The Perfect Complement to "Chenier's" Practical Math Dictionary:

PRESENTING:

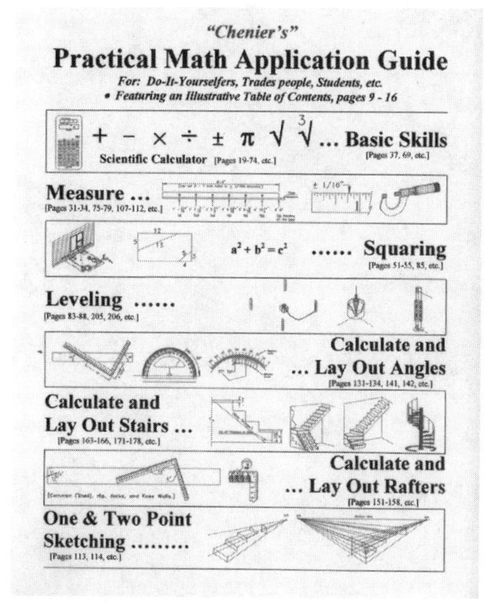

Front Cover

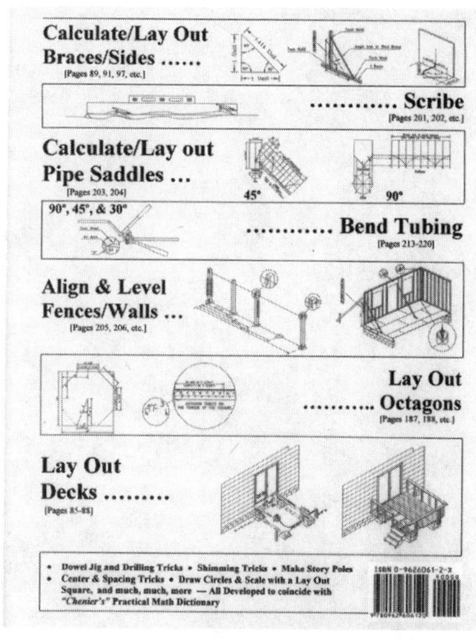

Back Cover

[256 Pages, 8 1/2″× 10 ¼″ with perforated pages and 3 hole drilled. ISBN: 0-9626061-2-X]

Information not depicted on the cover: (***These books can be used individually or as a set.***)

- Referenced to Chenier's Practical Math Dictionary, this book is designed to parallel and enhance any practical math class from general education through college level programs. Many of the math concepts are left out of traditional math books and are relevant to many different trades, occupations, do-it-yourselfers, home owners, home schools, etc.
- Includes: testing material, economical hands-on projects that simulate industry (use with sticks of wood, chalk lines, flip chart paper, etc.), the answers, and many different unique modules for projects, classroom situations, self-study, industry, etc. All proven in the classroom and on-the-job.
- Drilling Tricks, drill and tap charts (English and Metric), drill numbers, American Standard pipe chart, shimming tricks, draw circles with a lay out square, plus much, much, more.

For Further Information Contact: Chenier's Educational Enterprises, Inc.
3999 Co. 416 – 20th Rd. Gladstone, MI 49837
Phone: 906-786-1630 Fax: 906-786-8088
www.cheniermath.com

"Chenier's"

PRACTICAL MATH DICTIONARY

"The only book of its kind in America"

- ***ILLUSTRATIVE Table of Contents & English/ Metric Appendix ------ found only in this book.***
- ***UNIQUE INDEX ------ find problems instantly.***
- ***PROBLEMS RELATE to practical applications with self checking techniques, a unique feature.***
- ***DESCRIPTIVE GEOMETRY - lay out geometric figures from 3 to 8 sides, circles, ellipses, etc.***
- ***ILLUSTRATIVE TRIG -- formulas in pictures. Find the picture you need -- quickly and easily.***
- ***LEVELING TECHNIQUES - Learn how to use a water level, transits, plumb bob (level poles, fence post, your Christmas tree, etc.).***
- ***TRADE TRICKS -- Lay out/measuring secrets (draw circles with a square, lay out buildings, lay out machines, etc.).***
- ***PLUS many more features ....***

"A BOOK FOR EVERYONE"

3999 County 416-20th Road, Gladstone, MI 49837

Printed and bound in the United States of America.

6th Printing

International Standard Book Number: 0-9626061-1-1

PREFACE

PREFACE

This book is an extension of the Chenier Math Method, A Practical Math Dictionary and Workbook-Textbook. However, the content of this book is mainly for reference. Its size is designed to make it as versatile as possible and still give the reader the necessary tools to master basic mathematical concepts. All the basic mathematical concepts are designed with practical applications in mind. Hence, squaring techniques, leveling techniques, lay out techniques, and so on are included in this book. Many of these concepts are unique to this book and give the reader a totally new approach to learning and referencing mathematics.

Development of this book was through the perspective of both education and industry, evolving from many years of practical experiences, applications, and research. Education complements the world of work; therefore, this book is designed to be an easy-to-use resource for anyone involved in teaching or learning math skills. Industrial and commercial workers, apprentices, tradespeople, teachers, students, parents . . . all will find this volume helpful.

From the Table of Contents (unique because it is in both pictures and words), to the Tasks themselves, to the Appendix (also in pictures and in words), to the Glossary of Terms, and the Index, the book is a complete kit for anyone who needs a handy reference or wants to upgrade and become proficient in basic applied math skills. Its purpose is to serve as an asset in any home, at work, or in any mathematics, vocational, or apprentice program.

ACKNOWLEDGMENTS

ACKNOWLEDGMENTS

Foremost, I give thanks to my Lord and Savior, Jesus Christ for giving me the health, talent, and the right people to finish this gigantic project. In doing so I would like to acknowledge these people that have supported me in so many ways. I extend to each of them my appreciation and sincere thanks. First, I would like to acknowledge my family members: to my loving wife Joanne, to my loving sisters Betty Chenier and Carol Carlson, to my brother-in-law and true friend David Carlson. They were always there when I needed them for support, inspiration, and encouragement. I would also like to thank my brothers Donald Chenier and Jimmy Frazer for helping with the trigonometry part of this book, and my brother Pat just for being my friend and brother. Second, I would like to acknowledge my investors and friends: Al Atwood, my Editor, and his wife, Clare; Vicki Meyer, and Jim Chouinard, my computer programmers; Jim Almonroeder, marketing chairman, and colleague; Gary Olsen, corporate lawyer; Paul Paulson, CPA; Gary and Annis Bengston; Thomas and Dorothy Srock; Ronald and Patricia Carlson; Clayton and Dorlene Carlson; and Robert and Kathy LaRoche. Sadly, I extend my utmost appreciation and sympathy to the family of Jim Popelka, especially to his wife, Sue. Jim passed away October 10, 1996. He was an investor, former neighbor, and a true friend, indeed. He will be sadly missed.

I would also like to extend my appreciation and sincere thanks to my mother, Jeanette (Chenier) Frazer who encouraged me and who was always there for support and inspiration; to James Mitchell, my copyright lawyer, who guided me each step of the way; to Joe Larson for his illustrative work; to Karen Meiers, CPA for her assistance with our corporate bookkeeping; and finally to all my friends and relatives, too numerous to mention, who guided me through the years.

Grateful appreciation and thanks are also extended to Stanley Tools Division of The Stanley Works, New Britain, Connecticut; The L.S. Starrett Company, Athol, Massachusetts; Pat and Patricia Mitchell, The Hydrolevel Co., Ocean Springs, Mississippi; and The David White Company, Germantown, Wisconsin; for their photographs and line art.

DEDICATION

DEDICATION

I dedicate this book to my father Laurence Chenier, who passed away at the age of 40 in 1950, but who left me with an inspiration to follow all the days of my life.

Chenier Educational Enterprises would like to dedicate this book to the people of America to be used as a tool to help improve math skills in our homes, schools, and workplaces. We also would like to challenge everyone else to take up this call to help achieve this goal, and make our country a better place than we found it.

TABLE OF CONTENTS

TABLE OF CONTENTS

TABLE OF CONTENTS

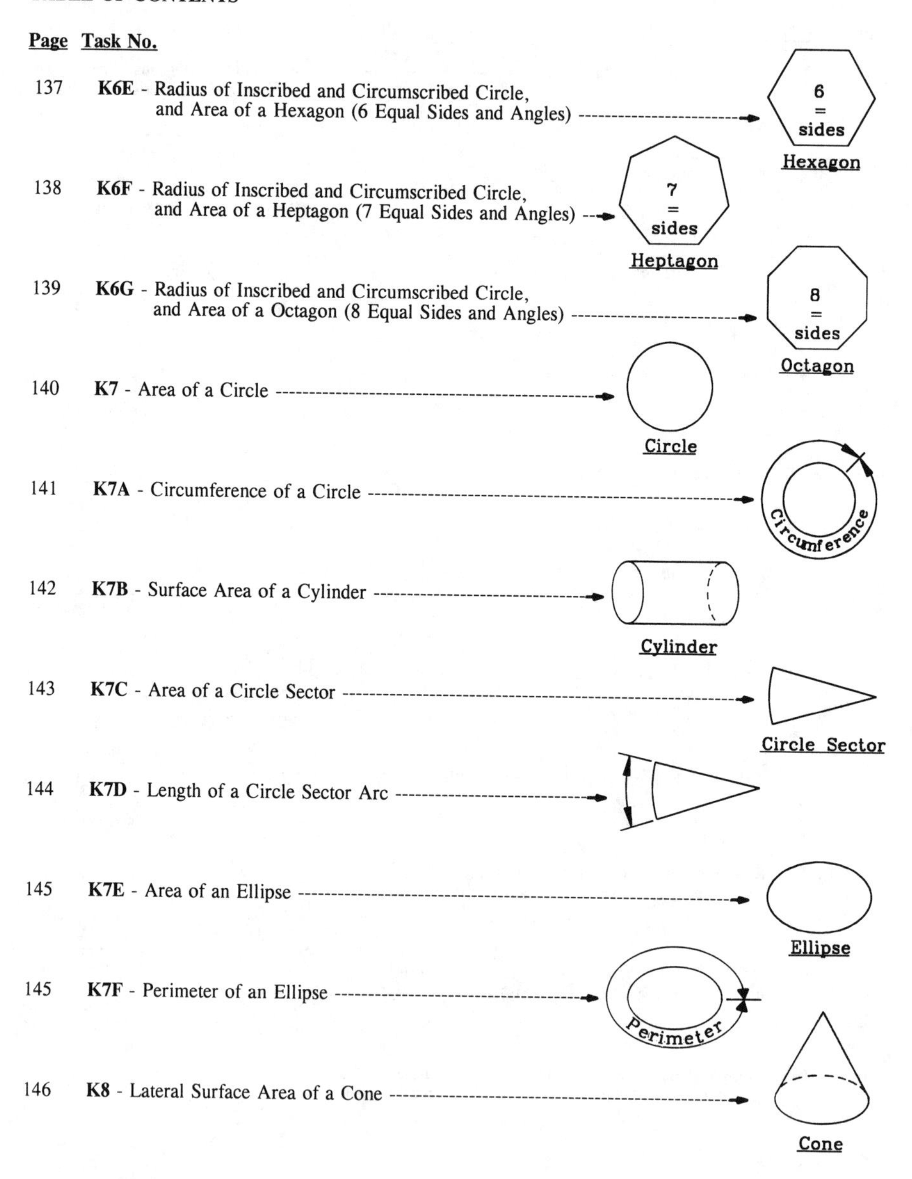

TABLE OF CONTENTS

TABLE OF CONTENTS

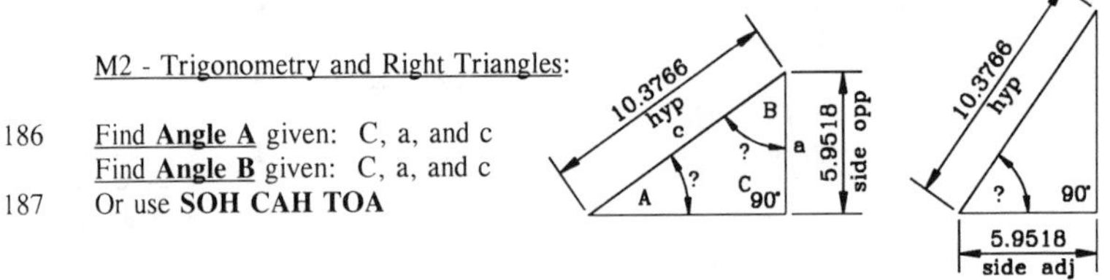

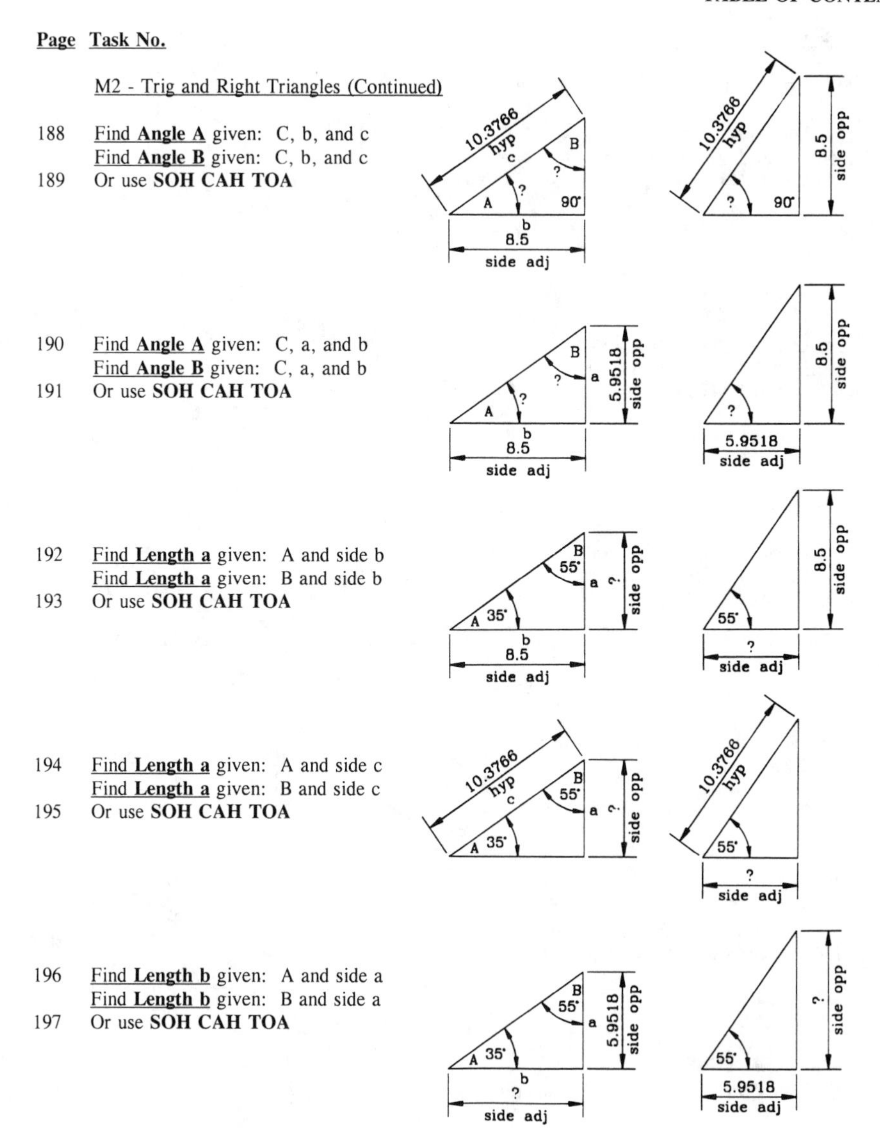

10.3766
hyp
c
A
B
?
90˚
b
8.5
side adj
side opp
5.9518
a
35˚
55˚

TABLE OF CONTENTS

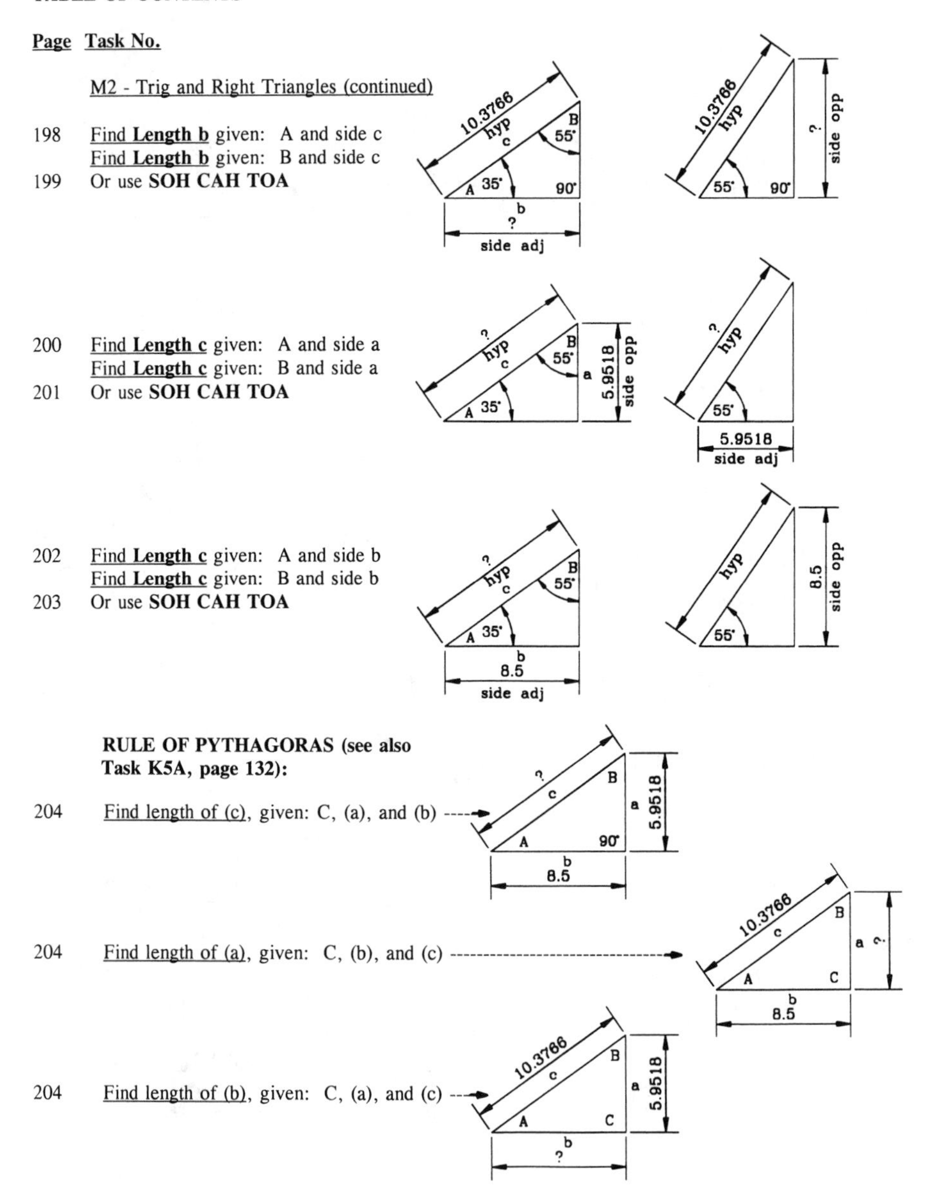

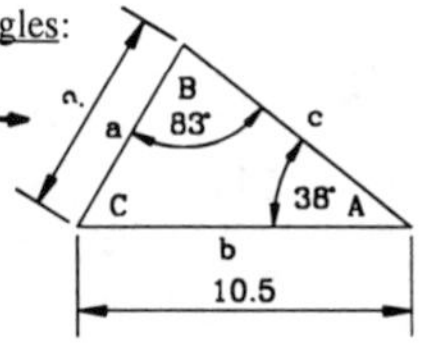

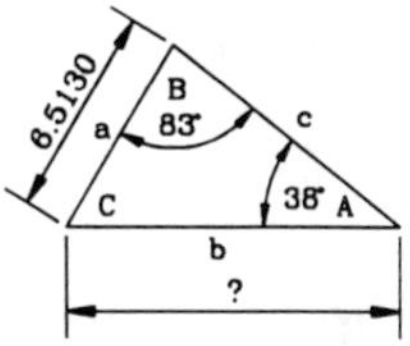

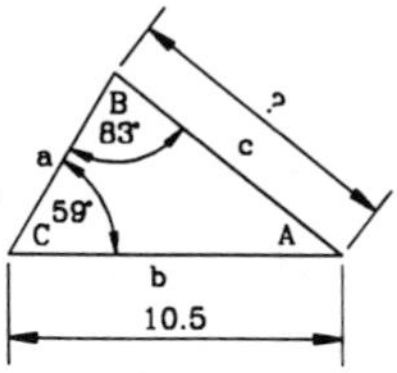

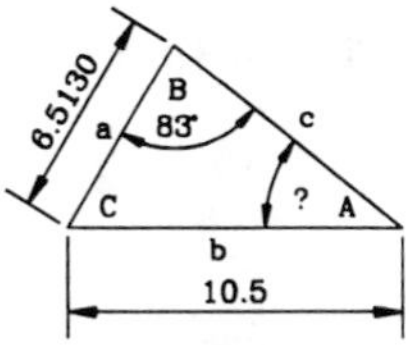

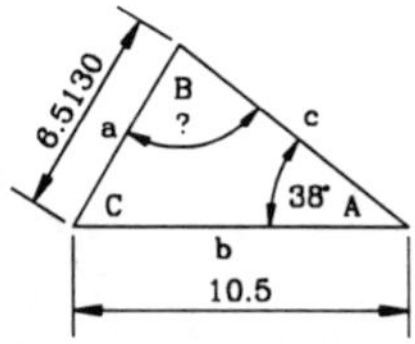

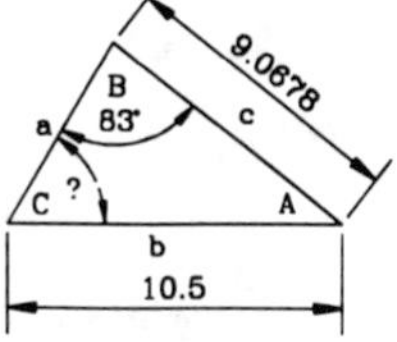

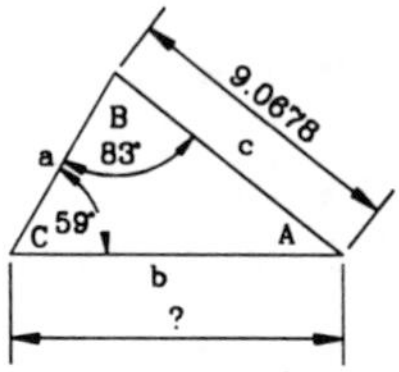

TABLE OF CONTENTS

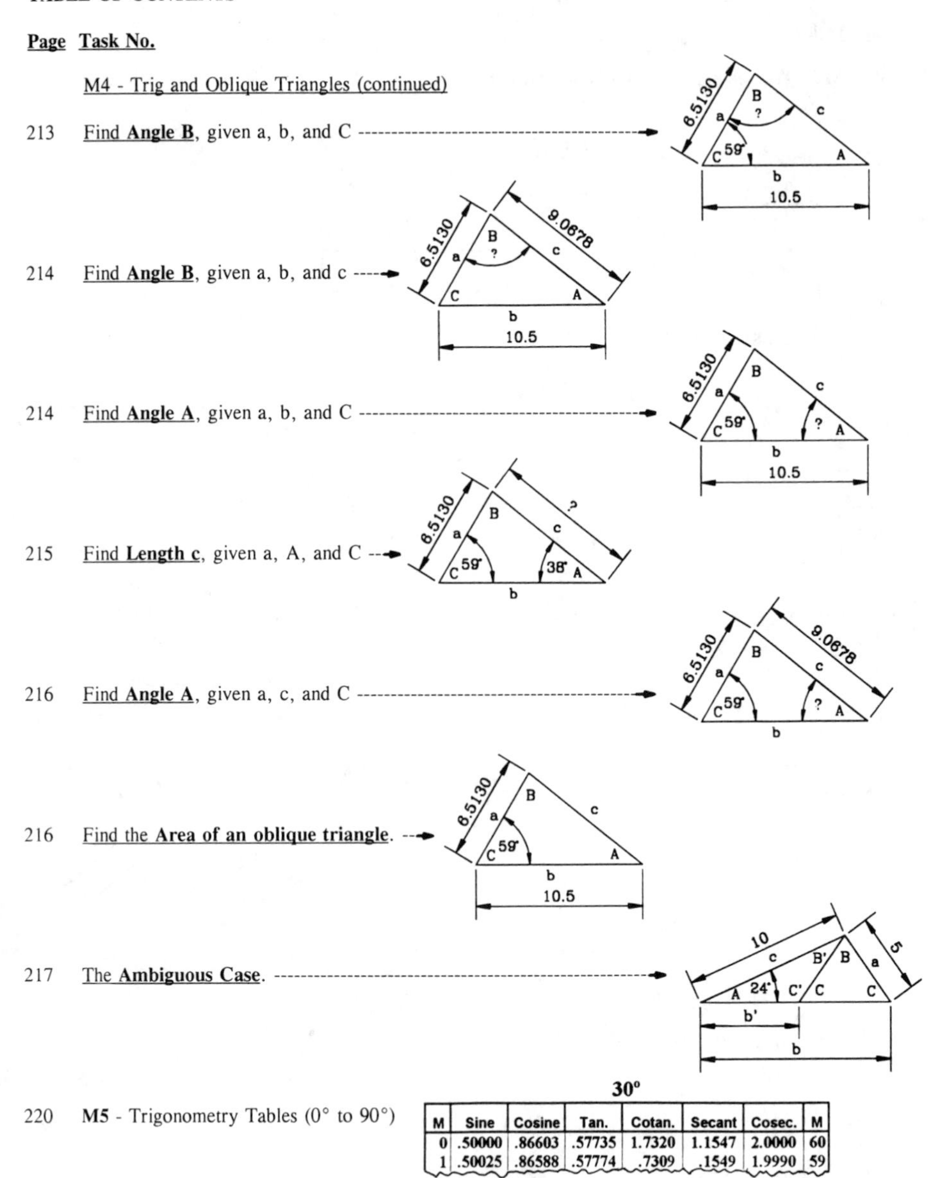

M	Sine	Cosine	Tan.	Cotan.	Secant	Cosec.	M
0	.50000	.86603	.57735	1.7320	1.1547	2.0000	60
1	.50025	.86588	.57774	.7309	.1549	1.9990	59

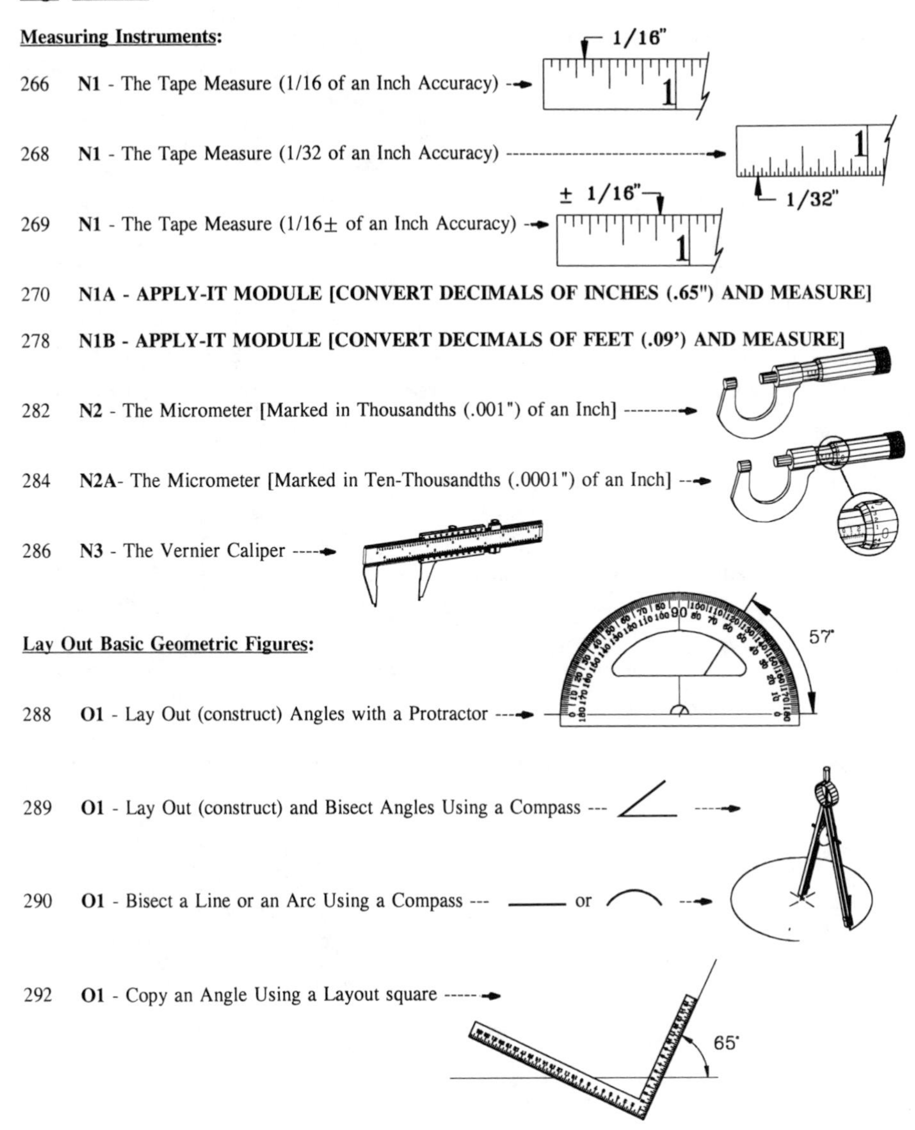

1/16"
1
1
1/32"
± 1/16"
1
57°
65°

TABLE OF CONTENTS

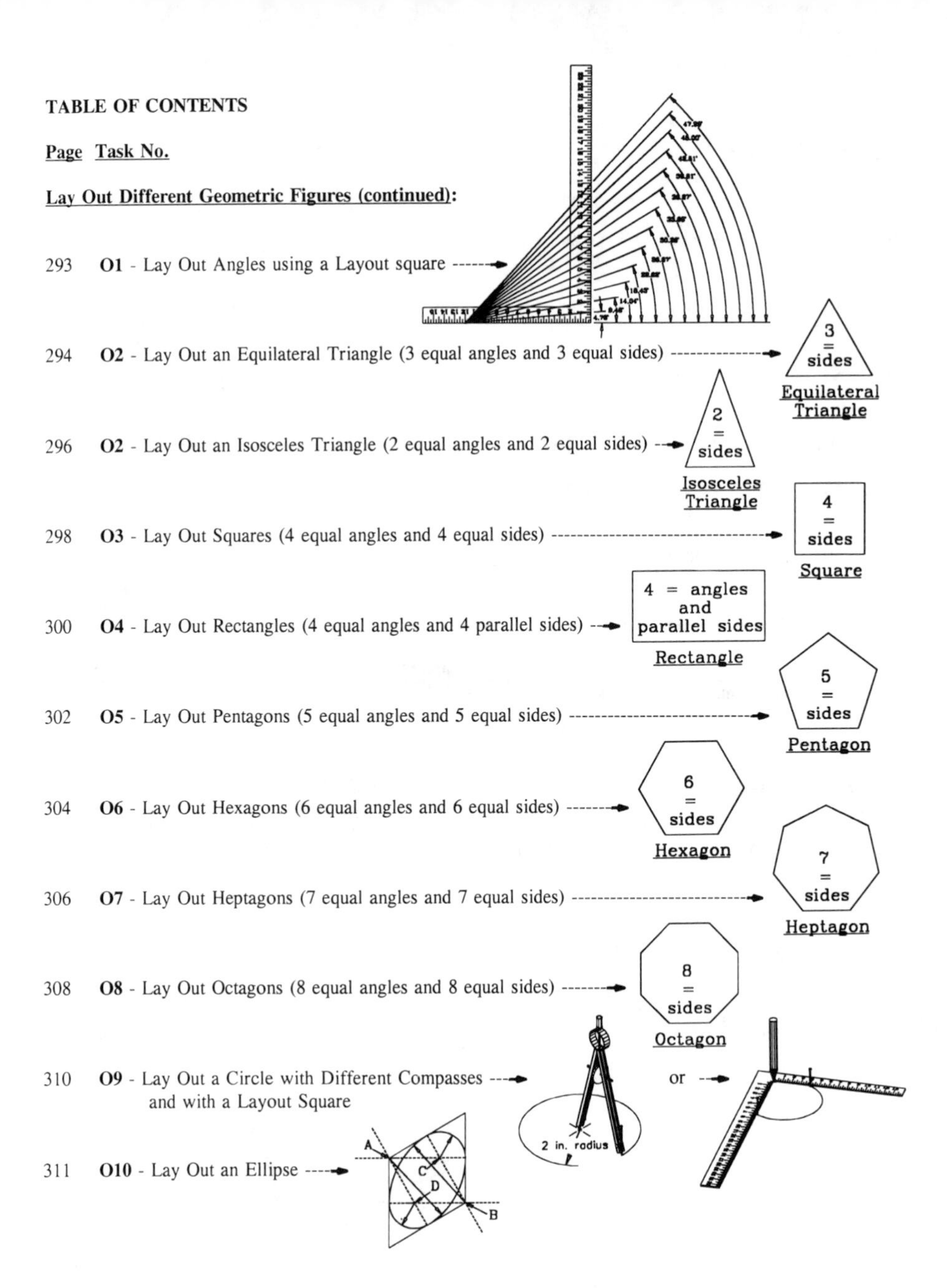

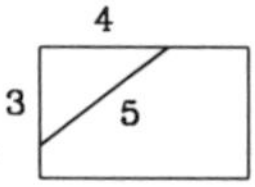

and

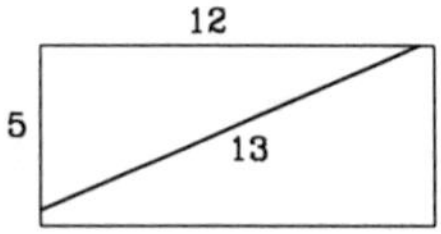

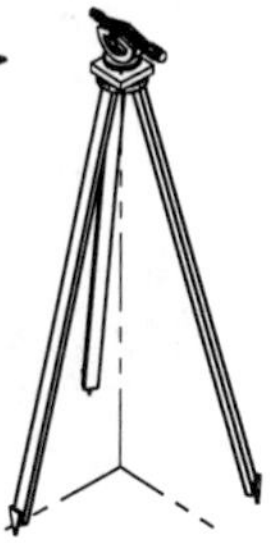

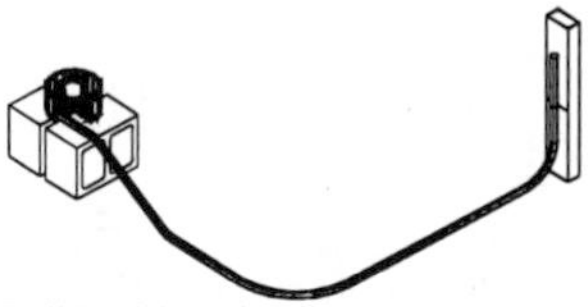

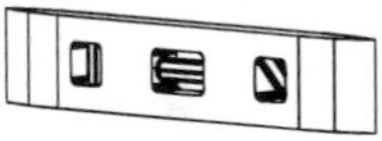

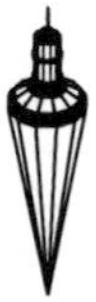

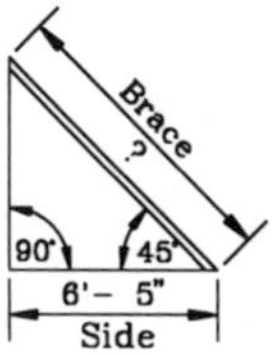

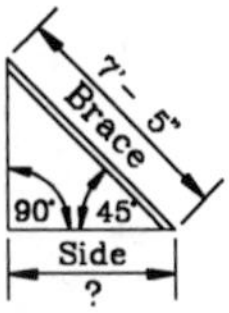

TABLE OF CONTENTS

Page

“This page intentionally blank.”

A1 ◂ Task No. Mathematics and Numeral Systems (WHOLE NUMBERS)

A whole number (integer) is the entire number. It is not a fraction or does not include a fraction of a number.

Mathematics is a group of such sciences as arithmetic, algebra, geometry, and trigonometry, which deals with quantities, sizes and forms, and their relationships to each other by use of numbers and symbols.

Numbers, strictly speaking, are ideas. **Numerals** are the symbols we use to express those number ideas. For example, to express the ideas of the number of legs a dog has, we might use the numeral 4, four, IV, or ////.

Arabic numerals are the most commonly used symbols to represent numbers. Any number can be represented by the proper combination of its ten symbols, or **digits**--0, 1, 2, 3, 4, 5, 6, 7, 8, and 9.

The decimal numeral system is the way we most often express numbers with the Arabic numerals. The word "decimal" comes from the Latin **decem**, meaning ten, which is the base of the decimal system. In a decimal numeral, the value of each digit depends on its place, or position, in the complete numeral. In the numeral 786, for example, the 6 is in the units, or ones, place; the 8 is in the tens place; and the 7 is in the hundreds place. In other words, 786 means (7 x 100) + (8 x 10) + (6 x 1). In the decimal system, the value of each place is 10 times greater than that of the place to its right.

A numeral system can have any base, not just the ten of the decimal system. The **binary numeral system** is commonly used in computer applications. It is a base-two system that uses only two digits, 0 and 1. In the binary system, the value of each place is twice that of the place to its right. Thus, in the binary numeral 1011, the 1 at the far right is in the ones place; the 1 to its left is in the twos place; the 0 is in the fours place; and the 1 to the left of that is in the eights place. In other words, binary 1011 means (1 x 8) + (0 x 4) + (1 x 2) + (1 x 1), which equals decimal 11. For everyday use, binary would be extremely cumbersome because of the number of places needed to express most numbers, Decimal 38, for example, requires six places in binary--100110. In electronic devices, however, binary is efficient because it utilizes the concept that a switch is either on or off (1 or 0).

The **Roman numeral system**, in common use until about 400 years ago, used seven letters of the Latin alphabet as its numerals-- I(1), V(5), X(10), L(50), C(100), D(500), M(1,000). There was no zero. It was a decimal system in the sense that numbers were written left to right, thousands first, hundreds next, then tens and finally ones. It did not make use of place values, however. Instead, the principles of addition and subtraction were used. In this system, the groupings of thousands, hundreds, tens, and ones are added to arrive at the number being written. Within each of those groupings, a numeral following itself or one of greater value is added to whatever precedes it. Thus, XXVIII is (10 + 10) + (5 + 1 + 1 + 1), or 28. Usually, 4s, 9s, and numbers beginning with 4 or 9 are written to show subtraction of a numeral from the larger numeral following it. Thus, IV is 4, XL is 40, CD is 400. A bar over a numeral means it should be multiplied by 1,000--$\overline{\text{V}}$ is 5,000, M$\overline{\text{V}}$ is 4,000, and $\overline{\text{XI}}$CMLXIV is 11,964. Roman numerals are still used occasionally for such purposes as numbering clock faces, listing outline topics, and showing dates on monuments and in some copyright notices.

Task No. ▶ A1

DECIMAL NUMERAL SYSTEM

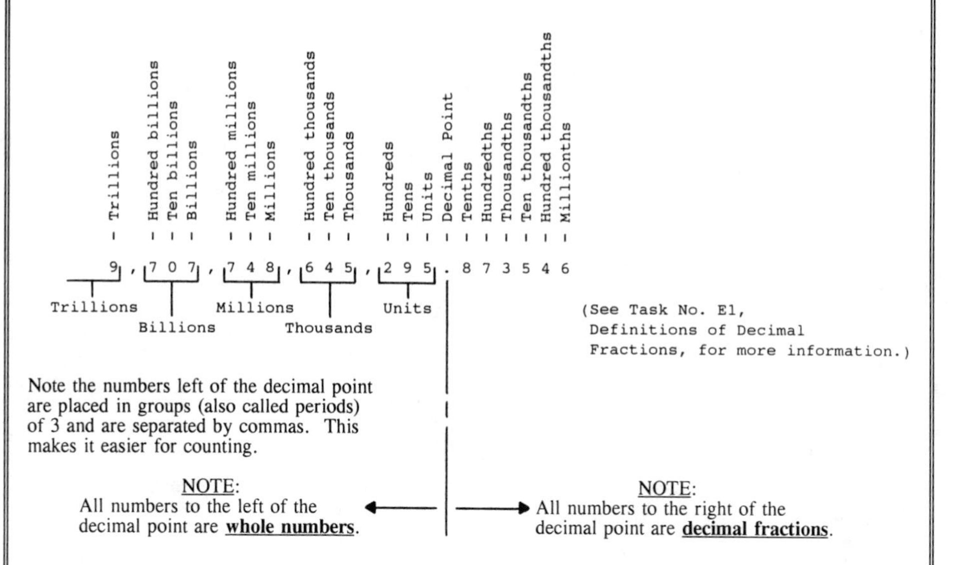

BINARY NUMERALS FOR DECIMAL 0 THROUGH 32

Binary	Decimal	Binary	Decimal	Binary	Decimal	Binary	Decimal
0	0						
1	1	1001	9	10001	17	11001	25
10	2	1010	10	10010	18	11010	26
11	3	1011	11	10011	19	11011	27
100	4	1100	12	10100	20	11100	28
101	5	1101	13	10101	21	11101	29
110	6	1110	14	10110	22	11110	30
111	7	1111	15	10111	23	11111	31
1000	8	10000	16	11000	24	100000	32

ROMAN NUMERALS

Arabic	1	2	3	4	5	6	7	8	9	10	11	12	13	14	15	16	17
Roman	I	II	III	IV	V	VI	VII	VIII	IX	X	XI	XII	XIII	XIV	XV	XVI	XVII

Arabic	18	19	20	30	40	50	60	70	80	90	100	200	300	400	500	600
Roman	XVIII	XIX	XX	XXX	XL	L	LX	LXX	LXXX	XC	C	CC	CCC	CD	D	DC

Arabic	700	800	900	1000	2000	3000	4000	5000
Roman	DCC	DCCC	CM	M	MM	MMM	M$\overline{V}$	$\overline{V}$

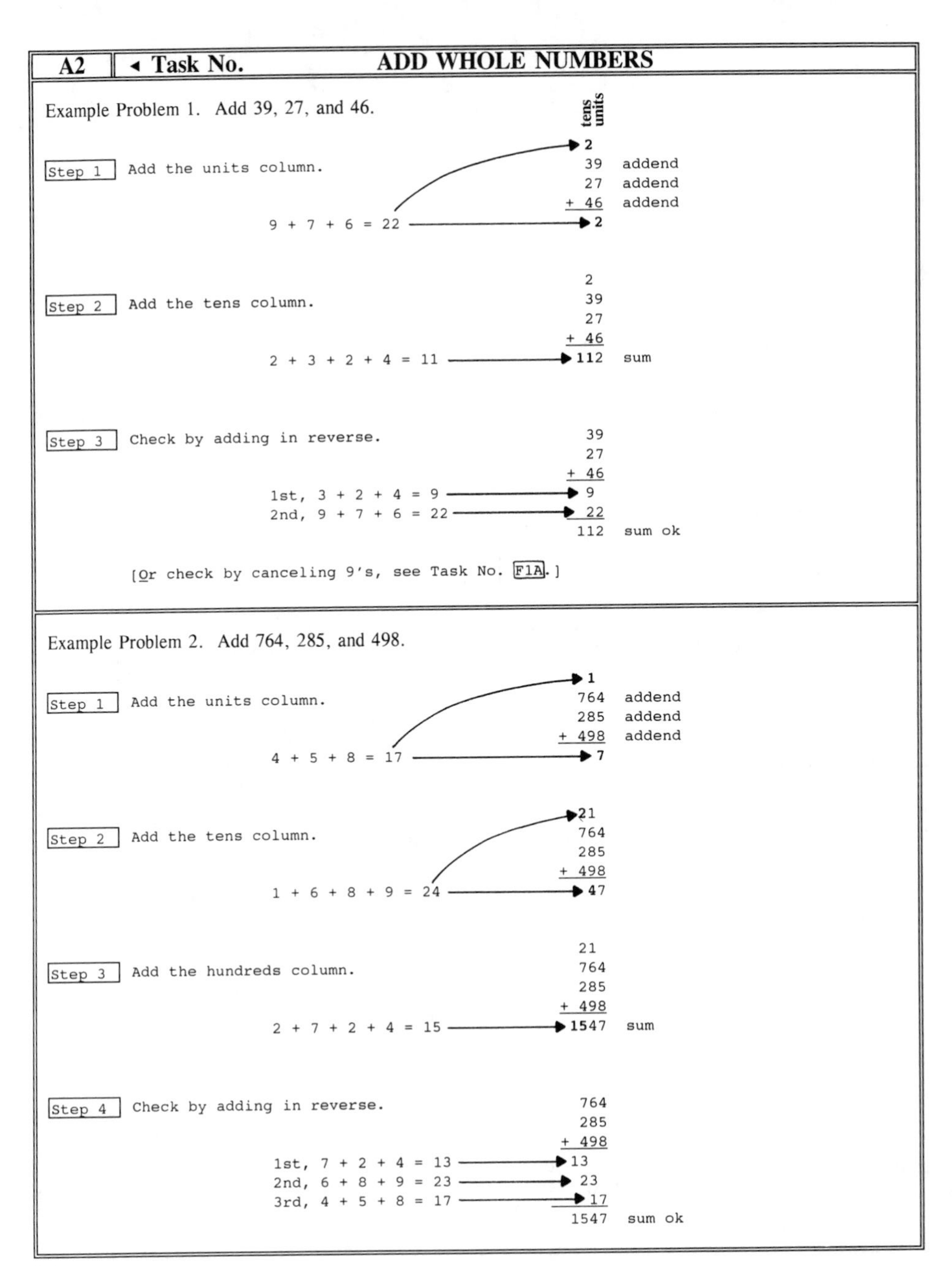

A2	◂ Task No.	ADD WHOLE NUMBERS

Example Problem 1. Add 39, 27, and 46.

Step 1 Add the units column.

```
              tens units
                  2
                 39   addend
                 27   addend
               + 46   addend
9 + 7 + 6 = 22 →  2
```

Step 2 Add the tens column.

```
                      2
                     39
                     27
                   + 46
2 + 3 + 2 + 4 = 11 → 112   sum
```

Step 3 Check by adding in reverse.

```
                          39
                          27
                        + 46
1st, 3 + 2 + 4 = 9  →      9
2nd, 9 + 7 + 6 = 22 →     22
                         112   sum ok
```

[Or check by canceling 9's, see Task No. F1A.]

Example Problem 2. Add 764, 285, and 498.

Step 1 Add the units column.

```
                     1
                   764   addend
                   285   addend
                 + 498   addend
4 + 5 + 8 = 17 →     7
```

Step 2 Add the tens column.

```
                        21
                       764
                       285
                     + 498
1 + 6 + 8 + 9 = 24 →    47
```

Step 3 Add the hundreds column.

```
                        21
                       764
                       285
                     + 498
2 + 7 + 2 + 4 = 15 → 1547   sum
```

Step 4 Check by adding in reverse.

```
                          764
                          285
                        + 498
1st, 7 + 2 + 4 = 13 →      13
2nd, 6 + 8 + 9 = 23 →     23
3rd, 4 + 5 + 8 = 17 →    17
                         1547   sum ok
```

SUBTRACT WHOLE NUMBERS

Task No. ▸ A3

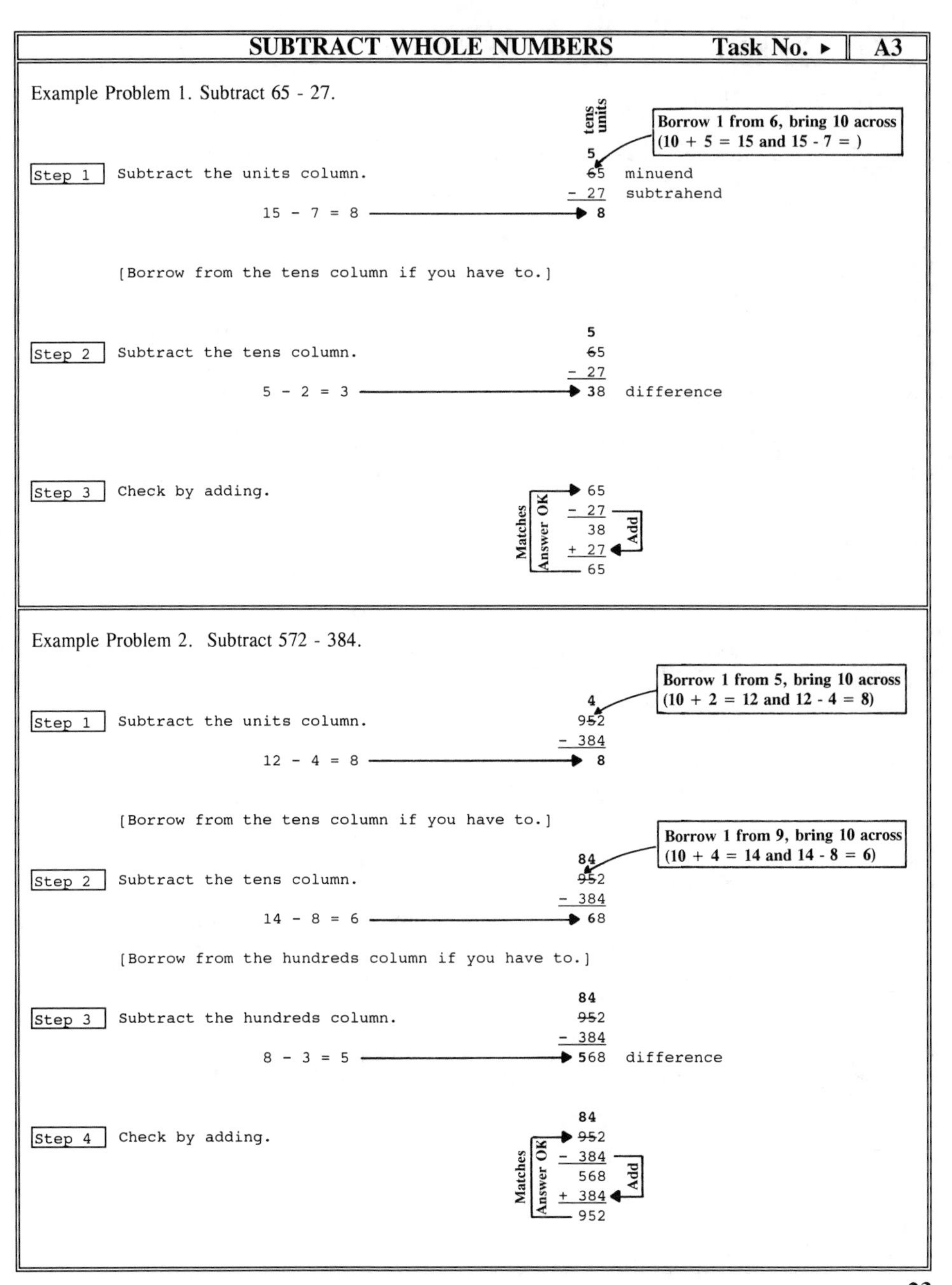

Example Problem 1. Subtract 65 - 27.

Step 1 Subtract the units column.

15 - 7 = 8

[Borrow from the tens column if you have to.]

Step 2 Subtract the tens column.

5 - 2 = 3

Step 3 Check by adding.

Example Problem 2. Subtract 572 - 384.

Step 1 Subtract the units column.

12 - 4 = 8

[Borrow from the tens column if you have to.]

Step 2 Subtract the tens column.

14 - 8 = 6

[Borrow from the hundreds column if you have to.]

Step 3 Subtract the hundreds column.

8 - 3 = 5

Step 4 Check by adding.

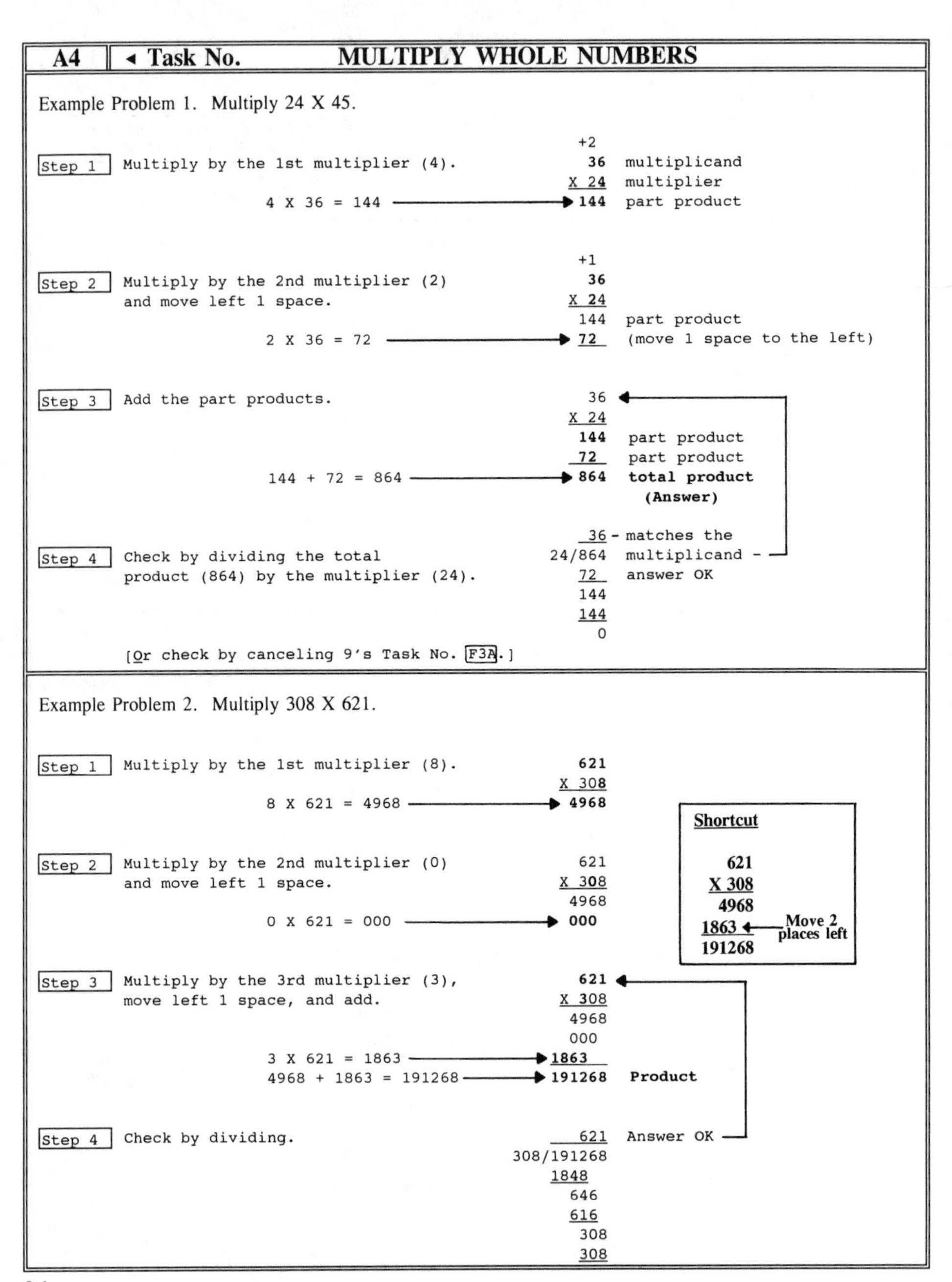

A4 ◂ Task No. MULTIPLY WHOLE NUMBERS

Example Problem 1. Multiply 24 X 45.

Step 1 Multiply by the 1st multiplier (4).

```
                       +2
                       36   multiplicand
                     X 24   multiplier
4 X 36 = 144  ──────▶ 144   part product
```

Step 2 Multiply by the 2nd multiplier (2) and move left 1 space.

```
                       +1
                       36
                     X 24
                      144   part product
2 X 36 = 72   ──────▶ 72    (move 1 space to the left)
```

Step 3 Add the part products.

```
                        36
                      X 24
                       144   part product
                       72    part product
144 + 72 = 864  ─────▶ 864   total product
                               (Answer)
```

Step 4 Check by dividing the total product (864) by the multiplier (24).

```
    36 - matches the
24/864   multiplicand -
    72   answer OK
    144
    144
      0
```

[Or check by canceling 9's Task No. F3A.]

Example Problem 2. Multiply 308 X 621.

Step 1 Multiply by the 1st multiplier (8).

```
                        621
                      X 308
8 X 621 = 4968  ────▶  4968
```

Step 2 Multiply by the 2nd multiplier (0) and move left 1 space.

```
                        621
                      X 308
                       4968
0 X 621 = 000   ────▶  000
```

Shortcut

```
   621
 X 308
  4968
1863 ◀── Move 2 places left
191268
```

Step 3 Multiply by the 3rd multiplier (3), move left 1 space, and add.

```
                                621
                              X 308
                               4968
                               000
3 X 621 = 1863          ────▶ 1863
4968 + 1863 = 191268    ────▶ 191268   Product
```

Step 4 Check by dividing.

```
       621   Answer OK
308/191268
    1848
     646
     616
      308
      308
```

DIVIDE WHOLE NUMBERS

Task No. ▸ A5

Example Problem 1. Divide 2608 ÷ 8.

Step 1 Divide into the 1st group (26) and subtract.

```
                3
   divisor  8/2608  dividend
             -24
               2
```

Step 2 Bring down another number (0), divide into the new group (20), and subtract.

```
      32
  8/2608
    24↓
     20
    -16
      4
```

Step 3 Bring down another number (8), divide into the new group (48), and subtract.

```
     326  quotient
  8/2608
    24
     20
     16↓
      48
     -48
       0
```

Step 4 Check by multiplying the divisor (8) by the quotient 326.

```
   326  (matches
   x 8   dividend)
  2608   Answer OK
```

Example Problem 2. Divide 268 ÷ 12.

Step 1 Divide into the 1st group (26) and subtract.

```
       2
  12/268
     -24
       2
```

Step 2 Bring down another number (8), divide into the new group (28), and subtract.

```
              22 4/12 quotient
          12/268
             24↓
              28
             -24
  Remainder    4
```

Step 3 To bring down another number (0) would be to make a decimal fraction, see division of of decimals, Task No. E5.

$22\frac{4}{12}$ or $22\frac{1}{3}$ reduced is a mixed number.

Step 4 Check by multiplying and adding the remainder.

```
               22
             x 12
               44
              22
              264
  Remainder  +  4
              268  Answer OK
```

A6 ◂ Task No. MISCELLANEOUS WHOLE NUMBERS, SYMBOLS, AND AVERAGES

Whole numbers can be either **prime** or **composite**. However zero or 1 is not a prime number or a composite number.

A **prime** number is a number that can be divided only by itself and 1 exactly. Some of the prime numbers are 2, 3, 5, 7, 11, 13, 17, 19, and so on.

Composite numbers are numbers made up of the product of two or more whole numbers called **factors** (the number 1 or zero is not a composite number). For example, the whole number 12 is composed of the factors 2 X 2 X 3; the whole number 45 is composed of the factors 3 X 3 X 5, and so on.
A **factor** is a number that can be divided into the composite number exactly.
For example, 2, 3, 4, and 6 are factors of 12.

Prime factors are factors that can be divided only by themselves and 1.
For example, 2 X 2 X 3 are prime factors of 12. Each composite number can have only one set of prime factors. Prime factors are the prime numbers that, after multiplication, make up the original number or composite number.

Denominate number is any number that applies to a specific unit of measure, such as 2 feet 1 inch, or 5 pounds 3 ounces, 2 bushels 1 peck, and so on. To calculate denominate numbers, see Task D1 through D6 .

Symbols used to calculate numbers:

+ sign indicates addition (2 + 1) means 2 plus 1.

For addition of whole numbers, see Task No. A2.

```
  2  addend
+ 1  addend
  3  sum
```

- sign indicates subtraction (8 - 5) means 8 minus 5.

For subtraction of whole numbers, see Task No. A3 .

```
  8  minuend
- 5  subtrahend
  3  difference or
     remainder
```

X sign indicates multiplication (6 X 7) means 6 times 7.
• raised dot indicates multiplication (6 • 7) means 6 times 7.

For multiplication of whole numbers, see Task No. A4 .

```
   7  multiplicand
 X 6  multiplier
  42  product
```

÷ sign indicates division (10 ÷ 2) means 10 divided by 2.
/ sign indicates division (10/2) means 10 ÷ 2 or 10 divided by 2.

For division of whole numbers, see Task No. A5 .

```
               5  quotient
divisor - 2/10  dividend
             10
              0  remainder
```

Note: The divisor always follows the division sign (÷) and/or (/). In this case, the divisor is 2.

= sign indicates equality (3 + 2) = 5 means 3 plus 2 equals 5.

Task No. ▸ A6

How to find averages:

RULES: To find an average, (1) find the sum of all the quantities, and (2) divide the sum by the number of quantities.

Example Problem 1. Find the average of 7, 11, 5, and 9.

Step 1 Find the sum of all the quantities.

Line up vertically and add, or use a calculator.

```
  7       2
 11       7
  5      11
+ 9       5
        + 9
         32  sum
```

Step 2 Divide the sum by the number of quantities.

Note: There are 4 quantities (7, 11, 5, and 9), therefore, divide by 4.

```
   8  average
4/32
  32
```

Step 3 Check by multiplying.

```
             8
           X 4
Answer OK   32
```

Example Problem 2. Find the average of 16, 27, 41, 18, and 24.

Step 1 Find the sum of all the quantities.

```
   2
  16
  27
  41
  18
+ 24
 126
```

Step 2 Divide the sum by the number of quantities.

Note: There are 5 quantities (16, 27, 41, 18, and 24), therefore, divide by 5.

```
   25.2  average
5/126.0
  10
   26
   25
    10
    10
```

Step 3 Check by multiplying.

```
 25.2
 X  5
126.0  Answer OK
```

B1	◂ Task No.	COMMON FRACTIONS

A **common fraction** is any number of equal parts of a whole, for example:

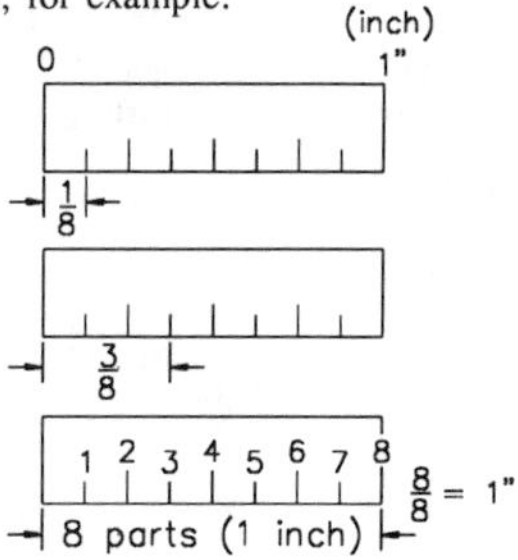

A whole inch can be separated into 8 equal parts. Each part is 1 of 8 parts and is written as 1/8 of an inch.

Three parts would be 3 of 8 parts or 3/8 of a whole inch.

eight parts would be 8 of 8 parts or 8/8 of a whole inch, therefore, the whole or 1 inch.

A Common Fraction

The top number is the numerator (tells how many parts are taken) → $\frac{7}{8}$

The bottom number is the denominator (tells into how many equal parts the whole is divided).

The line of a fraction indicates division, and the denominator is always the divisor.

Thus: $\frac{7}{8} = 8\overline{)7}$ or $7 \div 8$

The **terms** of a fraction are the numerator and the denominator -- together ---- $\frac{3}{8}$

In a **proper fraction**, the numerator is smaller than the denominator ---- $\frac{5}{8}$

In an **improper fraction**, the numerator is larger than the denominator ---- $\frac{9}{8}$

Or, the numerator is equal to the denominator ---- $\frac{8}{8}$

A **mixed number** is a whole number and a fraction written together ---- $3\frac{1}{4}$

In a **complex fraction**, a mixed number or a fraction appears in the numerator or the denominator or both, such as: ---- $\dfrac{\frac{5}{16}}{\frac{1}{8}}$, $\dfrac{6}{3\frac{1}{4}}$, or $\dfrac{\frac{1}{8}}{9\frac{1}{2}}$

RULES: **To change (reduce) a fraction to its lowest terms, divide both the numerator and the denominator by a common factor.**

Example: $\frac{6}{8} = \dfrac{\frac{3}{2/6}}{\frac{4}{2/8}} = \frac{3}{4}$ [Note that the common factor is 2 because 2 goes into both 6 and 8 evenly.]

When the numerator and the denominator are divided by the same number, the value of the fraction does not change.

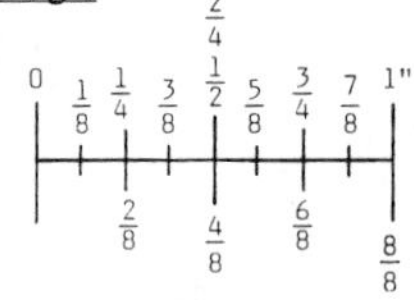

Note: $\frac{3}{4}$ is the same as $\frac{6}{8}$. Therefore, the value is the same.

RULES: **<u>To change an improper fraction</u>, divide the bottom number (denominator) into the top number (numerator). The result will be a mixed number or whole number.**

Examples: [$\frac{11}{8}$ = 8/11 = $1\frac{3}{8}$ (mixed number) $\frac{8}{8}$ = 8/8 = 1 (whole number)]

Note: One reason for reducing an improper fraction such as $\frac{11}{8}$, is that measuring-instrument parts (increments) are usually written and located in reduced form.

Example:

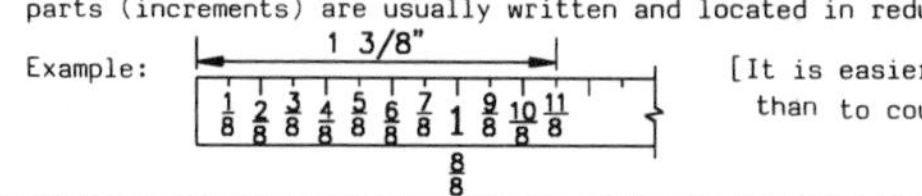

[It is easier to find $1\frac{3}{8}$ on a tape measure than to count out eleven $\frac{1}{8}$ parts.]

<u>To change a mixed number to an improper fraction</u>, multiply the whole number by the denominator in the fraction, then add this product to the numerator. <u>The result is an improper fraction</u>.

Example: $3\frac{5}{8}$ = 3 x 8 = 24 and 24 + $\frac{5}{8}$ = $\frac{29}{8}$ (improper fraction)

<u>To change a fraction to higher terms</u> (make an equivalent fraction) multiply both the numerator and the denominator by the same number.

Example: numerator 3 times 4 = 12 / denominator 4 times 4 = 16 [$\frac{12}{16}$ is an equivalent fraction of $\frac{3}{4}$ because their values are the same.]

When both the numerator and the denominator are multiplied by the same number, <u>the value of the fraction does not change</u>.

The lowest common denominator (L.C.D.) is the <u>lowest possible number</u> that can be divided evenly by the denominators of all given fractions.

Example:

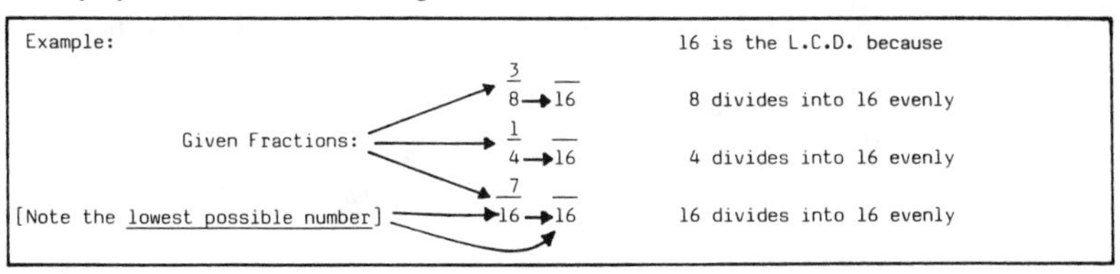

<u>To make equivalent fractions</u> with the L.C.D.: (1) divide the denominator of each given fraction into the L.C.D., (2) multiply that answer (quotient) by the numerator of the given fraction, and (3) place the answer (product) over the L.C.D. The new fractions are <u>equivalent fractions</u>.

Example:

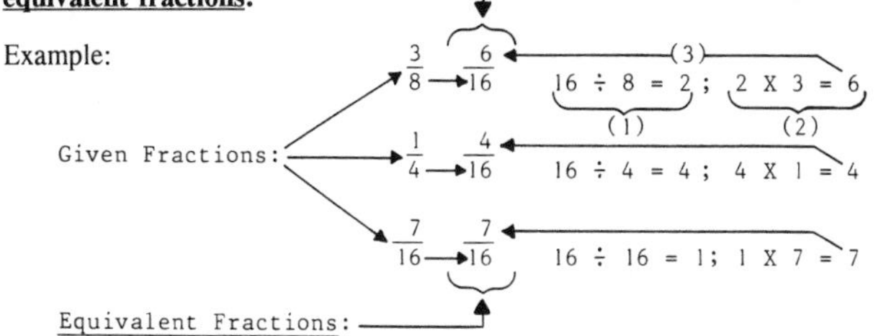

[See Task No. C1, page 39, for more information on equivalent fractions.]

B2	◂ Task No.	ADD COMMON FRACTIONS

Example Problem 1. Add $\frac{1}{4} + \frac{3}{8}$

Step 1 Find the lowest common denominator.

$$\begin{array}{rr} \frac{1}{4} & \frac{}{8} \\ + \frac{3}{8} & \frac{}{8} \\ \hline \end{array}$$

4 divides into 8 evenly
8 divides into 8 evenly } **8 L.C.D.**

Step 2 Make equivalent fractions with the lowest common denominator (L.C.D.) 8.

$$\begin{array}{rr} \frac{1}{4} & \frac{2}{8} \\ + \frac{3}{8} & \frac{3}{8} \\ \hline \end{array}$$

Step 3 Add the numerators.

$$\begin{array}{rr} \frac{1}{4} & \frac{2}{8} \\ + \frac{3}{8} & \frac{3}{8} \\ \hline \end{array} \qquad 2 + 3 = 5$$

(sum) $\frac{5}{8}$ Answer

Step 4 Check by subtracting.

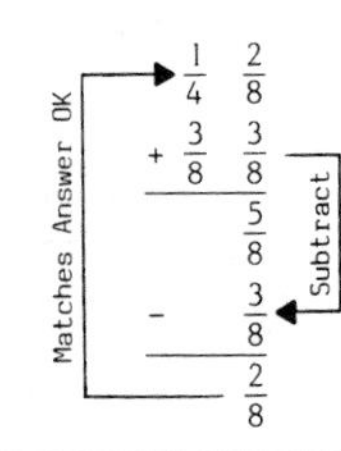

Example Problem 2. Add $\frac{5}{8} + \frac{9}{16}$

Step 1 Find the lowest common denominator.

$$\begin{array}{rr} \frac{5}{8} & \frac{}{16} \\ + \frac{9}{16} & \frac{}{16} \\ \hline \end{array}$$

8 divides into 16 evenly
16 divides into 16 evenly } **16 L.C.D.**

Step 2 Make equivalent fractions with the lowest common denominator (L.C.D.) 16.

$$\begin{array}{rr} \frac{5}{8} & \frac{10}{16} \\ + \frac{9}{16} & \frac{9}{16} \\ \hline \end{array}$$

Step 3 Add the numerators and reduce to lowest terms.

$$\frac{19}{16} = 16\overline{)19} = 1\frac{3}{16} \qquad \begin{array}{r} 16 \\ \hline 3 \end{array}$$

$$\begin{array}{rr} \frac{5}{8} & \frac{10}{16} \\ + \frac{9}{16} & \frac{9}{16} \\ \hline & \frac{19}{16} = 1\frac{3}{16} \end{array} \qquad 10 + 9 = 19$$

Answer

Step 4 Check by subtracting.

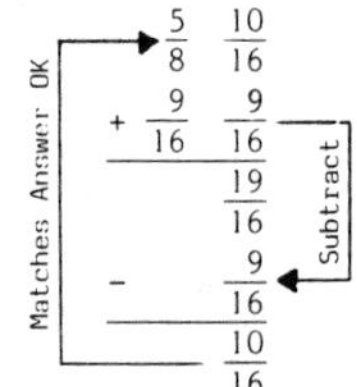

SUBTRACT COMMON FRACTIONS — Task No. ▸ B3

Example Problem 1. Subtract $\frac{3}{4} - \frac{1}{2}$

Step 1 — Find the lowest common denominator.

$$\frac{3}{4}\ \frac{\ }{4} \qquad -\frac{1}{2}\ \frac{\ }{4}$$

4 divides into 4 evenly
2 divides into 4 evenly } **4 L.C.D.**

Step 2 — Make equivalent fractions with the lowest common denominator (L.C.D.) 4.

$$\frac{3}{4} = \frac{3}{4} \qquad -\frac{1}{2} = \frac{2}{4}$$

Step 3 — Subtract the numerators.

$$\frac{3}{4} - \frac{2}{4}: \quad 3 - 2 = 1$$

(difference) $\frac{1}{4}$ Answer

Step 4 — Check by adding.

$$\frac{1}{4} + \frac{2}{4} = \frac{3}{4}$$

Add — Matches Answer OK

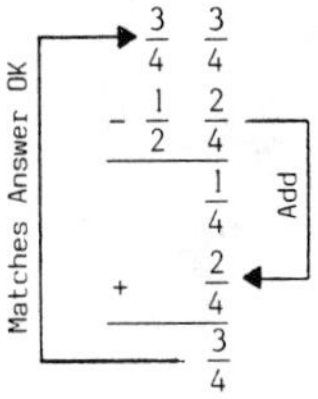

Example Problem 2. Subtract $\frac{7}{8} - \frac{3}{16}$

Step 1 — Find the lowest common denominator.

$$\frac{7}{8}\ \frac{\ }{16} \qquad -\frac{3}{16}\ \frac{\ }{16}$$

8 divides into 16 evenly
16 divides into 16 evenly } **16 L.C.D.**

Step 2 — Make equivalent fractions with the L.C.D. (16).

$$\frac{7}{8} = \frac{14}{16} \qquad -\frac{3}{16} = \frac{3}{16}$$

Step 3 — Subtract the numerators.

$$\frac{14}{16} - \frac{3}{16}: \quad 14 - 3 = 11$$

$\frac{11}{16}$ Answer

Step 4 — Check by adding.

$$\frac{11}{16} + \frac{3}{16} = \frac{14}{16}$$

Add — Matches Answer OK

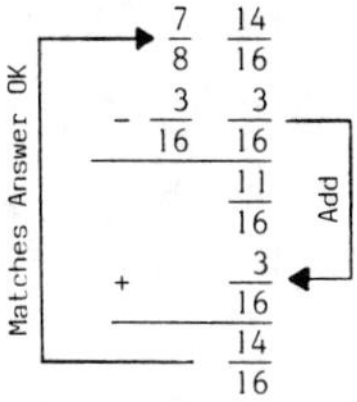

B4 ◂ Task No. MULTIPLY COMMON FRACTIONS

Example Problem 1. Multiply $\frac{1}{2} \times \frac{5}{8}$

Step 1 Reduce by canceling if you can. $\frac{1}{2} \times \frac{5}{8} =$ **[Neither fraction can be reduced]**

Step 2 Multiply straight across.

(1 X 5 = 5)

$\frac{1}{2} \times \frac{5}{8} = \frac{5}{16}$ Answer

(2 X 8 = 16)

Step 3 Reduce to lowest terms. $\frac{5}{16}$ is in lowest terms.

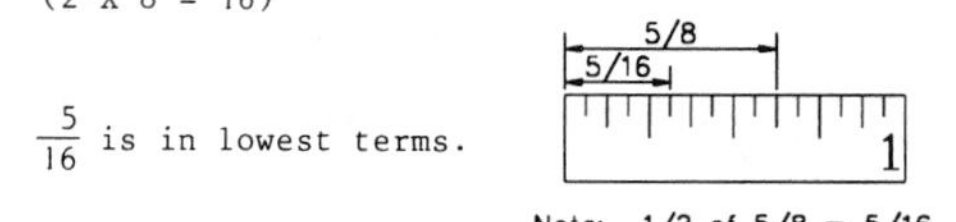

Note: 1/2 of 5/8 = 5/16

Step 4 Check by dividing.

$\frac{1}{2} \times \frac{5}{8} = \frac{5}{16}$

$\frac{5}{16} \div \frac{5}{8} = \frac{\not{5}^{1}}{\not{16}_{2}} \times \frac{\not{8}^{1}}{\not{5}_{1}} = \frac{1}{2}$

Matches Answer OK

Example Problem 2. Multiply $6 \times \frac{5}{16}$

Step 1 Reduce by canceling if you can. $\frac{\not{6}^{3}}{1} \times \frac{5}{\not{16}_{8}} =$ **[6 and 16 can be reduced by the common factor 2]**

Step 2 Multiply straight across.

(3 X 5 = 15)

$\frac{\not{6}^{3}}{1} \times \frac{5}{\not{16}_{8}} = \frac{15}{8} = 1\frac{7}{8}$

(1 X 8 = 8)

Step 3 Reduce to lowest terms.

$\frac{15}{8} = 8\overline{)15}$ → 1, $\frac{8}{7}$ $= 1\frac{7}{8}$ Answer

Step 4 Check by dividing.

$\frac{\not{6}^{3}}{1} \times \frac{5}{\not{16}_{8}} = \frac{15}{8}$

$\frac{15}{8} \div \frac{5}{16} = \frac{\not{15}^{3}}{\not{8}_{1}} \times \frac{\not{16}^{2}}{\not{5}_{1}} = \frac{6}{1}$

Matches Answer OK

DIVIDE COMMON FRACTIONS

Task No. ▸ B5

Example Problem 1. Divide $\frac{3}{4} \div 8$

Step 1 1st, invert the divisor.

$\frac{3}{4} \div \frac{8}{1}$ — Invert the divisor

$\frac{3}{4} \times \frac{1}{8}$

2nd, change the (÷) sign to (x) — Change the sign

Step 2 1st, cancel (if you can). [Cannot cancel]

2nd, multiply straight across. $\frac{3}{4} \times \frac{1}{8} = \frac{3}{32}$ Answer

Step 3 Reduce to lowest terms. $\frac{3}{32}$ is in lowest terms.

Step 4 Check by multiplying.

$\frac{3}{4} \div \frac{8}{1} = \frac{3}{32}$

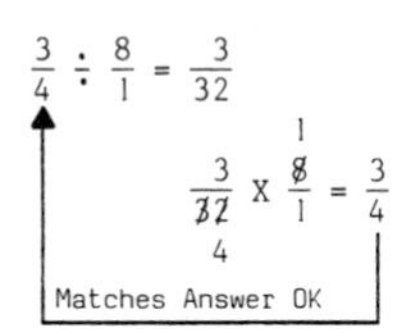

$\frac{3}{\not{32}_{4}} \times \frac{\not{8}^{1}}{1} = \frac{3}{4}$

Example Problem 2. Divide $\frac{15}{16} \div \frac{3}{4}$

Step 1 1st, invert the divisor.

$\frac{15}{16} \div \frac{3}{4}$ — Invert the divisor

$\frac{15}{16} \times \frac{4}{3}$

2nd, change the (÷) sign to (x) — Change the sign

Step 2 1st, cancel (if you can).

2nd, multiply straight across.

$\frac{\not{15}^{5}}{\not{16}_{4}} \times \frac{\not{4}^{1}}{\not{3}_{1}} = \frac{5 \times 1}{4 \times 1} = \frac{5}{4} = 1\frac{1}{4}$

Step 3 Reduce to lowest terms.

$\frac{5}{4} = 4\overline{)5} = 1\frac{1}{4}$ Answer ($\frac{4}{1}$)

Step 4 Check by multiplying.

$\frac{15}{16} \div \frac{3}{4} = \frac{5}{4}$

$\frac{5}{4} \times \frac{3}{4} = \frac{15}{16}$

Matches Answer OK

B6	◂ Task No.	CALCULATE COMPLEX FRACTIONS

In a complex fraction, a mixed number or a fraction appears in the numerator or the denominator or both, such as:

$$\frac{3\frac{1}{2}}{\frac{3}{4}}, \quad \frac{\frac{3}{8}+\frac{15}{16}}{\frac{25}{32}\div 5}, \quad \frac{10}{5\frac{1}{4}}, \quad \frac{\frac{3}{16}}{7}$$

REVIEW:
- The **line of a fraction means division**.
- The **bottom (denominator) is always the divisor**.
- The **divisor always follows the division (÷) sign**.

Note: Two of the most common applications of complex fractions in trade and industry are:

1) Converting a fractional part of a denominate number to a decimal.

Examples: $1\frac{1}{4}$ inches is $1\frac{1}{4}$ 12th parts of a foot or $\frac{1\frac{1}{4}}{12}$, a complex fraction. $12\overline{)1.25}$ = .104 feet as a decimal.

$2\frac{3}{8}$ ounces is $2\frac{3}{8}$ 16th parts of a pound or $\frac{2\frac{3}{8}}{16}$, a complex fraction. $16\overline{)2.375}$ = .1979 pound as a decimal.

2) Changing percents to decimals.

Examples: $\frac{1}{8}$% is $\frac{1}{8}$ of 100 parts or $\frac{\frac{1}{8}}{100}$, a complex fraction. $100\overline{).125}$ = .00125 as a decimal.

$15\frac{1}{2}$% is $15\frac{1}{2}$ of a 100 parts or $\frac{15\frac{1}{2}}{100}$, a complex fraction. $100\overline{)15.5}$ = .155 as a decimal.

Example Problem 1. Calculate $\frac{3\frac{1}{2}}{\frac{3}{4}}$

Step 1. Calculate the numerator ----- $\frac{3\frac{1}{2}}{\frac{3}{4}} \rightarrow \frac{\frac{7}{2}}{\ }$ [Extend the fraction line]

[Change $3\frac{1}{2}$ to an improper fraction and extend the fraction line.]

Step 2. Calculate the denominator --- $\frac{3\frac{1}{2}}{\frac{3}{4}} = \frac{\frac{7}{2}}{\frac{3}{4}}$

[$\frac{3}{4}$ cannot be changed; therefore, just bring across as shown.]

$\frac{3\frac{1}{2}}{\frac{3}{4}} = 3\frac{1}{2} \div \frac{3}{4}$ because the line of a fraction means divide (÷).

Step 3. Divide as with common fractions or mixed numbers.

The fraction line means divide.

The bottom is the divisor.

$$\frac{\frac{7}{2}}{\frac{3}{4}} = \frac{7}{2} \div \frac{3}{4}$$

[The divisor always follows the division (÷) sign.]

[Invert the divisor and change the sign.]

$$\frac{7}{\cancel{2}_1} \times \frac{\cancel{4}^2}{3} = \frac{14}{3} \text{ or } 4\frac{2}{3} \quad \text{Answer}$$

Step 4. Check the answer by reversing the calculations where possible, starting from the answer.

1st, check the answer against the numerator.

2nd, check the denominator.

3rd, check the numerator.

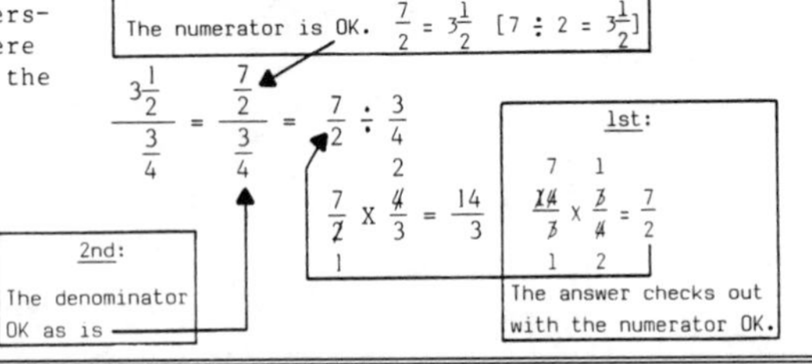

Example Problem 2. Calculate $\dfrac{\frac{3}{8}+\frac{15}{16}}{\frac{25}{32}\div 5}$

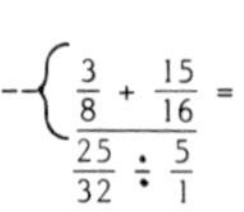

Step 1. Calculate the numerator

$$\frac{\frac{3}{8}+\frac{15}{16}}{\frac{25}{32}\div\frac{5}{1}} \qquad \frac{3}{8}=\frac{6}{16},\quad \frac{6}{16}+\frac{15}{16}=\frac{21}{16}$$

[Extend the fraction line]

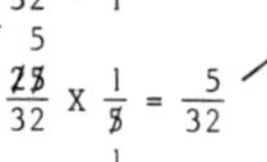

Step 2. Calculate the denominator

$$\frac{25}{32}\times\frac{1}{5}=\frac{5}{32} \qquad \frac{\frac{21}{16}}{\frac{5}{32}}$$

Step 3. Divide as with common fractions.

$$\frac{\frac{21}{16}}{\frac{5}{32}}=\frac{21}{16}\div\frac{5}{32}$$

$$\frac{21}{16}\times\frac{32}{5}=\frac{42}{5}=8\frac{2}{5} \quad \text{Answer}$$

Step 4. Check the answer by reversing the calculations where possible, starting from the answer.

1st, check the answer against the numerator.

2nd, check the denominator.

3rd, check the numerator.

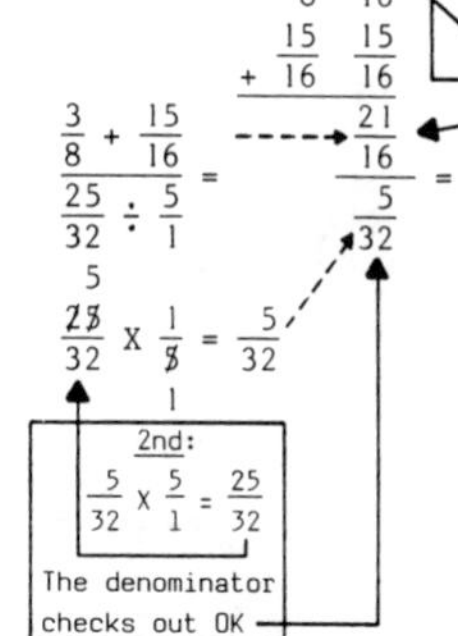

1st: $\frac{42}{5}\times\frac{5}{32}=\frac{21}{16}$ Answer checks out with the numerator OK.

2nd: $\frac{5}{32}\times\frac{5}{1}=\frac{25}{32}$ The denominator checks out OK

3rd: $\frac{21}{16}-\frac{15}{16}=\frac{6}{16}$ $[\frac{3}{8}+\frac{15}{16}=\frac{21}{16}]$ The numerator is OK.

B7	◂ Task No.	FACTORING

When adding and subtracting fractions with different denominators, it is necessary to find equivalent denominators for all the fractions. To make it easier, it is best to find the lowest common denominator (L.C.D.) of all the fractions (the reason is explained below). There are two popular ways to find the lowest common denominator of given fractions listed below; the **first method is to visually compare the given fractions**, and the **second method is to factor the given denominators**. Use the method that works best for you.

VISUALLY COMPARE THE GIVEN FRACTIONS TO FIND THE L.C.D.

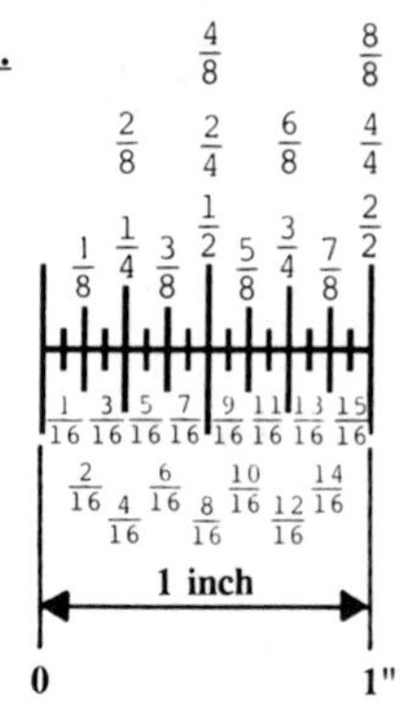

Illustrated to the right is an American English tape measure. Notice how equivalent fractions can be reduced by just observing where the fraction line lies. For example; 2/8 is the same as 4/16, and 4/16 is the same as 1/4 and so on.

This method can be used when adding and subtracting fractions that are quite common and easy to determine.

For example, find the L.C.D. of 1/4, 3/8, 9/16, and 23/32.

Step 1 List the fractions vertically as shown below.

$$\frac{1}{4}$$
$$\frac{3}{8}$$
$$\frac{9}{16}$$
$$\frac{23}{32}$$

Step 2 Notice that the largest denominator is divisible by the others; 32 is divisible by 16, 8, and 4. Therefore, list the largest denominator to the right of the given fractions as shown below.

$$\frac{1}{4} \quad \frac{\ }{32}$$
$$\frac{3}{8} \quad \frac{\ }{32}$$
$$\frac{9}{16} \quad \frac{\ }{32}$$
$$\frac{23}{32} \quad \frac{\ }{32}$$

Step 3 Make equivalent fractions with the L.C.D. as shown below.

$$\frac{1}{4} \quad \frac{8}{32}$$ [4 divided into 32 = 8; and (8 X 1 = 8)]

$$\frac{3}{8} \quad \frac{12}{32}$$ [8 divided into 32 = 4; and (4 X 3 = 12)]

$$\frac{9}{16} \quad \frac{18}{32}$$ [16 divided into 32 = 2; and (2 X 9 = 18)]

$$\frac{23}{32} \quad \frac{23}{32}$$

Task No. ▸ **B7**

FACTOR TO FIND THE L.C.D.

When adding fractions that are uncommon to each other such as 1/4 + 7/12 + 2/3 + 3/10, it is more difficult to determine the L.C.D. Listed below is a method that can be used to find the L.C.D. by factoring.

FACTOR BY CREATING A CHART:

Step 1 To find the L.C.D. by factoring, list all the denominators in a row horizontally as shown on the right ---------------▸

	4	12	3	10

Step 2 Find the smallest number that will divide evenly into one or more of the listed numbers. Write this number to the left as shown ---------------▸

2)	4	12	3	10

Step 3 Divide this number into the listed numbers and write these quotients on the line below as shown ---------------▸
[If the number does not divide evenly, write the original denominator.]

2)	4	12	3	10
	2	6	3	5

Step 4 Find the smallest number that will divide evenly into one or more of the new group of numbers and write these new quotients on the line below as shown. ---------------▸

2)	4	12	3	10
2)	2	6	3	5
	1	3	3	5

Step 5 Find the smallest number that will divide evenly into the new group of numbers and write these new quotients on the line below as shown. ---------------▸

2)	4	12	3	10
2)	2	6	3	5
3)	1	3	3	5
	1	1	1	5

Step 6 Repeat the above step one more time, until you have a row of ones.

2)	4	12	3	10
2)	2	6	3	5
3)	1	3	3	5
5)	1	1	1	5
	1	1	1	1

Step 7 The product of the divisors on the left is the lowest common denominator. 2 X 2 X 3 X 5 = 60

Step 8 Set up the given fractions and the lowest common denominators as shown, then make equivalent fractions out of the L.C.D., and add the numerators. Reduce if you can.

$$\frac{1}{4} \quad \frac{15}{60}$$
$$\frac{7}{12} \quad \frac{35}{60}$$
$$\frac{2}{3} \quad \frac{40}{60}$$
$$+\ \frac{3}{10} \quad \frac{18}{60}$$
$$\frac{108}{60} = 1\frac{48}{60} \text{ or } 1\frac{4}{5} \text{ reduced}$$

C1	◂ Task No.	MIXED NUMBERS

A **mixed number** is a whole number and a fraction written together.
For example, $3\frac{1}{4}$ is a mixed number.

The **whole number** is written first and shows how many complete units there are. For example, 3 means 3 whole units.

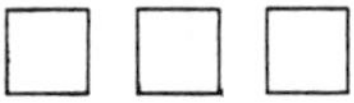

The **fraction** is written after the whole number and shows what part of a whole unit is taken. For example, $\frac{1}{4}$ is 1 of 4 parts of a whole unit.

Therefore, the mixed number $3\frac{1}{4}$ = 3 full units and $\frac{1}{4}$ of a unit.

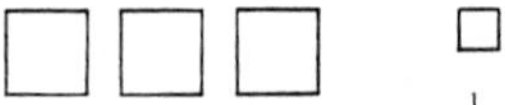

An **improper fraction** has a numerator larger than the denominator → $\frac{5}{4}$

Or a numerator equal to the denominator → $\frac{4}{4}$

RULES: To change an improper fraction, divide the bottom number (denominator) into the top number (numerator). The result will be a mixed number or whole number. Examples:

[$\frac{11}{8} = 8\overline{)11}$ = $1\frac{3}{8}$ (mixed number), with 1 in the quotient, 8 subtracted, remainder 3; $\frac{8}{8} = 8\overline{)8}$ = 1 (whole number), with 1 in the quotient, 8 subtracted]

> Note: One reason for reducing an improper fraction such as $\frac{11}{8}$, is that measuring-instrument parts (increments) are usually written and located in reduced form.
>
> Example: 1 3/8" — $\frac{1}{8}$ $\frac{2}{8}$ $\frac{3}{8}$ $\frac{4}{8}$ $\frac{5}{8}$ $\frac{6}{8}$ $\frac{7}{8}$ 1 ($\frac{8}{8}$) $\frac{9}{8}$ $\frac{10}{8}$ $\frac{11}{8}$
>
> [It is easier to find $1\frac{3}{8}$ on a tape measure than to count out eleven $\frac{1}{8}$ parts.]

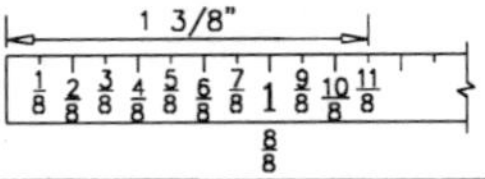

To change a mixed number to an improper fraction, multiply the whole number by the denominator in the fraction, then add this product to the numerator. The result is an improper fraction.

Example: $4\frac{3}{8} = 4 \times 8 = 32$ and $32 + \frac{3}{8} = \frac{35}{8}$ (improper fraction)

The lowest common denominator (L.C.D.) is the **lowest possible number** that can be divided evenly by the denominators of all given fractions.

> Example:
>
> Given Fractions: $\frac{3}{8} \rightarrow \frac{\ }{16}$, $\frac{1}{4} \rightarrow \frac{\ }{16}$, $\frac{7}{16} \rightarrow \frac{\ }{16}$
>
> [Note the lowest possible number]
>
> 16 is the L.C.D. because
> 8 divides into 16 evenly
> 4 divides into 16 evenly
> 16 divides into 16 evenly

To make **equivalent fractions** with the L.C.D., 1st divide the denominator of the given fractions into the L.C.D., 2nd multiply that answer (quotient) by the numerator of the given fraction, and 3rd place that answer (product) over the L.C.D. The new fractions are called **equivalent fractions**.

Example: Given fractions:

1st divide the bottom 8 into the bottom 16 = 2

2nd multiply the answer 2 by the top left 3 = 6

3rd place the answer 6 over the L.C.D. 16

$\frac{3}{8} = \frac{6}{16}$ (8 into 16 = 2; times)

[Calculate each given fraction as shown above and to the left.]

$\frac{3}{8}$ $\frac{6}{16}$ [16 ÷ 8 = 2; 2 X 3 = ⑥]

$\frac{1}{4}$ $\frac{4}{16}$ [16 ÷ 4 = 4; 4 X 1 = ④]

$\frac{7}{16}$ $\frac{7}{16}$ [16 ÷ 16 = 1; 1 X 7 = ⑦]

Subtraction of Mixed Numbers

Example: $6\frac{1}{4} - 2\frac{3}{8}$

How to borrow when needed:

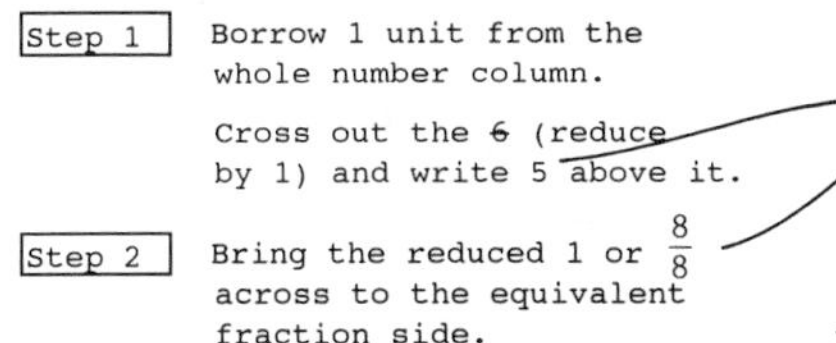

Step 1 — Borrow 1 unit from the whole number column.

Cross out the ~~6~~ (reduce by 1) and write 5 above it.

Step 2 — Bring the reduced 1 or $\frac{8}{8}$ across to the equivalent fraction side.

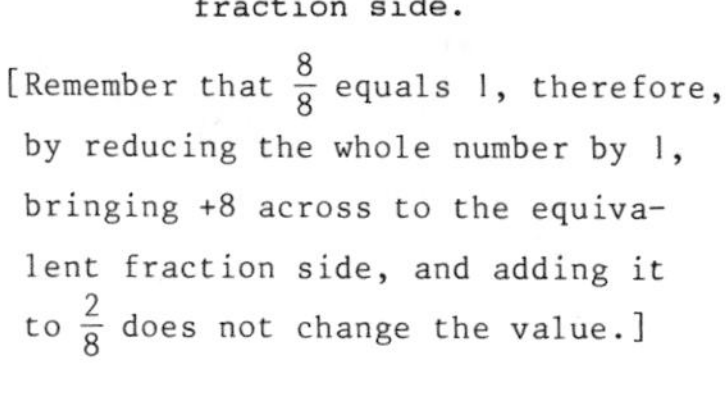

[Remember that $\frac{8}{8}$ equals 1, therefore, by reducing the whole number by 1, bringing +8 across to the equivalent fraction side, and adding it to $\frac{2}{8}$ does not change the value.]

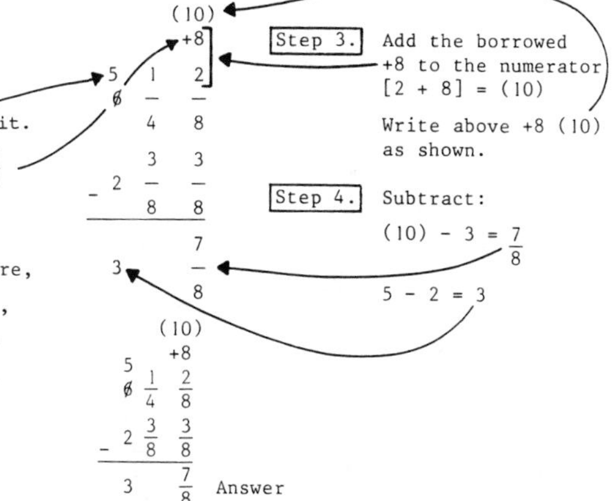

Step 3. — Add the borrowed +8 to the numerator [2 + 8] = (10)

Write above +8 (10) as shown.

Step 4. — Subtract:

(10) − 3 = $\frac{7}{8}$

5 − 2 = 3

$$\begin{array}{rcc} & & (10) \\ & & +8 \\ & 5 & \\ & \not{6}\ \frac{1}{4} & \frac{2}{8} \\ - & 2\ \frac{3}{8} & \frac{3}{8} \\ \hline & 3 & \frac{7}{8} \end{array}$$

Answer

Division of Mixed Numbers

Do's and Don'ts:

When calculating mixed numbers, always write down each step, especially after changing to improper fractions.

	DO		DON'T
Step 1.	$4\frac{1}{2} \div 2\frac{1}{4}$	Step 1.	$4\frac{1}{2} \div 2\frac{1}{4}$
Step 2.	$\frac{9}{2} \div \frac{9}{4}$	Step 3.	$\frac{9}{2} \times \frac{4}{9}$
Step 3.	$\frac{9}{2} \times \frac{4}{9}$		

Example: Divide

Don't skip Step 2.

(Skipping the 2nd step just increases your chances for errors.)

Get into the habit of writing down each step. If you don't, how will you be able to go back and check each step?

C2	◂ Task No.	ADD MIXED NUMBERS

Example Problem 1. Add $3\frac{1}{4} + 2\frac{5}{8}$

Step 1 Find the lowest common denominator (L.C.D.).

$$\begin{array}{rr} 3\frac{1}{4} & \frac{}{8} \\ +\ 2\frac{5}{8} & \frac{}{8} \end{array}$$

4 divides into 8 evenly; 8 divides into 8 evenly → 8 L.C.D.

Step 2 Make equivalent fractions with the L.C.D. (8).

$$\begin{array}{rr} 3\frac{1}{4} & \frac{2}{8} \\ +\ 2\frac{5}{8} & \frac{5}{8} \end{array}$$

Step 3 Add the numerators. $[\frac{2}{8} + \frac{5}{8} = \frac{7}{8}]$

$$\begin{array}{rrr} 3\frac{1}{4} & \frac{2}{8} & 2 \\ +\ 2\frac{5}{8} & \frac{5}{8} & +5 \\ & & 7 \\ \hline 5 & \frac{7}{8} & \end{array}$$

Add the whole numbers. 3 + 2 = 5 → $5\frac{7}{8}$ Answer

Step 4 Check by subtracting.

$$\begin{array}{rr} 3\frac{1}{4} & \frac{2}{8} \\ +\ 2\frac{5}{8} & \frac{5}{8} \\ \hline 5 & \frac{7}{8} \\ -\ 2 & \frac{5}{8} \\ \hline 3 & \frac{2}{8} \end{array}$$

Subtract; Matches Answer OK

Example Problem 2. Add $1\frac{5}{8}$, $2\frac{1}{2}$, and $4\frac{3}{16}$

Step 1 Find the lowest common denominator.

$$\begin{array}{rr} 1\frac{5}{8} & \frac{}{16} \\ 2\frac{1}{2} & \frac{}{16} \\ +\ 4\frac{3}{16} & \frac{}{16} \end{array}$$

8 divides into 16 evenly; 2 divides into 16 evenly; 16 divides into 16 evenly → 16 L.C.D.

Step 2 Make equivalent fractions with the L.C.D. (16).

Add the numerators (10 + 8 + 3 = 21), and add the whole numbers. 1 + 2 + 4 = 7 → 7

$$\begin{array}{rrr} 1\frac{5}{8} & \frac{10}{16} & 10 \\ 2\frac{1}{2} & \frac{8}{16} & 8 \\ +\ 4\frac{3}{16} & \frac{3}{16} & +3 \\ & & 21 \\ \hline 7 & \frac{21}{16} = 8\frac{5}{16} & \end{array}$$

Answer

Step 3 Reduce.

$$\frac{21}{16} = 16\overset{1}{\overline{)21}} = 1\frac{5}{16};\quad 1\frac{5}{16} + 7 = 8\frac{5}{16}$$

$$\begin{array}{r} 1 \\ 16\overline{)21} \\ \underline{16} \\ 5 \end{array}$$

Step 4 Check by adding the top two mixed numbers. Then add that sum to the third mixed number.

The results should match the step 2 answer before reducing.

$$\begin{array}{rr} 1\frac{5}{8} & \frac{10}{16} \\ 2\frac{1}{2} & \frac{8}{16} \\ +\ 4\frac{3}{16} & \frac{3}{16} \\ \hline 7 & \frac{21}{16} \end{array} \qquad \begin{array}{r} 1\frac{10}{16} \\ +\ 2\frac{8}{16} \\ \hline 3\frac{18}{16} \\ +\ 4\frac{3}{16} \\ \hline 7\frac{21}{16} \end{array}$$

Answer OK

SUBTRACT MIXED NUMBERS — Task No. ▸ C3

Example Problem 1. Subtract $6\frac{1}{2} - 2\frac{3}{8}$

$$\begin{array}{rr} 6\frac{1}{2} & \frac{\ }{8} \\ -\ 2\frac{3}{8} & \frac{\ }{8} \\ \hline \end{array}$$

2 divides into 8 evenly
8 divides into 8 evenly } **8 L.C.D.**

Step 1 Find the lowest common denominator.

$$\begin{array}{rr} 6\frac{1}{2} & \frac{4}{8} \\ -\ 2\frac{3}{8} & \frac{3}{8} \\ \hline \end{array}$$

Step 2 Make equivalent fractions with the lowest common denominator (L.C.D.) 8.

Step 3 Subtract the numerators. $[\frac{4}{8} - \frac{3}{8} = \frac{1}{8}]$

$$\begin{array}{rr} 6\frac{1}{2} & \frac{4}{8} \\ -\ 2\frac{3}{8} & \frac{3}{8} \\ \hline 4 & \frac{1}{8} \end{array}$$

$4 \rightarrow 4$, $3 \rightarrow -3$, $= 1$

Subtract the whole numbers. $6 - 2 = 4$ ----▸ 4 $\frac{1}{8}$

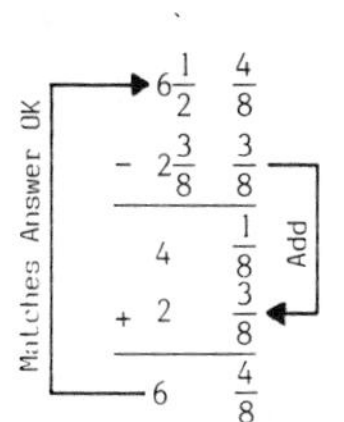

Step 4 Check by adding.

Example Problem 2. Subtract $14\frac{5}{16} - 5\frac{3}{4}$

$$\begin{array}{rr} 14\frac{5}{16} & \frac{\ }{16} \\ -\ 5\frac{3}{4} & \frac{\ }{16} \\ \hline \end{array}$$

16 divides into 16 evenly
4 divides into 16 evenly } **16 L.C.D.**

Step 1 Find the lowest common denominator.

$$\begin{array}{rr} 14\frac{5}{16} & \frac{5}{16} \\ -\ 5\frac{3}{4} & \frac{12}{16} \\ \hline \end{array}$$

Step 2 Make equivalent fractions with the L.C.D. (16).

(21) --▸ (21)
+16

$$\begin{array}{rr} \overset{13}{\not{14}}\frac{5}{16} & \frac{5}{16} \\ -\ 5\frac{3}{4} & \frac{12}{16} \\ \hline 8 & \frac{9}{16} \end{array}$$

$12 \rightarrow -12$, $= 9$ — Answer

Step 3 Borrow if you have to and subtract the numerators (21) - 12 = 9
Subtract the whole numbers. 13 - 5 = 8 ---▸ 8 $\frac{9}{16}$

(21)
+16

$$\begin{array}{rr} \overset{13}{\not{14}}\frac{5}{16} & \frac{5}{16} \\ -\ 5\frac{3}{4} & \frac{12}{16} \\ \hline 8 & \frac{9}{16} \\ +\ 5 & \frac{12}{16} \\ \hline 13 & \frac{21}{16} \end{array}$$

Add
Matches Answer OK

Step 4 Check by adding.

C4	◂ Task No.	MULTIPLY MIXED NUMBERS

Example Problem 1. Multiply $2\frac{3}{4} \times 2\frac{2}{3}$

$\left[2\frac{2}{3} = 2 \times 3 = 6 \text{ and } 6 + \frac{2}{3} = \frac{8}{3}\right]$

Step 1 Change to improper fractions.

$\left[2\frac{3}{4} = 2 \times 4 = 8 \text{ and } 8 + \frac{3}{4} = \frac{11}{4}\right]$

$2\frac{3}{4} \times 2\frac{2}{3} =$

$\frac{11}{4} \times \frac{8}{3} =$

Step 2 Cancel if you can and multiply straight across.

$(11 \times 2 = 22)$

$\frac{11}{\not{4}_1} \times \frac{\not{8}^2}{3} = \frac{22}{3} = 7\frac{1}{3}$ Answer

$(1 \times 3 = 3)$

Step 3 Reduce to lowest terms.

$\frac{22}{3} = 3\overline{)22}$ → 7, $\frac{21}{1}$ $= 7\frac{1}{3}$ Answer

Step 4 Check by dividing.

$\frac{11}{\not{4}_1} \times \frac{\not{8}^2}{3} = \frac{22}{3}$

$\frac{22}{3} \div \frac{8}{3} = \frac{\not{22}^{11}}{\not{3}_1} \times \frac{\not{3}^1}{\not{8}_4} = \frac{11}{4}$

Matches Answer OK

Example Problem 2. Multiply $4 \times 2\frac{5}{8}$

$\left[2\frac{5}{8} = 2 \times 8 = 16 \text{ and } 16 + \frac{5}{8} = \frac{21}{8}\right]$

Step 1 Change to improper fractions

$4 \times 2\frac{5}{8} =$

$\frac{4}{1} \times \frac{21}{8} =$

Step 2 Cancel if you can and multiply straight across.

$\frac{\not{4}^1}{1} \times \frac{21}{\not{8}_2} = \frac{21}{2} = 10\frac{1}{2}$ Answer

Step 3 Reduce to lowest terms.

$\frac{21}{2} = 2\overline{)21}$ → 10, $\frac{20}{1}$ $= 10\frac{1}{2}$ Answer

Step 4 Check by dividing.

$\frac{\not{4}^1}{1} \times \frac{21}{\not{8}_2} = \frac{21}{2}$

$\frac{21}{2} \div \frac{21}{8} = \frac{\not{21}^1}{\not{2}_1} \times \frac{\not{8}^4}{\not{21}_1} = \frac{4}{1}$

Matches Answer OK

DIVIDE MIXED NUMBERS — Task No. ▸ C5

Example Problem 1. Divide $7\frac{1}{2} \div 1\frac{1}{8}$

$7\frac{1}{2} = 7 \times \frac{1}{2} = 14$ and $14 + \frac{1}{2} = \frac{15}{2}$

$1\frac{1}{8} = 1 \times \frac{1}{8} = 8$ and $8 + \frac{1}{8} = \frac{9}{8}$

Step 1 — 1st, change to improper fractions

2nd, invert the divisor.

3rd, change the (÷) sign to (x).

$7\frac{1}{2} \div 1\frac{1}{8} =$

$\frac{15}{2} \div \frac{9}{8}$

$\frac{15}{2} \times \frac{8}{9}$

Step 2 — 1st, cancel (if you can).

2nd, multiply straight across.

$\frac{\cancel{15}^{5}}{\cancel{2}_{1}} \times \frac{\cancel{8}^{4}}{\cancel{9}_{3}} = \frac{20}{3} = 6\frac{2}{3}$ Answer

Step 3 — Reduce to lowest terms.

$\frac{20}{3} = 3\overline{)20}$ with 6, $\frac{18}{2}$ $= 6\frac{2}{3}$ Answer

Step 4 — Check by multiplying.

$\frac{15}{2} \div \frac{9}{8} = \frac{20}{3}$

$\frac{\cancel{20}^{5}}{\cancel{3}_{1}} \times \frac{\cancel{9}^{3}}{\cancel{8}_{2}} = \frac{15}{2}$

Matches Answer OK

Example Problem 2. Divide $2\frac{9}{16} \div 2$

$2\frac{9}{16} = 2 \times \frac{9}{16} = 32$ and $32 + \frac{9}{16} = \frac{41}{16}$

Step 1 — 1st, change to improper fractions.

2nd, invert the divisor.

3rd, change the (÷) sign to (x).

$2\frac{9}{16} \div 2 =$

$\frac{41}{16} \div \frac{2}{1}$

$\frac{41}{16} \times \frac{1}{2} =$

[cannot cancel]

Step 2 — 1st, cancel (if you can).

2nd, multiply straight across.

$\frac{41}{16} \times \frac{1}{2} = \frac{41}{32} = 1\frac{9}{32}$

Step 3 — Reduce to lowest terms.

$\frac{41}{32} = 32\overline{)41}$ with 1, $\frac{32}{9}$ $= 1\frac{9}{32}$ Answer

Step 4 — Check by multiplying.

$\frac{41}{16} \div \frac{2}{1} = \frac{41}{32}$

$\frac{41}{\cancel{32}_{16}} \times \frac{\cancel{2}^{1}}{1} = \frac{41}{16}$

Matches Answer OK

D1 ◄ Task No. DENOMINATE NUMBERS

A **Denominate number** is any number that applies to a specific unit of measure. For example, 5 ft. 2 in., 8 lb. 5 oz., and 2 hr. 5 min. are all denominate numbers.

A denominate number does not change in value when its form is changed. For example, 27" changed to feet and inches is 2'3". 27" and 2'3" are equal.

RULE: To change the form of a denominate number with small units to one with larger units, divide.

Example 1. Change 27 inches to feet.

Step 1 Divide 27" by 12 because there are 12" in 1 foot.

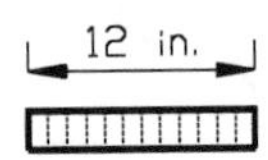


```
      2
12/27" = 2'3"
   24
    3
```

Example 2. Change 360 sq. in. to sq. ft

Step 1 Divide 360 sq. in. by 144 because there are 144 sq. in. in 1 sq. ft.

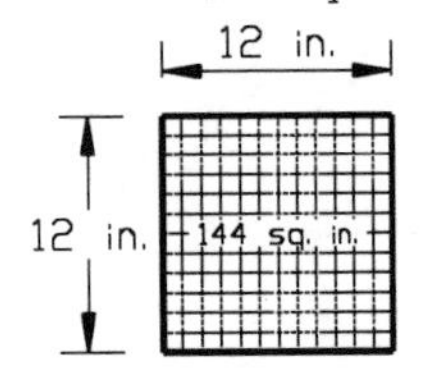


```
      2.5 sq. ft.
144/360
    288
     720
     720
```

Example 3. Change 135 cu. ft. to cu. yd.

Step 1 Divide 135 cu. ft. by 27 because there are 27 cu. ft. in 1 cubic yard.

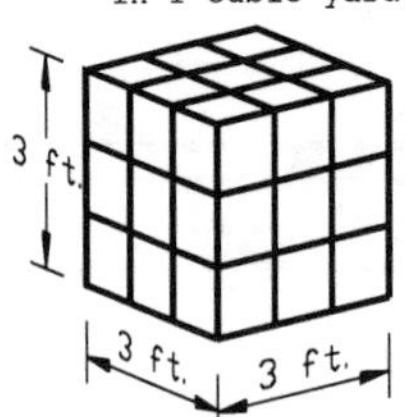


```
     5 cu. yd.
27/135
   135
     0
```

Example 4. Change 40 ounces to pounds.

Step 1 Divide 40 oz. by 16 because there are 16 oz. in 1 lb.

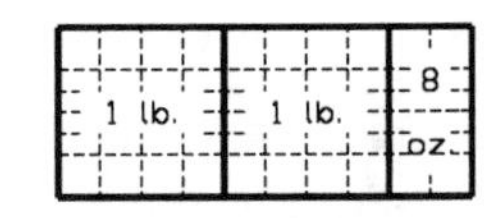


```
    2 lb. 8 oz.
16/40
   32
    8
```

RULE: To change the form of a denominate number with large units to one with small units, multiply.

Example 1. Change 4 yards to feet.

Step 1 Multiply by 3 because there are 3 ft. in 1 yd.

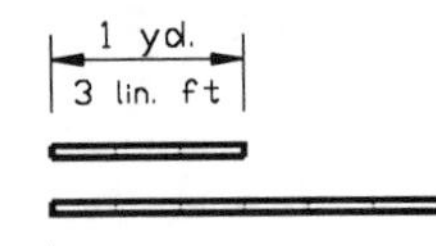


```
  4
X 3
 12 lin. ft.
```

Example 2. Change 6 1/2 feet to inches.

Step 1 Multiply by 12 because there are 12 inches in 1 ft.

```
   6.5
X   12
   130
   65
  78.0 lin. in.
```

Example 3. Change 4 3/4 pounds to ounces.

Step 1 Multiply by 16 because there there are 16 oz. in one lb.

```
   4.75
X    16
   2850
   475
  76.00 oz.
```

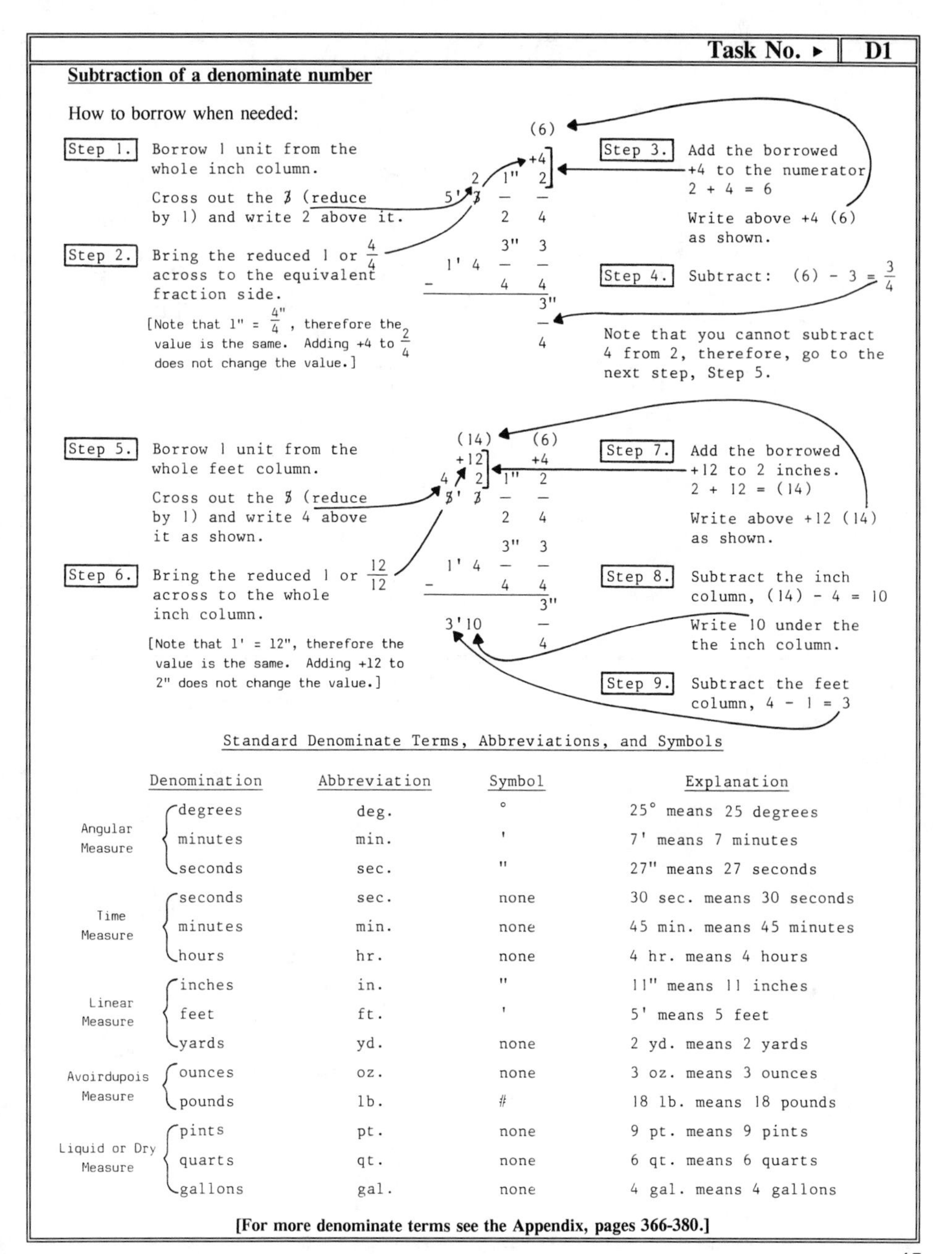

Task No. ▸ D1

Subtraction of a denominate number

How to borrow when needed:

Step 1. Borrow 1 unit from the whole inch column.

Cross out the 3 (reduce by 1) and write 2 above it.

Step 2. Bring the reduced 1 or $\frac{4}{4}$ across to the equivalent fraction side.

[Note that 1" = $\frac{4"}{4}$, therefore the value is the same. Adding +4 to $\frac{2}{4}$ does not change the value.]

Step 3. Add the borrowed +4 to the numerator 2 + 4 = 6

Write above +4 (6) as shown.

Step 4. Subtract: (6) − 3 = $\frac{3}{4}$

Note that you cannot subtract 4 from 2, therefore, go to the next step, Step 5.

Step 5. Borrow 1 unit from the whole feet column.

Cross out the 5 (reduce by 1) and write 4 above it as shown.

Step 6. Bring the reduced 1 or $\frac{12}{12}$ across to the whole inch column.

[Note that 1' = 12", therefore the value is the same. Adding +12 to 2" does not change the value.]

Step 7. Add the borrowed +12 to 2 inches. 2 + 12 = (14)

Write above +12 (14) as shown.

Step 8. Subtract the inch column, (14) − 4 = 10

Write 10 under the the inch column.

Step 9. Subtract the feet column, 4 − 1 = 3

Standard Denominate Terms, Abbreviations, and Symbols

	Denomination	Abbreviation	Symbol	Explanation
Angular Measure	degrees	deg.	°	25° means 25 degrees
	minutes	min.	'	7' means 7 minutes
	seconds	sec.	"	27" means 27 seconds
Time Measure	seconds	sec.	none	30 sec. means 30 seconds
	minutes	min.	none	45 min. means 45 minutes
	hours	hr.	none	4 hr. means 4 hours
Linear Measure	inches	in.	"	11" means 11 inches
	feet	ft.	'	5' means 5 feet
	yards	yd.	none	2 yd. means 2 yards
Avoirdupois Measure	ounces	oz.	none	3 oz. means 3 ounces
	pounds	lb.	#	18 lb. means 18 pounds
Liquid or Dry Measure	pints	pt.	none	9 pt. means 9 pints
	quarts	qt.	none	6 qt. means 6 quarts
	gallons	gal.	none	4 gal. means 4 gallons

[For more denominate terms see the Appendix, pages 366-380.]

D2	◄ Task No.	ADD DENOMINATE NUMBERS

Example Problem 1. Add $3'6\frac{1}{4}'' + 5'2\frac{1}{8}''$

Step 1 Find the lowest common denominator (L.C.D.).

$$\begin{array}{rr} 3'6\frac{1}{4}'' & \frac{\ }{8} \\ +\ 5'2\frac{1}{8}'' & \frac{\ }{8} \\ \hline \end{array}$$

4 divides into 8 evenly
8 divides into 8 evenly } **8 L.C.D**

Step 2 Make equivalent fractions with the lowest common denominator (L.C.D.) 8.

$$\begin{array}{rr} 3'6\frac{1}{4}'' & \frac{2}{8} \\ +\ 5'2\frac{1}{8}'' & \frac{1}{8} \\ \hline \end{array}$$

1 inch

$\frac{1}{4}$, $\frac{2}{8}$, $\frac{1}{8}$

Note: 1/4 and 2/8 are equivalent (equal)

Step 3 Add the numerators
2 + 1 = 3, and

add the whole inches
6 + 2 = 8

and feet 3 + 5 = 8

$$\begin{array}{rrl} 3'6\frac{1}{4}'' & \frac{2}{8} & \rightarrow 2 \\ +\ 5'2\frac{1}{8}'' & \frac{1}{8} & \rightarrow +1 \\ \hline 8'8 & \frac{3}{8}'' & 3 \end{array}$$

Answer

Step 4 Check by subtracting.

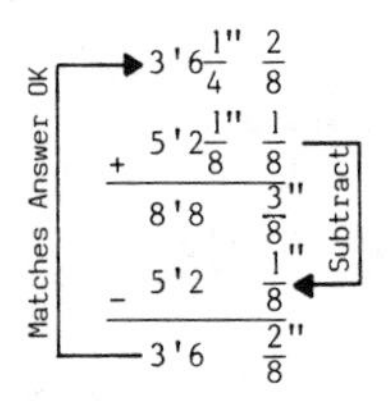

Example Problem 2. Add $1'1\frac{3}{4}''$, $2'0\frac{1}{2}''$, and $7'7\frac{5}{16}''$

Step 1 Find the L.C.D.

$$\begin{array}{rr} 1'1\ \frac{3}{4}'' & \frac{\ }{16} \\ 2'0\ \frac{1}{2}'' & \frac{\ }{16} \\ +\ 7'7\frac{5}{16}'' & \frac{\ }{16} \\ \hline \end{array}$$

4 divides into 16 evenly
2 divides into 16 evenly } **16 L.C.D.**
16 divides into 16 evenly

Step 2 Make equivalent fractions with the L.C.D. (16),

add the numerators
12 + 8 + 5 = 25, and

add the whole inches
1 + 0 + 7 = 8

and feet 1 + 2 + 7 = 10

$$\begin{array}{rrl} 1'1\ \frac{3}{4}'' & \frac{12}{16} & \rightarrow 12 \\ 2'0\ \frac{1}{2}'' & \frac{8}{16} & \rightarrow 8 \\ +\ 7'7\frac{5}{16}'' & \frac{5}{16} & \rightarrow +5 \\ \hline 10'8 & \frac{25}{16}'' & 25 \end{array}$$

$\frac{25}{16}'' = 10'9\frac{9}{16}''$ Answer

1 inch

$\frac{1}{2}$ $\frac{3}{4}$

$\frac{8}{16}$ $\frac{12}{16}$

3/4 and 12/16 are equivalent

1/2 and 8/16 are equivalent

Step 3 Reduce.

$$\frac{25}{16} = 16\overline{)25}^{\,1} = 1\frac{9}{16};\ 1\frac{9}{16}'' + 8'' = 9\frac{9}{16}''$$

(16/25: 1, −16, 9)

Step 4 Check by adding the top two mixed numbers. Then add that sum to the third mixed number.

The results should match the step 2 answer before reducing.

$$\begin{array}{rr} 1'1\ \frac{3}{4}'' & \frac{12}{16} \\ 2'0\ \frac{1}{2}'' & \frac{8}{16} \\ +\ 7'7\frac{5}{16}'' & \frac{5}{16} \\ \hline 10'8 & \frac{25}{16}'' \end{array}$$

$$\begin{array}{r} 1'1\frac{12}{16}'' \\ +\ 2'0\frac{8}{16}'' \\ \hline 1'1\frac{20}{16}'' \\ +\ 7'7\frac{5}{16} \\ \hline 10'8\frac{25}{16}''\ \text{OK} \end{array}$$

SUBTRACT DENOMINATE NUMBERS — Task No. ▸ D3

Example Problem 1. Subtract. $7'6\frac{1}{2}" - 3'4\frac{1}{8}"$

Step 1 Find the lowest common denominator (L.C.D.) and make equivalent fractions.

$$\begin{array}{rr} 7'6\frac{1}{2}" & \frac{4}{8} \\ -\ 3'4\frac{1}{8}" & \frac{1}{8} \\ \hline \end{array}$$

2 divides into 8 evenly
8 divides into 8 evenly } 8 L.C.D.

Step 2 Subtract the numerators (4 - 1 = 3), and the whole numbers:

inches (6 - 4 = 2)

feet (7 - 3 = 4)

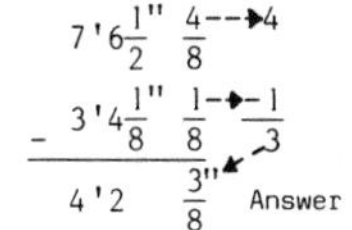

$$\begin{array}{rr} 7'6\frac{1}{2}" & \frac{4}{8} \\ -\ 3'4\frac{1}{8}" & \frac{1}{8} \\ \hline 4'2 & \frac{3}{8}" \end{array}$$

1 inch

$\frac{1}{2}$

$\frac{1}{8}$ $\frac{4}{8}$

Step 3 Check by adding.

$$\begin{array}{rr} 7'6\frac{1}{2}" & \frac{4}{8} \\ -\ 3'4\frac{1}{8}" & \frac{1}{8} \\ \hline 4'2 & \frac{3}{8}" \\ +\ 3'4 & \frac{1}{8}" \\ \hline 7'6 & \frac{4}{8}" \end{array}$$

Add

Matches Answer OK

Example Problem 2. Subtract $11'2\frac{1}{8}" - 6'3\frac{7}{16}"$

Step 1 Find the L.C.D. and make equivalent fractions.

$$\begin{array}{rr} 11'2\frac{1}{8}" & \frac{2}{16} \\ -\ 6'3\frac{7}{16}" & \frac{7}{16} \\ \hline \end{array}$$

8 divides into 16 evenly
16 divides into 16 evenly } 16 L.C.D.

Step 2 Borrow if you have to and subtract the numerators (18 - 7 = 11), and the whole numbers:

inches (13 - 3 = 10)

feet (10 - 6 = 4)

$$\begin{array}{rr} & (13)\quad (18) \\ & +12\quad +16 \\ & 10\ \ 1 \\ \not{1}\not{1}'\not{2}\frac{1}{8}" & \frac{2}{16} \\ -\ 6'3\frac{7}{16}" & \frac{7}{16} \\ \hline 4'10 & \frac{11}{16}" \end{array}$$

Answer

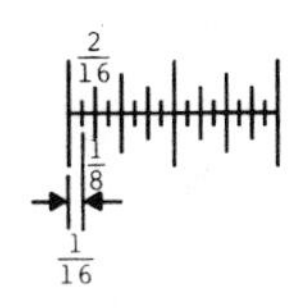

Step 3 Check by adding.

$$\begin{array}{rr} & (13)\quad (18) \\ & +12\quad +16 \\ & 10\ \ 1 \\ \not{1}\not{1}'\not{2}\frac{1}{8}" & \frac{2}{16} \\ -\ 6'3\frac{7}{16}" & \frac{7}{16} \\ \hline 4'10 & \frac{11}{16}" \\ +\ 6'\ 3 & \frac{7}{16}" \\ \hline 10'13 & \frac{18}{16}" \end{array}$$

Add

Matches Answer OK

D4 ◂ Task No. MULTIPLY DENOMINATE NUMBERS

Example Problem 1. Multiply 6'4" x 4'2".

Step 1 Change to all inches.

6'4" [6 X 12 (=72) + 4] = 76"
X 4'2" [4 X 12 (=48) + 2] = 50"

Step 2 Multiply:

76"
x 50"
3800 square inches (area)

Step 3 Convert to square feet.

1 sq. ft. = 144 sq. in., therefore divide by 144.

144/3800 = 26.38 sq. ft.

Step 4 Check by solving with mixed numbers.

$6'4" = [6\frac{4}{12} = 6\frac{1}{3}']$ $4'2" = [4\frac{2}{12} = 4\frac{1}{6}']$

$6\frac{1}{3}' \times 4\frac{1}{6}' = \frac{19}{3} \times \frac{25}{6} = \frac{475}{18} = 26.38$ sq. ft.

Or, solve by using decimal fractions.

$6\frac{1}{3}' = 6.33'$

$4\frac{1}{6}' = 4.16'$

```
   6.33'
 X 4.16'
   3798
   633
  2532
26.3328 sq. ft.
```

[For greater accuracy use 6.333 X 4.166 equals 26.383]

Example Problem 2. Multiply 3'5" x 5

Step 1 Change to all inches.

3'5" [3 X 12 (=36) + 5] = 41"
X 5

Step 2 Multiply:

41"
x 5
205 linear inches (length)

Step 3 Convert to linear feet.

```
    17'  = 17'1"
12/205
   12
    85
    84
     1
```

Step 4 Check by multiplying the multiplier by each specific unit.

3'5"
x 5
15'25" =17'1" Answer OK

Or, solve by using mixed numbers. $[3'5" = 3\frac{5}{12}']$ $3\frac{5}{12}' \times 5 =$

$\frac{41}{12} \times \frac{5}{1} = \frac{205}{12} = 17'1"$

Answer OK

DIVIDE DENOMINATE NUMBERS   Task No. ▸ D5

Example Problem 1. Divide 6'8" ÷ 4

Step 1 Change to all inches.

6'8" = [6 x 12 (=72) + 8] = 80"

Step 2 Divide: 80" ÷ 4 =

$$4\overline{)80} = 20" \quad (8,\ 0) = 1'8" \text{ Answer}$$

Convert 20" to feet.
Divide by 12".

$$12\overline{)20} = 1' \quad (12,\ 8) = 1'8"$$

Check by multiplying: 20 x4 = 80 **Answer OK**

Or see **Step 3** below.

Step 3 Check by dividing with mixed numbers. Or by dividing into each unit separately.

$$[6'8" = 6\frac{8}{12} \text{ or } 6\frac{2}{3}']$$

$$6\frac{2}{3} \div 4 =$$

$$\frac{20}{3} \div \frac{4}{1} =$$

$$\frac{20}{3} \times \frac{1}{4} = \frac{20}{12} \text{ or } 1'8" \quad \text{Answer OK}$$

$$4\overline{)6'\ 8"} = 1'\ 8" \quad \text{Answer OK}$$

4
2=+24"
32"
32"

[Decimal fractions can also be used.
Use one method to check the other method.]

Example Problem 2. Divide 5'6" ÷ 5

Step 1 Change to all inches.

5'6" = [5 x 12 (=60) + 6] = 66

Step 2 Divide: 66 ÷ 5

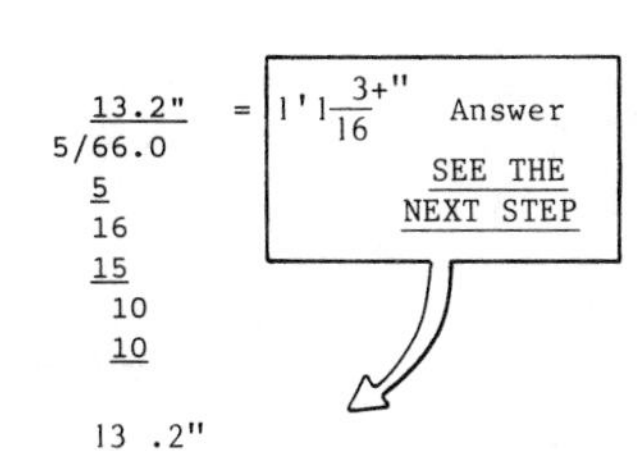

Step 3 Convert so you can read the dimension on a tape measure.

13 .2"
X16
3.2 = 3+/16"

$$\text{Reduce } (13") = 12\overline{)13} = 1'1" \quad (12,\ 1) \qquad (13)\frac{3+}{16}" = 1'1\frac{3+}{16}" \quad \text{Answer}$$

Step 4 Check by solving with mixed numbers. $[5'6" = 5\frac{6}{12} \text{ or } 5\frac{1}{2}']$

$$5\frac{1}{2}' \div 5 =$$

$$\frac{11}{2} \div \frac{5}{1} =$$

$$\frac{11}{2} \times \frac{1}{5} = \frac{11}{10} = 10\overline{)11} = 1.1' \text{ or } 1'1\frac{3+}{16}" \quad \text{Answer OK}$$

1 .1'
X12 for inches
1.2"
X16 for 16ths
3.2 = 3+/16"

D6	◂ Task No.	FIND ½ OF A DENOMINATE NUMBER

Shown below and on the next page are three different methods to find ½ of a denominate number. The method shown step by step is emphasized because, after some practice, it can be a quick and accurate way to do it on the job. **Apply-It Module #1, Task No. D7**, also demonstrates these methods and illustrates some practical applications for them. Use the methods that work best for you.

Example Problem 1. Find ½ of 8 ft. 4 in.

(ft. column | in. column)

Step 1 Find ½ of the feet column and bring down as shown. 8 ft. 4 in.

$[\frac{1}{2}$ of $8' = \frac{1}{2} \times \frac{8}{1} = \frac{4}{1}]$ -------------------- 4 ft.

Step 2 Find ½ of the inch column and bring down as shown.

$[\frac{1}{2}$ of $4'' = \frac{1}{2} \times \frac{4}{1} = \frac{2}{1}]$ ---------------------------- 2 in.

Step 3 Add Step 1 and Step 2.

```
(Step 1)  4 ft.
       +        2 in.  (Step 2)
          4 ft. 2 in. Answer
```

Step 4 Check by doubling the answer.

```
    4 ft. 2 in.
  + 4 ft. 2 in.
    8 ft. 4 in. Answer OK
```

Divide by 2.

```
   4'2"  Answer
2/8'4"
  8
    4
    4
```

Change to all inches and divide by 2.

```
8' = 96"          50"
   + 4"       2/100
   100"         10
                 0
                 0
             4
50" = 12/50 = 4'2" Answer
             48
              2
```

Example Problem 2. Find ½ of 5'3".

(ft. column | in. column)

Step 1 Find ½ of the feet column and bring down as shown. 5' 3"

$[\frac{1}{2}$ of $5' = \frac{1}{2} \times \frac{5}{1} = \frac{5}{2}$ or $2\frac{1}{2}$ or 2'6"] ------- 2'6"

Step 2 Find ½ of the inch column and bring down as shown.

$[\frac{1}{2}$ of $3'' = \frac{1}{2} \times \frac{3}{1} = \frac{3}{2}$ or $1\frac{1}{2}''$] ---------------- $1\frac{1}{2}''$

Step 3 Add Step 1 and Step 2.

(Step 1) 2'6"
\+ $1\frac{1}{2}''$ (Step 2)
$2'7\frac{1}{2}''$ Answer

Step 4 Check by doubling the answer.

$2'\ 7\frac{1}{2}''$
\+ $2'\ 7\frac{1}{2}''$
$4'14\frac{2}{2}'' = 4'15'' = 5'3''$ Answer OK

Divide by 2.

```
   2'  7.5" = 2'7 1/2" Answer
2/5'  3"
  4
  1=+12
     15
     14
      10
      10
```

Change to all inches and divide by 2.

```
5' = 60"       31.5
   + 3"      2/63
   63"         62
                10
                10
```

$31\frac{1}{2}'' = 2'7\frac{1}{2}''$ Answer

[Please note that after adequate practice, the first 3 steps can be done mentally. This is up to you. <u>However, Step 4 should always be written down to check the results.</u>]

Example Problem 3. Find ½ of 17'7 5/8"

ft. col. | in. col. | fraction column

17' 7 $\frac{5"}{8}$

Step 1. Find ½ of the feet column and bring down as shown.

$[\frac{1}{2}$ of 17' $= \frac{1}{2} \times \frac{17}{1} = \frac{17}{2} = 8\frac{1}{2}$ or 8'6"] ----- 8'6"

Step 2. a) Find ½ of the inch column and bring down as shown.

$[\frac{1}{2}$ of 7" $= \frac{1}{2} \times \frac{7}{1} = \frac{7}{2} = 3\frac{1}{2}"]$ ------------------ $3\frac{1}{2}"$

b) Find ½ of the fraction column and bring down as shown.

$[\frac{1}{2}$ of $\frac{5"}{8} = \frac{1}{2} \times \frac{5}{8} = \frac{5"}{16}]$ ------------------------ $\frac{5"}{16}$

Step 3. Add Step 1, and a) and b) of Step 2.

(Step 1) 8'6"
$3\frac{1"}{2}$ $\frac{8}{16}$ [a) Step 2]
+ $\frac{5"}{16}$ $\frac{5}{16}$ [b) Step 2]
8'9 $\frac{13"}{16}$ Answer

Step 4. Check by doubling the answer.

8' 9 $\frac{13"}{16}$
+ 8' 9 $\frac{13"}{16}$
16'18 $\frac{26"}{16} = 16'19\frac{10"}{16} = 17'7\frac{5"}{8}$ Answer OK

Divide by 2.

```
   8'   9.8125" = 8'9 13"/16
2/17'   7.625"       Answer
  16
   1 =+12
       19
       18
        16
        16
         2
         2
          5
          4
         10
         10
```

Change to all inches and divide by 2.

17' = 204"
+ $7\frac{5}{8}$
211.625"

105.8125" = $8'9\frac{13"}{16}$ Answer
2/211.625"

Example Problem 4. Find ½ of 10 lb. 7½ oz.

lb. column | oz. column | fraction column

10 lb. 7 $\frac{1}{2}$ oz.

Step 1. Find ½ of the pounds column and bring down as shown.

$[\frac{1}{2}$ of 10 lb. $= \frac{1}{2} \times \frac{10}{1} = \frac{10}{2} = 5$ lb.] --------- 5 lb.

Step 2. a) Find ½ of the ounce column and bring down as shown.

$[\frac{1}{2}$ of 7 oz. $= \frac{1}{2} \times \frac{7}{1} = \frac{7}{2} = 3\frac{1}{2}$ oz.] ----------------- $3\frac{1}{2}$ oz.

b) Find ½ of the fraction column and bring down as shown.

$[\frac{1}{2}$ of $\frac{1}{2}$ oz. $= \frac{1}{2} \times \frac{1}{2} = \frac{1}{4}$ oz.] ---------------------- $\frac{1}{4}$ oz.

Step 3. Add Step 1, and a) and b) of Step 2.

(Step 1) 5 lb.
$3\frac{1}{2}$ $\frac{2}{4}$ oz a)
+ $\frac{1}{4}$ $\frac{1}{4}$ oz. b)
5 lb.3 $\frac{3}{4}$ Answer

Step 4. Check by doubling the answer.

5 lb. $3\frac{3}{4}$ oz.
+ 5 lb. $3\frac{3}{4}$
10 lb. $6\frac{6}{4}$ oz. = 10 lb. $7\frac{2}{4}$ oz. or 10 lb. $7\frac{1}{2}$ oz. Answer OK

Divide by 2.

```
   5 lb. 3.75 oz.   Answer
2/10 lb. 7.5 oz.
  10
          7   [5 lb. 3 3/4 oz.]
          6
          15
          14
           10
           10
```

Change to all ounces and divide by 2.

10 lb. = 160 oz.
+ 7½ oz.
167.5 oz.

83.75 oz. = 5 lb. $3\frac{3}{4}$ oz.
2/167.5 oz.

FIND 1/2 OF A DENOMINATE NUMBER

The purpose of this instruction sheet is to illustrate a few methods of finding the center of objects by using different instruments. The methods shown below should be used in conjunction with a mathematical method as shown on the next page. The reason for this is to make sure that the center (dimension) is correct. Double checking builds confidence and helps eliminate errors.

Find the center of a shaft by using a center square.

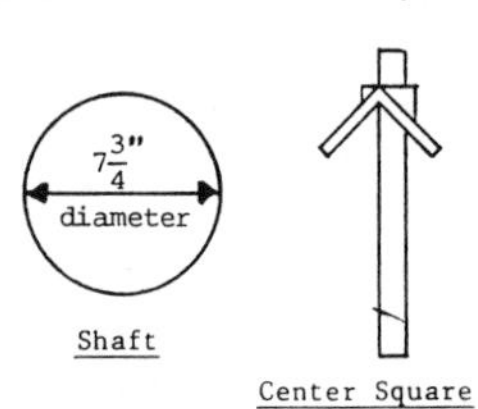

1st position the center square on the shaft and mark.

2nd reposition the center square on the shaft and mark.

Check by measuring different spots

$3\frac{7}{8}$" $3\frac{7}{8}$" $3\frac{7}{8}$"

3rd where the marks cross should be the correct center.

Find the center of a flat object by using a tape measure or rule.

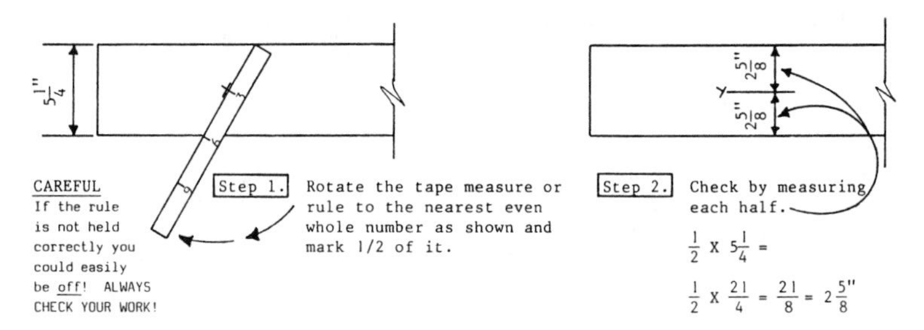

Find the center of different geometric figures by marking the diagonals or by using a compass and swinging arcs.

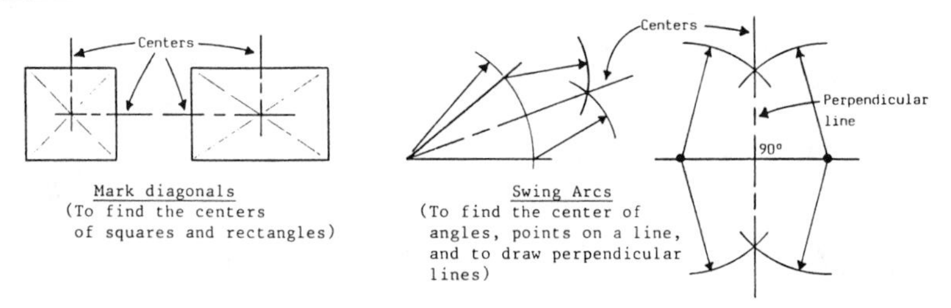

SHOWN ABOVE ARE JUST A FEW BASIC METHODS -- ALWAYS CHECK YOUR WORK BY USING ANOTHER METHOD AND BY MEASURING -- SEE THE MATHEMATICAL METHODS ON THE FOLLOWING PAGES.

Many times it is necessary to find the center of an object or find ½ of a portion. In doing so you are finding ½ of a denominate number. Listed below are different methods that can be used to achieve that goal. Use the method(s) that works best for you.

Example 1. Find 1/2 of 9'8 1/4"

METHOD 1

Step 1 Change to all inches.

$$9 \times 12 = 108$$
$$108 + 8\tfrac{1}{4} \qquad 9'8\tfrac{1}{4}" = 116\tfrac{1}{4}"$$

Step 2 Multiply by ½.

$$\tfrac{1}{2} \times 116\tfrac{1}{4}"$$
$$= \frac{1}{2} \times \frac{465}{4} = 8/465 = 58\tfrac{1}{8}"$$
$$40, \; 65, \; 64, \; 1$$
$$= 58\frac{1}{8}"$$

Step 3 Change to ft. and in.

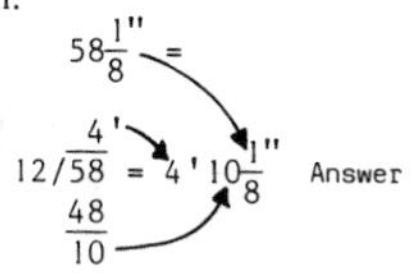

METHOD 2

Step 1 Divide by 2.

$$2/9' \; 8\tfrac{1}{4}" = 4' \; 10\tfrac{1}{8}" \quad \text{Answer}$$
$$8$$
$$1' = +12"$$
$$20$$
$$20$$
$$\frac{1}{4} \div \frac{2}{1} =$$
$$\frac{1}{4} \times \frac{1}{2} = \frac{1}{8}"$$

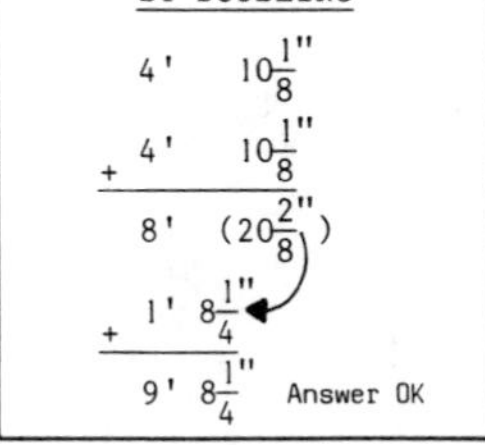

METHOD 3

(Split each column in half separately -- this is usually done mentally)

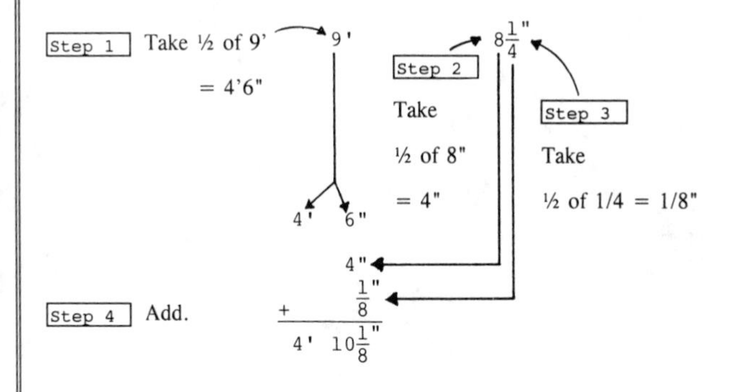

Example 2. Find ½ of 15 lb. 9 oz.

METHOD 1

Step 1 Change to all ounces.

[1 lb. = 16 oz.]

```
                      15 lb.
                    X 16
                      90
                     15
                     240 oz.
                       9 oz.
15 lb. 9 oz.  =      249 oz.
```

Step 2 Multiply by ½.

$\frac{1}{2} \times 249 =$

$\frac{1}{2} \times \frac{249}{1} = \frac{249}{2}$

$= 124\frac{1}{2}$ oz.

Step 3 Change 124½ oz. to lb. and oz. =

```
       7 lb. 12½ oz.
16/124½
   112
    12½ oz.
```

METHOD 2

Step 1 Divide by 2.

```
    7 lb.  12.5 oz.
2/15 lb.    9.0 oz.
  14
   1 =    +16 oz.
           25
           24
            1 0
            1 0
```

CHECK ALL THREE METHODS BY DOUBLING

```
   7 lb. 12 1/2 oz.
+  7 lb. 12 1/2 oz.
  14 lb. (24 2/2 oz.)
+  1 lb. 9 oz.
  15 lb. 9 oz.   Answer OK
```

METHOD 3

(Split each column in half separately -- this is usually done mentally)

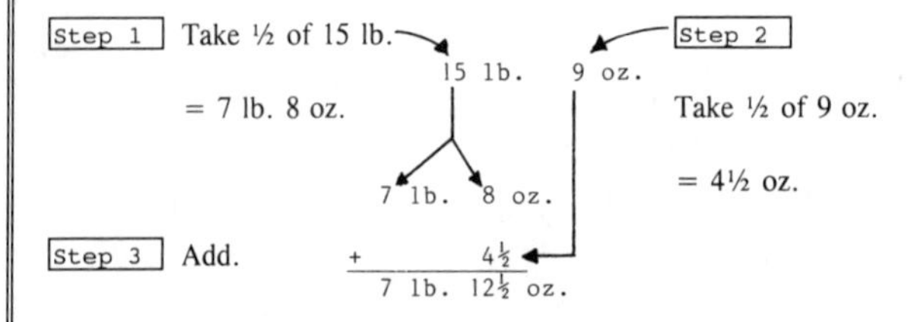

Another Example:

Example 3. Find the center of the object below, real fast.

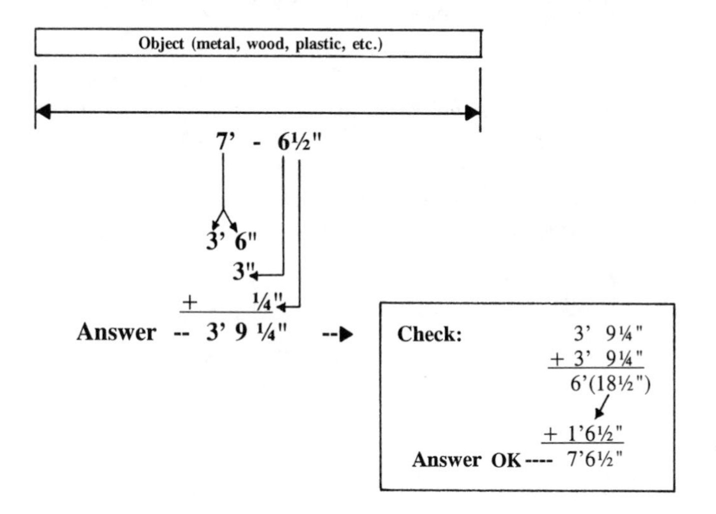

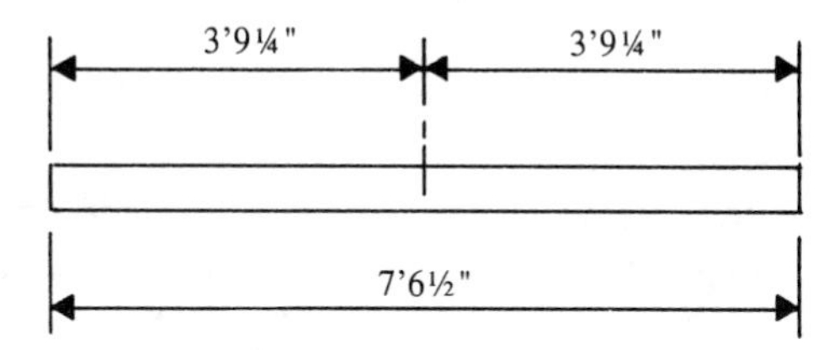

Note: **Another way to check this problem is to actually measure 3'9 1/4" each way to the center. The center mark should be in the correct place. If not, you did something wrong.**

E1 | ◂ Task No. | DECIMAL FRACTIONS

DECIMAL FRACTIONS

A decimal fraction is the end result of a common fraction, for example: common fraction $\frac{3}{4}$ equals 3 ÷ 4 or $4\overline{)3.00}$ = .75 decimal fraction.

RULES: **To change a common fraction to a decimal fraction divide the numerator of the fraction by the denominator of the fraction,**

for example: $\frac{N}{D}$ = N ÷ D or $D\overline{)N}$ = d.f. (decimal fraction)

The end result of an improper fraction is a whole number and a decimal fraction,

for example: $\frac{7}{4}$ equals 7 ÷ 4 or $4\overline{)7.00}$ = 1.75

All decimal fractions have a denominator made up of 10 or multiples of 10,

for example: $\frac{3}{4}$ = 3 ÷ 4 or $4\overline{)3.00}$ = .75 = $\frac{75}{100}$ [$\frac{75 \div 25}{100 \div 25} = \frac{3}{4}$] $100\overline{)75.00}$ = .75

[Note common factors]

Try any fraction and it works out $\frac{5}{8}$ = 5 ÷ 8 or $8\overline{)5.000}$ = .625 [$\frac{625 \div 125}{1000 \div 125} = \frac{5}{8}$] $1000\overline{)625.000}$ = .625

Illustrative examples:

$\frac{1}{10}$ = $10\overline{)1.0}$ = .1, 1 0 — read as one tenth

$\frac{1}{100}$ = $100\overline{)1.00}$ = .01, 1.00 — read as one hundredth

$\frac{1}{1000}$ = $1000\overline{)1.000}$ = .001, 1 000 — read as one thousandth

$\frac{1}{10000}$ = $10000\overline{)1.0000}$ = .0001, 1 0000 — read as one ten thousandth

By using the above examples, we can assume that to divide a whole number or decimal by 10 or multiples of 10, we can then move the decimal point to the left once for each zero in the divisor.

Examples:
348 ÷ 10 = 34.8 move the decimal to the left once.
348 ÷ 100 = 3.48 move the decimal to left twice.
348 ÷ 1000 = .348 move the decimal to the left three times
348 ÷ 10000 = .0348 move the decimal to the left four times

The reciprocal of division is multiplication. Therefore, to multiply a whole number or decimal by 10 or multiples of 10, move the decimal point to the right once for each zero in the multiplier.

Examples:
10 X 34.8 = 348 move the decimal to the right once
100 X 3.48 = 348 move the decimal to the right twice
1000 X .348 = 348 move the decimal to the right three times
10000 X .0348 = 348 move the decimal to the right four times

or:
10 X 246 = 2460 move once
100 X 246 = 24600 move twice
1000 X 246 = 246000 move three times

or:
10 X .005 = .05
100 X .005 = .5
1000 X .005 = 5

RULES: **<u>To carry a decimal a required number of places</u>, when after division there is a remainder, you must keep dividing, until you obtain the accuracy needed.**

Example: 89 ÷ 45 =

```
   1.9 for tenths          1.97 for hundredths      1.977 for thousandths
45/89.0        =       45/89.00        =        45/89.000
   45                     45                       45
   440                    440                      440
   405                    405                       405
    35                     350                      350
                           315                      315
                            35                       350
                                                     315
                                                      35
```

Note: Carrying the decimal to two or three places is usually accurate enough for most practical applications.

<u>To round off a decimal a required number of places</u>, do not change the digit being rounded off if the next digit is below 5 , or add 1 to the digit if the next digit is 5 or above. See the examples shown below.

round off 1.97 to one place (tenths) = 2.0 because 7 is greater than 5
round off 1.977 to two places (hundredths) = 1.98 because 7 is greater than 5
round off 1.93 to one place (tenths) = 1.9 because 3 is less than 5
round off 9.7386 to thousandths = 9.739 because 6 is greater than 5

<u>The above rule of rounding off decimals applies to repeating decimals too</u>:

```
                     .33333   The further the decimal is carried out the greater the accuracy
Example 1. 1/3 = 3/1.00000
                     9        .3 for tenths
                     10       .33 for hundredths
                      9       .333 for thousandths
                      10      (and so on)
```

<u>For greater accuracy, multiply by the fraction</u>.

For example: 1/3 X 24 = 8 --- .33 X 24 = 7.92 or .333 X 24 = 7.992 and so on.
(the further a decimal is carried out, the greater the accuracy).

```
                     .66666
Example 2. 2/3 = 3/2.00000
                     18       .6 for tenths
                     20       .66 for hundredths
                     18       .666 for thousandths
                      20      (and so on)
```

<u>For greater accuracy, multiply by the fraction</u>.

For example: 2/3 X 30 = 20 --- .66 X 30 = 19.8 or .666 X 30 = 19.98 and so on.
These illustrations illustrate the importance of carrying a decimal out further to obtain greater accuracy. Or whenever possible, use the fraction to reach the least tolerance possible.

E2	◂ Task No.	ADD DECIMAL FRACTIONS

RULE: To add decimal fractions, 1st line up the decimal points vertically (straight up and down), and 2nd add. The decimal point always goes to the right of the whole number.

Example Problem 1. Add 24, .005, and 7.9

[The decimal points must line up vertically -- straight up and down.]

Step 1 Set up vertically as shown.

```
 24
   .005
+ 7.9
```

[The decimal point is always to the right of the whole number.]

Step 2 Add as with whole numbers. **[Fill spaces with zeroes as shown.]**

```
   1st         2nd          3rd          4th
                             1            1
 24.000      24.000       24.000       24.000
   .005        .005         .005         .005
+ 7.900     + 7.900      + 7.900      + 7.900
               .905        1.905       31.905  sum
```

[Bring down .905] [4 + 7 = 11] [1 + 2 = 3]

Step 3 Check by adding in reverse.

```
 24.000
   .005
+ 7.900
```

[Or check by canceling 9's, see Task No. F1A .]

1st bring down the 2 → 2
2nd 4 + 7 = 11 → 11
3rd bring down the 905 → + .905
4th find the total sum → 31.905 OK

Example Problem 2. Add 7.63, 24.5, 17, and .48

[Vertically line up decimal points]

Step 1 Set up vertically as shown.

```
  7.63
 24.5
 17.
   .48
```

Step 2 Add as with whole numbers.

```
   1st        2nd        3rd        4th
     1        1 1       11 1       11 1
   7.63       7.63       7.63       7.63
  24.50      24.50      24.50      24.50
  17.00      17.00      17.00      17.00
    .48        .48        .48        .48
      1        .61       9.61      49.61  sum
```

[3 + 0 + 0 + 8 = 11] → 1
[1 + 6 + 5 + 0 + 4 = 16]
[1 + 7 + 4 + 7 = 19] **[1 + 2 + 1 = 4]**

Step 3 Check by adding in reverse.

```
  7.63
 24.50
 17.00
   .48
```

1st add 2 + 1 = 3 → 3
2nd add 7 + 4 + 7 = 18 → 18
3rd add 6 + 5 + 0 + 4 = 15 → 1.5
4th add 3 + 0 + 0 + 8 = 11 → 11
5th find the total sum → 49.61 OK

SUBTRACT DECIMAL FRACTIONS Task No. ▸ E3

RULE: To subtract decimal fractions, 1st line up the decimal points vertically and 2nd subtract. The decimal point always goes to the right of the whole number.

Example Problem 1. Subtract 27 - 18.05

[The decimal points must line up vertically -- straight up and down.]

Step 1 Set up vertically as shown.

```
  27.00
- 18.05
```

Step 2 Subtract as with whole numbers.

```
 1st            2nd            3nd

    9             6 9           16 9
  27.00          27.00          27.00
- 18.05        - 18.05        - 18.05
      5             95           8.95   Answer (difference)
[10 - 5 = 5]   [9 - 0 = 9]    [16 - 8 = 8]
```

Step 3 Check by adding.

Add the subtrahend to the difference, **Step 1**, and that sum should match the minuend, **Step 2**.

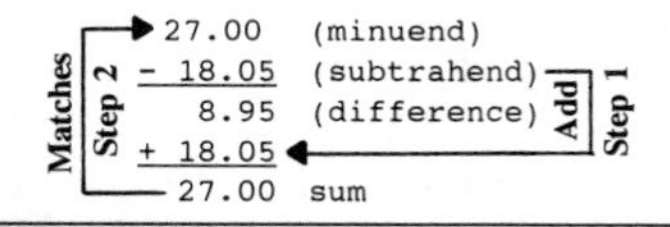

```
  27.00   (minuend)
- 18.05   (subtrahend)
   8.95   (difference)
+ 18.05
  27.00   sum
```


Example Problem 2. Subtract 4.56 - 1.87

[Vertically line up the decimal points]

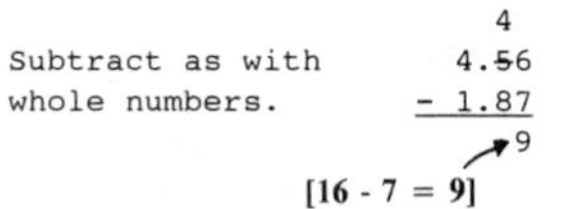

Step 1 Set up vertically as shown.

```
  4.56
- 1.87
```

Step 2 Subtract as with whole numbers.

```
 1st            2nd            3nd

    4              4            3 4
  4.56           4.56           4.56
- 1.87         - 1.87         - 1.87
     9            .69           2.69   Difference (answer)
[16 - 7 = 9]   [14 - 8 = 6]   [3 - 1 = 2]
```

Step 3 Check by adding.

Add the subtrahend to the difference, **Step 1**, and that sum should match the minuend, **Step 2**.

```
  4.56   (minuend)
- 1.87   (subtrahend)
  2.69   (difference)
+ 1.87
  4.56   sum
```

Matches / Step 2 / Add / Step 1

E4	◂ Task No.	MULTIPLY DECIMAL FRACTIONS

RULE: To multiply decimal fractions, 1st multiply as with whole numbers, 2nd from right to left count the number of decimal places in both the multiplier and the multiplicand, and 3rd move the decimal point in the product, from right to left the total number of places counted.

Example Problem 1. Multiply 25 X 2.05

```
Step 1   Set up          2.05  (multiplicand)    [Put larger number on top to have
         as shown.      X  25  (multiplier)       fewer numbers to multiply by.]

                         1st         2nd          3rd          4th

                          2           1
Step 2   Multiply as     2.05        2.05         2.05         2.05   2 places
         with whole     X  25       X  25        X  25        X  25  +0 places
         numbers.        1025        1025         1025         1025   2
                                     410        + 410          410
                                                  5125         51.25  Product

                                                         [Move the decimal point
                                                          2 places to the left.]

Step 3   Check the answer (product) by dividing.    (quotient)      2.05  Answer OK
                                                   (multiplier)  25/51.25 (product)
                                                                    50
         Divide the answer (product) by the multiplier.             125
         The quotient should match the multiplicand.                125

         [Or check by canceling 9's, see Task Number F3A]
```

Example Problem 2. Multiply 4.5 X 3.82

```
Step 1   Set up          3.82  (multiplicand)    [Put larger number on top to have
         as shown.      X 4.5  (multiplier)       fewer numbers to multiply by.]

                         1st         2nd          3rd          4th

                         4 1          3
Step 2   Multiply as     3.82        3.82         3.82         3.82   2 places
         with whole     X 4.5       X 4.5        X 4.5        X 4.5  +1 places
         numbers.        1910        1910         1910         1910   3
                                     1528       + 1528         1528
                                                 17190        17.190  Product

                                                         [Move the decimal point
                                                          3 places to the left.]

Step 3   Check the answer (product) by dividing.    (quotient)       3.82  Answer OK
                                                   (multiplier)  4.5/17.190 (product)
                                                                    135
         Divide the answer (product) by the multiplier.              369
         The quotient should match the multiplicand.                 360
                                                                      90
         [Or check by canceling 9's, see Task Number F3A]             90
```

DIVIDE DECIMAL FRACTIONS — Task No. ▸ E5

RULE: **To divide decimal fractions, 1st change the divisor to a whole number by moving the decimal point to the right, 2nd move the decimal point in the dividend the same number of places, and 3rd divide as with whole numbers.**

Example Problem 1. Divide 13.03 ÷ 025

Step 1 — Set up as shown. (divisor) .025/13.03 (dividend)

Step 2 — Make a whole number out of the divisor: ×025./13030.
[1st, make the divisor a whole number] **[2nd, move the decimal point in the dividend the same number of places.]**

Step 3 — Divide as with whole numbers Task Number A5.

```
1st            2nd            3rd            4th
                                             Answer (quotient)
      5             52            521           521.2
25/13030.      25/13030.      25/13030.      25/13030.0
 - 125          - 125          - 125          - 125
     53             53             53             53
                  - 50           - 50           - 50
                     3             30             30
                                 - 25           - 25
                                    5             50
                                                  50
```

Step 4 — Check the answer (quotient) by multiplying.

[Multiply the divisor in Step 3 times the quotient. The product should match the dividend in Step 3.]

```
(quotient)    521.2
(divisor)     X 25
              26060
              10424
(product)     13030.0
```

Matches dividend — Answer OK

Example Problem 2. Divide 28.44 ÷ 1.6

Step 1 — Set up as shown. (divisor) 1.6/28.44 (dividend)

Step 2 — Make a whole number out of the divisor: 1×6./284.4

Step 3 — Divide as with with whole numbers.

```
1st            2nd            3rd
      1             17            17.7   Answer (quotient)
16/284.4       16/284.4       16/284.4
 -16            - 16           - 16
  12              124            124
                - 112          - 112
                   12            124
                               - 112
                                  12
```

Step 4 — Check the answer (quotient) by multiplying.

[Multiply the divisor in Step 3 times the quotient. The product should match the dividend in Step 3.]

```
(quotient)     17.7
(divisor)      X 16
               1062
               177
              283.2
Remainder  +     12
(product)     284.4
```

Matches dividend — Answer OK

E6	◂ Task No.	CHANGE FRACTIONS TO DECIMALS (see also Task E7)

Review of Common Fractions

Numerator (top number) → $\frac{3}{8}$ ← Fraction line means divide, therefore, $\frac{3}{8} = 3 \div 8$ or $8/\overline{3}$ or $8/\overline{3.000}$ = .375

Denominator (bottom no.) is always the divisor

The divisor always follows the division sign.

[Thus, common fraction $\frac{3}{8}$ = .375, a decimal fraction.]

Example Problem 1. Change 5/16 inch to a decimal.

Step 1. Set up as shown. $\frac{5''}{16} = 5 \div 16$ or $16/\overline{5.}$ [The decimal point is always to the right of a whole number.]

Step 2. Divide as with decimal fractions task no. E5 .

```
      .3125   Answer
16/5.0000     [Therefore 5"/16 equals .3125"]
   4 8
     20
     16
      40
      32
       80
       80
```

```
Check by          .3125
multiplying.    X    16
                  18750
                  3125
Answer OK ---->  5.0000 = 5"/16
```

Example Problem 2. Change 5" to decimals of a foot.

Step 1. Set up as shown. $5'' = \frac{5'}{12}$ or $5 \div 12$ or $12/\overline{5.}$

[12" = 1 foot, therefore, 5" is $\frac{5}{12}$ of a foot.]

Step 2. Divide.

```
      .416'   Answer
12/5.000      [Therefore 5'/12 equals .416']
   4 8
     20
     12
      80
      72
       8
```

```
Check by
                  .416'
multiplying.    X   12
                   832
                  416
                 4.992"
      remainder  +   8
Answer OK ---->  5.000" = 5'/12
```

Example Problem 3. Change 3 1/2" to decimals of a foot.

Step 1. Set up as shown. $3\frac{1}{2}'' = \frac{3\frac{1}{2}}{12}$ [A complex fraction] or $3\frac{1}{2} \div 12$ or $12/\overline{3.5}$

[12" = 1 foot, therefore, $3\frac{1}{2}''$ is $\frac{3\frac{1}{2}}{12}$ of a foot.]

Step 2. Divide as with complex fractions task no. B6.

```
      .291'   Answer
12/3.500      [Therefore 3 1/2" equals .291']
   2 4
   1 10
   1 08
      20
      12
       8
```

```
Check by          .291'
multiplying.    X   12
                   582
                  291
                 3.492
      remainder  +   8
Answer OK ---->  3.500"
```

Other Examples

Change 1 oz. to decimals of a lb.

[16 oz. = 1 pound, therefore, 1 oz. is $\frac{1}{16}$ lb.]

$\frac{1}{16}$ lb. = $16/\overline{1.000}$ = .0625 lb.

Thus 1 oz. = $\frac{1}{16}$ lb. or .0625 lb.

Change 7 oz. to decimals of a lb.

[16 oz. = 1 pound, therefore, 7 oz. is $\frac{7}{16}$ lb.]

$\frac{7}{16}$ lb. = $16/\overline{7.0000}$ = .4375 lb.

Thus 7 oz. = $\frac{7}{16}$ lb. or .4375 lb.

Change $\frac{1}{2}$ oz. to decimals of a lb.

[16 oz. = 1 pound, therefore, $\frac{1}{2}$ oz. is $\frac{\frac{1}{2}}{16}$ lb.]

```
 1/2               .03125 lb.
 ---  lb. = 16/.5000
 16             48
                 20
                 16
                  40
                  32
                   80
                   80
```

Thus $\frac{1}{2}$ oz. = $\frac{\frac{1}{2}}{16}$ or .03125 lb.

[See also Task No. E9, N1A, and N1B for more information and application.]

(see also Task E6) **CHANGE DECIMALS TO FRACTIONS** **Task No. ▸ E7**

Review: **A decimal fraction is the end result of a common fraction,** thus $\frac{3}{8} = 8\overline{)3.000}$ = .375

Therefore, to change the form of a decimal back to a fraction, multiply (X) the denominator (divisor) by the decimal and place the product over the denominator.

```
 .375  (decimal)
X   8  (divisor)
3.000  (product)
8  (denominator)
```

RULE: To change a decimal fraction to a common fraction, multiply the decimal by the desired denominator.

```
Examples:  .0625"        .375"          .5"         .75"          .416'
          X   16 <---   X  8 <---     X 2 <---    X  4 <---     X  12 <--- desired denominators
            3750        3.000 = 3"/8  1.0 = 1"/2  3.00 = 3"/4     832
            625                                                   416
          1.0000 = 1"/16                                        4.992 = 5" or 5'/12 = 12/5.000 (.416')
```

Example Problem 1. Change 2.89" so you can read the dimension on an American (English) tape measure.

Step 1. Multiply the decimal fraction by the desired denominator.

[If you want 16ths, multiply by 16. For 32nds, multiply by 32. And so on.]

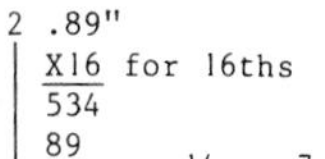

```
2 .89"                           2 .89"
  X16  for 16ths                   X64  for 64ths
  534                              356
  89                               534
 14.24 = 14/16 or 7+"/8           56.96 = 57-"/64
```

Step 2. Bring down the whole inches and add the fraction to it. → $2\frac{7+''}{8}$ → $2\frac{57-''}{64}$

Example Problem 2. Change 5.62' so you can read the dimension on an American (English) tape measure.

Step 1. Multiply the decimal fraction by the desired denominator -- 12, to find inches.

Step 2. Multiply the decimal fraction of inches by the desired denominator -- 16, to find 16ths.

```
5 .62'
  X12  for 12ths (because 12" = 1 foot)
  124
  62
 7 .44"
   X16  for 16ths
   264
   44
   7.04 = 7+"/16
```

Step 3. Bring down the whole feet and the whole inches, and add to the fraction: → $5'7\frac{7+''}{16}$ Answer

Check the answer by converting back to a decimal.

```
5'7 7"/16 = 7.4375/12

      .6197
12/7.4375
   7 2
     23
     12
     117
     108
5.62'
```

Example Problem 3. Change 2.9 lb. to pounds and ounces.

Step 1. Multiply the decimal fraction by the desired denominator -- 16, to find ounces.

```
2 .9 lb.
 X16  for oz. because 16 oz. = 1 lb.
 14.4 oz.
```

Step 2. Bring down the whole pounds and add the oz. to it. → 2 lb. 14.4 oz.

[See also Task No. E9, N1A, and N1B for more information and application.]

E8 ◄ Task No. FIND AVERAGES

How to find Averages:

RULES: **To find averages, (1) find the sum of all the quantities, and (2) divide the sum by the number of quantities.**

Example Problem 1. Find the average of 75, 245, 96, and 90.

Check by adding in reverse

Step 1 — Find the sum of all the quantities.

```
  75              75
 245             245
  96              96
+ 90            + 90
 506  sum        2
                 29
                 16
                 506  sum OK
```

Step 2 — Divide the sum by the number of quantities.

Note: There are four quantities (75, 245, 96, and 90), therefore, divide by 4.

```
   126.5   average
 4/506.0
   4
   10
    8
    26
    24
     20
     20
```

Step 3 — Check by multiplying.

```
 126.5
   X 4
 506.0  Answer OK
```

Note: Another efficient method of checking long addition problems, is to use a calculator (preferably with a tape). Use whatever method works best for you. However, get in the habit of checking your work. This will build up confidence and minimize errors.

Example Problem 2. Find the average of 245.09, 32.42, 58, 147.5, 309.36, and 187.16.

Check by adding in reverse

Step 1 — Find the sum of all the quantities.

```
  245.09              245.09
   32.42               32.42
   58.00               58.00
  147.50              147.50
  309.36              309.36
+ 187.16            + 187.16
  979.53  sum         7
                      24
                       38
                        1.3
                          23
                      979.53  Answer OK
```

Task No. ▸ E8

Step 2 — Divide the sum by the number of quantities.

Note: There are six quantities, therefore, divide by 6.

```
  163.255  average
6/979.530
  6
  37
  36
   19
   18
    15
    12
     33
     30
      30
      30
```

Step 3 — Check by multiplying.

```
163.255
X     6
979.530  answer OK
```

CONVERT FRACTIONS TO DECIMALS AND CONVERT DECIMALS TO FRACTIONS

This module demonstrates converting common fractions to decimal fractions and, vice-versa, converting decimal fractions to common fractions. It seems that most people have trouble converting. Hopefully this module will help clarify the procedure.

CONTENTS OF SAMPLE CONVERSION PROBLEMS INCLUDED IN THIS MODULE

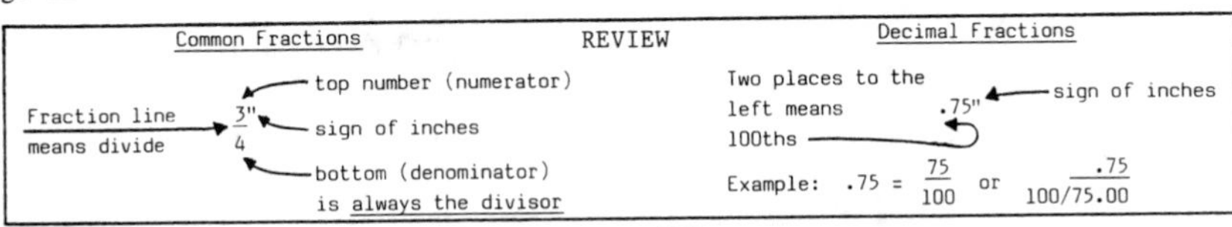

E6 Convert Fractions to decimals

Rule: To convert fractions to decimals, you must divide the bottom (denominator into the top (numerator)

Problem 1. Convert $\frac{3''}{4}$ to a decimal fraction.

Step 1. Divide $4\overline{)3} = 4\overline{)3.00}$ = .75

```
     .75
 4/3.00
   2 8
    20
    20
```

Bottom always the divisor

Therefore: Common Fraction $\frac{3''}{4}$ equals .75"

E7 Convert Decimals to Fractions*

Rule: To convert decimals to fractions, you must multiply the decimal by the desired denominator (lower number of the fraction desired).

Problem 2. Convert .75" to a common fraction with a denominator (bottom number) of 4, 8, 16, 32, or 64.

Step 1. Multiply by the desired denominator.

```
 .75"    .75"    .75"     .75"     .75"
 X 4     X 8     X 16     X 32     X 64
3.00    6.00     450      150      300
                  75      225      450
                12.00    24.00    48.00
```

$\frac{3}{4}$ $\frac{6}{8}$ $\frac{12}{16}$ $\frac{24}{32}$ $\frac{48}{64}$

Note: All 5 values are the same.

Therefore: Decimal Fraction .75" equals $\frac{3''}{4}$

*You can also change decimals to fractions by using ratio and proportion, but it takes more time and is more confusing. Example: If you want to find out how many 16ths are in .75", you could change .75 to 75/100 and set up an equation.

Thus: $\frac{75}{100} \times \frac{X}{16}$ or $(100X = 75 \times 16)$ → or $100X = 1200$

75:100 :: X:16 or $(100X = 75 \times 16)$

$$\frac{\cancel{100}X}{\cancel{100}} = \frac{1200}{100}$$

$$X = 12 \text{ or } \frac{12}{16} \text{ or } \frac{3''}{4}$$

Note: All the above means the same as $.75 \times 16 = 12.00$ or $\frac{12}{16}$ or $\frac{3}{4}$ because .75 is the end result of $\frac{75}{100}$ or $100\overline{)75.00}$. Therefore, you save the step of dividing by 100.

E6 Convert Fractions to Decimals

Problem 2. Convert 9" to decimals of a foot.

(Note: 12" = 1 foot and $\frac{12}{12} = 1$)

Step 1. Convert 9" to a fraction.

9" is 9 12th parts of a ft. or $\frac{9}{12}$ of a ft.

Common fraction $\frac{9}{12}$ equals $\frac{3}{4}$

Step 2. Divide $4/\overline{3} = 4/\overline{3.00}$ = .75

28
20
20

Therefore: 9" = fraction $\frac{3'}{4}$ or decimal .75'

E7 Convert Decimals to Fractions

Problem 2. Convert .75' to inches and to a common fraction of a foot.

Step 1. Multiply by the desired denominator.

.75' ← indicates feet
X 12 ← 12" = 1 foot — desired denominator
150
75
9.00" = $\frac{9}{12}$ or $\frac{3'}{4}$

Note: When you multiply decimals of a foot by 12 (inches), that answer (product) becomes inches -- make sure to identify with the inch sign (") as shown.

Problem 3. Convert 2 3/8" to a decimal in inches and a decimal in feet.

Step 1. Change $2\frac{3"}{8}$ to decimals of an inch.

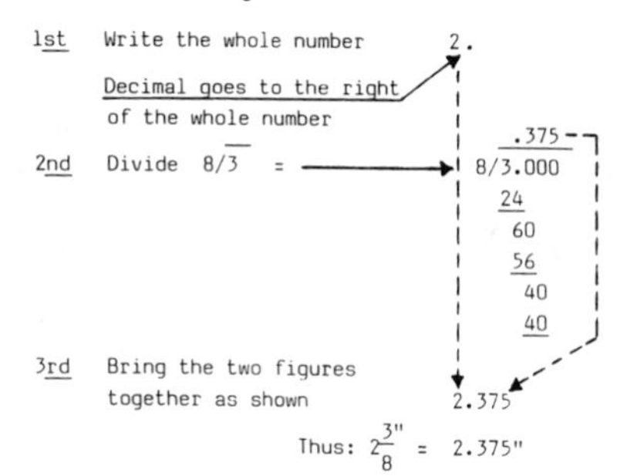

Thus: $2\frac{3"}{8}$ = 2.375"

Step 2. Convert $2\frac{3"}{8}$ to decimals of a foot.

1st Write $2\frac{3"}{8}$ as a decimal 2.375"

2nd Place 2.375 over 12 $\frac{2.375}{12}$

(Note: 12" = 1 foot, therefore, 2.375" is 2.375 parts of a foot or $\frac{2.375}{12}$)

3rd Divide 12/2.375 = .1979 Thus: $2\frac{3"}{8}$ = .1979'

Problem 3. Convert .1979 to inches and fractions of an inch.

Step 1. Multiply by the desired denominator.

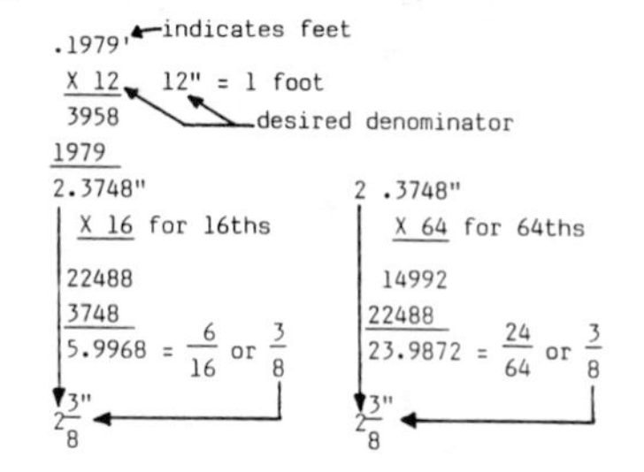

Note: Multiply the decimal only by 16 or 64 -- 2 is already whole inches.

NOTE

The reason the problems are back to back (the same problems for fractions and the same problem for decimals) is to show you how they interrelate. When you calculate a specific problem, that problem can be self-checked by doing the opposite (reciprocal) calculation.

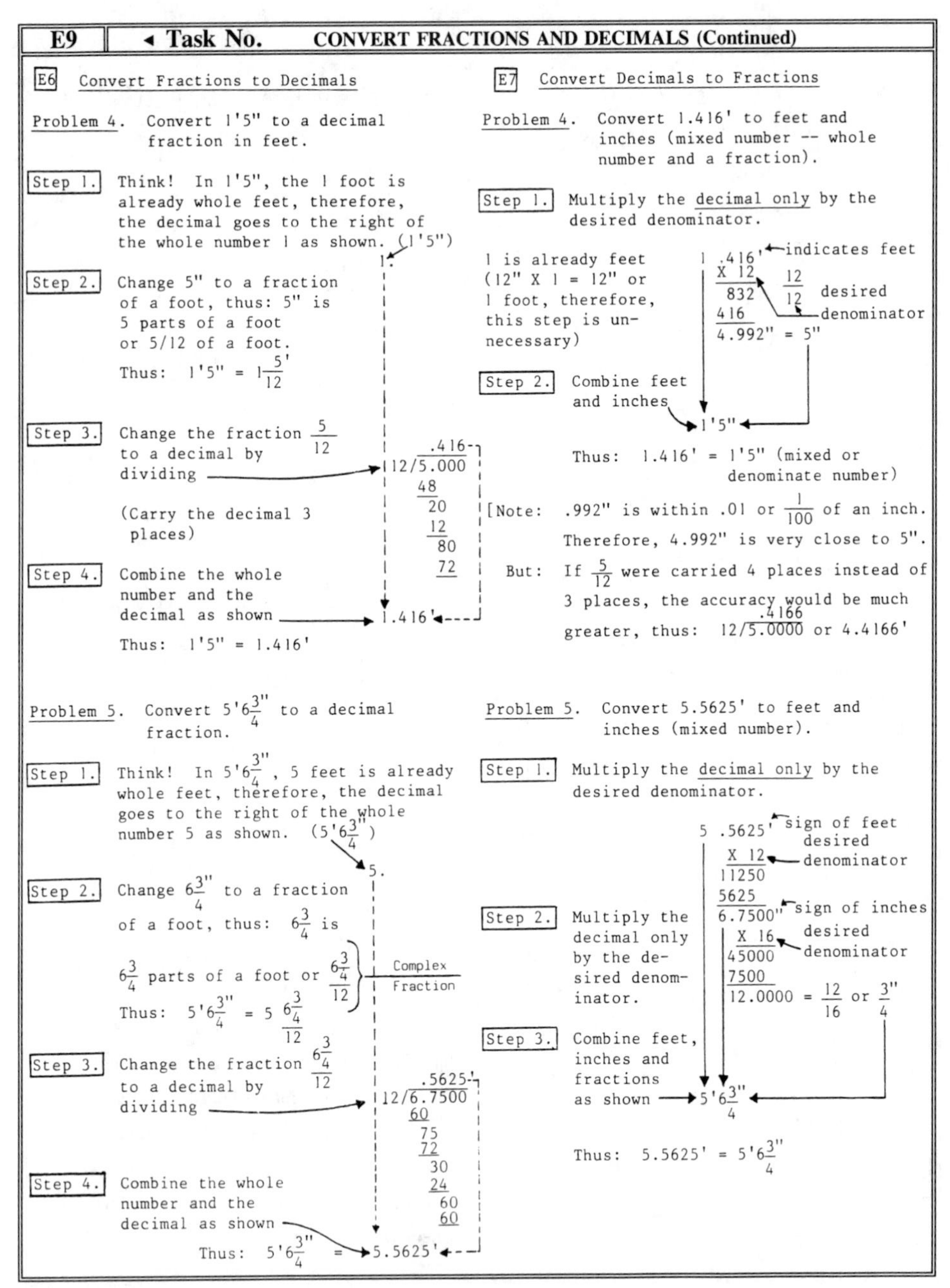

E9 ◄ **Task No.** **CONVERT FRACTIONS AND DECIMALS (Continued)**

E6 Convert Fractions to Decimals

Problem 4. Convert 1'5" to a decimal fraction in feet.

Step 1. Think! In 1'5", the 1 foot is already whole feet, therefore, the decimal goes to the right of the whole number 1 as shown. (1'5") → 1.

Step 2. Change 5" to a fraction of a foot, thus: 5" is 5 parts of a foot or 5/12 of a foot.

Thus: $1'5" = 1\frac{5}{12}'$

Step 3. Change the fraction $\frac{5}{12}$ to a decimal by dividing → 12/5.000 = .416 (48, 20, 12, 80, 72)

(Carry the decimal 3 places)

Step 4. Combine the whole number and the decimal as shown → 1.416'

Thus: 1'5" = 1.416'

E7 Convert Decimals to Fractions

Problem 4. Convert 1.416' to feet and inches (mixed number -- whole number and a fraction).

Step 1. Multiply the decimal only by the desired denominator.

1 is already feet (12" X 1 = 12" or 1 foot, therefore, this step is unnecessary)

1 .416' ← indicates feet
X 12 ← $\frac{12}{12}$ desired denominator
832
416
4.992" = 5"

Step 2. Combine feet and inches → 1'5"

Thus: 1.416' = 1'5" (mixed or denominate number)

[Note: .992" is within .01 or $\frac{1}{100}$ of an inch. Therefore, 4.992" is very close to 5".

But: If $\frac{5}{12}$ were carried 4 places instead of 3 places, the accuracy would be much greater, thus: 12/5.0000 = .4166 or 4.4166'

Problem 5. Convert $5'6\frac{3}{4}"$ to a decimal fraction.

Step 1. Think! In $5'6\frac{3}{4}"$, 5 feet is already whole feet, therefore, the decimal goes to the right of the whole number 5 as shown. ($5'6\frac{3}{4}"$) → 5.

Step 2. Change $6\frac{3}{4}"$ to a fraction of a foot, thus: $6\frac{3}{4}$ is $6\frac{3}{4}$ parts of a foot or $\frac{6\frac{3}{4}}{12}$ — Complex Fraction

Thus: $5'6\frac{3}{4}" = 5\ \frac{6\frac{3}{4}}{12}$

Step 3. Change the fraction $\frac{6\frac{3}{4}}{12}$ to a decimal by dividing → 12/6.7500 = .5625 (60, 75, 72, 30, 24, 60, 60)

Step 4. Combine the whole number and the decimal as shown

Thus: $5'6\frac{3}{4}"$ = 5.5625'

Problem 5. Convert 5.5625' to feet and inches (mixed number).

Step 1. Multiply the decimal only by the desired denominator.

5 .5625' ← sign of feet
X 12 ← desired denominator
11250
5625
6.7500" ← sign of inches

Step 2. Multiply the decimal only by the desired denominator.

X 16 ← desired denominator
45000
7500
12.0000 = $\frac{12}{16}$ or $\frac{3}{4}"$

Step 3. Combine feet, inches and fractions as shown → $5'6\frac{3}{4}"$

Thus: $5.5625' = 5'6\frac{3}{4}"$

PRACTICAL APPLICATION OF CHANGING FRACTIONS AND DECIMALS

[NOTE: The examples shown below and on the next two pages demonstrate how to change various types of denominate numbers from fractions to decimals and vice-versa from decimals to fractions.]

More Feet and Inch Examples:

1. Change $\frac{7}{8}''$ to decimals of an inch = .875"

$\frac{7''}{8} = 8\overline{)7.000}$ = .875"

2. Change 5" to decimals of a foot = .416' or .417'

$5'' = \frac{5'}{12} = 12\overline{)5.000}$ = .416' or .4166' or .417'

3. Change $\frac{3}{4}''$ to decimals of a foot = .0625'

$\frac{3''}{4} = \frac{\frac{3}{4}}{12}$ or $\frac{.75}{12}$ or $12\overline{).7500}$ = .0625'

4. Change $3\frac{1}{2}''$ to decimals of a foot = .291'

$3\frac{1}{2}'' = \frac{3\frac{1}{2}}{12}$ or $\frac{3.5}{12}$ or $12\overline{)3.500}$ = .291'

5. Change $7\frac{1}{16}''$ to decimals of an inch = 7.0625"

$7\frac{1''}{16} = 16\overline{)1.0000}$ = .0625" = 7.0625"

6. Change $4'2\frac{1}{2}''$ to decimals of a foot = 4.208'

$4'2\frac{1}{2}'' = \frac{2\frac{1}{2}}{12}$ or $\frac{2.5}{12}$ or $12\overline{)2.500}$ = .208'

= 4.208'

7. Change .6875" to 16ths = $\frac{11''}{16}$

```
 .6875"
X   16
 41250
 6875
11.0000 = 11/16
```

8. Change 2.71" to $\pm\frac{1}{16}$ of an inch = $2\frac{11+''}{16}$

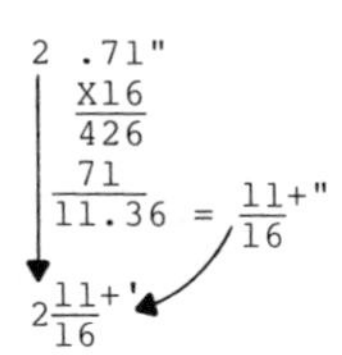

9. Change .43' to inches and 16ths = $5\frac{1+}{8}''$

```
     .43
   X 12
      86
     43
  5  .16"
     X16
      96
     16
  2.56 = 2+"/16  or  1+"/8

  5 1+"/8
```

10. Change 3.073' to feet, inches and 16ths = $3'0\frac{7}{8}''$

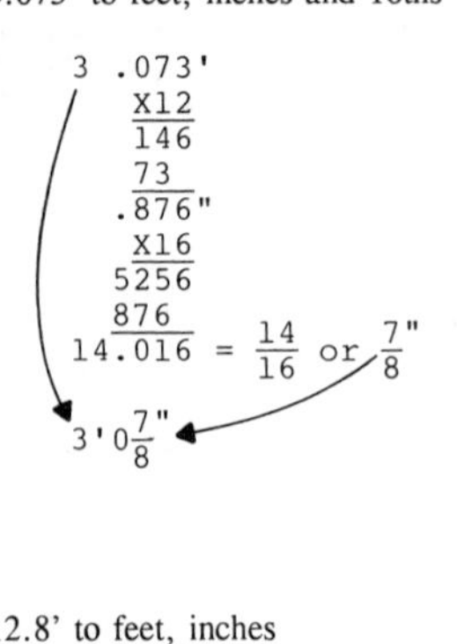

11. Change 24.625" to feet, inches and and 16ths = $24\frac{5}{8}''$ or $2'0\frac{5}{8}''$

```
  24 .625"
       X16
      3750
      625
    10.000 = 10/16  or  5/8

  24 5"/8  or  2'0 5"/8
```

12. Change 12.8' to feet, inches and 16ths = $12'9\frac{9+}{16}''$

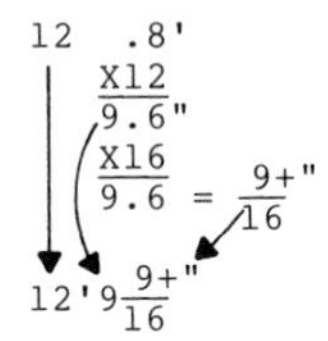

Change Pound and Ounce Examples:

1. Change 2.5 pounds to pounds and ounces = <u>40 ounces</u> or <u>2 lb. 8 oz.</u>.

```
    2.5 lb.
  X 16
    150
    25
    40.0 oz. or 2 lb. 8 oz.
```

2. Change .7 pounds to ounces = <u>11.2 ounces</u>.

```
     16
  X  .7
   11.2 oz.
```

3. Change 3/4 pounds to ounces = <u>12 ounces</u>.

$\frac{3}{\not{4}} \times \frac{\not{16}^{\,4}}{1} = 12$ oz.

<u>Or</u> .75 X 16 = 12 oz.

4. Change 68 ounces to pounds = <u>4 1/4 pounds</u>. or <u>4 lb. 4 oz.</u>.

```
        4.25  or  4 1/4 lb.
  16/68.00
```

5. Change 3 3/4 tons to pounds
   = 7,500 pounds .

```
  2000
X 3.75
 10000
14000
6000
7500.00 pounds
```

6. Change 5,750 pounds to tons
   = 2 tons 1750 pounds or 2.875 tons (2 7/8 tons)

```
          2.875 tons
2000/5750
```

Change Other Miscellaneous Denominate Numbers:

1. Change 45.5 degrees to degrees and minutes = 45 degrees and 30 minutes .

```
  60
X .5
30.0 minutes
```

2. Change 24 degrees and 12 minutes to to degrees and decimals of a degree
   = 24.2 degrees .

$\frac{12}{60} = \frac{1}{5}$ or .2 degrees

3. Change 13.275 degrees to degrees, minutes, and seconds = 13 degrees, 16 minutes, and 30 seconds .

```
13   .275 degrees
     X 60 minutes
16   .500 minutes
      X60 seconds
30   .000 seconds
13°16'30"
```

Read: 13 degrees 16 minutes 30 seconds

4. Change 77.09 degrees to degrees, minutes, and seconds = 77 degrees, 5 minutes, and 24 seconds .

```
77    .09 degrees
     X 60 minutes
5     .40 minutes
     X 60 seconds
24    .00 seconds
77°5'24"
```

Read: 77 degrees 5 minutes 24 seconds

[SEE THE APPENDIX FOR OTHER KINDS OF DENOMINATE NUMBERS]

F1 ◂ Task No. **CHECK ADDITION BY RE-ADDING**

Example
Problem 1. Add and check the problem on the right.

```
 7124
  129
+ 999
```

Step 1. Add from right to left.

```
  112
 7124
  129
+ 999
 8252  Answer
```

The answer should be 8252.

← [Add right to left]

[How do you know for sure that the answer is correct?

Do the next step to make sure!]

Step 2. Check by re-adding in the opposite direction left to right.

```
 7124
  129
+ 999
```

1st bring down the 7 ------→ 7
2nd add 1 + 1 + 9 = 11 ------→ 11
3rd add 2 + 2 + 9 = 13 ------→ 13
4th add 4 + 9 + 9 = 22 ------→ + 22
5th find the total sum ------→ 8252 **Matches the above answer, therefore, the answer should be OK**

[Add left to right]

Example
Problem 2. Add and check 127, 483, 927, 645, and 113.

Step 1. Set up as shown and add from right to left.

```
   22
  127
  483
  927
  645
+ 113
 2315  Answer
```

[Add right to left]

Step 2. Check by re-adding in the opposite direction left to right.

```
  127
  483
  927
  645
+ 113
```

1st add 1 + 4 + 9 + 6 + 1 = 21 ------→ 21
2nd add 2 + 8 + 2 + 4 + 1 = 17 ------→ 17
3rd add 7 + 3 + 7 + 5 + 3 = 25 ------→ 25
4th find the total sum ------→ 2295 **Doesn't match the above answer, therefore, something is wrong. Do it over until the answer is correct.**

CHECK ADDITION BY CANCELING NINES   Task No. ▸ F1A

Canceling nines is an especially effective way to check addition, because it allows you to check your work step by step.

One principle of the method is that if you add digit to digit in a row of figures, then add the digits in that sum to each other, and so on, you will end up with a single digit greater than zero.

Example:			
7124:	7+1+2+4 = 14:	1+4 = 5	
129:	1+2+9 = 12:	1+2 = 3	
999:	9+9+9 = 27:	2+7 = 9	

Another principle of the method is that if you cancel out any digits in a row that do add up to nine, the single-digit "sum" for that line will end up the same as if you added all the digits -- with one exception.

Example:		
7124:	~~7~~ 1 ~~2~~ 4 :	1+4 = 5
129:	1 2 ~~9~~ :	1+2 = 3
999:	~~9~~ ~~9~~ ~~9~~ :	0 -- **[This is the exception. To cancel nines, the single-digit result has to be greater than zero. So leave one 9 in the row.]**
	~~9~~ ~~9~~ 9 :	9

RULE: The sum of the canceling-nines digits in the addends should equal the canceling-nines digit of the sum.

(addend)	7124:	~~7~~ 1 ~~2~~ 4 :	1+4 = 5	5
(addend)	129:	1 2 ~~9~~ :	1+2 = 3	or 3
(addend)	+ 999:	~~9~~ ~~9~~ 9 :	9	~~9~~
(sum)	8252:	8 ~~2~~ ~~5~~ ~~2~~ :	17: 1+7 = ⑧*	

~~7~~ 1 ~~2~~ 4	-5
1 2 ~~9~~	-3
+ ~~9~~ ~~9~~ 9	-9
8 ~~2~~ ~~5~~ ~~2~~	⑧

matches
OK

[*Circle makes it easier to remember the sum has to be matched.]

The ⑧ from the sum equals the 8 resulting from the addends, so we know the answer is not wrong and is probably right.

Canceling nines is not a fail-safe method (an 8522 answer to the problem above would have yielded the same result), but it is a quick way to double check. If the results do not match, at least you know something is probably wrong with the calculation.

Another method of checking addition is to check with a calculator with a tape.* However, many times this will not be readily available for you. Therefore, it is to your benefit to use one of the methods shown on this and the preceding pages.

*If you check addition, especially with many addends, on a calculator without a tape, a wrong number could be plugged in and you would not know it. [Try adding a long list of addends, such as a grocery list on a calculator without a tape, and you will find that it is very easy to get mixed up.]

F2	◂ Task No.	CHECK SUBTRACTION BY ADDING

Example Problem 1. Subtract and check the problem on the right.

```
  743  (minuend)
- 295  (subtrahend)
```

Step 1. Subtract as with whole numbers, Task No. A3.

```
   6 3
  7̸4̸3
- 295
  448  (difference)
```

Step 2. Add the subtrahend to the difference.

```
  743
- 295  (subtrahend) ─┐
  448  (difference)  │
+ 295  ◄─────────────┘
```

Step 3. Check by matching that sum with the minuend. If the sum matches the minuend, the answer should be correct.

```
Matches / Answer OK
 ┌──►  743  (minuend)
 │   - 295
 │     448  Answer
 │   + 295
 └───  743  (sum)
```

Example Problem 2. Subtract and check the problem on the right.

```
  9604  (minuend)
- 2846  (subtrahend)
```

Step 1. Subtract.

```
  8 5 9
  9̸6̸0̸4
- 2846
  6768  (difference)
```

Step 2. Add the subtrahend to the difference.

```
  9604
- 2846 ─┐
  6768  │ Add
+ 2846 ◄┘
```

Step 3. Check by matching that sum with the minuend.

Note: The sum does not match the minuend, therefore, redo until correct.

```
Does Not Match / Redo
 ┌──►  9604  (minuend)
 │   - 2846
 │     6768  Answer Wrong
 │   + 2846
 └───  9614  (sum)
```

```
Subtract:   9604        Check:  ┌──► 9604
          - 2846                │  - 2846 ─┐
            6758  Answer   Matches  6758   │ Add
                                │  + 2846 ◄┘
                                └─── 9604  Answer OK
```

CHECK SUBTRACTION BY CANCELING NINES   Task No. ▸ F2A

Probably the most effective and the quickest method to use for checking subtraction is by adding as shown on the preceding page and/or by using a calculator with a tape on it*. Canceling nines is still another method that can be used. Therefore, it is included as an alternative.

[NOTE: Cancel nines in the subtrahend, minuend, and the difference in the same manner as you cancel nines in each row of numbers (addends) and the sum for addition. Refer to Task Number F1A, if needed.]

RULE: The sum of the canceling-nines digits in the difference, and the subtrahend is equal to the canceling-nines digit in the minuend.

```
                          equal
(minuend)       743  (5)*<--------------------+
(subtrahend)  - 295--7 \                       |
(difference)    448--7 /  7+7 = 14: 1+4 = 5
```

[*Circle makes it easier to remember the minuend has to be matched.]

Example:

```
                                                     equal
(minuend)       743:   7+4+3 = 14:   1+4 = (5)<------------------+
(subtrahend)  - 295:     2+5 =  7:          7 \                   |
(difference)    448:   4+4+8 = 16:   1+6 =  7 /  7+7 = 14: 1+4 = 5
```

The (5) from the minuend equals the 5 resulting from the subtrahend and the difference, so we know the answer is not wrong, and is probably right.

Canceling nines is not a fail-safe method (a 484 answer to the problem above would have yielded the same result), but it is a quick way to double check. If the results do not match, at least you know something is probably wrong with the calculation.

*If you check subtraction with a calculator without a tape, a wrong number could be plugged in and you would not know it. And, since a calculator with a tape may not be readily available to you, it is to your benefit to use one of the methods shown on this and the preceding page.

F3 ◂ Task No. CHECK MULTIPLICATION BY DIVIDING

Example
Problem 1. Multiply and check the problem on the right.

```
  7256   (multiplicand)
X 213    (multiplier)
```

Step 1. Multiply as with whole numbers, Task No. A4 .

```
                                                 7256     (multiplicand) <--+
                                                X 213     (multiplier)      |
1st 3 X 7256 = ----------------------------->   21768     (part product)    |
2nd 1 X 7256 = ----------------------------->   7256      (part product)    | Matches
3rd 2 X 7256 = -------------------------->    +14512      (part product)    | Answer OK
4th Add all the part products ------------>   1545528     (total product)   |
                                                                            |
```

Step 2. Check by dividing the total product by the multiplier. If the quotient matches the multiplicand, the answer should be correct.

```
        7256  (quotient) ---------------------+
213/1545528
    1491
     545
     426
     1192
     1065
      1278
      1278
```

Example
Problem 2. Multiply and check the problem on the right.

```
 3073 <---------+
 X 48           |
```

Step 1. Multiply.

```
  3073          |
  X 48          | Does Not Match
 24584          | Redo
12282           |
147404          |
```

Step 2. Check by dividing the product by the multiplier. NOTE: Does not match, therefore, multiply again and check again.

```
   3070.91 -----+
48/147404
```

Step 3. Redo

```
                 Matches Answer OK
            +--------------------------------+
            v                                |
Multiply:   3073               Check:        3073
            X 48                      48/147504
           24584
          12292
          147504  Answer
```

Note: The above method can be very time consuming, especially if you don't have a calculator with a tape on it. If you have a wrong answer, you have to do all steps over again. The example on the next page shows you how to correct a multiplication problem step by step.

CHECK MULTIPLICATION BY CANCELING NINES — Task No. ► F3A

Canceling nines is an especially effective way to check multiplication, because it allows you to check your work, step by step.

[NOTE: Cancel nines in the multiplicand, multiplier, part products, and the total product in the same manner as you cancel nines in each row of numbers (addends) and the sum for addition. Refer to Task No. F1A, if needed.]

RULE: **The product of the canceling-nines digits in the multiplier and the multiplicand is equal to the canceling-nines digit in the total product as well as the canceling-nines result of the partial products.**

```
(multiplicand)      7256   -2 }  2 X 6 = 12:   1 + 2 = 3
(multiplier)      X  213   -6 }
(part product)     21768   -6 }
(part product)      7256   -2 }  6+2+4 = 12:   1 + 2 = 3
(part product)    14512    -4 }      equal
(total product)   1545528  *(3)      equal
```

[*Circle makes it easier to remember the total product has to be matched.]

```
Example:    7256:          7 2 5+6 = 11: 1+1 = 2    2 X 6 = 12:     1+2 = 3
           X 213:            2+1+3 =  6        6
           21768:        2 1 7 6 8 =  6        6
            7256 :         7 2 5+6 = 11: 1+1 = 2    6+2+4 = 12:     1+2 = 3
          14512  :         1+4 5 1+2 = 4       4           equal
         1545528:    1 5 4 5+5+2 8 = 12: 1+2 = (3)         equal
```

Or, step by step:

```
  7256 x 2 = 6          7256 x 2 = 2          7256 x 2 = 4
X    3               X    13               X   213
 21768  -6            21768  -6             21768  -6
                       7256  -2              7256  -2
                                           14512   -4
```

Canceling nines is not a fail-safe method (a 1555428 answer to the problem above would have yielded the same result), but it is a quick way to double check. If the results do not match, at least you know something is probably wrong with the calculation.

Another method of checking multiplication is to check with a calculator with a tape.* However, many times this will not be readily available for you. Therefore, it is to your benefit to use one of the methods shown on this and the preceding page.

*If you check multiplication with a calculator without a tape, a wrong number could be plugged in and you would not know it.

F4 ◂ Task No. **CHECK DIVISION BY MULTIPLYING**

Example Problem 1. Divide and check the problem on the right.

(divisor) 56/44184 (dividend)

Step 1. Divide as with whole numbers, Task No. A5.

```
                 789   (quotient)
(divisor) 56/44184     (dividend)
             392
              498
              448
               504
               504
    (remainder)  0
```

Step 2. Check by multiplying the quotient by the divisor and adding any remainder to the product. If the total matches the quotient, the answer should be correct.

```
                 789   (quotient)
              X   56   (divisor)
                4734
               3945
               44184
    (remainder)    0
    (total)    44184
```

Matches Answer OK

Example Problem 2. Divide and check the problem on the right.

24/2763

Step 1. Divide.

```
         115 3/24  or  115 1/8   Answer (reduced)
    24/2763
       24
        36
        24
        123
        120
          3
```

Step 2. Check by multiplying the quotient by the divisor, plus the remainder.

```
                   115   (quotient)
                X   24   (divisor)
                   460
                  230
(product)         2760
(remainder)      +   3
(total)           2763
```

Matches Answer OK

CHECK DIVISION BY CANCELING NINES Task No. ▸ F4A

Probably the most effective and the quickest method to use for checking division is by multiplying as shown on the preceding page, and/or by using a calculator with a tape on it*. Canceling nines is still another method that can be used. Therefore, it is included as an alternative way to check division. Use the methods that work best for you.

[NOTE: Cancel nines in the divisor, dividend, and the quotient in the same manner as you cancel nines in each row of numbers (addends) and the sum for addition. Refer to Task Number F1A, if needed.]

RULE: The product of the canceling-nines digit in the quotient and the divisor, plus the remainder (if there is one) is equal to the canceling-nines digit in the dividend.

```
                      789 -6    (quotient)
(divisor)   2-  56/44184  ③*   (dividend)
                   392
                    498
                    448
                     504
                     504
```

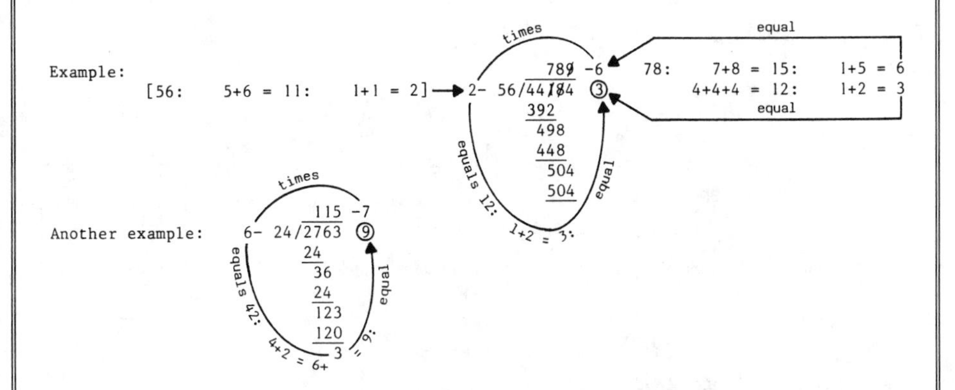

Canceling nines is not a fail-safe method (a 798 answer to the first example above would have yielded the same result), but it is a quick way to double check. If the results do not match, at least you know something is probably wrong with the calculation.

*If you check division with a calculator without a tape, a wrong number could be plugged in and you would not know it. And, since a calculator with a tape may not be readily available to you, it is to your benefit to use one of the methods on this and the preceding page.

G1 | Task No. CHANGE A PERCENT TO A DECIMAL (Whole and Mixed Numbers)

REVIEW OF PERCENTAGE:

Percent is any part of 100 parts. For example 12% is 12 of 100 parts or $\frac{12}{100}$ parts, $2\frac{1}{2}$% is $2\frac{1}{2}$ of 100 parts or $\frac{2\frac{1}{2}}{100}$ parts, or $\frac{3}{4}$% is $\frac{3}{4}$ part of 100 parts or $\frac{\frac{3}{4}}{100}$ parts.

A hundred percent is 100 of 100 parts or 1 full unit, thus 100% is $\frac{100}{100}$ or 1 full unit.

The sign of percent is %.

The percent sign means: o ← The two zeros stand for 100; / ← The line of a fraction means divide. THEREFORE, TO CHANGE A PERCENT TO A DECIMAL, DIVIDE THE PERCENT BY 100 (or move the decimal point to the left 2 places) AND REMOVE THE PERCENT SIGN. For example, 75% = .75

RULE: To change a percent to a decimal, divide the percent by 100 (or move the decimal to the left 2 places) and remove the percent sign (%).

Example Problem 1. Change 12% to a decimal.

Step 1. Place the percent (12%) over 100; $\frac{12\%}{100}$

Step 2. Remove the percent sign ($\frac{12\not\%}{100}$) and divide by 100 ------------> $\frac{12}{100}$ = 12 ÷ 100 or 100/12.00

```
     .12  Answer
100/12.00
    10 0
     2 00
     2 00
```

Or ------------ remove the percent sign 12~~%~~

and move the decimal point to the left .12 two places

Check the Answer by Multiplying

```
  .12
X 100
12.00 + % = 12%
```

Answer OK

Example Problem 2. Change 2 1/2% to a decimal.

Step 1. 1st, place the percent ($2\frac{1}{2}$%) over 100 ——> $\frac{2\frac{1}{2}\%}{100} = \frac{2.5\%}{100}$

2nd, change $2\frac{1}{2}$% to a decimal

Step 2. Remove the percent sign ($\frac{2.5\not\%}{100}$) and divide by 100 ------------> $\frac{2.5}{100}$ = 2.5 ÷ 100 or 100/2.500

```
     .025  Answer
100/2.500
    2 00
      500
      500
```

Or change the percent to a decimal 2 1/2% = 2.5%

remove the percent sign, 2.5~~%~~

and move the decimal point to the left .025 two places.

Check the Answer by Multiplying

```
 .025
X 100
2.500 + % = 2 1/2%
```

Answer OK

CHANGE A PERCENT TO A DECIMAL (Fraction Only) — Task No. ▸ G1

Example Problem 3. Change 3/4% to a decimal.

Step 1. 1st, place the percent ($\frac{3}{4}\%$) over 100 ⟶ $\frac{\frac{3}{4}\%}{100} = \frac{.75\%}{100}$

2nd, change $\frac{3}{4}\%$ to a decimal

Step 2. Remove the percent sign ($\frac{.75\%}{100}$)

and divide by 100 ⟶ $\frac{.75}{100} = .75 \div 100$ or $100\overline{).7500}$ = .0075 Answer

```
   .0075  Answer
100/.7500
    700
     500
     500
```

Or change the percent to a decimal, 3/4% = .75%

remove the percent sign, .75~~%~~

and move the decimal point to the left .0075 two places.

Check the Answer by Multiplying

.0075
X 100
.7500 + % = $\frac{3}{4}\%$
Answer OK

CHANGE A DECIMAL TO A PERCENT — Task No. ▸ G2

RULE: To change a decimal to a percent, multiply the decimal by 100 (or move the decimal point to the right 2 places) and add the percent sign (%).

Example Problem 1. Change .32 to a percent.

Step 1. Multiply the decimal --------▸ .32
by 100 --------------------▸ X 100
32.00

[Note that moving the decimal two places to the right is the same as multiplying by 100.]

Step 2. Add the percent sign. 32 + % = 32% Answer

Check the Answer by Dividing

```
      .32 Answer OK
100/32.00
    30 0
     2 00
     2 00
```

Example Problem 2. Change .085 to a percent.

Step 1. Multiply the decimal --------▸ .085
by 100 --------------------▸ X 100
8.500

Step 2. Add the percent sign. 8.5 + % = 8.5% or $8\frac{1}{2}\%$ Answer

Check the Answer by Dividing

```
     .085 Answer OK
100/8.500
    8 00
      500
      500
```

Example Problem 3. Change .005 to a percent.

Step 1. Multiply the decimal --------▸ .005
by 100 --------------------▸ X 100
.500

Step 2. Add the percent sign. .5 + % = .5% or $\frac{1}{2}\%$ Answer

Check the Answer by Dividing

```
     .005 Answer OK
100/.500
     500
```

Example Problem 4. Change 2.5 to a percent.

Step 1. Multiply the decimal ------▸ 2.5
by 100 ------------------▸ X 100
250.0

Step 2. Add the percent sign. 250 + % = 250% Answer

Check the Answer by Dividing

```
     2.5  Answer OK
100/250
```

G3 ◂ Task No. CHANGE A FRACTION TO A PERCENT

RULE: To change a fraction to a percent, (1) change the fraction to a decimal, (2) multIply by 100, and (3) add the percent sign (%).

Example Problem 1. Change 1/4 to a percent.

Step 1. Change the fraction $\frac{1}{4}$ to a decimal: $\frac{1}{4} = 1 \div 4$ or $4\overline{)1.00}$ = .25 (8, 20, 20)

[To change a fraction to a decimal, divide by the denominator of the fraction.]

Step 2. Multiply by 100. ⟶ .25 X 100 = 25.00

Step 3. Add the percent sign. 25 + % = 25% Answer

Or change the fraction $\frac{1}{4}$ to a decimal, ⟶ .25

move the decimal pt. two places to the right, x25.

and add the percent sign (%). 25%

To Check the Answer, Reduce by Canceling

$25\% = \frac{25}{100}$

$\frac{25 \div 25}{100 \div 25} = \frac{1}{4}$ Answer OK

Example Problem 2. Change 5/4 to a percent.

Step 1. Change the fraction $\frac{5}{4}$ to a decimal, thus $\frac{5}{4} = 5 \div 4$ or $4\overline{)5.00}$ = 1.25 (4, 1 0, 8, 20, 20)

Step 2. Multiply by 100, ⟶ 1.25 X 100 = 125.00

Step 3. Add the percent sign. 125 + % = 125% Answer

Or change the fraction $\frac{5}{4}$ to a decimal, ⟶ 1.25

move the decimal pt. two places to the right, 1x25.

and add the percent sign (%). 125%

To check the Answer, Reduce by Canceling

$125\% = \frac{125}{100}$

$\frac{125 \div 25}{100 \div 25} = \frac{5}{4}$ Answer OK

G4 ◂ Task No. THE PERCENTAGE FORMULA (FORMULA 1)

Formula 1

RULE: To find the Percentage of a number, multiply the Base times (x) the Rate.
FORMULA: P = BR (Note the Rate or percent must be changed to a decimal before calculating.)

Example Problem 1. Find 4% of $26.

Percentage	?
Base	$26
Rate (percent)	4%

Step 1. Use the formula ------------▸ P = BR

Step 2. Substitute the known values into the formula -----------▸ P = $26 X 4%

Step 3. Change the Rate or percent to a decimal -----------▸ $4\% = \frac{4}{100} =$ $100\overline{)4.00}$ = .04 (4 00)

P = 26 X .04

RULE: To change a percent to a decimal, divide the percent by 100 (or move the decimal point to the left 2 places) and remove the percent sign (%).

Step 4. Calculate the formula (do the operations indicated) --------▸

$$\begin{array}{r} 26 \\ \times\ .04 \\ \hline \$1.04 \end{array}$$

P = \$1.04 Answer

Check by substituting the answer for the Percentage and calculating.

1.04 = 26R

1.04 ÷ 26 = .04

.04 = 4% Answer OK

Example Problem 2. 17 3/4% of 125 is what number?

Step 1. Use the formula ------------▸ P = BR

Step 2. Substitute the known values into the formula -----------▸ $P = 125 \times 17\frac{3}{4}\%$

Percentage = ?
Base = 125
Rate (percent) = $17\frac{3}{4}\%$

Step 3. Change the Rate or percent to a decimal -----------▸ $17\frac{3}{4}\% = \frac{17.75}{100} = 100\overline{)17.75}$ = .1775

P = 125 X .1775

Step 4. Calculate the formula (do the operations indicated) --▸

$$\begin{array}{r} .1775 \\ \times\ 125 \\ \hline 8875 \\ 3550 \\ 1775 \\ \hline 22.1875 \end{array}$$

P = 22.1875 Answer

Check by substituting for P and calculating.

22.1875 = 125R

22.1875 ÷ 125 = .1775

.1775 = 17 3/4%

Answer OK

Example Problem 3. Find 1/2% of 66.

Step 1. Use the formula ------------▸ P = BR

Step 2. Substitute: ----------------▸ $P = 66 \times \frac{1}{2}\%$

Percentage = ?
Base = 66
Rate (percent) = $\frac{1}{2}\%$

Step 3. Change the Rate or percent to a decimal ----------▸ $\frac{1}{2}\% = \frac{\frac{1}{2}}{100}$ or $\frac{.5}{100}$ or $100\overline{).500}$ = .005; 500

P = 66 X .005

Step 4. Calculate: -----------------▸

$$\begin{array}{r} 66 \\ \times\ .005 \\ \hline .330 \end{array}$$

P = .330 Answer

Check by substituting for P and calculating.

.33 = 66R

.33 ÷ 66 = .005

.005 = 1/2% Answer OK

G4 ◂ Task No. THE PERCENTAGE FORMULA (FORMULA 2)

Formula 2

RULE: To find the Rate of a number, divide the Percentage by the Base.
FORMULA: R = P/B [Note the answer (quotient) must be changed to a percent.]

Example Problem 1. What percent of 35 is 7? **[Can also be read: 7 is what percent of 35?]**

Step 1. Use the formula ----------------→ $R = \frac{P}{B}$

Step 2. Substitute the known values into the formula ----------------→ $R = \frac{7}{35}$

Percentage = 7
Base = 35
Rate (percent) = ?

Step 3. Calculate the formula (do the operations indicated) -----------→ R = 7 ÷ 35

R = .2 (quotient)

```
      .2
35/7.0
   7 0
```

Step 4. Change the quotient to a percent --------------------→ .2

```
   .2
X 100
20.00 + % = 20%
```

RULE: To change a decimal to a percent, multiply the decimal by 100 (or move the decimal point to the right 2 places) and add the percent sign (%).

R = 20% Answer

Check by substituting the answer for the Rate (percent) and calculating.

What __% of 35 is 7?

20% of 35 is 7.

```
  35
X .20      .20 X 35 = 7
 7.00 -----→ 7 = 7  Answer OK
```

Example Problem 2. $30.00 is what percent of $86.00? **[Can also be read: What percent of $86.00 is $30.00.]**

Step 1. Use the formula ----------------→ $R = \frac{P}{B}$

Step 2. Substitute: --------------------→ $R = \frac{30.10}{86}$

Percentage = 30.10
Base = 86
Rate (percent) = ?

Step 3. Calculate: ---------------------→ R = 30.10 ÷ 86

R = .35 (quotient)

```
      .35
86/30.10
   25 8
    4 30
    4 30
```

Step 4. Change the quotient to a percent -------------------→ .35

```
  .35
X 100
35.00 + % = 35%
```

R = 35% Answer

Check by substituting for R and calculating.

$30.10 is __% of $86.00?

$30.10 is 35% of $86.00.

30.10 = .35 X 86

30.10 = 30.10 Answer OK

Formula 3

RULE: To find the Base of a number, divide the Percentage by the Rate.
FORMULA: B = P/R (Note the Rate or percent must be changed to a decimal before calculating.)

Example Problem 1. 15% of what number is 12. **[Can also be read: 12 is 15% of what number?]**

Step 1. Use the formula ----------------▶ $B = \frac{P}{R}$

Step 2. Substitute the known values into the formula ---------------▶ $B = \frac{12}{15\%}$

Percentage = 12
Base = ?
Rate (percent) = 15%

Step 3. Change the Rate or percent to a decimal ---------▶ $15\% = \frac{15}{100} =$ 100/15.00 = .15 (10 0; 5 00; 5 00)

$B = \frac{12}{.15}$

RULE: To change a percent to a decimal, divide the percent by 100 (or move the decimal point to the left 2 places) and remove the percent sign (%).

Step 4. Calculate the formula (do the operations indicated) ---------▶ $B = 12 \div .15$

B = .15/12.00 = 80 Answer (12 0; 0; 0)

Check by substituting the answer for the Base and calculating.
15% of __ is 12?
15% of 80 is 12.
.15 X 80 = 12 (.15 X 80 = 12.00)
12 = 12 ◀---- Ans. OK

Example Problem 2. $2.60 is 4% of what number? **[Can also be read: 4% of what number is $2.60?]**

Step 1. Use the formula ----------------▶ $B = \frac{P}{B}$

Step 2. Substitute: --------------------▶ $B = \frac{2.60}{4\%}$

Percentage = 2.60
Base = ?
Rate (percent) = 4%

Step 3. Change the Rate or percent to a decimal ----------▶ $4\% = \frac{4}{100} =$ 100/4.00 = .04 (4 00)

$B = \frac{2.60}{.04}$

Step 4. Calculate: --------------------▶ $B = 2.60 \div .04$

B = .04/2.60. = 65 Answer (2 4; 20; 20)

Check by substituting for B and calculating.
$2.60 is 4% of __?
$2.60 is 4% of $65.
2.60 = .04 X 65
2.60 = 2.60 Answer OK

BASICS OF PERCENTAGE

The purpose of this module is to illustrate why and how percentage works, and to assist you in solving the basic percentage problems encompassed in the world of work.

Review of Percentage

Percent is any part of 100 parts. A hundred percent is 100 parts of 100 parts or 1 full unit -------- thus $\frac{100}{100} = 1$.

The sign of percent is %.

The percent sign means (o ← The two zeros stand for 100; / ← The line of a fraction means divide) THEREFORE, TO CHANGE A PERCENT TO A DECIMAL, DIVIDE BY 100.

Examples:

A 100% would be 100 parts of 100 parts

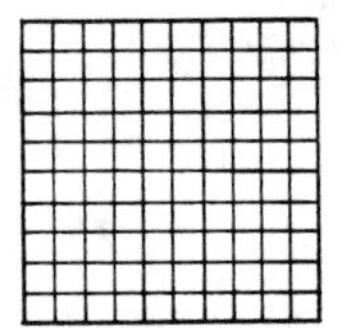

or $\frac{100}{100}$ parts

or $100\overline{)100}$ = 1 full part

1% would be 1 part of 100 parts

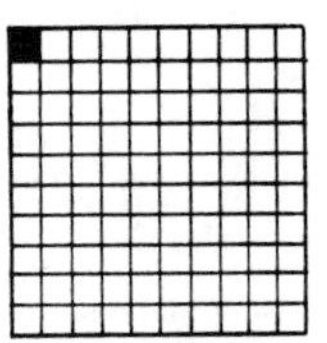

or $\frac{1}{100}$ parts

or $100\overline{)1.00}$ = .01 or 1%

$\frac{3}{4}$% would be $\frac{3}{4}$ of 1 part of 100 parts

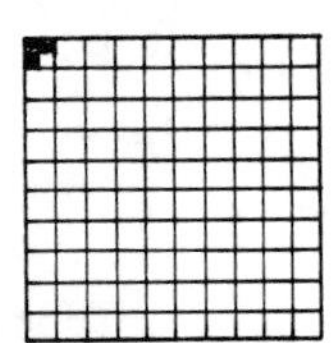

or $\frac{\frac{3}{4}}{100}$ parts

or $100\overline{).7500}$ = .0075 or $\frac{3}{4}$%

4% would be 4 parts of 100 parts

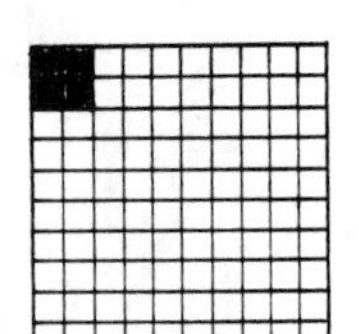

or $\frac{4}{100}$ parts

or $100\overline{)4.00}$ = .04 or 4%

[Notice when solving percent problems the denominator will always be 100, THEREFORE TO CHANGE A PERCENT TO A DECIMAL, REMOVE THE % SIGN AND MOVE THE DECIMAL TWO PLACES TO THE LEFT.] THUS: 4% = .04

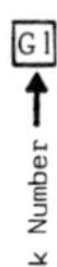

G1 TO CHANGE A PERCENT TO A DECIMAL, DIVIDE THE PERCENT BY 100 (or move the decimal point to the left two places) AND REMOVE THE PERCENT SIGN (%).

If you divide by 100 to change a percent to a decimal, then to change a decimal to a percent, do the exact opposite and multiply by 100.

Percent to decimal <u>divide by 100</u> — Decimal to percent <u>multiply by 100</u>

Example: 4% = $\frac{4}{100}$ or $100\overline{)4.00}$ = .04 (or 4% = .04% = .04) — Remove the percent sign

```
   .04
 X 100
  4.00 + % = 4%
```

Add the percent sign

G2 TO CHANGE A DECIMAL TO A PERCENT MULTIPLY BY 100 (or move the decimal point to the right two places) AND ADD THE PERCENT SIGN (%)*.

Examples: Change .01 to a percent =

```
   .01
 X 100
  1.00 + % = 1%
```

(or .01 = 01. + % = 1%)

Change .0075 to a percent =

```
  .0075
  X 100
  .7500 + % = .75% or 3/4%
```

(or .0075 = 00.75 + % = .75% or $\frac{3}{4}$%)

Change .235 to a percent =

```
    .235
   X 100
  23.500 + % = 23.5% or 23 1/2%
```

(or .235 = 23.5 + % = 23.5% or $23\frac{1}{2}$%)

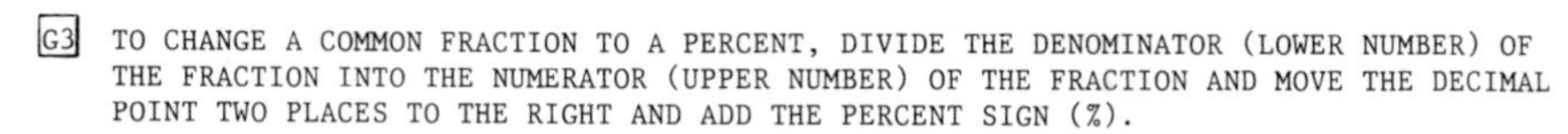

G3 TO CHANGE A COMMON FRACTION TO A PERCENT, DIVIDE THE DENOMINATOR (LOWER NUMBER) OF THE FRACTION INTO THE NUMERATOR (UPPER NUMBER) OF THE FRACTION AND MOVE THE DECIMAL POINT TWO PLACES TO THE RIGHT AND ADD THE PERCENT SIGN (%).

Example: Change the common fraction $\frac{7}{8}$ to a percent.

Divide 8 into 7 = $8\overline{)7.000}$ = .875 = 87.5 + % = 87.5% or $87\frac{1}{2}$% [Move decimal 2 places to the right]

<u>Or</u> multiply by 100 =

```
   .875
  X 100
 87.500 + % = 87 1/2%
```

[Note: $\frac{7}{8}$ is a proper fraction because the numerator (7) is smaller than the denominator (8).]

Example: Change the common fraction $\frac{9}{6}$ to a percent.

Divide 6 into 9 = $6\overline{)9.0}$ = 1.5 = 150. + % = 150% [Move decimal 2 places to the right]

<u>Or</u> multiply by 100 =

```
   1.5
 X 100
 150.0 + % = 150%
```

[Note: $\frac{9}{6}$ is an improper fraction because the numerator (9) is larger than the denominator (6).]

[*Another way to look at it -- to change a decimal to a common fraction multiply by the desired denominator -- <u>the denominator in percent is always 100, therefore, multiply by 100.</u>]

Definitions involved with calculating percent problems.

Definitions: Percent or Rate is the number of parts taken.

Example: 2% of 25 = ? (The percent or rate is 2%, .02 or $\frac{2}{100}$ parts)

Base is the original number that the rate or percent is applied to. Example: 2% of 25 = ? (the base is 25)

Percentage is the result of multiplying the percent or the rate times the base. Example: 2% of 25 = ?

.02 X 25 = .50 (.5 is the percentage)

The Percentage Formula (the three different percentage formulas)

RULE: **The Percentage equals the Base times the Rate.**
FORMULA: **P = BR**

By applying basic algebra to the above formula -- the three different percentage formulas listed below can be solved.

Formula 1. **The base and the rate are known, therefore, it is necessary to find the percentage.**

Example: Formula: P = BR [Base = 25 Rate = 2% or .02]

Substitute: P = 25 X 2%

Calculate: P = .5 [25 X .02 = .50]

Formula 2. **The percentage and the base are known, therefore, it is necessary to find the rate.**

Example: Original formula: P = BR

New formula: $R = \frac{P}{B}$ [Percentage = .5] [Base = 25]

Substitute: $R = \frac{.5}{25}$

Calculate: R = .02 [25/.50] (quotient .02; 50)

or R = 2%

Note in the formula P = BR to solve for the Rate, divide both sides by the Base.

Example: $\frac{P}{B} = \frac{\not{B}R}{\not{B}}$

Formula 3. **The percentage and the rate are known, therefore, it is necessary to find the base.**

Example: Original formula: P = BR

New formula: $B = \frac{P}{R}$ [Percentage = .5] [Rate = 2% or .02]

Substitute: $B = \frac{.5}{.02}$

Calculate: B = 25 [.02/.50] (quotient 25; 4, 10, 10)

Note in the formula P = BR to solve for the Base, divide both sides by the Rate.

Example: $\frac{P}{R} = \frac{B\not{R}}{\not{R}}$

Practical Applications of Percentage

1) Write 8% as a decimal = .08

$8\% = \frac{8}{100}$ or $100\overline{)8.00}$ = .08 (800)

or 8% = .08

2) Write $\frac{3}{4}\%$ as a decimal = .0075

$\frac{3}{4}\% = \frac{\frac{3}{4}}{100}$ or $.\frac{75}{100}$ = $100\overline{).7500}$ = .0075

or $\frac{3}{4}\%$ = .75% = .00.75

3) Write $\frac{1}{2}$ as a percent = 50%

$\frac{1}{2} = 2\overline{)1.0}$ = .5 (10) = 100 x .5 = 50.0 + % = 50%

or .50. + % = 50%

4) Write .10 as a percent = 10%

.10 = 100 x .10 = 10.00 + % = 10%

or .10. + % = 10%

5) Find 6% of 206 = 12.36

P = BR

P = .06 X 206 — 206 x .06 = 12.36

P = 12.36

6) 9% of what number is 7.65 = 85

P = BR or $B = \frac{P}{R}$

$\frac{7.65}{.09}$ = B = $.09\overline{)7.65}$ = 85 (72, 45, 45)

7) 75.68 is what percent of 344? 22%

P = BR or $R = \frac{P}{B}$

$\frac{75.68}{344}$ = R = $344\overline{)75.68}$ = .22 = .22.%

22% = R

8) Write 175% as a decimal = 1.75

$175\% = \frac{175}{100}$ or $100\overline{)175.00}$ = 1.75

or 1.75

9) Write 2.25 as a percent = 225%

2.25 X 100 = 225.00 + % = 225%

10) Write $2\frac{1}{8}\%$ as a decimal = .02125

$2\frac{1}{8}\% = \frac{2\frac{1}{8}}{100}$ or $100\overline{)2.125}$ = .02125

11) Write .0025 as a percent = 1/4%

.0025 X 100 = .2500 + % = $\frac{1}{4}\%$

12) Find $9\frac{1}{2}\%$ of \$62.25 = \$5.91

\$62.25 X .095 = 31125, 56025 = \$5.91375

13) Write .0025 as a percent = .25%

.0025 X 100 = .2500 + % = .25%

14) Find $\frac{3}{4}\%$ of 28 = .21

$\frac{3}{4}\%$ = .75%

.0075 X 28 = 600, 150 = .2100

H1 ◂ Task No. **POWERS AND ROOTS**

POWERS OF NUMBERS:

The **square** of a number is a common example of a **power** of a number. A number written with a raised ^{2} immediately following it is to be multiplied by itself, in other words, raised to the second power, or squared.

Examples: 2^2 means 2 x 2, or 4

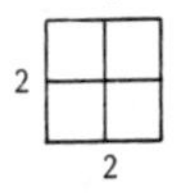

5^2 means 5 x 5, or 25

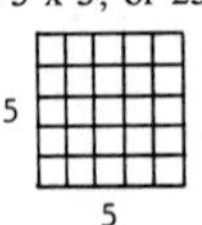

The **power** of a number, then, indicates the number of times that number is a factor when multiplying it by itself. The power is shown by the raised number, which is called an **exponent**. In 3 x 3, for example, 3 is multiplied by itself just once, but it is a factor twice. Therefore, we can write 3 x 3 as 3 with an exponent of 2: 3^2 (read "three squared" or "three to the second power"). Three to the third power is wrriten 3^3, usually read as "three cubed," and means 3 x 3 x 3, or 27. Three to the fourth power is written 3^4 (read "three to the fourth") and means 3 x 3 x 3 x 3, or 81.

Examples: $3^3 = 3 \times 3 \times 3 = 27$

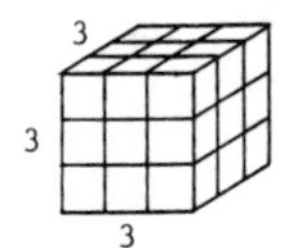

$3^4 = 3 \times 3 \times 3 \times 3 = 81$

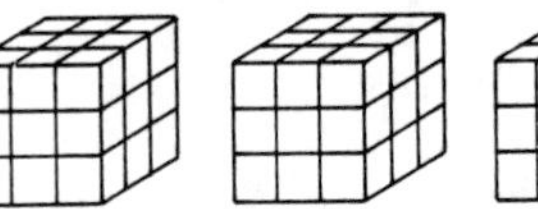

As with other kinds of calculations, when powers are indicated in conjunction with parentheses ( ), brackets [ ], or braces { }, do the innermost calculations first. Compare these examples step by step.

	Example 1	Example 2
	$[7 = (2 + 3)^2]^2$	$[(7 + 2) + 3^2]^2$
Step 1	$[7 + (5)^2]^2$	$[(9) + 3^2]^2$
Step 2	$[7 + (5 \times 5)]^2$	$[9 + (3 \times 3)]^2$
Step 3	$[7 + 25]^2$	$[9 + 9]^2$
Step 4	$[32]^2 = 1024$	$[18]^2 = 324$

ROOTS OF NUMBERS:

The **root** of a number is whatever number taken as a factor a given number of times yields the original number. The **square root** of a number is the number which taken as a factor twice, that is, multiplied by itself, yields the original number. The square root of 9 is 3 because 3 multiplied by itself yields 9. The square root of 100 is 10 for the same reason -- 10 x 10 = 100. The **cube root** of a number is whatever number taken as a factor three times yields the original number. The cube root of 8 is 2 because 2 x 2 x 2 = 8. In another example, the fourth root of 81 is 3 because 3 x 3 x 3 x 3 = 81.

The **radical sign** $\sqrt{}$ indicates square root. When the radical sign is placed over a number such as $\sqrt{36}$, the square root of that number is to be found. When a raised number is placed in the radical sign, as $\sqrt[3]{}$, the indicated root of that number is to be found. For example, $\sqrt[3]{8}$ means find the cube root of 8, or 2 [2 x 2 x 2 = 8]; $\sqrt[4]{16}$ means find the fourth root of 16 or 2 [2 x 2 x 2 x 2 = 16]. The raised number placed in the radical sign is called the **index** of the root. When no raised number is placed in the radical sign, the index is understood to be 2.

When the radical sign $\sqrt{\frac{49}{64}}$ is placed over a fraction, find the square root of both the numerator (top number) and the denominator (bottom number) individually.
Example:

$$\sqrt{\frac{49}{64}} = \frac{\sqrt{49}}{\sqrt{64}} = \frac{7}{8}$$

[To find the square root of a number see Task No. H2 on the next two pages.]

SCIENTIFIC NOTATION:

Talking about powers of numbers leads to another term called scientific notation. **Scientific notation** is used to express very large and very small numbers in terms of the powers of 10. Example: 1,000,000 is equal to 1 x 10 X 10 X 10 X 10 X 10 X 10. An easier (shorter) way to write this number is 1×10^6. To write a decimal fraction such as .00010, the same principal is true. However, the exponent of 10 would be negative to indicate dividing by 10. Example: .0001 is equal to $1 \div 10 \div 10 \div 10 \div 10$, or 1×10^{-4}.

To change any large number to scientific notation, count to the left or right of the decimal point until you are to the left or right of the first whole number.

Examples: 1,200,000 equals 1.2×10^6
.00000082 equals 8.2×10^{-7}

RULE: When adding or subtracting numbers using scientific notation (without the use of a scientific calculator), change the numbers so they have the same exponent of 10 by moving the decimal point, then add or subtract.

Example 1. Add $(9.7 \times 10^6) + (4.3 \times 10^7)$ =

$$\begin{array}{r} 4.3 \times 10^7 \\ + \underline{.97 \times 10^7} \\ \text{Answer } 5.27 \times 10^7 \end{array} \qquad \text{Or} \qquad \begin{array}{r} 9.7 \times 10^6 \\ + \underline{43.0 \times 10^6} \\ 52.7 \times 10^6 \end{array}$$

Example 2. Subtract $(5.2 \times 10^3) - (1.3 \times 10^{-1})$ =

$$\begin{array}{r} 5.20000 \times 10^3 \\ - \underline{.00013 \times 10^3} \\ \text{Answer } 5.19987 \times 10^3 \end{array}$$

RULE: When multiplying numbers using scientific notation (without the use of a scientific calculator), calculate the numbers and the exponents of 10 separately. 1st multiply the numbers and, 2nd add the exponents.

Example 3. Multiply $(1.25 \times 10^5) \times (6.25 \times 10^4)$ = 1st 1.25, **Multiply** → X 6.25; 2nd 10^5, 10^4 — **Add the exponents [(5 + 4) = 9]**

Answer 7.8125×10^9 Or 7,812,500,000

RULE: When dividing numbers using scientific notation (without the use of a scientific calculator), calculate the numbers and the exponents of 10 separately. 1st divide the numbers, and 2nd subtract the exponents.

Example 4. Divide $(2.5 \times 10^4) \div (6.25 \times 10^{-2})$ = 2nd $\frac{2.5 \times 10^4}{6.25 \times 10^{-2}}$ — **Subtract the exponents [4-(-2) = 6]**

Divide 1st $\frac{2.5}{6.25} = .4$

Answer $.4 \times 10^6$ or 4×10^5 or 400,000

[Note: Most scientific notation problems are solved by using a scientific calculator.]

H2	◂ Task No.	FIND SQUARE ROOTS

REVIEW: The radical sign $\sqrt{}$ indicates square root. $\sqrt{25}$ means the square root of 25. 5 X 5 = 25; therefore, the root is 5.

Example Problem 1. Find the square root of 4624. **[What number times itself equals 4624?]**

Step 1. From right to left separate the number into groups of two's and draw a straight, light vertical line as shown.

$\sqrt{46|24.}$ [The decimal point always goes to the right of the whole number.] 2 spaces left

Step 2. 1st, find the largest square root (number times itself) that will go into the first group, (6 X 6 = 36), and place as shown.

[Note that the root 7 would not work because 7 X 7 = 49 (too big); therefore, the root has to be 6.]

2nd place the root 6 above the 1st group as shown.

3rd subtract 46 - 36 = 10 and place 10 as shown.

Step 3. 1st, bring down another group.

2nd, double the root 6: 6 + 6 = 12 → 12?

X ?

3rd, place blanks and question marks, and set up as shown.

[Note the 2nd part of this step is the most important part but also the most forgotten in finding square roots.]

Step 4. 1st, using 12? and ?, replace the question marks with a number that when multiplied will go into the new group 1024. Place 1024 as shown.

[Note that if 9 were used, instead of 8, the number would be too big.] 129 X 9 = 1161

6 + 6 = 128, X 8 = 1024

2nd, place the number 8 above the 2nd group as shown.

3rd, subtract 1024 - 1024 = 0; this is as far as you can go. The answer is 68.

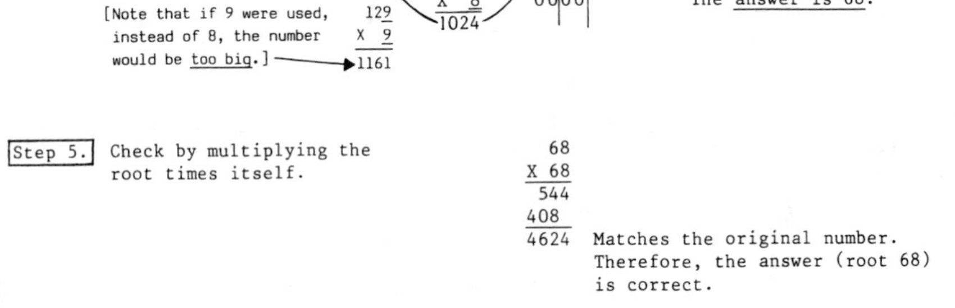

Step 5. Check by multiplying the root times itself.

68 X 68 = 544 + 408 = 4624 Matches the original number. Therefore, the answer (root 68) is correct.

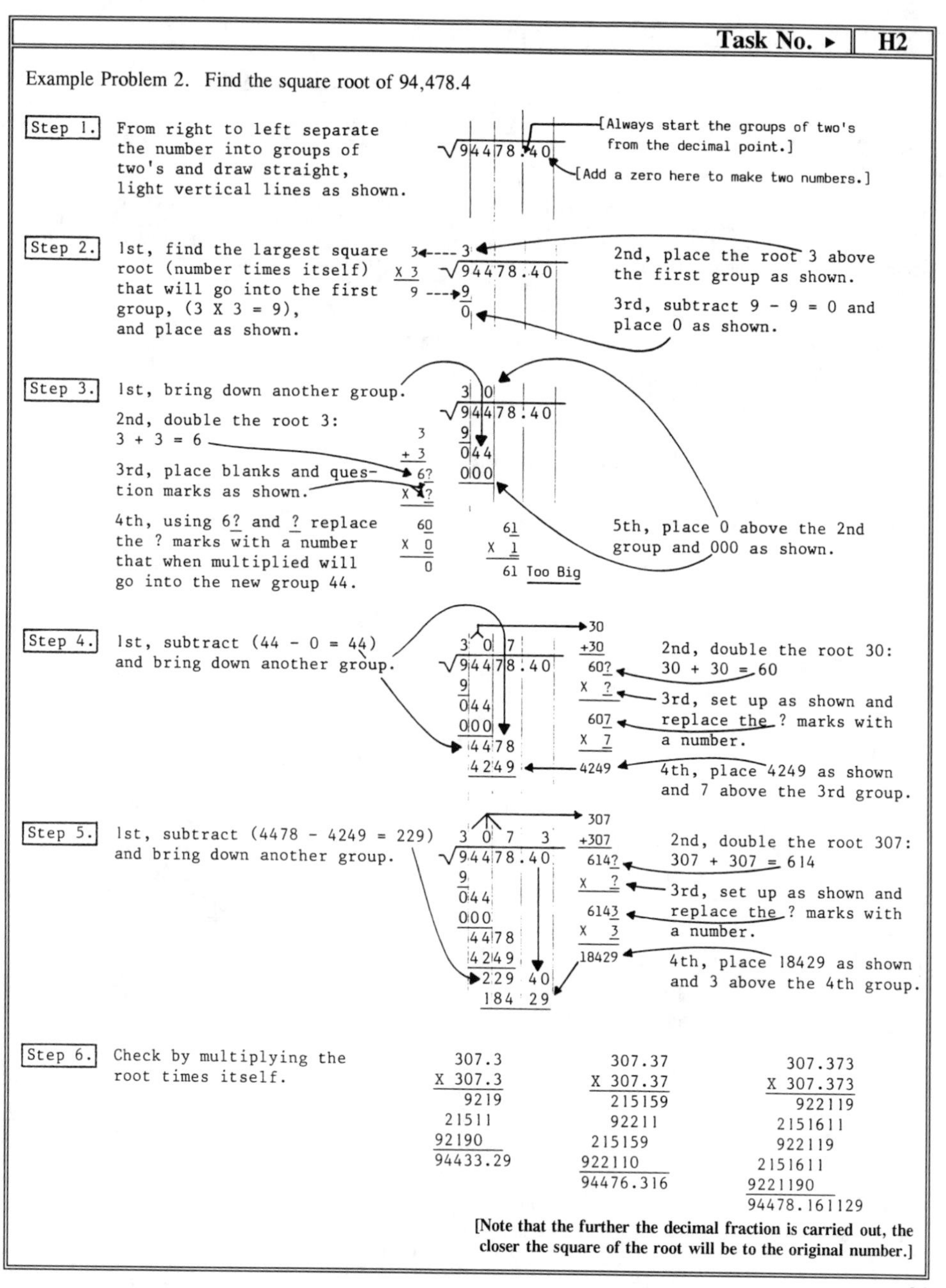
Task No. ▸ H2
Example Problem 2. Find the square root of 94,478.4
Step 1.
From right to left separate the number into groups of two's and draw straight, light vertical lines as shown.
√94478.40
[Always start the groups of two's from the decimal point.]
[Add a zero here to make two numbers.]
Step 2.
1st, find the largest square root (number times itself) that will go into the first group, (3 X 3 = 9), and place as shown.
2nd, place the root 3 above the first group as shown.
3rd, subtract 9 - 9 = 0 and place 0 as shown.
Step 3.
1st, bring down another group.
2nd, double the root 3: 3 + 3 = 6
3rd, place blanks and question marks as shown.
4th, using 6? and ? replace the ? marks with a number that when multiplied will go into the new group 44.
61 Too Big
5th, place 0 above the 2nd group and 000 as shown.
Step 4.
1st, subtract (44 - 0 = 44) and bring down another group.
2nd, double the root 30: 30 + 30 = 60
3rd, set up as shown and replace the ? marks with a number.
4th, place 4249 as shown and 7 above the 3rd group.
Step 5.
1st, subtract (4478 - 4249 = 229) and bring down another group.
2nd, double the root 307: 307 + 307 = 614
3rd, set up as shown and replace the ? marks with a number.
4th, place 18429 as shown and 3 above the 4th group.
Step 6.
Check by multiplying the root times itself.
307.3 X 307.3 = 94433.29
307.37 X 307.37 = 94476.316
307.373 X 307.373 = 94478.161129
[Note that the further the decimal fraction is carried out, the closer the square of the root will be to the original number.]

H2A	◂ Task No.	APPLY-IT MODULE

APPLICATION OF SQUARE ROOTS, AND THE 3-4-5 AND 5-12-13 SQUARING METHODS

Before assembling or building practically any object, it must be laid out true in dimension and square (at 90° angles) before actually starting the construction of it. Otherwise, the parts that make up the object would not fit properly.

This page and the next page demonstrate why the Rule of Pythagoras ($a^2 + b^2 = c^2$), and a shortcut (trick of the trades) using the largest combination of 3-4-5 actually work for squaring up objects. There are times that you will have to use the Rule of Pythagoras. However, for most practical applications the largest combination of 3-4-5 will work and be accurate enough. The pages following these two pages demonstrate how to apply these principles and some examples of practical problems you may encounter.

THE RULE OF PYTHAGORAS: [Task No. K5A] **In any right triangle, the square of the hypotenuse is equal to the sum of the squares of the other two sides.**

FORMULA: $a^2 + b^2 = c^2$

c a 90° b

How to use the Rule of Pythagoras

Example Problem 1. Use the Rule of Pythagoras and square roots to prove the 3-4-5 squaring method.

c 3 90° 4

Step 1. Substitute the known values of the problem into the formula.

Formula: $a^2 + b^2 = c^2$

Substitute: $3^2 + 4^2 = c^2$

Step 2. Calculate the known values and complete the formula.

Calculate:

$$\begin{array}{r} 3 \\ \underline{\times 3} \\ 9 \end{array} + \begin{array}{r} 4 \\ \underline{\times 4} \\ 16 \end{array} = c^2$$

$$25 = c^2$$

$$\sqrt{25} = c$$

5 3 90° 4

Step 3. Find the square root of 25. (What number times itself equals 25)

$$\begin{array}{r} ? \\ \underline{\times ?} \\ 25 \end{array} \quad \begin{array}{r} 5 \\ \underline{\times 5} \\ 25 \end{array} \quad 5 = c, \ \sqrt{25}$$

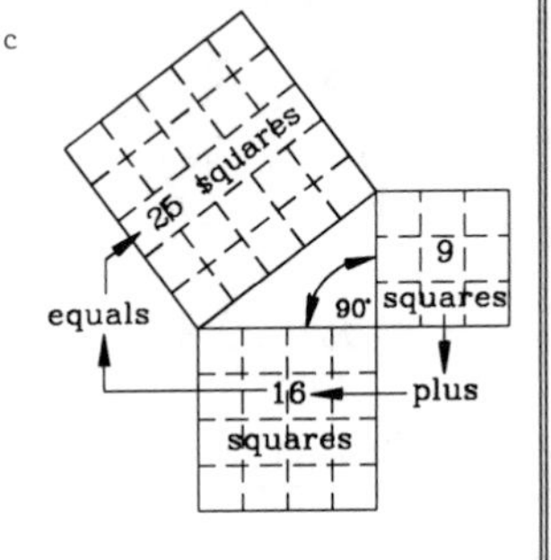

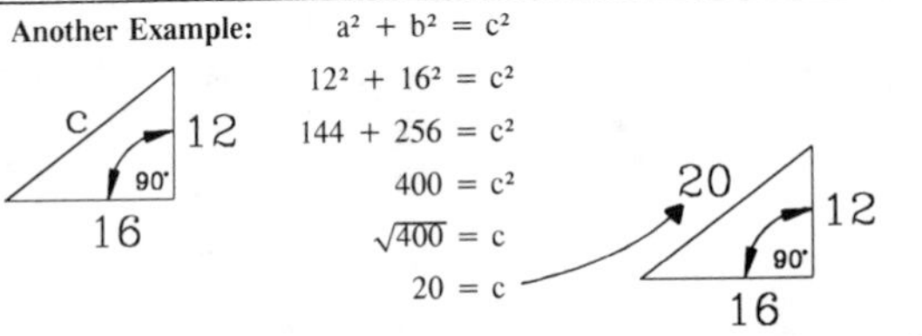

Another Example:

$$a^2 + b^2 = c^2$$
$$12^2 + 16^2 = c^2$$
$$144 + 256 = c^2$$
$$400 = c^2$$
$$\sqrt{400} = c$$
$$20 = c$$

Try any combination of 3-4-5 and the formula will work. The examples on this page demonstrate the Rule of Pythagoras and why any combination of 3-4-5 will work.

Another combination that can be used to square an object, is 5-12-13. Plug it into the rule above and try it. See Task Number P1 , for more information on using 5-12-13 for squaring.

5 13 12

Example

Problem 2. Use the Rule of Pythagoras and square roots to find the exact hypotenuse of a right triangle.

Find the exact hypotenuse of the right triangle on the right so you can read the dimension on an American (English) tape measure 1/16± of an inch.

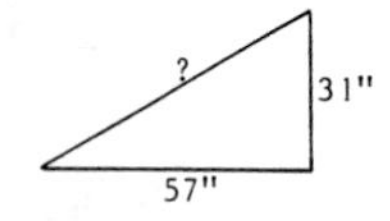

YOU WILL NOTE BELOW, THAT THERE IS A LOT OF EXTRA WORK INVOLVED IN FINDING THE EXACT HYPOTENUSE -- HOWEVER THERE WILL BE TIMES THAT YOU WILL HAVE TO DO IT THIS WAY.

Step 1. Substitute the known values of the problem into the formula.

Formula

$a^2 + b^2 = c^2$

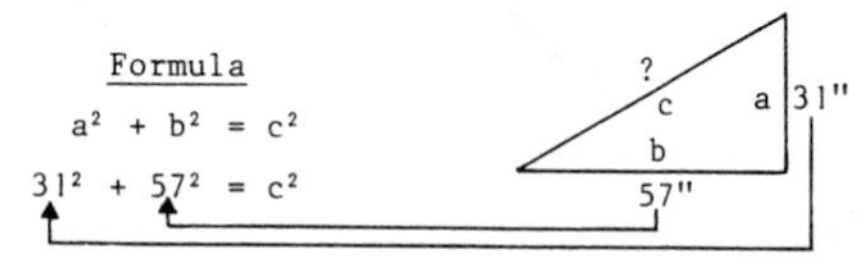

Step 2. Calculate the known values and complete the formula.

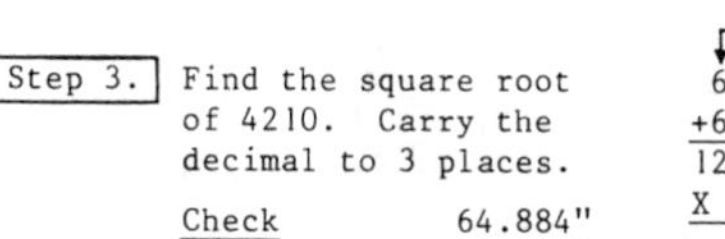

$a^2 + b^2 = c^2$

$31^2 + 57^2 = c^2$

$961 + 3249 = c^2$

$4210 = c^2$

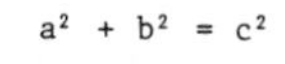

Step 3. Find the square root of 4210. Carry the decimal to 3 places.

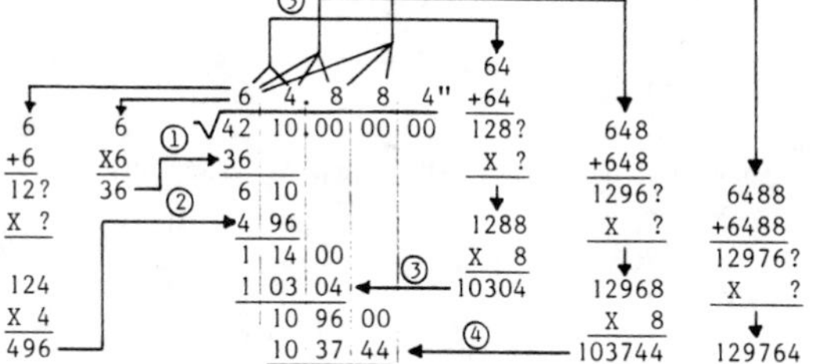

Check

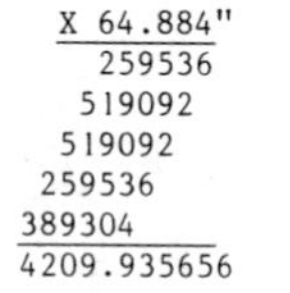

The answer is very close to the original number 4210. Therefore, the answer must be correct.

Step 4. Convert from a decimal to a common fraction so you can read the dimension on a tape measure.

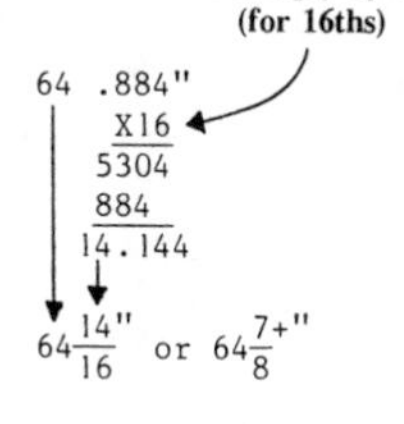

Equals $5'4\frac{7+}{8}$"

How to convert if the answer 64.884 were feet instead of inches.

Multiply by 12 (for inches)

64 .884′
X12
1768
884
10.608"
X16
3648
608
9.728 = 9/16+" or 5/8-"

64′ 10 5/8-"

Example
Problem 3. Use the Rule of Pythagoras and square roots to find the exact diagonal of a square.

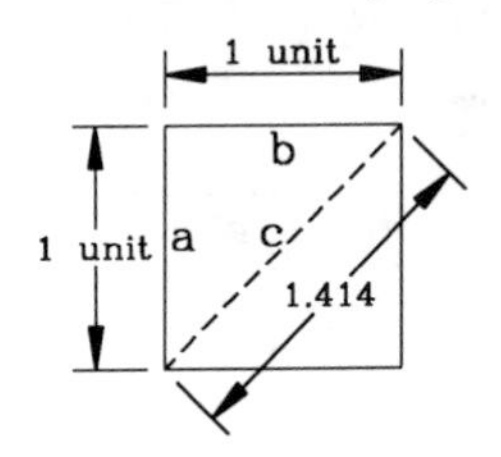

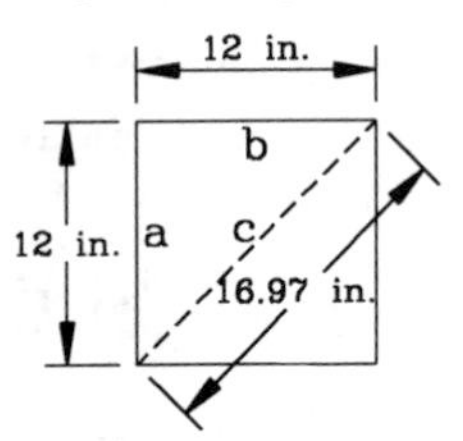

Step 1 $a^2 + b^2 = c^2$

Step 2 $1^2 + 1^2 = c^2$

$1 + 1 = c^2$

$2 = c^2$

Step 3 $\sqrt{2} = c^2$

$1.414 = c^2$

Step 1 $a^2 + b^2 = c^2$

Step 2 $12^2 + 12^2 = c^2$

$144 + 144 = c^2$

$288 = c^2$

Step 3 $\sqrt{288} = c^2$

$16.97 \text{ in.} = c^2$

Example
Problem 4. Use the Rule of Pythagoras and square roots to find the exact diagonal of a rectangle.

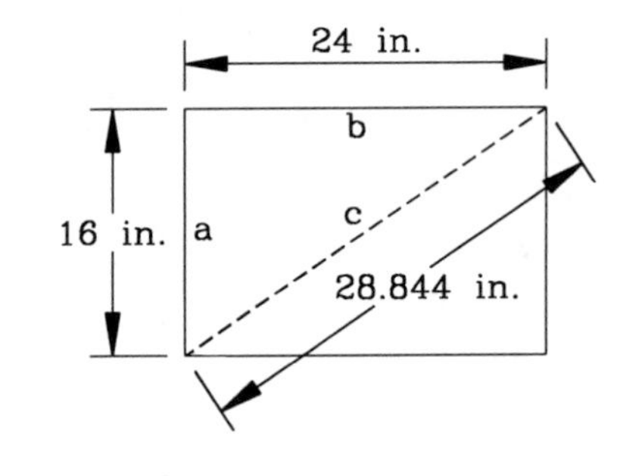

Step 1 $a^2 + b^2 = c^2$

$16^2 + 24^2 = c^2$

Step 2 $256 + 576 = c^2$

$832 = c^2$

Step 3 $\sqrt{832} = c$

28.84 inches = c

Step 4

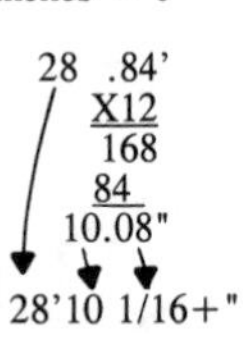

Example
Problem 5. Use the Rule of Pythagoras and square roots to find the exact diagonal of a building (rectangle).

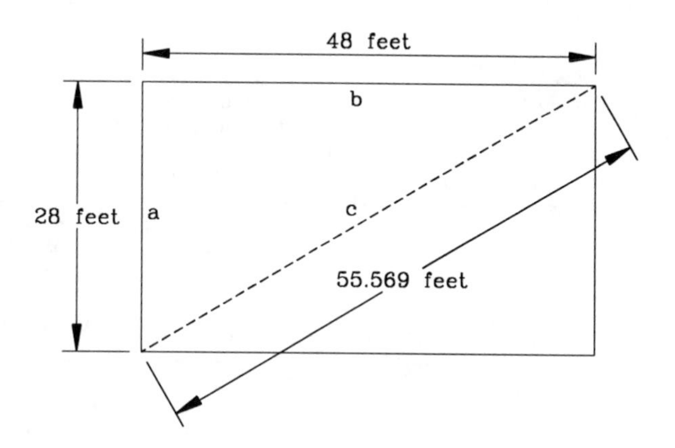

Step 1 $a^2 + b^2 = c^2$

$28^2 + 48^2 = c^2$

Step 2 $784 + 2304 = c^2$

$3088 = c^2$

Step 3 $\sqrt{3088} = c$

$55.569 = c$

Step 4 55 .569'

X12
1138
569
6.828"

55'6 13/16"

The example on this page and the examples on the preceding pages demonstrate the application of square roots. To actually lay out a square or a rectangle, see Task No. O3 and O4.

I1	◂ Task No.	RATIO

A **ratio** is the quotient of two numbers of the same kind. "Of the same kind" means both numbers are either abstract numbers, such as 40 and 10, or both numbers are expressed in the same units of measure, such as 40 feet and 10 feet.

The two numbers being compared to find a ration are the **terms** of the ratio. They often are written with a colon between the first term and the second term, such as 40:10, which is read "40 to 10", but they can also be written with the line of a fraction, such as 40/10. In either case, the first term (the numerator) is to be divided by the second term (the denominator) in order to find the ratio.

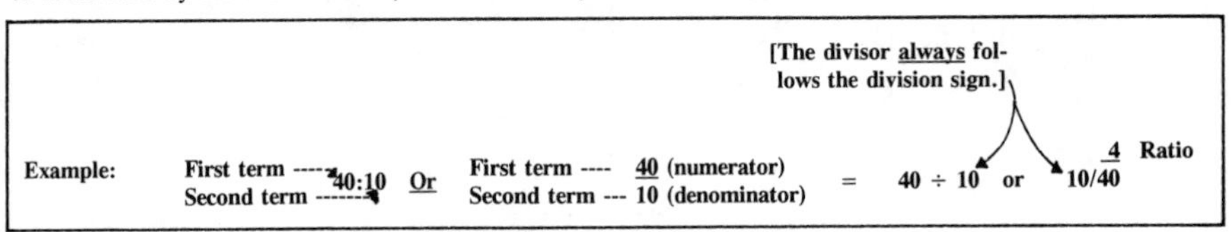

The ratio of 40 to 10, written as 40:10, is 4 (the quotient of 40 divided by 10). The ratio of $40 to $10 is also 4. And the ratio of 40' to 10' is 4. But there is no such thing as a ratio of $40 to 10', because dollars and feet are not the same units of measure and, therefore, the numbers are not of the same kind. You can, of course, determine a ratio of two objects of different material as long as they are expressed in the same units of measure. For example, to find the ratio of 60 inches of steel pipe to an 8-foot piece of lumber, you could convert the 60 inches to 5 feet (60 ÷ 12) to put both numbers in the same units of measure. The ratio would be 5 to 8, 5:8, or 5 divided by 8, or simply 5/8.

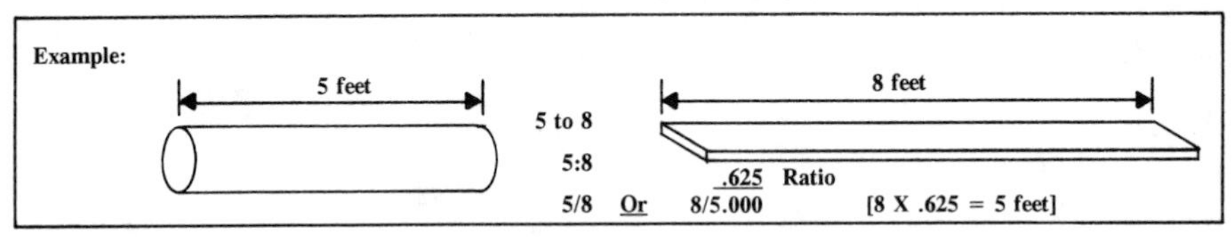

In all cases, a ratio itself is an abstract number. It has no units of measure. One of the most common examples of a ratio used in everyday mathematics is pi (π), the 3.1416 ratio of the circumference of a circle to its diameter. It applies no matter whether the circle is measured in feet, inches, rods, or whatever. Ratio even finds its way into law. A U.S. flag, for example, is to have a 1.9 ratio of length to width -- whatever its size, its length should be 1.9 times its width.

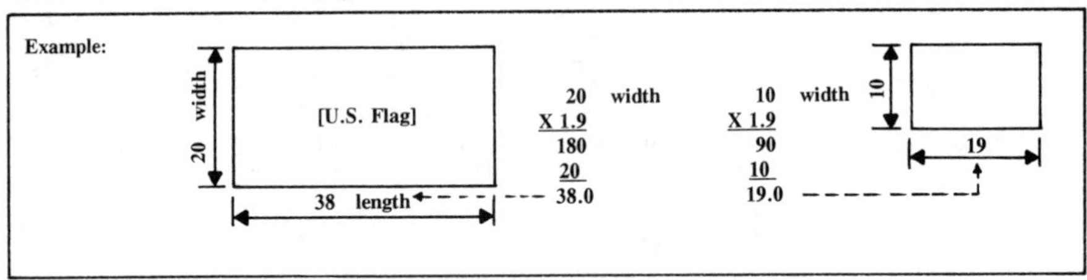

PROPORTION — Task No. ▸ I2

A **proportion** is an equality of two ratios. For example, the ratio of 5 to 10 is equal to the ratio of 10 to 20. Those two ratios form a proportion written $\frac{5}{10} = \frac{10}{20}$ or 5:10 = 10:20 or 5:10 :: 10:20. Whichever way it is written, that proportion is read "5 is to 10 as 10 is to 20." In a proportion, the product of the **means** is equal to the product of the **extremes**.

Example: The **extremes** are the outside terms of a proportion --------▶ extemes

5 : 10 = 10 : 20

The **means** are the inside terms of a proportion -------------▶ means

To prove a proportion, set up an equation -----------▶ 5 X 20 = 10 X 10

and calculate --▶ 100 = 100

[Note that both ratios equal each other.]

DIRECT PROPORTION — Task No. ▸ I3

A **direct proportion** is one in which the order of the ratios is the same.

For example, small is to LARGE as small is to LARGE.

4 is to 16 as X is to 24

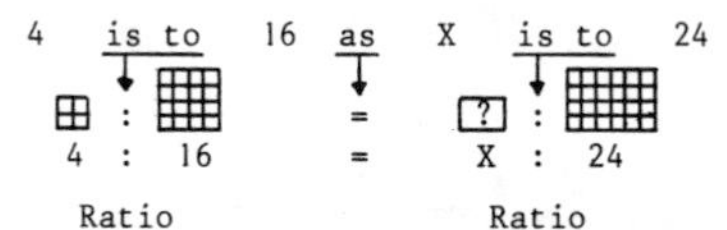

4 : 16 = X : 24

Ratio Ratio

or large is to small as large is to small

is to as is to ?

In a proportion, the product of the means is equal to the product of the extremes.

Example problem 1. Solve for X in the proportion 4:16 = X:24

Step 1. Set up as shown.

Extremes
4 : 16 = X : 24
Means

Or set up this way and cross multiply.

$\frac{4}{16} \times \frac{X}{24}$ = 16 (X) X = 4 X 24

16X = 96

Step 2. Make into an equation ---------▶ 16(X)X = 4 X 24 [The product of the means equals the product of the extremes.]

and calculate --------------------▶ 16X = 96

Step 3. Solve for X.

(To find X, divide both sides by 16)

$\frac{16X}{16} = \frac{96}{16}$ 16/96 = 6

X = 6 Answer

Step 4. Check by substituting for X and calculating.

16 x 6 = 4 X 24

96 = 96 Answer OK because both sides match.

I3 ◂ Task No. DIRECT PROPORTION (continued)

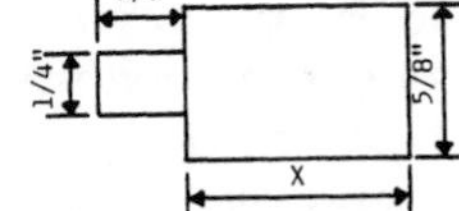

Example Problem 2. The dimensions for the machined part on the right are proportional. What is dimension X?

Step 1. Set up as shown. $\frac{1}{4}:\frac{3}{8} = \frac{5}{8}:X$

Step 2. Make into an equation ------------→ $\frac{1}{4}(X)X = \frac{3}{8} X \frac{5}{8}$

and calculate ------------→ $\frac{1}{4}X = \frac{15}{64}$

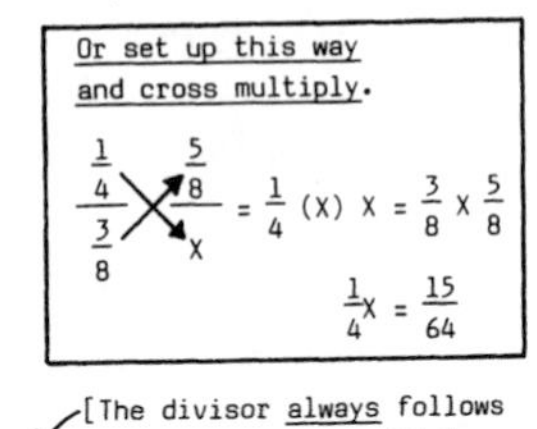

Step 3. Solve for X.

$\frac{\frac{1}{4}X}{\frac{1}{4}} = \frac{\frac{15}{64}}{\frac{1}{4}}$

numerator — $\frac{15}{64}$; denominator — $\frac{1}{4}$ } complex fraction

[Note the bottom (denominator) of a fraction is always the divisor.]

$X = \frac{\frac{15}{64}}{\frac{1}{4}} = \frac{15}{64} \div \frac{1}{4} =$ [The divisor always follows the division sign (÷).]

$\frac{15}{\cancel{64}_{16}} X \frac{\cancel{4}^{1}}{1} = \frac{15}{16}$ Answer

$[X = \frac{15}{16}]$

Step 4. Check by substituting for X and calculating. $\frac{1}{4} X \frac{15}{16} = \frac{15}{64}$

$\frac{15}{64} = \frac{15}{64}$ Answer OK

I4 ◂ Task No. INDIRECT PROPORTION

An **indirect proportion** is one in which the order of the ratios is indirect (inverse). For example, LARGE is to small as small is to LARGE.

D	is to	d	=	r	is to	R
↓	↓	↓		↓	↓	↓
20"	:	5"	=	540	:	135

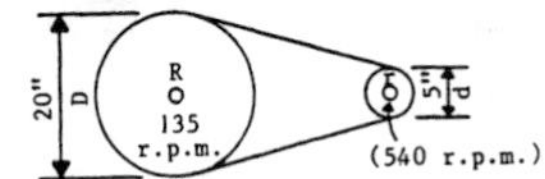

In a proportion, the product of the means is equal to the product of the extremes. (Note that this rule is the same for both direct and indirect proportion.)

RULE: Pulley diameters are indirectly (inversely) proportional to their r.p.m.
FORMULA: D:d = r:R

Example Problem 1. What are the r.p.m. of the little pulley?

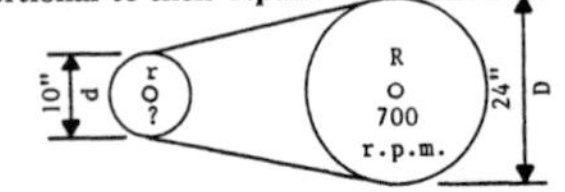

Step 1. Use the formula ------------→ D:d = r:R

Step 2. Substitute the known values into the formula ----------→ 24:10 = r:700

Step 3. Make into an equation ----→ 10 X r = 24 X 700
and calculate ------------→ 10r = 16800

Or use the formula:

(D:d = r:R) = (DR = dr) — extremes: D and R; means: d and r — Equality of products of the means and the extremes.

Substitute: 24 X 700 = 10r

Calculate: 16800 = 10r

Solve for r and calculate: $\frac{16800}{10} = \frac{\cancel{10}r}{\cancel{10}}$ or $\frac{1680}{10/16800}$

Answer: 1680 = r

Step 4. Solve for the unknown r and calculate

(To find r, divide both sides by 10.)

$$\frac{\not{1}\not{0}r}{\not{1}\not{0}} = \frac{16800}{10}$$

$$r = \frac{16800}{10} \text{ or } 10\overline{)16800} = 1680$$

```
    1680
10/16800
   10
    68
    60
     80
     80
      0
      0
```

Answer r = 1680 r.p.m.

Step 5. Check by substituting for r and calculating.

10 X 1680 = 16800
16800 = 16800 Answer OK because both sides match.

RULE: The r.p.m. of gears are indirectly (inversely) proportional to their teeth.
FORMULA: T:t = r:R

Example Problem 2. What are the r.p.m. of the big gear?

Step 1. Use the formula → T:t = r:R

Step 2. Substitute the known values into the formula → 45:36 = 90:R

Step 3. Make into an equation → 45 X R = 36 X 90
and calculate → 45R = 3240

Step 4. Solve for the unknown R and calculate

(To find R, divide both sides by 45.)

$$\frac{\not{4}\not{5}R}{\not{4}\not{5}} = \frac{3240}{45}$$

$$R = \frac{3240}{45} \text{ or } 45\overline{)3240} = 72$$

```
     72
45/3240
   315
    90
    90
```

Answer R = 72 r.p.m.

Step 5. Check by substituting for R and calculating.

72 X 45 = 3240
3240 = 3240 Answer OK

Or use the formula:

extremes: (T:t = r:R) = (TR = tr) — means
Equality of products of the means and the extremes

Substitute: 45R = 36 X 90

Calculate: 45R = 3240

Solve for R and calculate: $\frac{\not{4}\not{5}R}{\not{4}\not{5}} = \frac{3240}{45}$ or $45\overline{)3240} = 72$

Answer: R = 72 r.p.m.

I5	◂ Task No.	APPLY-IT MODULE

DIRECT AND INDIRECT PROPORTION

The purpose of this module is to demonstrate the difference between direct and indirect (or inverse) proportion. The main difference between direct and indirect proportion is the order in which the different quantities are placed in each specific ratio. For direct proportion the quantities in each ratio are placed in the same order to each other. For indirect (or inverse) proportion the quantities in each ratio are placed in a different or inverse order to each other. Keep in mind that both types of proportion are solved the same way once the proper order is found. The examples below, hopefully, will help clarify each type for you.

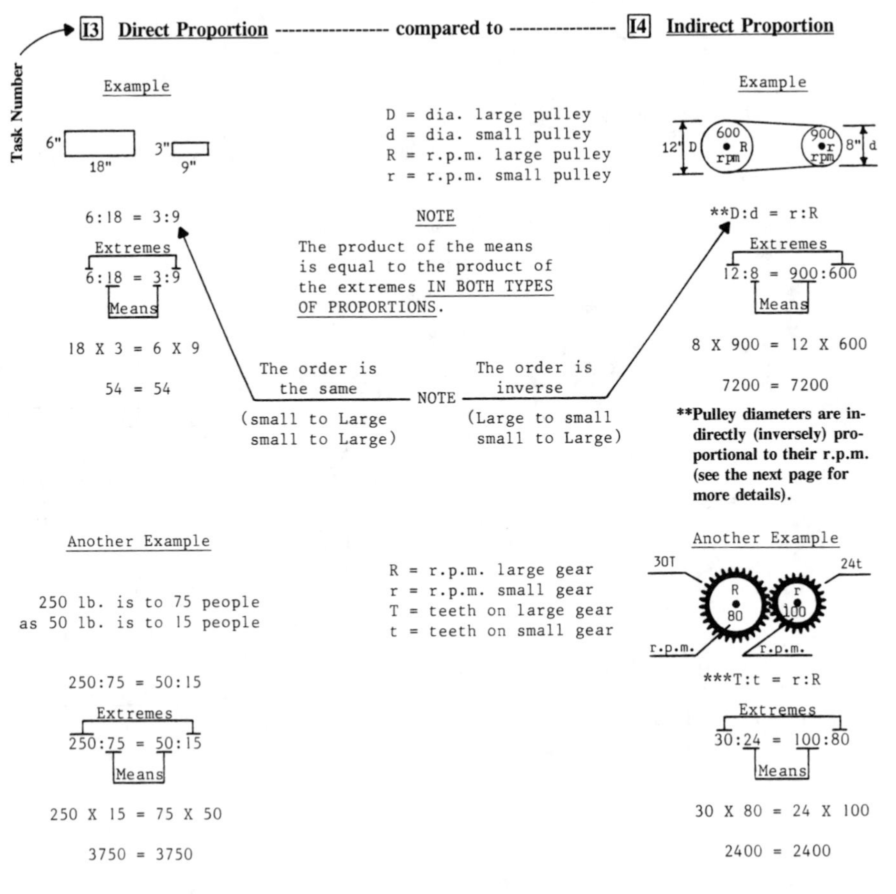

***The r.p.m. of gears are indirectly (inversely) proportional to the number of their teeth. See the next page for more details.

Pulley Diameters and r.p.m.

RULE: Pulley diameters are indirectly (inversely) proportional to their r.p.m.

From the above statement comes the proportion: D is to d as r is to R

Take the proportion --------- D:d = r:R and multiply the <u>means</u> by the <u>extremes</u>

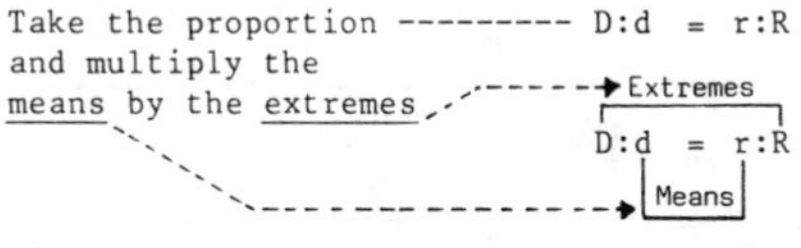

Example

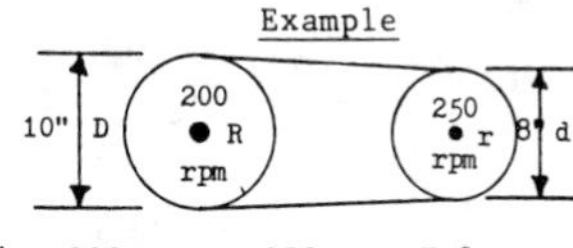

Thus you get the formula: -----▸DR = dr

10" X 200 rpm = 250 rpm X 8

2000 = 2000 checks out

<u>Therefore</u>
Using the above formula we can solve for each term as follows.

RULE: When two pulleys are connected by a belt or a chain, the large pulley diameter times its r.p.m. equals the small pulley diameter times its r.p.m.

> **<u>NOTE</u>**
>
> **All the information on this page is to illustrate how the formula works. <u>However</u>, probably the earliest and least confusing way to find pulley sizes and speeds (r.p.m.) is to draw a sketch of the pulleys and their r.p.m. and use the proportion D:d = r:R and substitute. <u>Note</u> how easy the formulas are to develop and use when you have an understanding of basic math and basic algebra. <u>Use the method that works best for you</u>.**

<u>Solve for D</u>

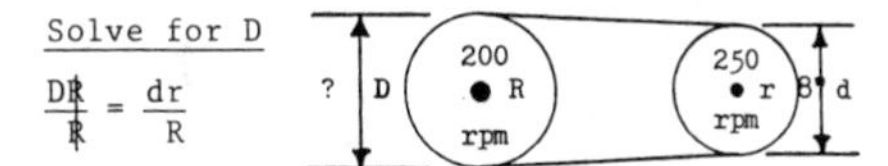

$\frac{DR}{R} = \frac{dr}{R}$ (R cancelled)

$D = \frac{dr}{R}$ $D = \frac{8 \text{ X } 250}{200}$ $D = \frac{2000}{200}$ $D = 10$

<u>Solve for R</u>

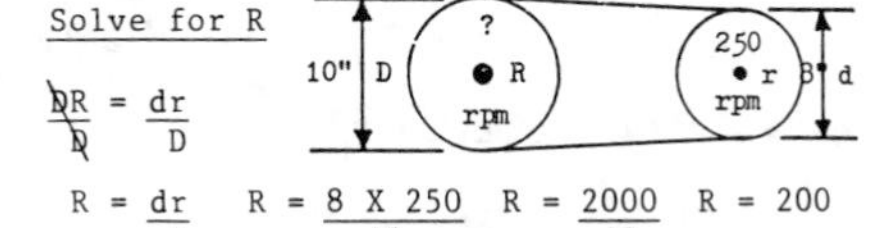

$\frac{DR}{D} = \frac{dr}{D}$ (D cancelled)

$R = \frac{dr}{D}$ $R = \frac{8 \text{ X } 250}{10}$ $R = \frac{2000}{10}$ $R = 200$

<u>Solve for d</u>

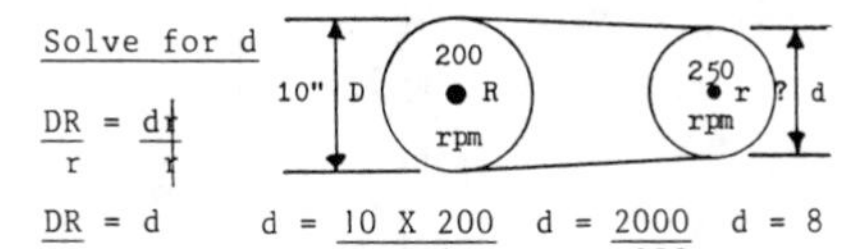

$\frac{DR}{r} = \frac{dr}{r}$ (r cancelled)

$\frac{DR}{r} = d$ $d = \frac{10 \text{ X } 200}{250}$ $d = \frac{2000}{250}$ $d = 8$

<u>Solve for r</u>

$\frac{DR}{d} = \frac{dr}{d}$ (d cancelled)

$\frac{DR}{d} = r$ $r = \frac{10 \text{ X } 200}{8}$ $r = \frac{2000}{8}$ $r = 250$

RULE: The r.p.m. of gears are indirectly (inversely) proportional to their number of teeth. From this rule comes the proportion: T:t = r:R

Therefore we can also get the formula: TR = tr

Extremes
T:t = r:R
Means

Other formulas that come from the formula TR = tr: $T = \frac{tr}{R}$ $R = \frac{tr}{T}$ $\frac{TR}{r} = t$ $\frac{TR}{t} = r$

[The formula TR = tr comes from the rule: **The number of <u>T</u>eeth multiplied by the <u>R</u>.p.m. of the large gear is equal to the number of <u>t</u>eeth multiplied by the <u>r</u>.p.m. of the small gear.**]

Practical Application of Direct Proportion

Example 1. Direct proportion can be used to solve right triangle problems that are proportional in size to each other.

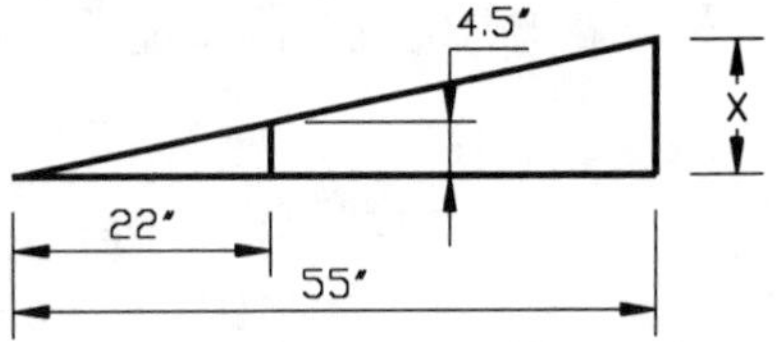

Proportion:

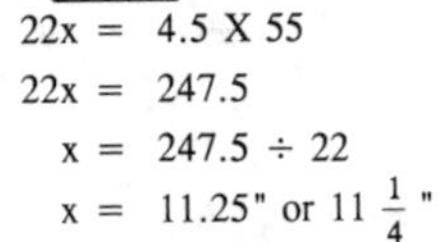

$22:4.5 = 55:x$

$22x = 4.5 \times 55$

$22x = 247.5$

$x = 247.5 \div 22$

$x = 11.25" \text{ or } 11\frac{1}{4}"$

Example 2. Direct proportion can be used to find unknown heights when certain values are known.

Proportion:

$12:10 = 50:x$

$12x = 10 \times 50$

$12x = 500$

$x = 41.66'$

$12/\overline{500.00}$ = 41.66

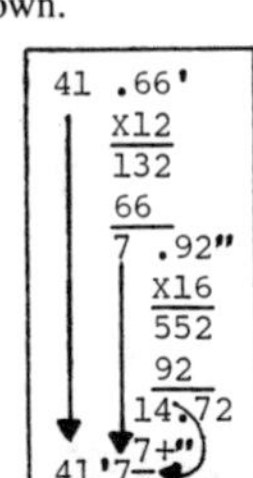

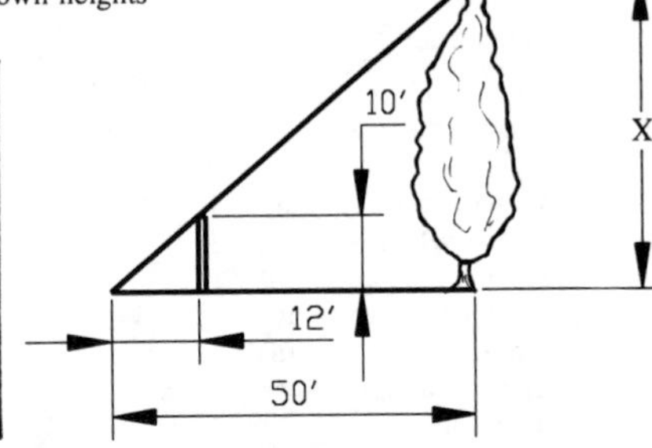

Example 3. Direct proportion can be used to find rectangles and squares that are proportional in size to each other.

Proportion:

$13:x = 31:11$

$31x = 13 \times 11$

$31x = 143$

$x = 4.61"$

$31/\overline{143.00}$ = 4.61

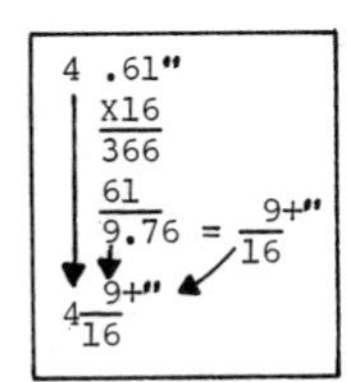

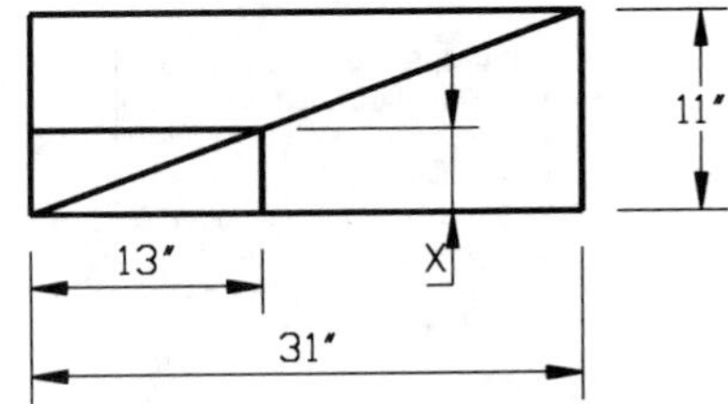

Example 4. Direct proportion can be used to find parts that are proportional in size to each other.

Proportion:

$X:1\frac{1}{4} = \frac{3}{8}:\frac{5}{16}$

$\frac{5}{16}x = 1\frac{1}{4} \times \frac{3}{8}$

$\frac{5}{16}x = \frac{5}{4} \times \frac{3}{8}$

$\frac{5}{16}x = \frac{15}{32}$

$x = \frac{\frac{15}{32}}{\frac{5}{16}}$

$x = \frac{15}{32} \div \frac{5}{16}$

$x = \frac{\cancel{15}^{3}}{\cancel{32}_{2}} \times \frac{\cancel{16}^{1}}{\cancel{5}_{1}}$

$x = \frac{3}{2}$

$x = 1\frac{1}{2}$

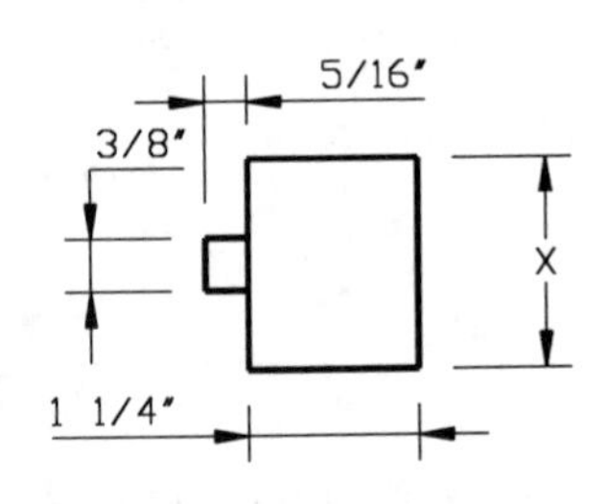

Example 5. Direct proportion can be used to calculate the gable end studs of a given building.

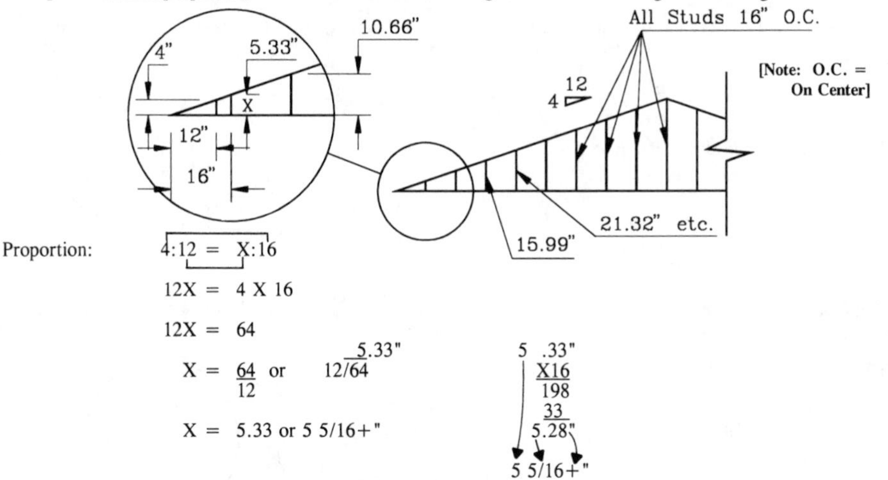

Proportion: 4:12 = X:16

12X = 4 X 16

12X = 64

X = $\frac{64}{12}$ or 12/64 = 5.33"

X = 5.33 or 5 5/16+"

5 .33" X16 = 198, 33, 5.28" → 5 5/16+"

Example 6. Direct proportion can be used to find the thickness of shims to line up motors with shafts, pumps, drives, etc., when certain factors are known.

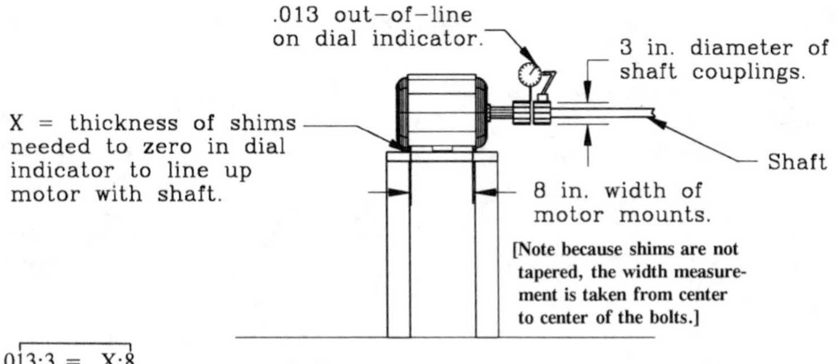

Proportion: .013:3 = X:8

3X = .013 X 8

3X = .104

X = $\frac{.104}{3}$ or 3/.104 = .0346

X = .0346 inches (thickness of shims)

[The examples included on these two pages are just a few different ways direct proportion can be used. Once you start implementing these examples, you will find more examples and different ways to use direct proportion. The next three pages include examples of indirect proportion applications.]

Practical Application of Indirect proportion

Example 1. Indirect proportion can be used to solve pulley diameter problems when the r.p.m. of the pulleys and one of the two pulley diameters is known.

Formula: **DR = dr**

Proportion:

D:d = r:R

14:d = 980:700

980d = 14 X 700

980d = 9800 $\quad 980\overline{)9800}^{\,10}$

d = 10" diameter

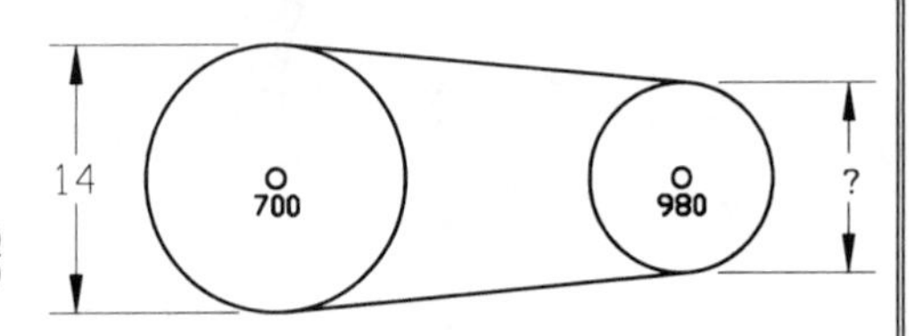

Example 2. Indirect proportion can be used to solve pulley r.p.m. problems when two pulley diameters and the r.p.m. of one of the two pulleys is known.

Formula: **DR = dr**

Proportion:

D:d = r:R

15:13.5 = 600:R

15R = 13.5 X 600

15R = 8100

R = 540

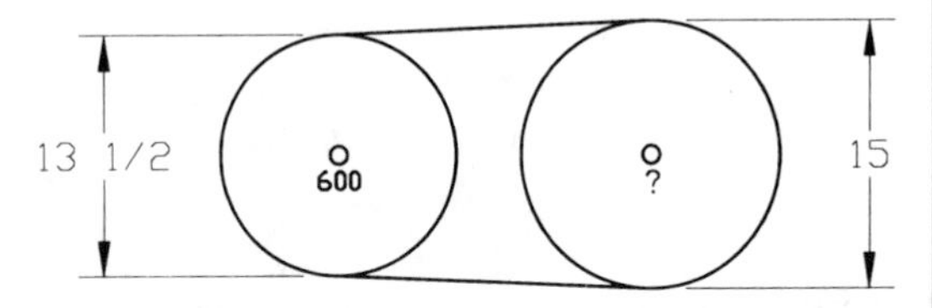

Example 3. Indirect proportion can be used to find the number of teeth in a gear when the r.p.m. of each gear and the number of teeth of one of the two gears is known.

Formula: **TR = tr**

Proportion:

T:t = r:R

48:t = 280:245

280t = 48 X 245

280t = 11760

t = 42

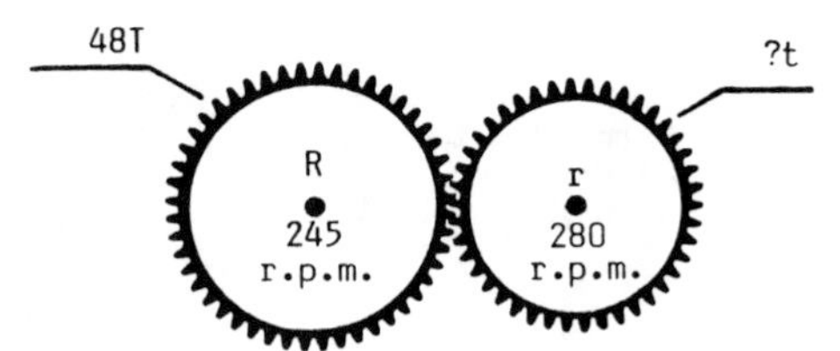

Example 4. Indirect proportion can be used to find the r.p.m. of a gear when the number of teeth on each gear and the r.p.m. of one of the gears is known.

Formula: **TR = tr**

Proportion:

T:t = r:R

64:50 = 480:R

64R = 50 X 480

64R = 24000

R = 375

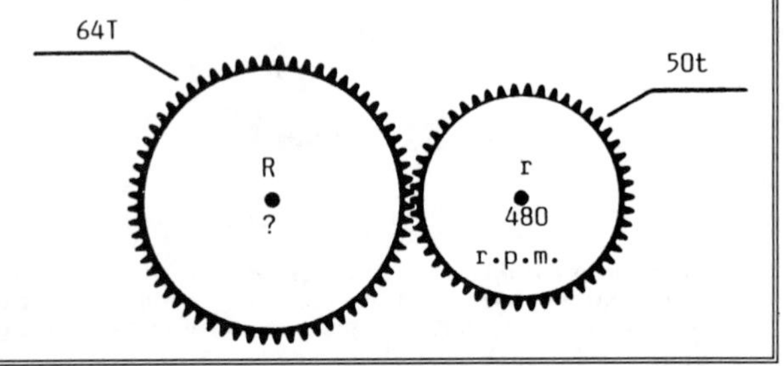

PULLEY TRAINS

A pulley train is when a driven pulley is connected to a series of pulleys via a belt or chain.

RULE: The r.p.m. of the first small pulley times the sum of all the small pulley diameters equals the r.p.m. of the last large pulley times the sum of all the large pulley diameters.

FORMULA: $rdd_1d_2 = RDD_1D_2$

Label pulleys as follows:

r = r.p.m small pulley	R = r.p.m last large pulley
d = diameter 1st small pulley	D = diameter 1st large pulley
d_1 = diameter 2nd small pulley	D_1 = diameter 2nd large pulley
d_2 = diameter 3rd small pulley	D_2 = diameter 3rd large pulley
d_3 = diameter 4th small pulley, etc.	D_3 = diameter 4th small pulley, etc.

By using basic algebra each term in the above formula can be solved as follows:

FORMULAS:

$$r = \frac{RDD_1D_2}{dd_1d_2} \qquad d = \frac{RDD_1D_2}{rd_1d_2}$$

$$d_1 = \frac{RDD_1D_2}{rdd_2} \qquad d_2 = \frac{RDD_1D_2}{rdd_1}$$

$$R = \frac{rdd_1d_2}{DD_1D_2} \qquad D = \frac{rdd_1d_2}{RD_1D_2}$$

$$D_1 = \frac{rdd_1d_2}{RDD_2} \qquad D_2 = \frac{rdd_1d_2}{RDD_1}$$

Example 1. Solve for the last large pulley in the pulley train shown below.

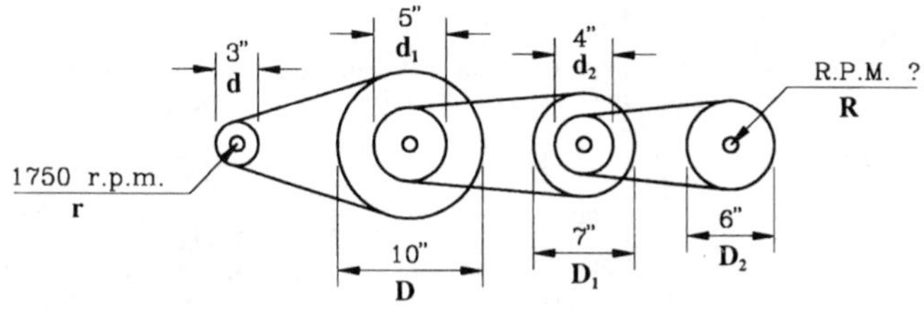

FORMULA: $R = \frac{rdd_1d_2}{DD_1D_2}$

SUBSTITUTE: $R = \frac{1750 \times 3 \times 5 \times 4}{10 \times 7 \times 6}$

CALCULATE: $R = \frac{(1750 \times 3) \times (5 \times 4)}{(10 \times 7) \times 6}$

$R = \frac{5250 \times 20}{70 \times 6}$

$R = \frac{105000}{420}$

R = 250 R.P.M.

Example 2. Using the same example, solve for d_2.

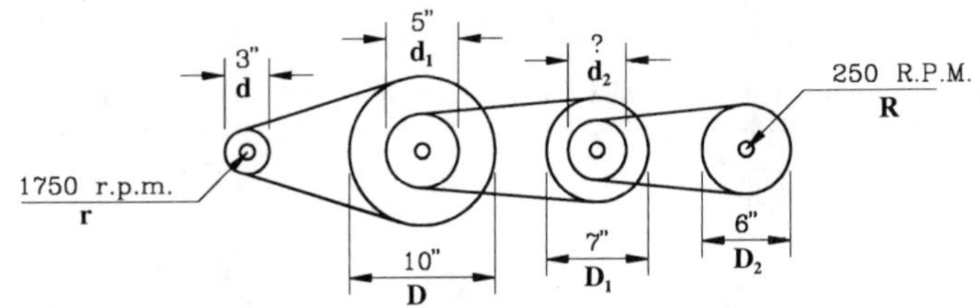

FORMULA: $d_2 = \frac{RDD_1D_2}{rdd_1}$

SUBSTITUTE: $d_2 = \frac{250 \times 10 \times 7 \times 6}{1750 \times 3 \times 5}$

CALCULATE: $d_2 = \frac{(250 \times 10) \times (7 \times 6)}{(1750 \times 3) \times 5}$

$d_2 = \frac{2500 \times 42}{5250 \times 5}$

$d_2 = \frac{105000}{26250}$

$d_2 = 4$ inches

GEAR TRAINS

A gear train is when a driven gear is connected to a series of gears.

RULE: The r.p.m. of the first small gear times the sum of teeth on all the small gears equals the R.P.M. of the last large gears times the sum of teeth on all the large gears.

FORMULA: $rtt_1t_2 = RTT_1T_2$

Label gears as follows:

r = r.p.m small gear
t = diameter 1st small gear
t_1 = diameter 2nd small gear
t_2 = diameter 3rd small gear
t_3 = diameter 4th small gear, etc.

R = r.p.m last large gear
T = diameter 1st large gear
T_1 = diameter 2nd large gear
T_2 = diameter 3rd large gear
T_3 = diameter 4th small gear, etc.

By using basic algebra each term in the above formula can be solved as follows:

FORMULAS:

$r = \frac{RTT_1T_2}{tt_1t_2}$ $t = \frac{RTT_1T_2}{rt_1t_2}$

$t_1 = \frac{RTT_1T_2}{rtt_2}$ $t_2 = \frac{RTT_1T_2}{rtt_1}$

$R = \frac{rtt_1t_2}{TT_1T_2}$ $T = \frac{rtt_1t_2}{RT_1T_2}$

$T_1 = \frac{rtt_1t_2}{RTT_2}$ $T_2 = \frac{rtt_1t_2}{RTT_1}$

Example 1. Solve for the R.P.M. of the last large gear shown below.

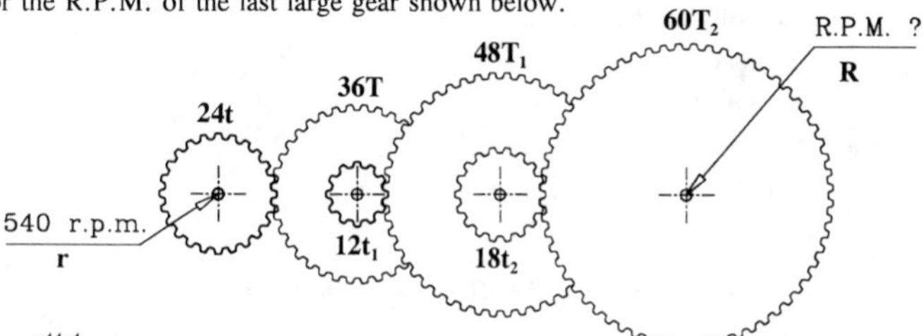

FORMULA: $\mathbf{R} = \dfrac{rtt_1t_2}{TT_1T_2}$

SUBSTITUTE: $\mathbf{R} = \dfrac{540 \text{ X } 24 \text{ X } 12 \text{ X } 18}{36 \text{ X } 48 \text{ X } 60}$

CALCULATE: $\mathbf{R} = \dfrac{(540 \text{ X } 24) \text{ X } (12 \text{ X } 18)}{(36 \text{ x} 48) \text{ X } 60}$

$\mathbf{R} = \dfrac{12960 \text{ X } 216}{1728 \text{ X } 60}$

$\mathbf{R} = \dfrac{2799360}{103680}$

R = 27 R.P.M.

Example 2. Using the same example, solve for $\mathbf{T_1}$.

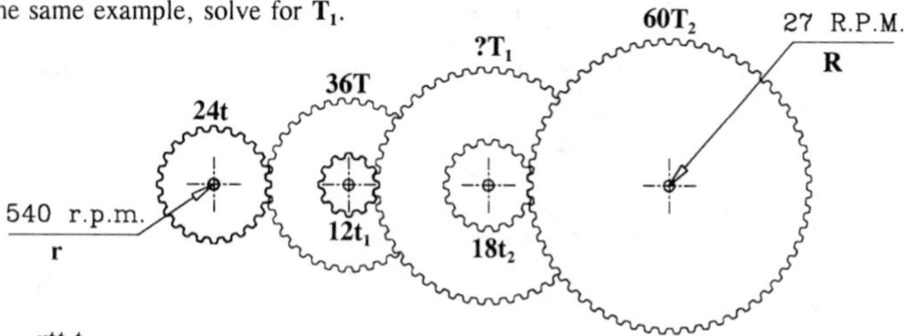

FORMULA: $\mathbf{T_1} = \dfrac{rtt_1t_2}{RTT_2}$

SUBSTITUTE: $\mathbf{T_1} = \dfrac{540 \text{ X } 24 \text{ X } 12 \text{ X } 18}{27 \text{ X } 36 \text{ X } 60}$

CALCULATE: $\mathbf{T_1} = \dfrac{(540 \text{ X } 24) \text{ X } (12 \text{ X } 18)}{(27 \text{ x } 36) \text{ X } 60}$

$\mathbf{T_1} = \dfrac{12960 \text{ X } 216}{972 \text{ X } 60}$

$\mathbf{T_1} = \dfrac{2799360}{58320}$

$\mathbf{T_1}$ = 48 TEETH

J1	◂ Task No.	**BASIC ALGEBRA** (Equations and Formulas)

Algebra is a part of mathematics that uses letters to represent numbers in the forms of equations and formulas calculated according to the rules of arithmetic.

An **equation** is a statement that says two number expressions are equal. Each side of the equal sign (=) is equal to the other.

Example: 2A = 24 is an equation.

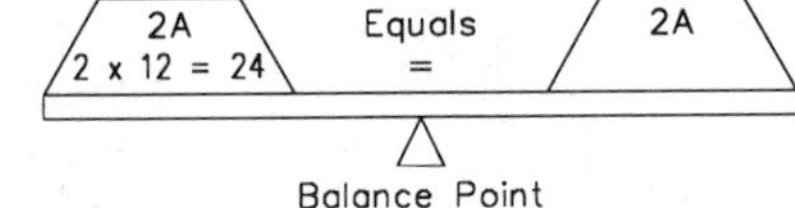

[In this example, 2A has to be 2 X 12 because the left side of the equation must equal the right side of the equation (2 X 12 = 24).]

Since both sides of an equation must be equal (balance out), whatever you do to one side you must do to the other side. When both sides of an equation are either added, subtracted, multiplied, or divided by the same quantity, the results again form an equation.

[See Addition of Equations Task No. J2 , Subtraction of Equations Task No. J3 , Multiplication of Equations Task No. J4 , and Division of Equations Task No. J5 on the next eight pages.]

A **formula** is a shortened version of a specific mathematical rule.

Substitution is the changing of letters or symbols in a formula or equation to numbers.

Example 1. **RULE: Area equals length times width**

FORMULA: A = lw

Substitution: A = 8 X 6

Note: 8 is substituted for the length (l) and 6 is substituted for the width (w).

8
Length
Width
6

Example 2. **RULE: Circumference for a circle equals pi times the diameter**

FORMULA: C = πd

Substitution: C = 3.1416 X 12

Note: 3.1416 is substituted for pi (π) and 12 is substituted for the diameter (d). π is a Greek letter equal to 3.14 or 3 1/7. However, for greater accuracy use 3.1416.

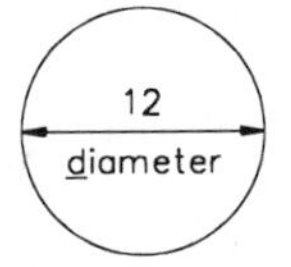

The four basic operation signs in formulas and equations are used in the following manner:

When you have to add, the plus sign (+) is used.

When you have to subtract, the minus sign (-) is used.

When you have to multiply, the times sign (X), raised dot (•), or no sign is used between a number next to a letter.

Examples: 4 X 4 = 16
4 • 4 = 16
4A = 4 X A or 4•A
AB = A X B or A•B

When you have to divide, rather than the division sign (÷), the line of a fraction is usually used.

Example: $\frac{A}{3}$ indicates division

Task No. ▸ J1

Those four basic operations -- addition, subtraction, multiplication, and division -- can be performed with both positive and negative numbers. A positive number is any number greater than zero and is written with or without a positive sign, such as +9, or 9. A negative number is any number less than zero and is written with a negative sign, such as -9.

To add numbers with the same sign, add their absolute values and give the answer that same sign. In other words, the sum of positive numbers is positive, and the sum of negative numbers is negative.

To add numbers with different signs, subtract the smaller absolute value from the larger absolute value, and give the answer the sign of the number with the larger absolute value.

To subtract signed numbers, change the sign of the subtrahend (the number being subtracted) and add according to the rules above. Subtraction problems can be changed to addition problems because the result of subtracting a positive number is the same as that of adding the corresponding negative number --

$$(+15) - (+7) = (+15) + (-7) = (+8)$$

$$(-15) - (+7) = (-15) + (-7) = (-22)$$

and the result of subtracting a negative number is the same as that of adding the corresponding positive number --

$$(+15) - (-7) = (+15) + (+7) = (+22)$$

$$(-15) - (-7) = (-15) + (+7) = (-8)$$

To multiply signed numbers, multiply the absolute values. If the numbers being multiplied have the same sign, the product is positive. If the numbers have different signs, the product is negative*.

To divide signed numbers, divide using the absolute values. If the numbers have the same sign, the quotient is positive. If the numbers have different signs, the quotient is negative*.

*In other words:
(+) X or ÷ (+) = (+)
(+) X or ÷ (-) = (-)
(-) X or ÷ (+) = (-)
(-) X or ÷ (-) = (+)

The table below illustrates these rules for working with signed numbers:

	SAME SIGNS		DIFFERENT SIGNS	
	Both Positive	Both Negative	Greater Positive	Greater Negative
ADD	+2 +2 +5	−3 −2 −5	+3 −2 +1 or 1	−3 +2 −1
SUBTRACT	+3 −(+2) +1	−3 −(−2) −1	+3 −(−2) +5 or 5	−3 −(+2) −5
MULTIPLY	(+3)X(+2) = +6 or 3 X 2 = 6	(−3)X(−2) = +6 or (−3)X(−2) = 6	(+3)X(−2) = −6	(−3)X(+2) = −6
DIVIDE	$\frac{+9}{+3} = +3$ or $\frac{9}{3} = 3$	$\frac{-9}{-3} = +3$ or $\frac{-9}{-3} = 3$	$\frac{+9}{-3} = -3$	$\frac{-9}{+3} = -3$

J2	◂ Task No.	ADDITION AND EQUATIONS

RULE: Whatever calculation is done to one side of an equation, the same calculation must be done to the other side.

To solve an equation showing addition, calculate by doing the opposite operation; that is, <u>subtract an equal quantity from both sides</u>. The results will be the value of the unknown.

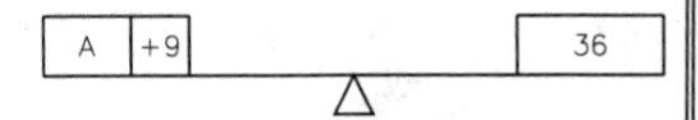

Example Problem 1. In the equation $A + 9 = 36$, solve for A.

Step 1. Use the equation as shown ------------ $A + 9 = 36$

Step 2. Addition is indicated. Therefore, get the unknown (A) by subtraction. Subtract 9 from both sides ----------

$$\begin{array}{rcl} A + 9 &=& 36 \\ \underline{-9} && \underline{-9} \end{array}$$

Step 3. Calculate the equation (do the operations indicated) ---------------

$$\begin{array}{rcl} A + 9 &=& 36 \\ \underline{-9} && \underline{-9} \\ A \quad 0 && 27 \end{array}$$

$A = 27$ Answer

Step 4. Check by substituting the value of the unknown (A) and compute the equation.

$A + 9 = 36$

$(27) + 9 = 36$

$36 = 36$ Answer OK because both sides are equal.

27 +9 (36) — 36 (36)

Example Problem 2. In the equation $a + b = c$, solve for b.

Step 1. Use the equation --------------------- $a + b = c$

Step 2. Addition is indicated. Therefore, get the unknown (b) by subtraction. Subtract a from both sides ----------

$$\begin{array}{rcl} a + b &=& c \\ \underline{-a} && \underline{-a} \end{array}$$

Step 3. Calculate the equation (do the operations indicated) ---------------

$$\begin{array}{rcl} a + b &=& c \\ \underline{-a} && \underline{-a} \\ 0 \quad b && c - a \end{array}$$

$b = c - a$ Answer

Step 4. Check by substituting the value of the unknown (b) and compute the equation.

$a + b = c$

$a + (c - a) = c$

$c = c$ Answer OK because both sides are equal.

Note: The example below is used to demonstrate practical application of addition and equations. See also the next 2 pages for subtraction and equations, as one is the reciprocal (opposite) of the other.

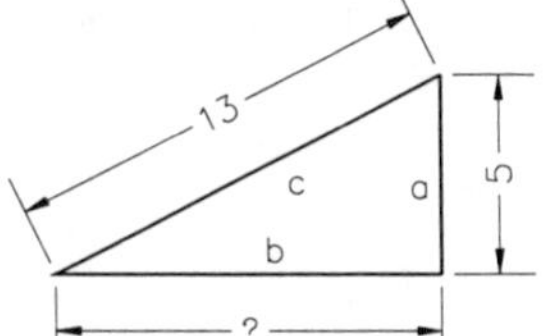

Example Problem 3. Solve for b in the right triangle on the right.

Step 1. Use the formula (equation)* -------- $a^2 + b^2 = c^2$

Step 2. Addition is indicated. Therefore, get the unknown (b^2) by subtraction. Subtract a^2 from both sides ------------------

$$\begin{array}{rcl} a^2 + b^2 & = & c^2 \\ \underline{-a^2} & & \underline{-a^2} \end{array}$$

Step 3. Calculate the formula (do the operations indicated) ------------

$$\begin{array}{rcl} a^2 + b^2 & = & c^2 \\ \underline{-a^2} & & \underline{-a^2} \\ 0 \quad b^2 & = & c^2 - a^2 \end{array}$$

Revised formula ----- $b^2 = c^2 - a^2$

Step 4. Substitute the known values into the formula ---------------------- $b^2 = 13^2 - 5^2$

Step 5. Calculate the formula (do the operations indicated) ----------------- $b^2 = (13 \text{ X } 13) - (5 \text{ X } 5)$

$b^2 = 169 - 25$

$b^2 = 144$

Step 6. Solve for (b) by finding the square root of 144 ---------------- $b = \sqrt{144}$

$b = 12$ Answer

```
 1      1<---- 1 2
+1     X1    /144
 22     1 ---> 1
X 2            44
 44 -------->  44
```

[To find the square root of a number, see Task No. H2.]

Step 7. Check by substituting the value of the unknown (b^2) and compute the equation.

$b^2 = c^2 - a^2$

$(12^2) = 169 - 25$

$144 = 144$ Answer OK because both sides are equal.

*$a^2 + b^2 = c^2$ is the Rule of Pythagoras. See Task No. K5A.

J3	◂ Task No.	SUBTRACTIONS AND EQUATIONS

RULE: **Whatever calculation is done to one side of an equation, the same calculation must be done to the other side.**

To solve an equation showing subtaction, calculate by doing the opposite (reciprocal) operation; that is, <u>add an equal quantity to both sides</u>. The results will be the value of the unknown.

Example Problem 1. In the equation B - 7 = 19, solve for B.

-7 19 b

Step 1. Use the equation as shown ------------ B - 7 = 19

Step 2. Subtraction is indicated. Therefore, get the unknown (B) by addition. Add 7 to both sides ------------------

$$B - 7 = 19$$
$$\quad +7 \quad +7$$

Step 3. Calculate the equation (do the operations indicated) -----------------

$$B - 7 = 19$$
$$\quad +7 \quad +7$$
$$B \quad 0 \quad 26$$

B = 26 Answer

Step 4. Check by substituting the value of the unknown (B) and compute the equation.

$$B - 7 = 19$$
$$(26) - 7 = 19$$
$$19 = 19$$

(19) (19) -7 19 26

Answer OK because both sides are equal.

Example Problem 2. In the equation b = c - a, solve for c.

Step 1. Use the equation --------------------- b = c - a

Step 2. Subtraction is indicated. Therefore, get the unknown (c) by addition. Add a to both sides -----------------

$$b = c - a$$
$$+a \quad +a$$

Step 3. Calculate the equation (do the operations indicated) ---------------

$$b = c - a$$
$$+a \quad +a$$
$$a + b \quad c \quad 0$$

a + b = c Answer

Step 4. Check by substituting the value of the unknown (c) and compute the equation.

$$b = c - a$$
$$b = (a + b) - a$$

b = b Answer OK because both sides are equal.

Note: The example below is used to demonstrate practical application of subtraction and equations. See also the preceding 2 pages for addition and equations, as one is the reciprocal (opposite) of the other.

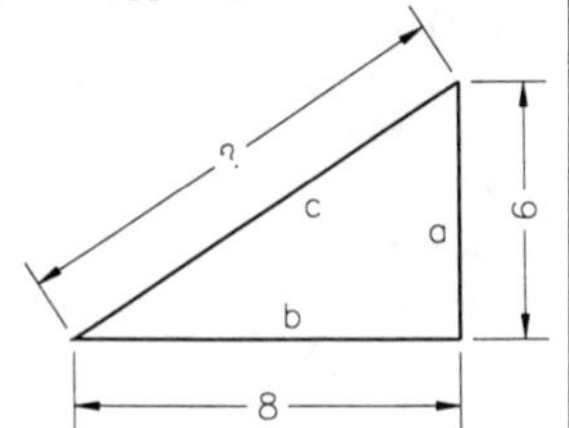

Example Problem 3. Solve for c in the right triangle on the right.

Step 1. Use the formula ------------------------ $b^2 = c^2 - a^2$

Step 2. Subtraction is indicated. Therefore, get the unknown (c^2) by addition. Add a^2 to both sides -------

$$\begin{array}{rcl} b^2 & = & c^2 - a^2 \\ \underline{+a^2} & & \underline{+a^2} \end{array}$$

Step 3. Calculate the formula (do the operations indicated) -----------------

$$\begin{array}{rcrr} b^2 & = & c^2 & - a^2 \\ \underline{+a^2} & & & \underline{+a^2} \\ a^2 + b^2 & & c^2 & 0 \end{array}$$

Revised formula ---- $a^2 + b^2 = c^2$*

Step 4. Substitute the known values into the formula ------------------ $8^2 + 6^2 = c^2$

Step 5. Calculate the formula (do the operations indicated) --- $(8 \times 8) + (6 \times 6) = c^2$

$$64 + 36 = c^2$$

$$100 = c^2$$

Step 6. Solve for (c) by finding the square root of 100 --------------- $\sqrt{100} = c$

$10 = c$ Answer

```
 1     1 <---- 1 0
+1    X1     / 1 00
 20    1 ----> 1
X 0              00
 00 -----------> 00
```

[To find the square root of a number, see Task No. H2.]

Step 7. Check by substituting the value of the unknown (c^2) and compute the equation.

$$a^2 + b^2 = c^2$$

$$64 + 36 = (10^2)$$

$100 = 100$ Answer OK because both sides are equal.

*$a^2 + b^2 = c^2$ is the Rule of Pythagoras. See Task No. K5A.

J4	◂ Task No.	MULTIPLICATION AND EQUATIONS

RULE: Whatever calculation is done to one side of an equation, the same calculation must be done to the other side.

To solve an equation showing multiplication, calculate by doing the opposite (reciprocal) operation; that is, <u>divide each side by an equal quantity</u>. The results will be the value of the unknown.

Example Problem 1. In the equation 6A = 84, solve for A.

6 x A | 84

Step 1. Use the equation as shown ------------- $6A = 84$

Step 2. Multiplication is indicated. Therefore, get the unknown (A) by division. Divide both sides by 6 ---------------------------------- $\frac{6A}{6} = \frac{84}{6}$

Step 3. Calculate the formula (do the operations indicated) ----------------- $\frac{\not{6}A}{\not{6}} = \frac{86}{6}$

[86 ÷ 6 = 14]

$A = 14$ Answer

Step 4. Check by substituting the value of the unknown (A) and compute the equation.

$6A = 84$

$6 \text{ X } (14) = 84$

$84 = 84$ Answer OK because both sides are equal.

6 x 14 (84) | 84 (84)

Example Problem 2. In the equation A = lw, solve for l.

By using division and equations, the formula A = lw can be changed into the other formulas shown below.

$\frac{A}{w} = \frac{l\not{w}}{\not{w}}$ [divide each side by w] $l = \frac{A}{w}$

$\frac{A}{l} = \frac{\not{l}w}{l}$ [divide each side by l] $w = \frac{A}{l}$

The next 3 pages show how to solve problems using these formulas.

Step 1. Use the equation --------------------- $A = lw$

Step 2. Multiplication is indicated. Therefore, get the unknown (l) by division. Divide both sides by w ---------------------------------- $\frac{A}{w} = \frac{lw}{w}$

Step 3. Calculate the formula (do the operations indicated) ----------------- $\frac{A}{w} = \frac{l\not{w}}{\not{w}}$

$\frac{A}{w} = l$ Answer

Step 4. Check by substituting the value of the unknown (l) and compute the equation.

$A = lw$

$A = (\frac{A}{\not{w}}) \text{ X } \frac{\not{w}}{1}$

$A = A$ Answer OK because both sides are equal.

Note: The example below is used to demonstrate practical application of multiplication and equations. See also the next 2 pages for division and equations, as one is the reciprocal (opposite) of the other.

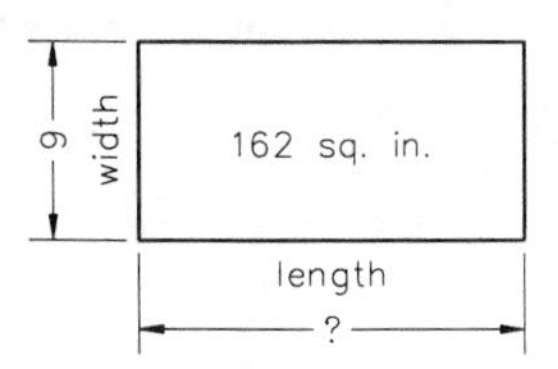

Example Problem 3. Solve for l in the rectangle on the right.

Note: The rule for finding the area of a rectangle is <u>Area</u> equals the length times the width. Therefore, the formula is -- A = lw.

Step 1. Use the formula (equation) ------------- $A = lw$

Step 2. Multiplication is indicated. Therefore, get the unknown (l) by division. Divide both sides by w ----------------------------------- $\frac{A}{w} = \frac{lw}{w}$

Step 3. Calculate the formula (do the operations indicated) ------------------ $\frac{A}{w} = \frac{l\not{w}}{\not{w}}$

Revised formula ------- $\frac{A}{w} = l$

Step 4. Substitute the known values into the formula --------------------- $\frac{162}{9} = l$

Step 5. Calculate the formula (do the operations indicated) ------- $[162 \div 9] = l$

$18 = l$ Answer

Step 6. Check by substituting the value of the unknown (l) and compute the equation.

$\frac{A}{w} = l$

$\frac{162}{9} = 18$

$18 = 18$ Answer OK because both sides are equal.

[For another example of multiplication and equations, see the bottom right page of Task No. J5 .]

J5	◂ Task No.	DIVISION AND EQUATIONS

RULE: Whatever calculation is done to one side of an equation, the same calculation must be done to the other side.

To solve an equation showing division, calculate by doing the opposite (reciprocal) operation that is, multiply each side by an equal quantity. The results will be the value of the unknown.

Example Problem 1. In the equation Y/9 = 12, solve for Y.

$\frac{Y}{9}$ | 12

Step 1. Use the equation as shown --------------- $\frac{Y}{9} = 12$

Step 2. Division is indicated. Therefore, get the unknown (Y) by multiplication. Multiply both sides by 9 ----- $9 \cdot \frac{Y}{9} = 12 \cdot 9$

Step 3. Calculate the formula (do the operations indicated) ----------------- $\not{9} \cdot \frac{Y}{\not{9}} = 108$

$Y = 108$ Answer

Step 4. Check by substituting the value of the unknown (Y) and compute the equation.

$\frac{Y}{9} = 12$

$\frac{(108)}{9} = 12$

$\frac{108}{9}$ (12) | 12 (12)

$12 = 12$ Answer OK because both sides are equal.

Example Problem 2. In the equation w = A/l, solve for l.

Step 1. Use the equation ------------------------ $w = \frac{A}{l}$

Step 2. Division is indicated. Therefore, get the unknown (l) by multiplication. Multiply both sides by l ----- $l \cdot w = \frac{A}{l} \cdot l$

Step 3. Calculate the formula (do the operations indicated) ----------------- $lw = \frac{A}{\not{l}}\not{l}$

$lw = A$

Step 4. Multiplication is indicated. Therefore, divide both sides by w ------------ $\frac{lw}{w} = \frac{A}{w}$

Step 5. Calculate the formula ------------------ $\frac{l\not{w}}{\not{w}} = \frac{A}{w}$

$l = \frac{A}{w}$ Answer

Or, invert each side of the equation and multiply each side by A.

$w = \frac{A}{l}$

$\frac{1}{w} = \frac{l}{A}$

$\frac{A}{1} \cdot \frac{1}{w} = \frac{l}{\not{A}} \cdot \frac{\not{A}}{1}$

$\frac{A}{w} = l$

Step 6. Check by substituting the value of the unknown (l) and compute the equation.

$w = \frac{A}{l}$

$w = \frac{A}{(\frac{A}{w})} = \frac{A}{1} \div \frac{A}{w}$

$\frac{\not{A}}{1} \times \frac{w}{\not{A}} = w$

$w = w$ Answer OK because both sides are equal.

Note: The example below is used to demonstrate practical application of division and equations. See also the preceding 2 pages for multiplication and equations, as one is the reciprocal (opposite) of the other.

Example Problem 3. Solve for w in the rectangle on the right.

width ?
168 sq. in.
length
24

Step 1. Use the formula (equation) ---------- $l = \frac{A}{w}$

Step 2. Division is indicated. Therefore, get the unknown (w) by multiplication. Multiply both sides by w -- $w \cdot l = \frac{A}{w} \cdot w$

Step 3. Calculate the formula (do the operations indicated) -------------- $wl = \frac{A}{\not w}\not w$

$wl = A$

Step 4. Multiplication is indicated, therefore, divide both sides by l -------- $\frac{wl}{l} = \frac{A}{l}$

Step 5. Calculate the formula -------------- $\frac{w\not l}{\not l} = \frac{A}{l}$

Revised formula ----- $w = \frac{A}{l}$

Step 6. Substitute the known values into the formula -------------------- $w = \frac{168}{24}$

Step 7. Calculate the formula --------------- $w = 168 \div 24$

$w = 7$ Answer

Step 8. Check by substituting the value of the unknown (w) and compute the equation.

[The answer is OK because both sides are equal in each formula.]

$l = \frac{A}{w}$ Or $w = \frac{A}{l}$

$24 = \frac{168}{7}$ Or $7 = \frac{168}{24}$

$24 = 24$ Or $7 = 7$

The formulas used in example problems 2 and 3, come from the formula A = lw. They are used to illustrate how basic formulas can be changed to solve for different unknowns. See the preceding 2 pages if needed. Use the formula(s) that work best for you.

For another example of how formulas can be changed to solve for different unknowns, see the formula used for solving electricity problems shown below. The formula is known as Ohm's Law.

Ohm's Law
Basic Formula

[solve for I] --------- $I = \frac{E}{R}$

Change the Form to Other Formulas:

$RI = \frac{E}{\not R}\not R$ [solve for E] --------- $E = RI$

$\frac{E}{I} = \frac{R\not I}{\not I}$ [solve for R] --------- $R = \frac{E}{I}$

I = amperes (intensity of current)

E = volts (electromotive force)

R = ohms (resistance)

BASIC FORMULAS
(The Application of Algebra)

A formula is a shortened version of a mathematical rule. In it, symbols such as the first letter of important words in the rule are used. For example, to find the area of a rectangle or square ------------------------------------

-- the rule is: The area equals the length times the width.

-- the formula is: A = l w

1 X w	l • w	l w

Each of the above means the same ------- length X width

Task Number → K2

Example Problem 1. How many square feet of material are needed to cover a rectangle 3 ft. by 9 ft.? __________.

How many square yards? ____________________.

Step 1. The rule is: Area equals the length times the width. Therefore, the formula is ----- A = lw

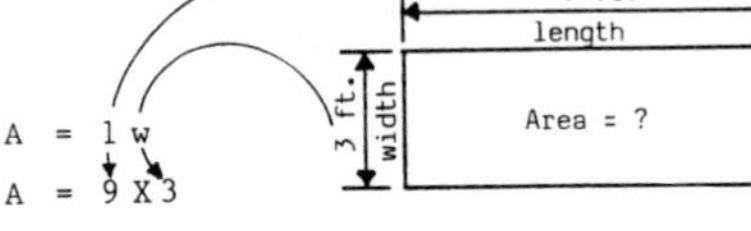

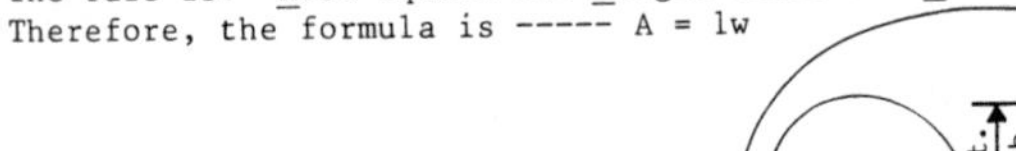

Step 2. Substitute the known values into the formula -------------- A = l w, A = 9 X 3

Step 3. Calculate the formula (do the operations indicated) --------- A = 9 X 3 = 27

[lw means l X w, therefore, multiply]

Thus the answer is ----------- **A = 27 square feet**

Step 4. Convert to square yards. **A = 3 square yards**

1 square yard =

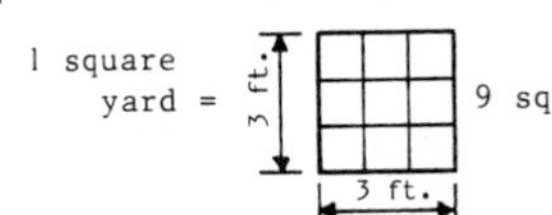

9 sq. ft., therefore in 27 sq. ft. there are --------------- 3 sq. yd. ($27 \div 9 = 3$)

Example
Problem 2. How many cubic feet of material are needed to fill a rectangular solid 3 ft. X 9 ft. X 2 ft.? __________.

How many cubic yards? ______________________________.

Step 1. The rule is: Volume equals the length X width X height.
Therefore, the formula is --------- V = lwh

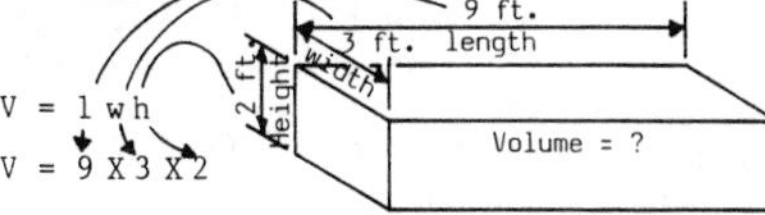

Step 2. Substitute the known values into the formula ------------------ V = l w h
V = 9 X 3 X 2

Step 3. Calculate the formula (do the operations indicated) ------------- V = (9 X 3) X 2

V = 27 X 2

Thus the answer is --------------- V = 54 cu. ft.

Step 4. Convert to cubic yards. V = 2 cu. yd.

1 cubic yard = 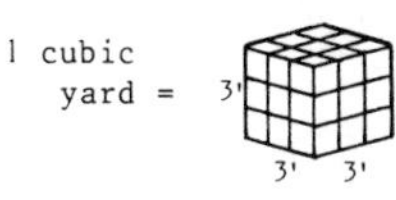

27 cu. ft., therefore in 54 cu. ft. there are -------------- 2 cu. yd.

$$\begin{array}{r} 2 \\ 27\overline{)54} \\ \underline{54} \end{array}$$

K1 Plane Geometry (Area)

L1 Solid Geometry (Volume)

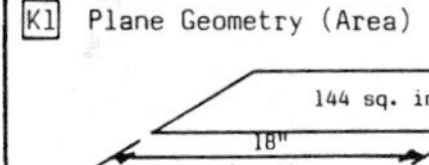

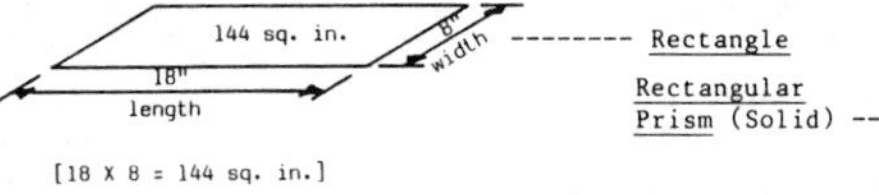

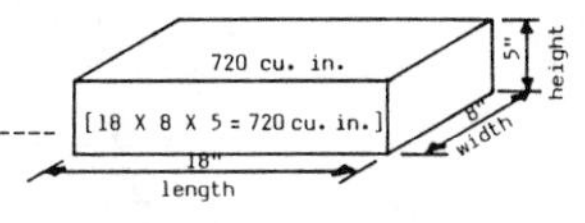

[18 X 8 = 144 sq. in.]

Notice when finding area there are two dimensions, the length and the width. Therefore, to find the Area multiply the length times the width. Use the formula: A = lw

When changing square inches to square feet, divide by 144 because a square foot is 12 inches by 12 inches or 144 sq. in.

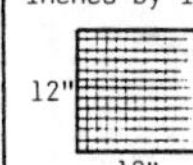

[12 X 12 = 144 sq. in. or 1 sq. ft.]

When changing square feet to square yards, divide by 9 because a square yard is 3 feet by 3 feet or 9 sq. ft.

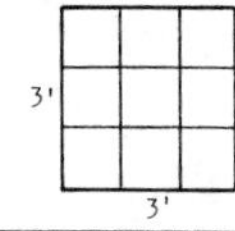

[3 X 3 = 9 sq. ft. or 1 sq. yd.]

Notice when finding volume there are three dimensions, the length, the width, and the height. Therefore, to find the Volume multiply the length times the width times the height. Use the formula: V = lwh

When changing cubic inches to cubic feet, divide by 1728 because a cubic foot is 12 inches by 12 inches by 12 inches or 1728 cu. in.

[12 X 12 X 12 = 1728 cu. in. or 1 cu. ft.]

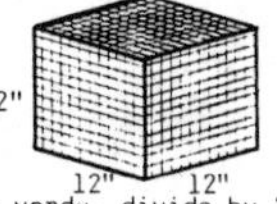

When changing cubic feet to cubic yards, divide by 27 because a cubic yard is 3 feet by 3 feet by 3 feet or 27 cu. ft.

[3 X 3 X 3 = 27 cu. ft. or 1 cu. yd.]

3'
3'
3'

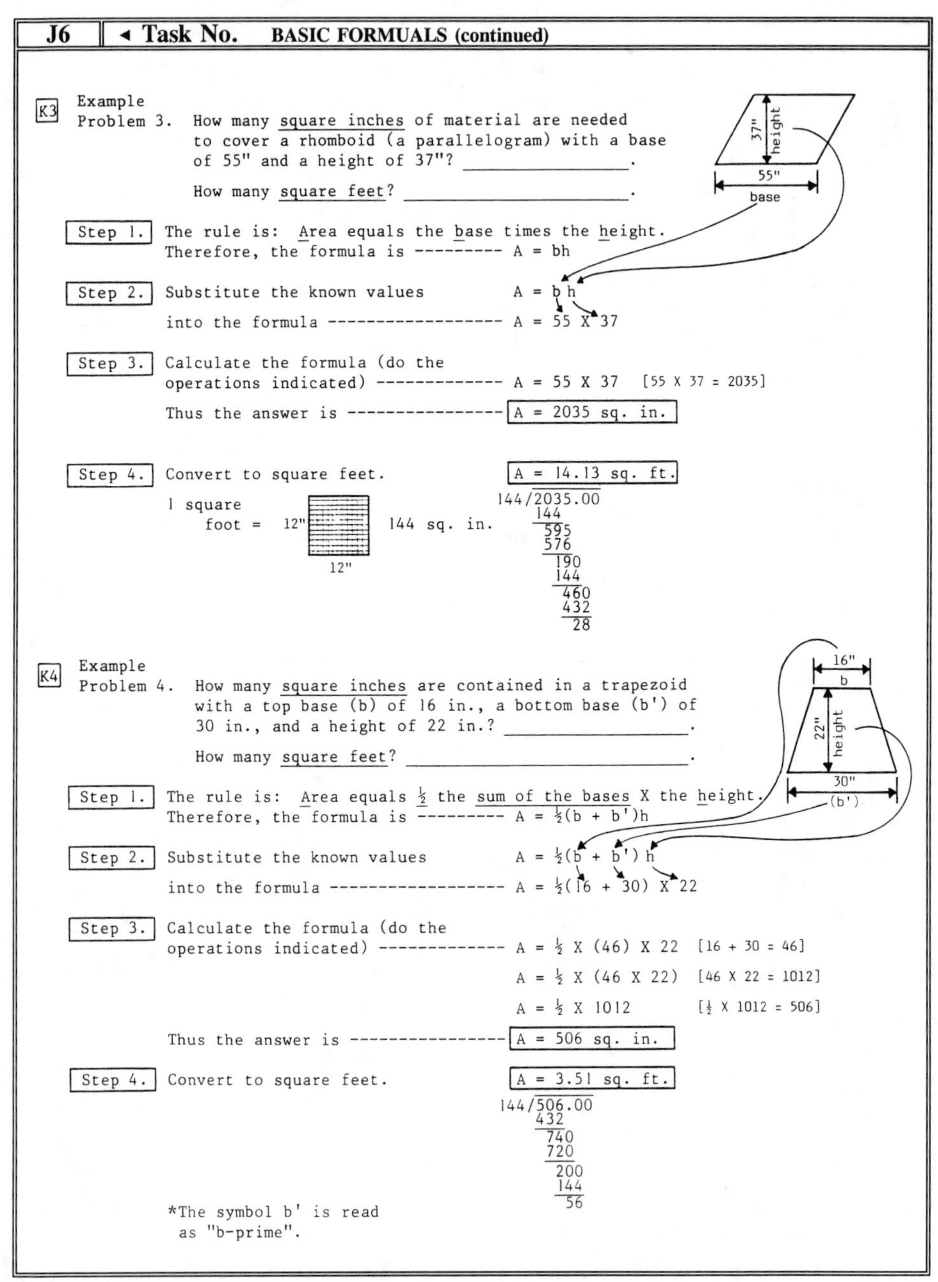

J6	◂ Task No.	BASIC FORMUALS (continued)

K3 Example Problem 3. How many square inches of material are needed to cover a rhomboid (a parallelogram) with a base of 55" and a height of 37"? ________________.

How many square feet? ________________________.

Step 1. The rule is: Area equals the base times the height. Therefore, the formula is --------- A = bh

Step 2. Substitute the known values into the formula ------------------ A = bh
A = 55 X 37

Step 3. Calculate the formula (do the operations indicated) ------------- A = 55 X 37 [55 X 37 = 2035]

Thus the answer is ---------------- A = 2035 sq. in.

Step 4. Convert to square feet. A = 14.13 sq. ft.

1 square foot = 12" × 12" 144 sq. in.

```
144/2035.00
    144
     595
     576
      190
      144
       460
       432
        28
```

K4 Example Problem 4. How many square inches are contained in a trapezoid with a top base (b) of 16 in., a bottom base (b') of 30 in., and a height of 22 in.? ________________.

How many square feet? ________________________.

Step 1. The rule is: Area equals ½ the sum of the bases X the height. Therefore, the formula is --------- A = ½(b + b')h

Step 2. Substitute the known values into the formula ------------------ A = ½(b + b') h
A = ½(16 + 30) X 22

Step 3. Calculate the formula (do the operations indicated) ------------- A = ½ X (46) X 22 [16 + 30 = 46]

A = ½ X (46 X 22) [46 X 22 = 1012]

A = ½ X 1012 [½ X 1012 = 506]

Thus the answer is ---------------- A = 506 sq. in.

Step 4. Convert to square feet. A = 3.51 sq. ft.

```
144/506.00
    432
     740
     720
      200
      144
       56
```

*The symbol b' is read as "b-prime".

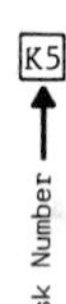

Example Problem 5. How many square inches of material are needed to cover a triangle with a base of 24 inches and a height of 18 inches? ______________.

How many square feet? ______________.

Step 1. The rule is: Area equals ½ the base times the height. Therefore, the formula is --------- $A = \frac{1}{2}bh$

Step 2. Substitute the known values into the formula ------------------ $A = \frac{1}{2}bh$; $A = \frac{1}{2} \times 24 \times 18$

Step 3. Calculate the formula (do the operations indicated) ------------- $A = \frac{1}{2} \times (24 \times 18)$ [$24 \times 18 = 432$]

$A = \frac{1}{2} \times 432$ [$\frac{1}{2} \times \frac{432}{1} = 216$]

Thus the answer is ---------------- **A = 216 sq. in.**

Step 4. Convert to square feet. **A = 1.5 sq. ft.**

1 square foot = 12" × 12" = 144 sq. in., therefore in 216 sq. in. there are ---------------- 1.5 sq. ft.

$$\begin{array}{r} 1.5 \\ 144\overline{)216.0} \\ \underline{144} \\ 720 \\ \underline{720} \end{array}$$

K7 **Example Problem 6.** How many square feet of material are needed to cover a circle that has a radius of 4.5 ft.? ______________.

How many square yards? ______________.

NOTE: pi or π = 3.1416

Step 1. The rule is: Area equals pi times the radius squared. Therefore, the formula is: $A = \pi r^2$ or $A = .7854d^2$

$A = \pi r^2$
$A = 3.1416(\frac{d}{2})^2$
$A = \frac{3.1416d^2}{4}$
$A = .7854d^2$
($\frac{d}{2}$ is the same as the radius)

Step 2. Substitute: $A = \pi r^2$; $A = 3.1416 \times 4.5^2$ | $A = .7854\ d^2$; $A = .7854 \times 9^2$

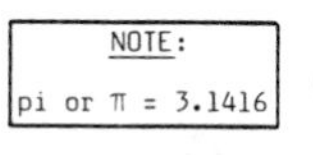

Step 3. Calculate: $A = 3.1416 \times 20.25$ | $A = .7854 \times 81$

The answer is -- **A = 63.617440 sq. ft.** | **A = 63.6174 sq. ft.**

[Note both answers are the same, except for the number of decimal points carried out.]

Step 4. Convert to square yards. **A = 63.61744/9 = 7.0686 sq. yd.**

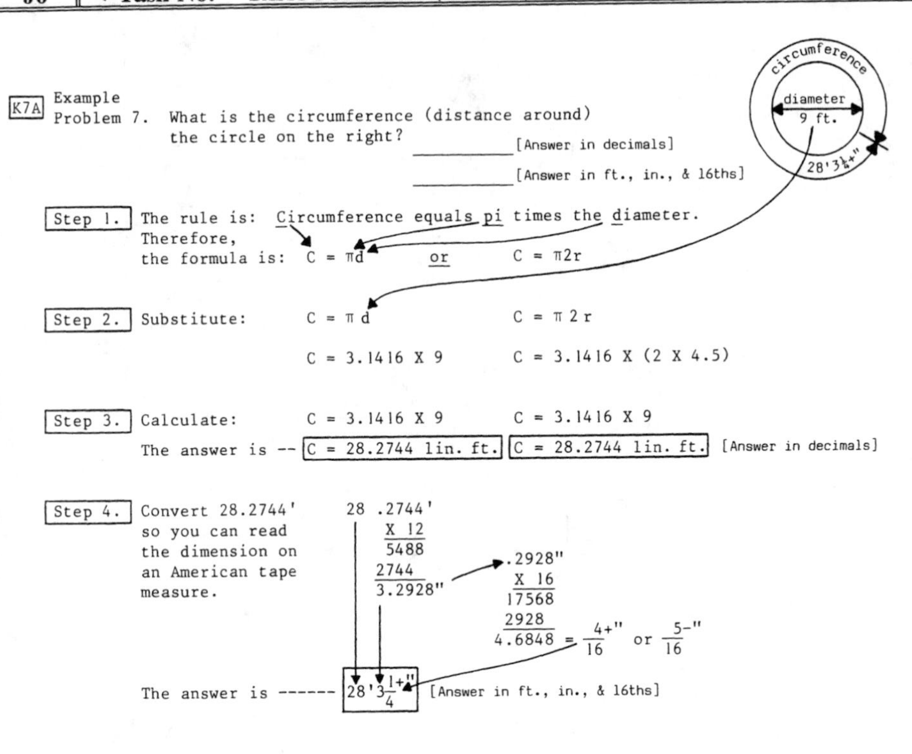

NOTE:

The circumference of the circle above is 28.2744 linear feet. However that dimension cannot be read on an American (English) tape measure. Therefore, the decimal has to be converted to feet, inches, and fractions of an inch (usually 16ths for most practical applications) before it can be measured. To convert 28.2744 feet so you can read the dimension on a tape measure refer to Task Number E7 or review Task Numbers N1 and N1A, pages 266 - 273.

A note about the formulas in this math apply-it module: The formulas listed in the preceding 4 pages are common formulas that are used quite frequently. It would be difficult, indeed, to list all the formulas there are. Even the most difficult formulas are comparatively easy once you have a basic understanding of mathematics. Many more formulas are listed throughout this book for easy reference.

How to manipulate the formula " Circumference = pi times the diameter ($C = \pi d$)" to find the diameter when the circumference is known. To measure the diameter of a run of a large pipe would be impossible without cutting the pipe and/or by using a huge caliper (in most cases this is impractical). Therefore to find the diameter, this formula is especially helpful and fairly easy to use once it has been practiced a few times.

Example
Problem 8. The circumference of the pipe shown on the right is 15.75". Find the diameter. To find the "diameter", manipulate the formula $C = \pi d$ as shown below.

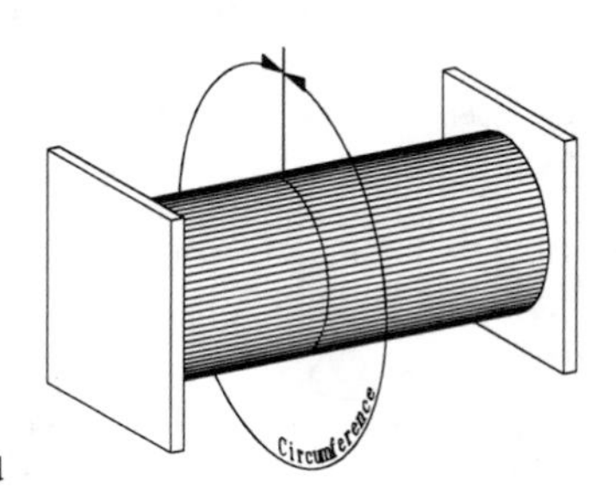

Note: Circumference = The distance around a circle.

Step 1. Use the basic formula: Also see Task K7A in the book.

$$C = \pi d$$

Step 2. Substitute the known values into the formula as shown.

$$15.75" = 3.1416 \text{ X } (d)$$

Step 3. Get the unknown (d) by itself. Therefore do the opposite as indicated. Divide both sides by 3.1416. Note (d) will now be by itself after canceling.

$$\frac{15.75"}{3.1416} = \frac{3.1416 \text{ X } (d)}{3.1416}$$

Step 4. Now you have the formula to find the diameter when the circumference is known.

$$\frac{15.75"}{3.1416} = d$$

Step 5. Calculate. $15.75 \div 3.1416 =$ To divide decimal fractions, see Task No. E5 in the book.

$$5.013" = d$$
$$5" = d$$

[For most practical purposes, in this case, 5" would be close enough.]

K1	◂ Task No.	PLANE GEOMETRY (Surface Measure)

Plane geometry (also known as surface measure) is the part of mathematics that deals with plane (flat) figures and lines.

Lines

Figures in plane geometry are made up of lines. A straight line is the shortest distance between two points. It has neither thickness or width.

Other types of lines are the curved line and the broken line.

Two lines that are equally distant from each other at all points are called parallel lines.

Types of Lines

Straight Line

Curved Line

Broken Line

Parallel Lines

Angles

When two lines meet or cross each other, the openings between them are called angles.

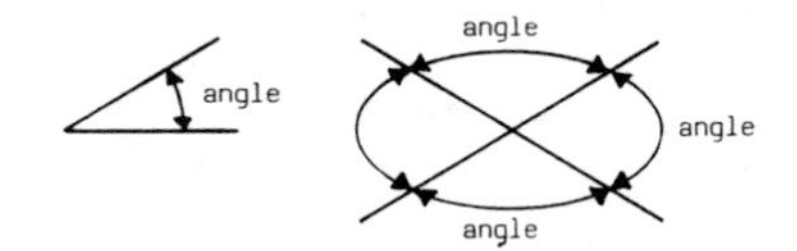

[For more information on angles, see pages 150 and 170.]

Plane Geometric Figures

When straight or curved lines on the same plane enclose a specific amount of surface, a plane figure is formed.

A polygon is a plane figure made up of straight lines with closed sides.

A Few Types of Polygons

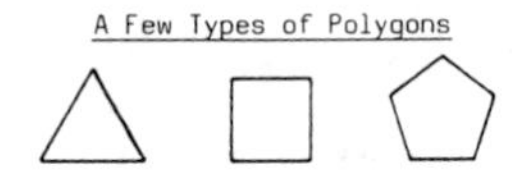

[For more information on polygons, see pages 152-154.]

A circle is a closed plane curve, all points of which are equidistant from the center point.

[For more information on circles, see pages 154 and 155.]

An ellipse is a closed plane curve with two focuses and shaped so that the sum of the distances from any point on the curve to the two focuses is always equal.

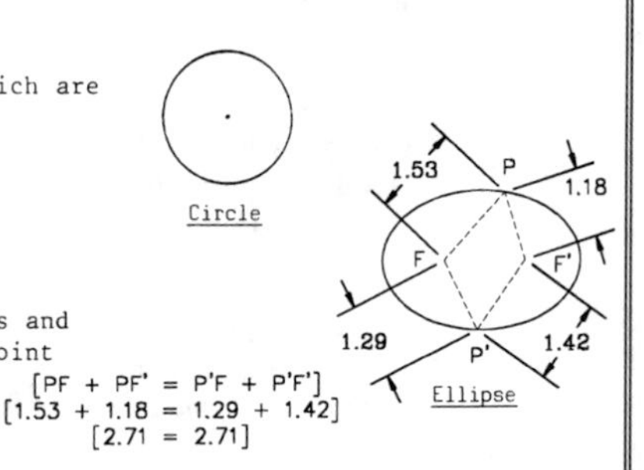

[For more information on ellipses, see pages 145, 155, and 311.]

Measuring Plane Geometric Figures

The measurement of the line length (distance) of lines and different geometric figures is called linear measurement. In America, the most common system of linear measure is the English system of measure. Common units of measure in this system are the inch and its fractions, the foot, the yard, and the mile.

[For more information on linear measure, see pages 270, 278, and 366.]

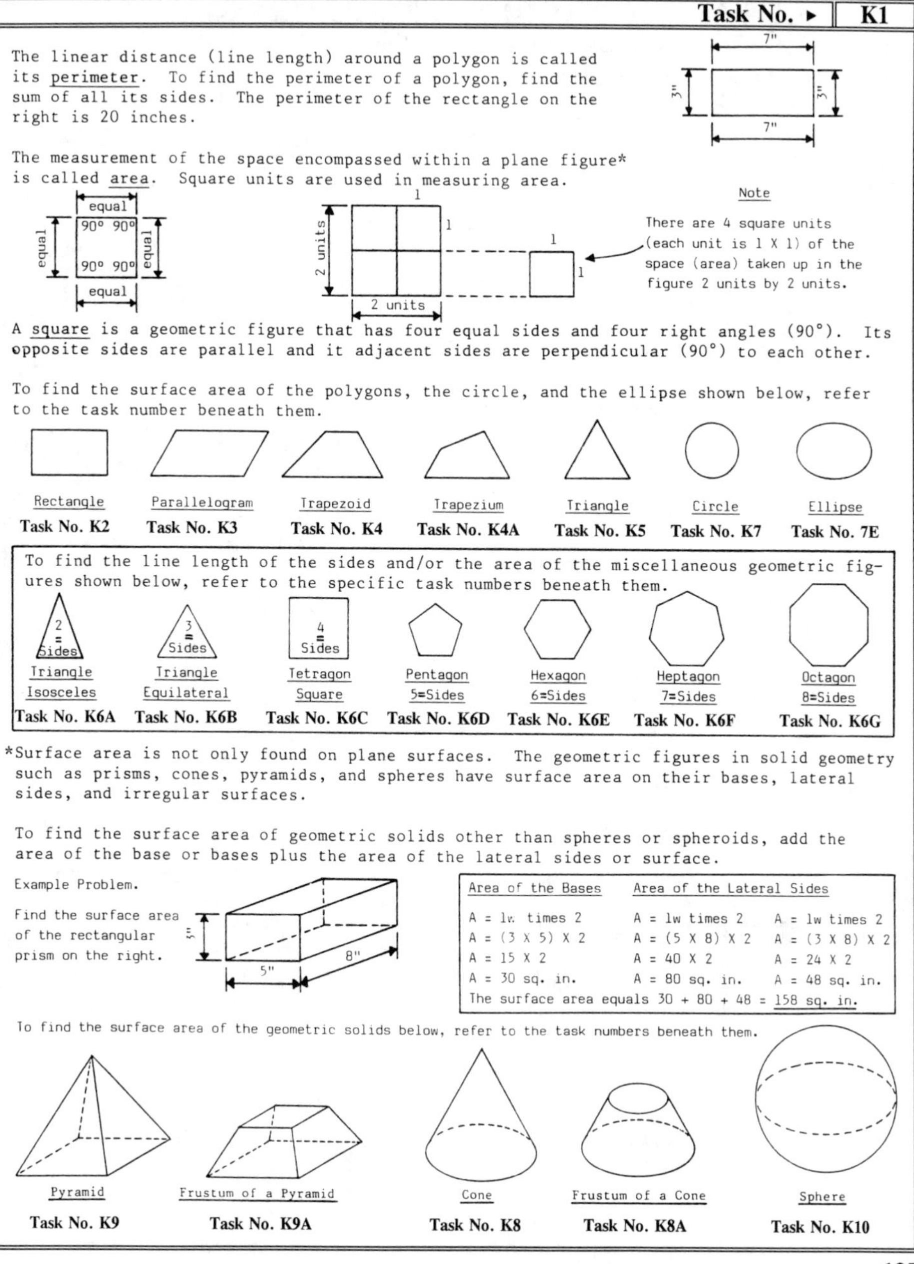

Task No. ▸ K1

The linear distance (line length) around a polygon is called its perimeter. To find the perimeter of a polygon, find the sum of all its sides. The perimeter of the rectangle on the right is 20 inches.

The measurement of the space encompassed within a plane figure* is called area. Square units are used in measuring area.

Note

There are 4 square units (each unit is 1 X 1) of the space (area) taken up in the figure 2 units by 2 units.

A square is a geometric figure that has four equal sides and four right angles (90°). Its opposite sides are parallel and it adjacent sides are perpendicular (90°) to each other.

To find the surface area of the polygons, the circle, and the ellipse shown below, refer to the task number beneath them.

To find the line length of the sides and/or the area of the miscellaneous geometric figures shown below, refer to the specific task numbers beneath them.

*Surface area is not only found on plane surfaces. The geometric figures in solid geometry such as prisms, cones, pyramids, and spheres have surface area on their bases, lateral sides, and irregular surfaces.

To find the surface area of geometric solids other than spheres or spheroids, add the area of the base or bases plus the area of the lateral sides or surface.

Example Problem.

Find the surface area of the rectangular prism on the right.

Area of the Bases	Area of the Lateral Sides	
A = lw times 2	A = lw times 2	A = lw times 2
A = (3 X 5) X 2	A = (5 X 8) X 2	A = (3 X 8) X 2
A = 15 X 2	A = 40 X 2	A = 24 X 2
A = 30 sq. in.	A = 80 sq. in.	A = 48 sq. in.

The surface area equals 30 + 80 + 48 = 158 sq. in.

To find the surface area of the geometric solids below, refer to the task numbers beneath them.

K2	◂ Task No.	AREA OF A SQUARE OR RECTANGLE

To find the area of a square or rectangle, multiply the length by the width.

RULE: Area equals length times width.

FORMULA: A = lw

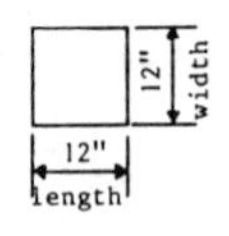

Example
Problem 1. Find the area of the square on the right.

Step 1. Use the formula ----------▶ $A = lw$ or $A = s^2$

Step 2. Substitute the known values into the formula ---------▶ $A = 12 \times 12$ $A = 12^2$

Step 3. Calculate the formula (do the operations indicated) ----▶

```
A =   12          A =   12
    X 12              X 12
      24                24
     12                12
     144 sq. in.       144 sq. in.
```

The answer is ----------------▶ 144 sq. in. 144 sq. in.

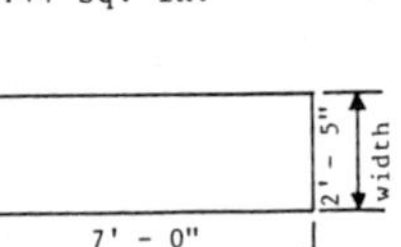

Example
Problem 2. Find the area of the rectangle on the right.

Step 1. Use the formula ----------▶ $A = lw$

Step 2. Substitute the known values into the formula ---------▶ $A = 2\frac{5}{12} \times 7$

Change to feet*
(solve with decimal fractions)

$2'5" = 2\frac{5}{12}$ feet

Step 3. Calculate the formula (do the operations indicated) ---▶ $A = 2\frac{5}{12} \times \frac{7}{1}$

$$A = \frac{29}{12} \times \frac{7}{1} = \frac{203}{12}$$

```
    16.91
12/203.00
```

The answer is ----------▶ A = 16.91 sq. ft.

Step 4. Convert to square inches or square yards.

```
  16.91                  1.87 sq. yd.
X  144               9/16.91
  6764                 9
 6764                  79
1691                   72
2535.04 sq. in.         71
                        63
                         8
```

***Note that dimensions must be changed to the same units of measure before multiplying.**

Change to feet
(solve with decimal fractions

$2\frac{5}{12}$ = 2.416 or 2.42

Step 1. A = lw
Step 2. A = 2.42 X 7
Step 3. A = 2.42

```
  2.42
   X 7
 16.94 sq. ft.
```

[NOTE: 2.416 is more accurate than 2.41. However, for most practical purposes when finding areas, 2 places is far enough. 2.416 = 2.42 rounded.]

Change to inches
(solve with whole numbers)

2'5" = 29" and 7'0" = 84"

Step 1. A = lw
Step 2. A = 84 X 29
Step 3. A =

```
   84
 X 29
  756
 168
 2436 sq. in.
```

Step 4. A =

```
       16.91 sq. ft.
144/2436.00
```

AREA OF A PARALLELOGRAM — Task No. ▸ K3

To find the area of a parallelogram, multiply the base by the height.

RULE: Area equals the base times the height.

FORMULA: A = bh **[The same formula is used for the rhombus or the rhomboid.]**

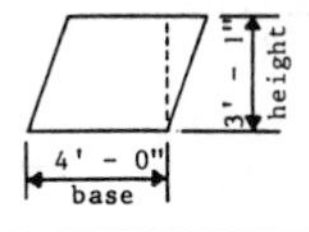

Example Problem. Find the area of the parallelogram on the right.

Step 1. Use the formula ------------▸ $A = bh$

Step 2. Substitute the known values into the formula* ----------▸ $A = 4 \times 3\frac{1}{12}$

Step 3. Calculate the formula (do the operations indicated) ------▸ $A = \frac{4}{1} \times \frac{37}{12} = \frac{148}{12}$

$A = 12.33$ sq. ft.

Step 4. Convert to square inches or square yards.

```
  12.33                    1.37 sq. yd.
 X 144                9/12.33
  4932                  9
 4932                   33
1233                    27
1775.52 Sq. in.          63
```

Change to feet
(solve with decimal fractions)

$3\frac{1}{12} = 3.083$ or 3.08

Step 1. $A = bh$
Step 2. $A = 4 \times 3.08$
Step 3. $A = 3.08$

```
  3.08
   X 4
 12.32 sq. ft.
```

Change to inches
(solve with whole numbers)

4'0 = 48" and 3'1" = 37"

Step 1. $A = bh$
Step 2. $A = 48 \times 37$
Step 3. $A = 1776$ sq. in.

```
  48
X 37
 336
144
1776
```

Step 4. $A = 1776 \div 144 = 12.33$ sq. ft.

AREA OF A TRAPEZOID — Task No. ▸ K4

To find the area of a trapezoid, multiply 1/2 the sum of the bases by the height.

RULE: Area equals 1/2 the sum of the bases times the height.

FORMULA: A = 1/2(b + b′)h

2' - 4"
base'
6' - 0"
height
4' - 10"
base

Example Problem. Find the area of the trapezoid on the right.

Step 1. Use the formula ------------▸ $A = \frac{1}{2}(b + b')h$

Step 2. Substitute the known values into the formula -----------▸ $A = \frac{1}{2}(4\frac{5}{6} + 2\frac{1}{3})6$

Change to feet*
(solve with mixed numbers)

$2'4" = 2\frac{4}{12}$ or $2\frac{1}{3}$ feet

$4'10" = 4\frac{10}{12}$ or $4\frac{5}{6}$ feet

Step 3. Calculate the formula (do the operations indicated) ------▸ $A = \frac{1}{2} \times (7\frac{1}{6}) \times 6$

$2\frac{1}{3}\ \frac{2}{6}$

$+\ 4\frac{5}{6}\ \frac{5}{6}$

$6\ \frac{7}{6} = 7\frac{1}{6}$

$7\frac{1}{6} \times 6$

$\frac{43}{\not 6} \times \frac{\not 6^{1}}{1} = 43$

$A = \frac{1}{2} \times \frac{43}{1}\ \ \frac{43}{2}$

$43 \div 2 = 21.5$ or $A = 21\frac{1}{2}$ sq. ft.

Change to inches
(solve with whole numbers)

2'4" = 28" and 4'10" = 58"

Step 1. $A = \frac{1}{2}(28 + 58)72$
Step 2. $A = \frac{1}{2} \times (86) \times 72$
Step 3. $A = \frac{1}{2} \times 6192$
Step 4. $A = 3096$ sq. in.
Step 5. $A = 3096.0 \div 144 = 21.5$ sq. ft.

[Note: When you calculate the problem by using two different methods, you are checking the results.]

***Note that the dimensions must be changed to the same units of measure before multiplying.**

K4A ◂ Task No. AREA OF A TRAPEZIUM

To find the area of a trapezium, find the sum of the area of the triangle ($A = \frac{1}{2}bh$) plus the area of the trapezoid [$A = \frac{1}{2}(b + b')h$].

FORMULAS:

$A = \frac{1}{2}bh$ ------------------ Area of a triangle, Task No. K5.
$A = \frac{1}{2}(b + b')h$ ----------- Area of a trapezoid, Task No. K4.
$A = [\frac{1}{2}bh] + [\frac{1}{2}(b + b')h$ -- Area of a trapezium, Task No. K4A.

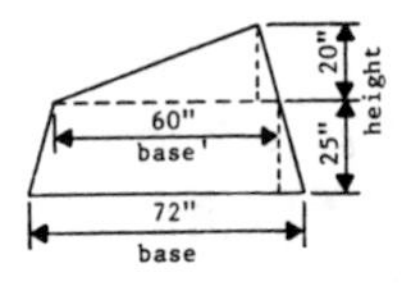

Example
Problem. Find the area of the trapezium on the right.

Step 1. Use the formula --------------- $A = [\frac{1}{2}bh] + [\frac{1}{2}(b + b')h]$

Step 2. Substitute the known values into the formula -------------- $A = [\frac{1}{2} \times 60 \times 20] + [\frac{1}{2} \times (72 + 60) \times 25]$

Step 3. Calculate the formula (do the operations indicated) --------- $A = [\frac{1}{2} \times (60 \times 20)] + [\frac{1}{2} \times (72 + 60) \times 25]$

$A = [\frac{1}{2} \times 1200] + [\frac{1}{2} \times 132 \times 25]$

$A = 600 + [\frac{1}{2} \times (132 \times 25)]$

$A = 600 + [\frac{1}{2} \times 3300]$

$A = 600 + 1650$

$A = 2250$ sq. in. Answer

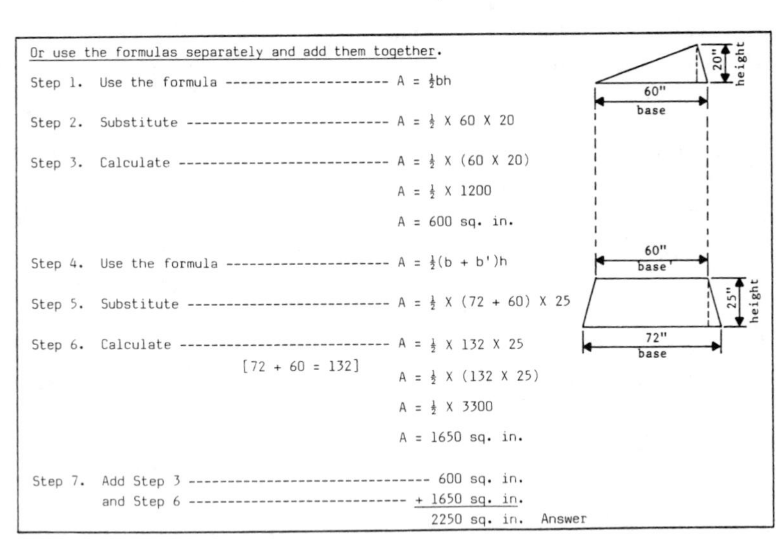

Or use the formulas separately and add them together.

Step 1. Use the formula --------------------- $A = \frac{1}{2}bh$

Step 2. Substitute -------------------------- $A = \frac{1}{2} \times 60 \times 20$

Step 3. Calculate --------------------------- $A = \frac{1}{2} \times (60 \times 20)$

$A = \frac{1}{2} \times 1200$

$A = 600$ sq. in.

Step 4. Use the formula --------------------- $A = \frac{1}{2}(b + b')h$

Step 5. Substitute -------------------------- $A = \frac{1}{2} \times (72 + 60) \times 25$

Step 6. Calculate --------------------------- $A = \frac{1}{2} \times 132 \times 25$

[72 + 60 = 132]

$A = \frac{1}{2} \times (132 \times 25)$

$A = \frac{1}{2} \times 3300$

$A = 1650$ sq. in.

Step 7. Add Step 3 ------------------------------ 600 sq. in.
and Step 6 ---------------------------- + 1650 sq. in.
2250 sq. in. Answer

AREA OF A TRIANGLE

Task No. ▸ K5

To find the area of a triangle, multiply the base by the height and divide that product by 2.

RULE: Area equal 1/2 the base times the height.

FORMULA: A = 1/2bh

height: 5' - $7\frac{1}{2}$"
base: 3' - 6"

Example Problem. Find the area of the triangle on the right.

Step 1. Use the formula ——▸ $A = \frac{1}{2}bh$

Step 2. Substitute the known values into the formula ——▸ $A = \frac{1}{2} \times 3.5 \times 5.625$

Change to feet*
(solve with decimal fractions)

$3'6" = 3\frac{6}{12}$ or $3\frac{1}{2}$ or 3.5 feet

$5'7\frac{1}{2}" = 5\frac{7\frac{1}{2}}{12}$ or $\frac{7.5}{12}$ or 5.625 feet

Step 3. Calculate the formula (do the operations indicated) ——▸ $A = \frac{1}{2} \times (3.5 \times 5.625)$

```
  5.625
 X 3.5
 28125
16875
19.6875
```

$2 \overline{)19.6875}$ = 9.84375

$A = \frac{1}{2} \times 19.6875$ sq. ft.

A = 9.84375 sq. ft.

A = 9.84 sq. ft. (rounded off to 2 places)

Change to inches
(solve with whole numbers)

3'6" = 42" and $5'7\frac{1}{2}"$ = 67.5"

Step 1. $A = \frac{1}{2}bh$

Step 2. $A = \frac{1}{2} \times 42 \times 67.5$

Step 3. $A = \frac{1}{2} \times (42 \times 67.5)$

$A = \frac{1}{2} \times 2835$

A = 1417.5 sq. in.

Step 4. A = $144 \overline{)1417.500000}$ = 9.84375 sq. ft.

NOTE: By re-arranging the formula $A = \frac{1}{2}bh$, you can find the base when the height and the area are known.

Example Problem. Using the same triangle as above, but knowing only the area and the height, find the base.

Step 1. Use the formula ——▸ $A = \frac{1}{2}bh$

Step 2. Solve for b ——▸ $\frac{A}{\frac{1}{2}h} = \frac{\frac{1}{2}bh}{\frac{1}{2}h}$ [$\frac{A}{\frac{1}{2}h} = b$ New Formula]

[Divide both sides by $\frac{1}{2}h$]

Step 3. Substitute the known values into the formula ——▸ $\frac{9.84}{.5 \times 5.625} = b$

Step 4. Calculate the formula (do the operations indicated) ——▸ $\frac{9.84}{2.8125} = b$

[.5 X 5.625 = 2.8125]

[9.84 ÷ 2.8125 = 3.498]

3.498 = b

3.5' = b

3'6" = b

Find the height when the area and the base are known.

$A = \frac{1}{2}bh$

$\frac{A}{.5b} = \frac{\frac{1}{2}bh}{\frac{1}{2}b}$

$\frac{A}{.5b} = h$

$\frac{9.84}{.5 \times 3.5} = h$

$\frac{9.84}{1.75} = h$

5.623' = h

$5'7\frac{1}{2}"$ = h

[NOTE: To find the area of a triangle having three known sides with the height unknown, see Hero's formula, page 216.]

K5A ◂ Task No. RULE OF PYTHAGORAS AND THE RIGHT TRIANGLE

The Rule of Pythagoras: **In any right triangle the altitude squared plus the base squared is equal to the hypotenuse (c) squared.**

FORMULA: $a^2 + b^2 = c^2$

Example Problem. Substitute the sides of the right triangle on the right into the Rule of Pythagoras.

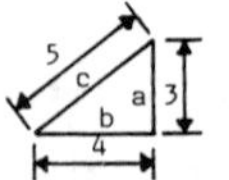

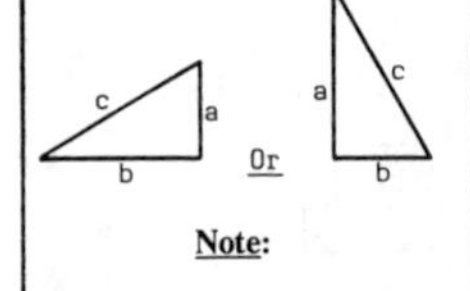

Note:

The proper way to label the sides of a right triangle is a is the altitude (side 90° to the base), b is the base (side it seems to rest on), and c is the hypotenuse (side opposite the right angle). Also see the note on Task Number K6B, page 134, and pages 171-206.

Step 1. Use the formula -------▶ $a^2 + b^2 = c^2$

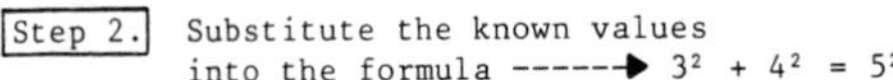

Step 2. Substitute the known values into the formula ------▶ $3^2 + 4^2 = 5^2$

Step 3. Calculate the formula (do the operations indicated) ---▶ $(3 \times 3) + (4 \times 4) = (5 \times 5)$

$9 + 16 = 25$

The formula checks out ----▶ $25 = 25$

Listed below are the three different formulas that are derived from the Rule of Pythagoras and how each one is used. [Refer to Basic Algebra Task No. J1, J2, and J3 if more information is needed.]

Formula:		$[a^2 + b^2 = c^2$, solve for c$]$	$[a^2 + b^2 = c^2$, solve for b$]$	$[a^2 + b^2 = c^2$, solve for a$]$
Step 1.	Use the formula:	$a^2 + b^2 = c^2$	$b^2 = c^2 - a^2$	$a^2 = c^2 - b^2$
Step 2.	Substitute:	$3^2 + 4^2 = c^2$	$b^2 = 5^2 - 3^2$	$a^2 = 5^2 - 4^2$
Step 3.	Calculate:	$9 + 16 = c^2$	$b^2 = 25 - 9$	$a^2 = 25 - 16$
		$25 = c^2$	$b^2 = 16$	$a^2 = 9$
		$\sqrt{25} = c$	$b = \sqrt{16}$	$a = \sqrt{9}$
		$5 = c$	$b = 4$	$a = 3$

Another Example. Find the hypotenuse (c) of the right triangle on the right.

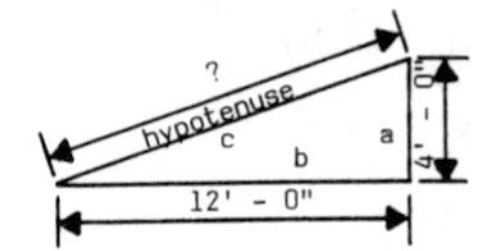

Step 1. Use the formula -------▶ $a^2 + b^2 = c^2$

Step 2. Substitute: $4^2 + 12^2 = c^2$

Step 3. Calculate: $(4 \times 4) + (12 \times 12) = c^2$

$16 + 144 = c^2$

$160 = c^2$

$\sqrt{160} = c$

$12.65' = c$ [Answer rounded off 2 decimal places.]

Step 4. Convert 12.65' so you can read the dimension on an American (English) tape measure.

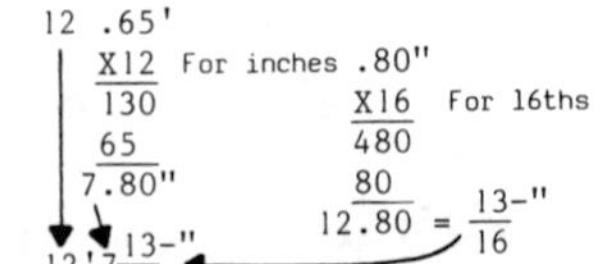

12 .65'

X12 For inches .80"

130 X16 For 16ths

65 480

7.80" 80

12.80 = 13-"/16

Answer $12' 7 \frac{13}{16}"$

ISOSCELES TRIANGLE

To solve the isosceles triangle below, the Rule of Pythagoras and the formula for the area of a triangle are used. However to solve the triangles encompassed within the other polygons (3 equal sides to 8 equal sides) that follow, trigonometry is used along with the Rule and the formula for the area of a triangle. Trigonometry is used to illustrate how the formulas came about, and to assure you that they do indeed work.

Should you need to calculate the sides of obtuse triangles (having one angle greater than 90°) and acute triangles (all angles less than 90°) other than with the formula A = ½bh, refer to the trigonometry part of this book. All the sines, cosines, tangents, and so on, along with the formulas for the different types of triangles are included. The advantage of solving triangles with trigonometry is that any type of triangle can be solved as long as one side and two angles, or two sides and one angle are known, whereas with the Rule of Pythagoras only right triangles can be solved and two sides have to be known.

ISOSCELES TRIANGLE

An isosceles triangle is a triangle that has two equal sides.

Example Problem . Find the height of the isosceles triangle on the right.

Step 1. Use the formula ----> $a^2 + b^2 = c^2$

Step 2. Solve for a --------> $a^2 = c^2 - b^2$

Step 3. Substitute: $a^2 = 8^2 - 1.5^2$

Step 4. Calculate: $a^2 = (8 \times 8) - (1.5 \times 1.5)$

$a^2 = 64 - 2.25$

$a^2 = 61.75$

$a = \sqrt{61.75}$

a = 7.86" height

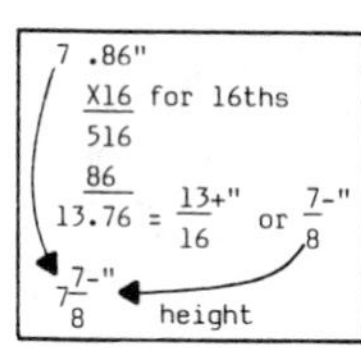

[Find the area now that the height is known.]

Step 1. Use the formula ----> $A = \frac{1}{2}bh$

Step 2. Substitute: $A = \frac{1}{2} \times 3 \times 7.86$

Step 3. Calculate: $A = \frac{1}{2} \times (3 \times 7.86)$

$A = \frac{1}{2} \times 23.58$

A = 11.79 sq. in.

[Or, find the height using the formula $A = \frac{1}{2}bh$ if the area and the base are known.]

Step 1. Use the formula ----> $\frac{A}{\frac{1}{2}b} = h$ $\left[\frac{A}{\frac{1}{2}b} = \frac{\frac{1}{2}bh}{\frac{1}{2}b}\right]$

Step 2. Substitute: $\frac{11.79}{1.5} = h$

Step 3. Calculate: 7.86" = h

[11.79 ÷ 1.5 = 7.86:]

K6B	◂ Task No.	EQUILATERAL TRIANGLE

Note: **The proper way to label the 3 sides and the 3 angles of a right triangle: A is the angle opposite the altitude, B is the angle opposite the base b, and C is the angle opposite the hypotenuse c. Oblique triangles are labeled in a similar way even though they have no right angles.**

EQUILATERAL TRIANGLES

An equilateral triangle is a triangle that has three equal sides.

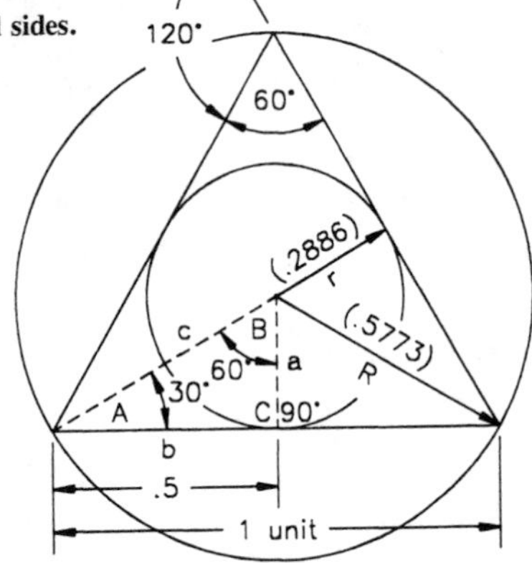

Formulas used to solve equilateral triangle problems:

r = .289S (radius of inscribed circle)

R = .577S (radius of circumscribed circle)

A = .433S² (Area of equilateral triangle)

S = 1.155h (length of the side)

Illustrated below is how the above formulas were determined and checked.

To find little r (a) when a side is 1, use the
Formula: a = b X tan A(30°)
Substitute: a = .5 X .5773
Calculate: a = .2886

Therefore to find little r for any side use the
Formula: r = .289S (rounded off to 3 decimal places)

To find big R (c) when a side is 1, use the
Formula: c = a X secant B(60°)
Substitute: c = .2886 X 2
Calculate: c = .5773

Therefore to find big R for any side use the
Formula: R = .577S (rounded off to 3 decimal places)

To find the Area (A) when a side is 1, use the
Formula: A = ½bh
Substitute: A = (½ X 1) X .866
Calculate: A = ½ X .866
A = .433 sq. units

Therefore to find the Area (A) for any side use the formula
Formula: A = lw
Substitute: .433 = S X S
Calculate: .433 = S^2
Formula: A = .433S²

The height of an equilateral triangle with a side of 1 is .866. Therefore, the ratio of any equilateral triangle would be 1 to .866; 1/.866 or 1.155 times the side. Therefore to find the Side of any equilateral triangle multiply 1.155 times the height. Use the
Formula: S = 1.155h

NOTE: The Rule of Pythagoras can be used to check the above out.

Formula:	$a^2 + b^2 =$	c^2
Substitute:	$.2886^2 + .5^2 =$	$.5773^2$
Calculate:	.0832 + .25 =	.3332
Answer OK	.3332 =	.3332

[See Task No. O2 to lay out an equilateral triangle]

SQUARE (Tetragon)
(4 Equal Sides and Angles)

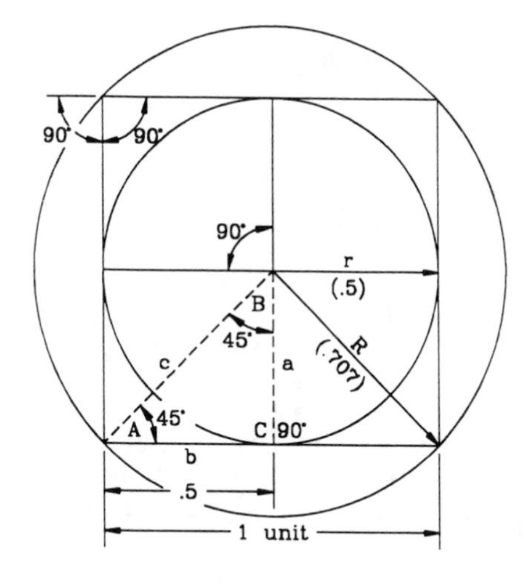

Formulas used to solve square problems:

S = Side
r = .5S (radius of inscribed circle)
R = .707S (radius of circumscribed circle)
A = lS^2 (area)

Illustrated below is how the above formulas were determined and checked.

To find little r (a) when a side is 1, use the
Formula: a = b X tan A(45°)
Substitute: a = .5 X 1
Calculate: a = .5
Therefore to find little r for any side, use the
Formula: r = .5S

To find big R (c) when a side is 1, use the
Formula: c = a X secant B(45°)
Substitute: c = .5 X 1.4142
Calculate: c = .7071
Therefore to find big R for any side, use the
Formula: R = .707S

To find the Area (A) when a side is 1, use the
Formula: A = lw
Substitute: A = 1 x 1
Calculate: A = 1 sq. unit
Therefore to find the Area (A) for any side, use the
Formula: A = l X S X S <u>or</u> **A = lS^2**

NOTE: The Rule of Pythagoras can be used to check all the formulas shown above.

Formula:	**$a^2 + b^2 = c^2$**
Substitute:	$.5^2 + .5^2 = .707^2$
Calculate:	.25 + .25 = .4998
Answer OK:	.5 = .5

[See Task No. O3 to lay out a square or Task No. O4 to lay out a rectangle]

PENTAGON
(5 Equal Sides and Angles)

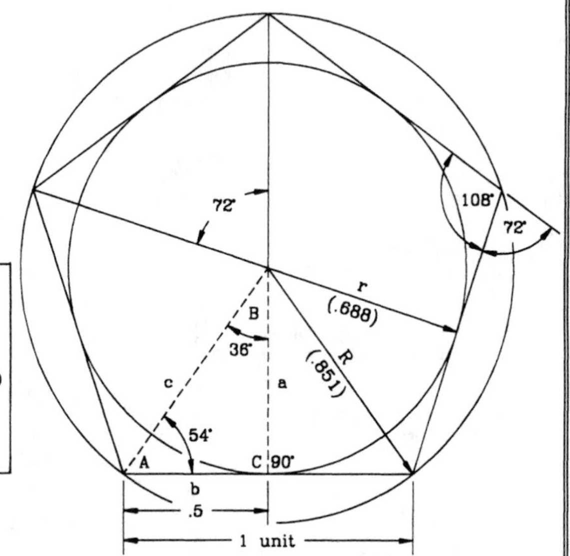

Formulas used to solve pentagon problems:

S = Side
r = .688S (radius of inscribed circle)
R = .851S (Radius of circumscribed circle)
A = 1.72S² (Area)

Illustrated below is how the above formulas were determined and checked.

To find little r (a) when a side is 1, use the
Formula: a = b X tan A (54°)
Substitute: a = .5 X 1.3764
Calculate: a = .6882
Therefore to find little r for any side, use the
Formula: r = .688S

To find big R (c) when a side is 1, use the
Formula: c = a X secant B (36°)
Substitute: c = .6882 X 1.2361
Calculate: c = .850684
Therefore to find big R for any side, use the
Formula: R = .851S

To find the Area (A) when a side is 1, use the
Formula: A = ½bh X 5 (5 triangles)
Substitute: A = ½ X 1 X .688 X 5
Calculate: A = (½ X 1) X (.688 X 5)
A = ½ X 3.44
A = 1.72 sq. units
Therefore to find the Area (A) for any side, use the
Formula: A = 1.72 X S X S or **A = 1.72S²**

NOTE: The Rule of Pythagoras can be used to check all the formulas shown above.

Formula:	$a^2 + b^2 = c^2$
Substitute:	$.6882^2 + .5^2 = .8506^2$
Calculate:	.4736 + .25 = .7235
Answer OK:	.7236 = .7235

[See Task No. O5 to lay out a pentagon]

HEXAGON
(6 Equal Sides and Angles)

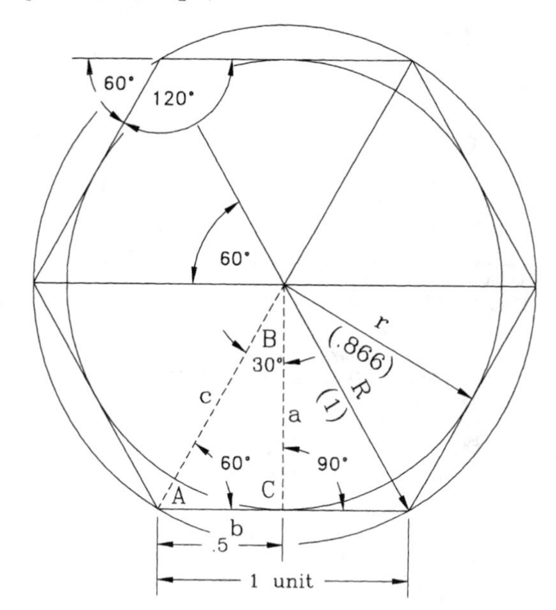

Formulas used to solve hexagon problems:

S = Side
r = .866S (radius of inscribed circle)
R = 1S (Radius of circumscribed circle)
A = 2.598S² (Area)

Illustrated below is how the above formulas were determined and checked.

To find little r (a) when a side is 1, use the
Formula: a = b X tan A (60°)
Substitute: a = .5 X 1.732
Calculate: a = .866
Therefore to find little r for any side, use the
Formula: r = .866s

To find big R (c) when a side is 1, use the
Formula: c = a X secant B (30°)
Substitute: c = .866 X 1.1547
Calculate: c = .9999
Therefore to find big R for any side, use the
Formula: R = 1S

To find the Area (A) when a side is 1, use the
Formula: A = ½bh X 6 (6 triangles)
Substitute: A = ½ X 1 X .866 X 6
Calculate: A = (½ X 1) X (.866 X 6)
A = ½ X 5.196
A = 2.598 sq. units
Therefore to find the Area (A) for any side, use the
Formula: 2.598 X S X S or **2.598S²**

NOTE: The Rule of Pythagoras can be used to check all the formulas shown above.

Formula:	**a² + b² =**	**c²**
Substitute:	.866 + .5² =	1²
Calculate:	.7499 + .25 =	1
Answer OK	----- .9999 =	1

[See Task No. O6 to lay out a hexagon]

HEPTAGON
(7 Equal Sides and Angles)

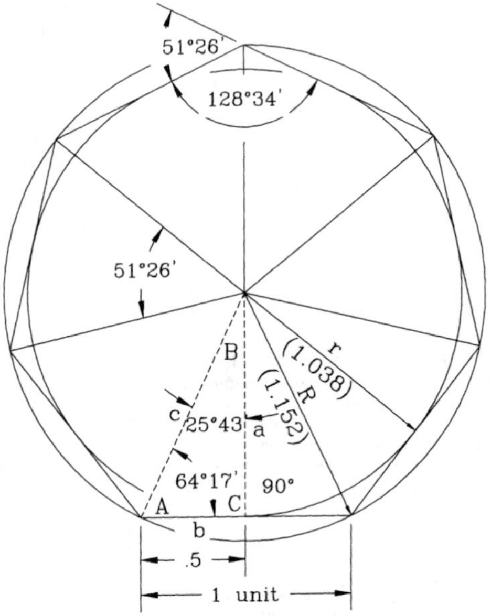

Formulas used to solve heptagon problems:

S = Side
r = 1.038S (radius of inscribed circle)
R = 1.152S (Radius of circumscribed circle)
A = 3.633S² (Area)

Illustrated below is how the above formulas were determined and checked.

51°26'
128°34'
51°26'
B
r
(1.038)
R
(1.152)
c
25°43
a
64°17'
90°
A
C
b
.5
1 unit

To find little r (a) when a side is 1, use the
Formula: a = b X tan A (64° 17')
Substitute: a = .5 X 2.0763
Calculate: a = 1.0381
Therefore to find little r for any side, use the
Formula: r = 1.038S

To find big R (c) when a side is 1, use the
Formula: c = a X secant B (25° 43')
Substitute: c = 1.038 X 1.1099
Calculate: c = 1.152
Therefore to find big R for any side, use the
Formula: R = 1.152S

To find the Area (A) when a side is 1, use the
Formula: A = ½bh X 7 (7 triangles)
Substitute: A = ½ X 1 X 1.0381 X 7
Calculate: A = (½ X 1) X (1.0381 X 7)
A = ½ X 7.2667
A = 3.633 sq. units
Therefore to find the Area (A) for any side, use the
Formula: 3.633 X S X S or **A = 3.633S²**

NOTE:	The Rule of Pythagoras can be used to check all the formulas shown above.	Formula:	$a^2 + b^2 =$	c^2
		Substitute:	$1.038^2 + .5^2 =$	1.152^2
		Calculate:	1.077 + .25 =	1.327
		Answer OK	------ 1.327 =	1.327

[See Task No. O7 to lay out a heptagon]

OCTAGON
(8 Equal Sides and Angles)

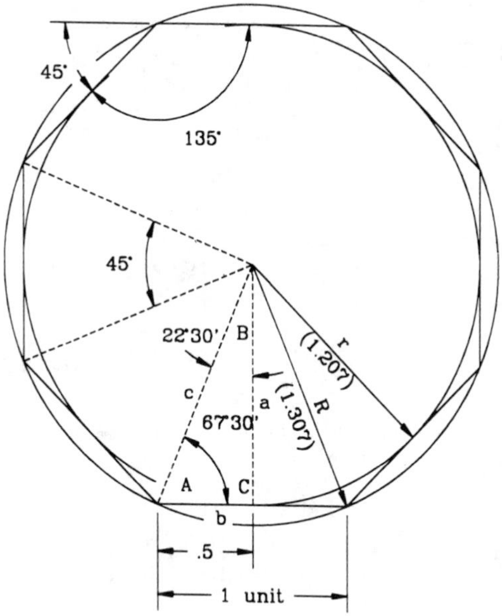

Formulas used to solve octagon problems:

S = Side
r = 1.207S (radius of inscribed circle)
R = 1.307S (Radius of circumscribed circle)
A = 4.828S² (Area)

Illustrated below is how the above formulas were determined and checked.

To find little r (a) when a side is 1, use the
Formula: a = b X tan A (67° 30′)
Substitute: a = .5 X 2.4142
Calculate: a = 1.2071

Therefore to find r for any side, use the
Formula: r = 1.207S

To find big R (c) when a side is 1, use the
Formula: c = a X secant B (22° 30′)
Substitute: c = 1.2071 X 1.0824
Calculate: c = 1.3065

Therefore to find big R for any side, use the
Formula: R = 1.307S

To find the Area (A) when a side is 1, use the
Formula: A = ½bh X 8 (8 triangles)
Substitute: A = ½ X 1 X 1.207 X 8
Calculate: A = (½ X 1) X (1.207 X 8)
A = ½ X 9.656
A = 4.828 sq. units

Therefore to find the Area (A) for any side, use the
Formula: 4.828 X S X S or **A = 4.828S²**

NOTE: The Rule of Pythagoras can be used to check all the formulas above.

Formula:	$a^2 + b^2$	=	c^2
Substitute:	$1.207^2 + .5^2$	=	1.3065^2
Calculate:	1.4568 + .25	=	1.7069
Answer OK	1.7068	=	1.7069 (answer very close)

[See Task No. O8 to lay out an octagon]

K7	◂ Task No.	AREA OF A CIRCLE

To find the area of a circle, multiply pi (π) by the radius squared.

RULE: Area equals pi times the radius squared.

FORMULA: $A = \pi r^2$

Example Problem. Find the area of the circle on the right.

Step 1. Use the formula --------→ $A = \pi r^2$

Step 2. Substitute the known values into the formula -------→ $A = 3.1416 \times 4.25^2$

Change to feet*
(solve with decimal fractions)

$4'3" = 4\frac{3}{12}$ or $4\frac{1}{4}$ or 4.25 feet

Step 3. Calculate the formula (do the operations indicated) --→ $A = 3.1416 \times 18.0625$

```
   4.25
 X 4.25
   2125
    850
  1700
18.0625
```

$A = 56.74515$ sq. ft.

$A = 56.75$ sq. ft.

*Note that dimensions must be changed to the same units of measure before multiplying.

Or use the formula:

$A = .7854d^2$

$A = .7854 \times 8.5^2$

$A = .7854 \times 72.25$

$A = 56.74515$ sq. ft.

$A = \pi r^2$

$A = 3.1416 \left(\frac{d}{2}\right)^2$

$A = \frac{3.1416d^2}{4}$

$A = .7854d^2$

($\frac{d}{2}$ is the same as the radius)

Change to inches
(solve with decimal fractions)

4'3" = 51"

Step 1. $A = \pi r^2$

Step 2. $A = 3.1416 \times 51^2$

Step 3. $A = 3.1416 \times 2601$

$A = 8171.3016$ sq. in.

Step 4. $A = 144\overline{)8171.3016}$ = 56.74515 sq. ft.

NOTE: By re-arranging the formula $A = \pi r^2$, you can find the radius if the area is known.

Example Problem. Using the same circle as above, but knowing only the area, find the radius.

Step 1. Use the formula ----→ $A = \pi r^2$

Step 2. Solve for r ---------→ $\frac{A}{\pi} = \frac{\not{\pi} r^2}{\not{\pi}}$ [$\frac{A}{\pi} = r^2$ New Formula]

[Divide both sides by π (3.1416).]

Step 3. Substitute the known values into the formula --------→ $\frac{56.75}{3.1416} = r^2$

Step 4. Calculate the folmula (do the operations indicated) ---→ $18.064 = r^2$

[56.76 ÷ 3.1416 = 18.067] $\sqrt{18.064} = r$

$4.25' = r$

Step 5. Convert 4.25' to inches and fractions.

```
4  .25'
   X12
    50
   25
  3.00"
4'3"  radius
```

Find the diameter using the formula $.7854d^2$ when the area is known.

$A = .7854d^2$

$\frac{A}{.7854} = \frac{\not{.7854}d^2}{\not{.7854}}$

$\frac{56.75}{.7854} = d^2$

$72.26 = d^2$

$\sqrt{72.26} = d$

$8.5 = d$

$8'6" = d$

CIRCUMFERENCE OF A CIRCLE — Task No. ▸ K7A

To find the circumference (distance around) of a circle, multiply pi (π) by the diameter.

RULE: Circumference equals pi times the diameter.

FORMULA: $C = \pi d$

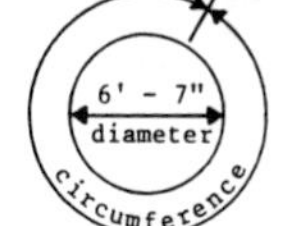

Example Problem. Find the circumference of the circle on the right.

Step 1. Use the formula ---------------▸ $C = \pi d$

Step 2. Substitute the known values into the formula ---------------▸ $C = 3.1416 \times 6.583$

Change to feet*
(solve with decimal fractions)

$6'7" = 6\frac{7}{12}$ or 6.583 feet

Step 3. Calculate the formula (do the operations indicated) --------▸

```
   3.1416
 X  6.583
    94248
   251328
  157080
 188496
20.6811528
```

The answer is ---------------▸ 20.6811528 ft.

Change to inches
(solve with decimal fractions)

6'7" = 79"

Step 1. $C = \pi d$

Step 2. $C = 3.1416 \times 79$

Step 3. $C = 3.1416 \times 79$

$C = 248.1864$ lin. in.

Step 4. $C = 12\overline{)248} = 20'8\frac{1}{8}+"$; 24; 8

Step 4. Convert 20.68' so you can read the dimension on an American (English) tape measure.

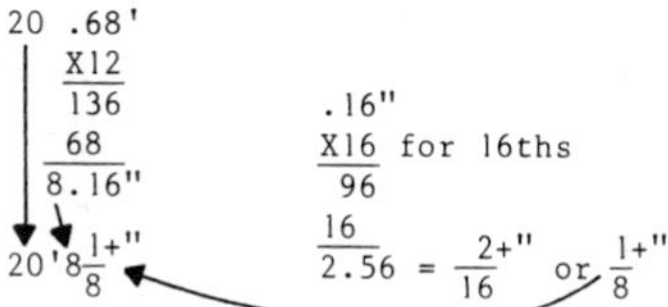

[Note the difference in accuracy when different values are used for pi and the diameter. The decimal fraction on each should be rounded off to obtain the accuracy that you need. Three places is usually close enough for most practical applications of linear measure.]

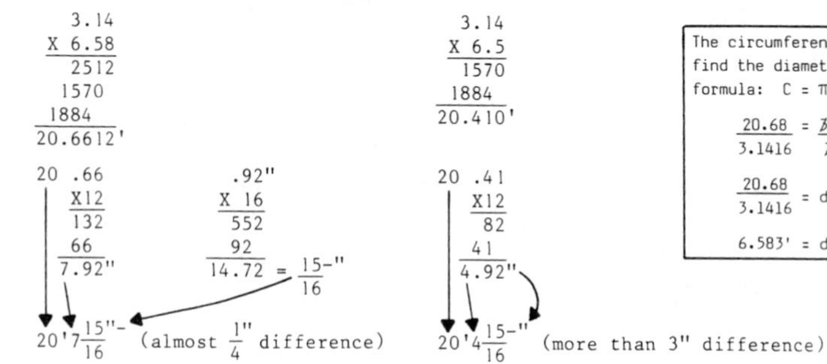

The circumference is 20.68', find the diameter. Use the formula: $C = \pi d^2$

$\frac{20.68}{3.1416} = \frac{\cancel{3.1416}d}{\cancel{3.1416}}$

$\frac{20.68}{3.1416} = d$

$6.583' = d$

*Note that the dimensions must be changed to the same units of measure before multiplying.

K7B	◂ Task No.	SURFACE AREA OF A CYLINDER

To find the cylindrical surface area of a cylinder multiply the circumference ($C = \pi d$) of the circle times the height of the cylinder.

RULE: Surface area equals pi times the diameter times the height.

FORMULA: $S = \pi dh$

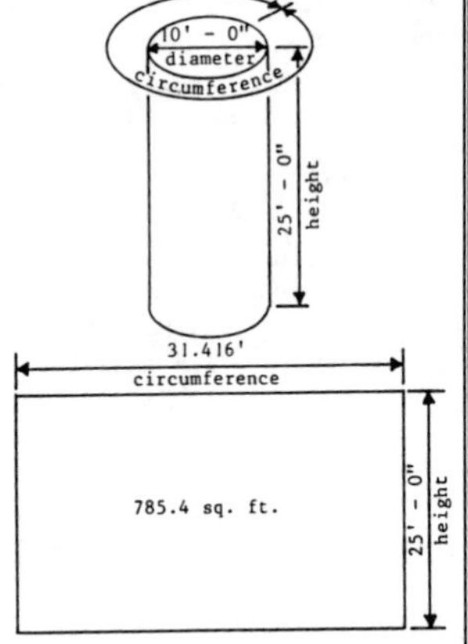

Example Problem. Find the surface area of the cylinder on the right.

Step 1. Use the formula ------------→ $S = \pi dh$

Step 2. Substitute the known values into the formula ------------→ $S = 3.1416 \times 10 \times 25$

Step 3. Calculate the formula (do the operations indicated) --------→ $S = (3.1416 \times 10) \times 25$

$S = 31.416 \times 25$

$S = 785.4$ sq. ft.

Step 4. Convert to square yards -----→ $S = 9\overline{)785.40}$ = 87.26 sq. yd.

To find the total surface area of a cylinder, find the cylindrical surface area and add the areas of the circles at each end.

In the example above, the area of each end of the cylinder is 78.54 sq. ft.

$A = \pi r^2$

$A = 3.1416 \times 5^2$

$A = 3.1416 \times 25 = 78.54$

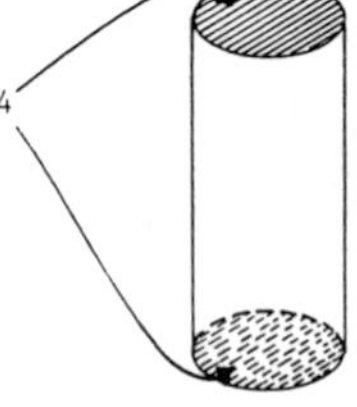

Twice 78.54 is 157.08. Add that to the cylindrical surface.

```
  785.40
+ 157.08
  942.48 sq. ft.
```

Converting to square yards, 942.48/9 = 104.72 sq. yd.

[To find the volume of a cylinder, see Task No. L7, page 164.]

AREA OF A CIRCLE SECTOR — Task No. ▸ K7C

A circle sector is a geometric figure that includes two equal radiuses and one part of the circumference the arc, of a circle.

To find the area of a circle sector, use the formula for finding the area of a circle ($A = \pi r^2$), divide by the number of degrees in a circle (360), and multiply that product by the number of degrees in the circle sector.

RULE: Area equals pi times the radius squared divided by 360 times the number of degrees in the circle sector.

FORMULAS: $A = \dfrac{\pi r^2 n}{360}$ or $\dfrac{3.1416 r^2 n}{360}$ or $A = .0087r^2n$

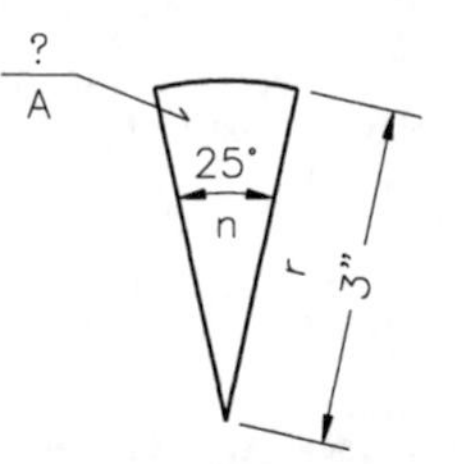

Example Problem. Find the area of the circular sector on the right.

Step 1. Use the formula --------▸ $A = .0087r^2n$

Step 2. Substitute the known values into the formula -------▸ $A = .0087 \times 3^2 \times 25$

Step 3. Calculate the formula (do the operations indicated) --▸ $A = .0087 \times (9 \times 25)$

$A = .0087 \times 225$

$A = 1.957$ sq. in.

$A = 1.96$ sq. in. (rounded)

> Area of a circle sector in terms of the diameter.
>
> $A = .7854d^2$ (formula for the area of a circle)
>
> $A = \dfrac{.7854d^2n}{360}$ or $A = .00218d^2n$
>
> Step 1. $A = .00218d^2n$
>
> Step 2. $A = .00218 \times 6^2 \times 25$
>
> Step 3. $A = .00218 \times (36 \times 25)$
>
> $A = .00218 \times 900$
>
> $A = 1.96$ sq. in.

USING BASIC ALGEBRA TO SOLVE FOR r AND TO SOLVE FOR n.

Given: $A = 1.96$

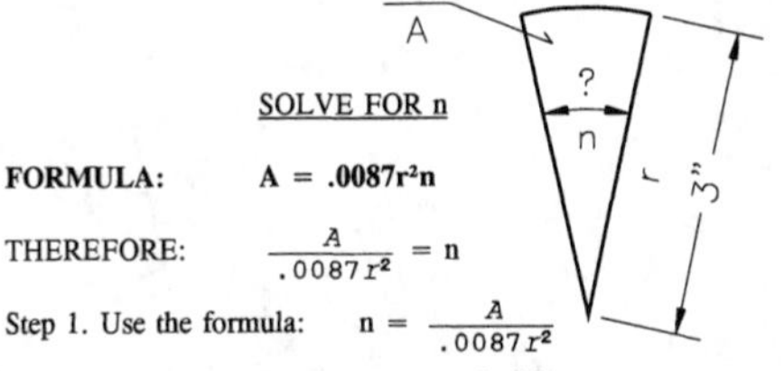

SOLVE FOR n

FORMULA: $A = .0087r^2n$

THEREFORE: $\dfrac{A}{.0087r^2} = n$

Step 1. Use the formula: $n = \dfrac{A}{.0087r^2}$

Step 2. Substitute: $n = \dfrac{1.96}{.0087 \times 3^2}$

Step 3. Calculate: $n = \dfrac{1.96}{.0087 \times 9}$

$.0783\overline{)1.9600.}$ = 25.

$n = \dfrac{1.96}{.0783}$

$n = 25$

SOLVE FOR r

FORMULA: $A = .0087r^2n$

THEREFORE: $\dfrac{A}{.0087n} = r^2$

Step 1. Use the formula: $r^2 = \dfrac{A}{.0087n}$

Step 2. Substitute: $r^2 = \dfrac{1.96}{.0087 \times 25}$

Step 3. Calculate: $r^2 = \dfrac{1.96}{.2175}$

$.2175\overline{)1.9600.}$ = 9.

$r^2 = 9$

$r = \sqrt{9}$

$r = 3$

K7D ◂ Task No. LENGTH OF A CIRCLE SECTOR ARC

To find the length of a circular sector arc, use the formula for finding the circumference of a circle ($C = \pi d$), divide by the number of degrees in a circle (360), and multiply that product by the number of degrees in the circle sector.

RULE: length equals pi times 2r (2r = the diameter), divided by 360, times the number of degrees in the circle sector.

FORMULAS: $\ell = \frac{\pi 2rn}{360}$ or $\ell = \frac{3.1416rn}{180}$ or $\ell = .01745rn$

Example Problem. Find the length of the circle sector on the right.

Step 1. Use the formula ------------→ $\ell = .01745rn$

Step 2. Substitute the known values into the formula ------------→ $\ell = .01745 \times 3 \times 25$

Step 3. Calculate the formula (do the operations indicated) -------→ $\ell = .01745 \times (3 \times 25)$

$\ell = .01745 \times 75$

$\ell = 1.308$ lin. in.

Step 4. Convert 1.308" so you can read the dimensions on an American (English) tape measure.

```
 1.308"
  X 16
  1848
  308
 5.928  =  6/16"  or  3/8"
```

The answer is ----------------→ $1\frac{3}{8}$" (linear inches)

USING BASIC ALGEBRA TO SOLVE FOR r AND TO SOLVE FOR n.

Given $\ell = 1.308$

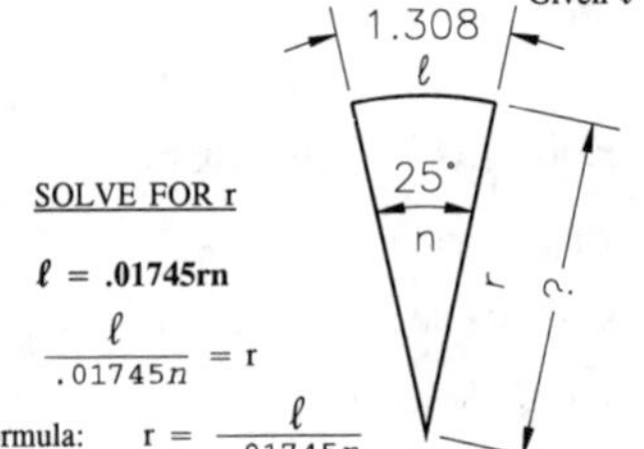

SOLVE FOR r

FORMULA: $\ell = .01745rn$

THEREFORE: $\frac{\ell}{.01745n} = r$

Step 1. Use the formula: $r = \frac{\ell}{.01745n}$

Step 2. Substitute: $r = \frac{1.308}{.01745 \times 25}$

Step 3. Calculate: $r = \frac{1.308}{.43625}$

$r = 2.9982$ or 3

SOLVE FOR n

FORMULA: $\ell = .01745rn$

THEREFORE: $\frac{\ell}{.01745r} = n$

Step 1. Use the formula: $n = \frac{\ell}{.01745r}$

Step 2. Substitute: $n = \frac{1.308}{.01745 \times 3}$

Step 3. Calculate: $n = \frac{1.308}{.05235}$

$n = 24.9857$ or 25°

AREA OF AN ELLIPSE — Task No. ▸ K7E

To find the area of an ellipse, multiply pi (π) times ½ of the short diameter times ½ of the long diameter.

RULE: Area equals pi times ½ of the short diameter (a) times ½ of the long diameter (b).

FORMULA: $A = \pi ab$

Example Problem. Find the area of the ellipse on the right.

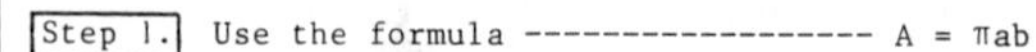

Step 1. Use the formula ------------------ $A = \pi ab$

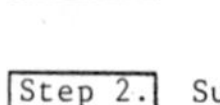

Step 2. Substitute the known values into the formula ----------------- $A = 3.14 \times 7 \times 9$

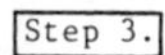

Step 3. Calculate the formula (do the operations indicated) ------------ $A = 3.14 \times (7 \times 9)$

$A = 3.14 \times 63$

$A = 197.82$ sq. in.

PERIMETER OF AN ELLIPSE — Task No. ▸ K7F

To find the approximate perimeter of an ellipse, multiply pi (π) times the sum of ½ of the short diameter times ½ of the long diameter.

RULE: Perimeter equals pi times the sum of ½ of the short diameter (a) plus ½ of the long diameter (b).

FORMULA: $P = \pi(a + b)$

Example Problem. Find the approximate perimeter of the ellipse on the right.

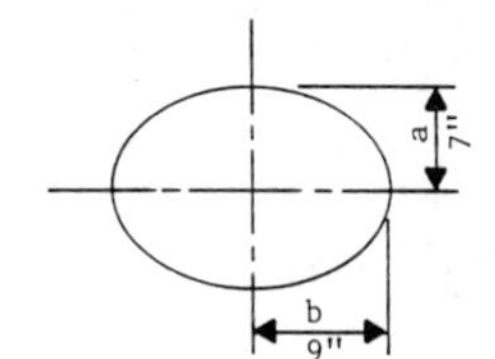

Step 1. Use the formula ------------------ $P = \pi(a + b)$

Step 2. Substitute the known values into the formula ----------------- $P = 3.14 \times (7 + 9)$

Step 3. Calculate the formula (do the operations indicated) ------------ $P = 3.14 \times (7 + 9)$

$P = 3.14 \times 16$

$P = 50.24$ linear inches approximately

[Note for greater accuracy use 3.1416 for pi.]

K8 ◂ Task No. LATERAL SURFACE AREA OF A CONE

To find the lateral surface area of a cone, multiply pi (π) times the diameter of the base times the slant height and divide by 2.

RULE: Lateral surface area equals pi times the diameter of the base times the slant height ÷ 2.

FORMULA: $L^s = \pi d s^h / 2$

Example Problem. Find the lateral surface area of the cone on the right.

Step 1. Use the formula ------------------- $L^s = \frac{\pi d s^h}{2}$

Step 2. Substitute the known values into the formula ------------------ $L^s = \frac{3.14 \times 24 \times 30}{2}$

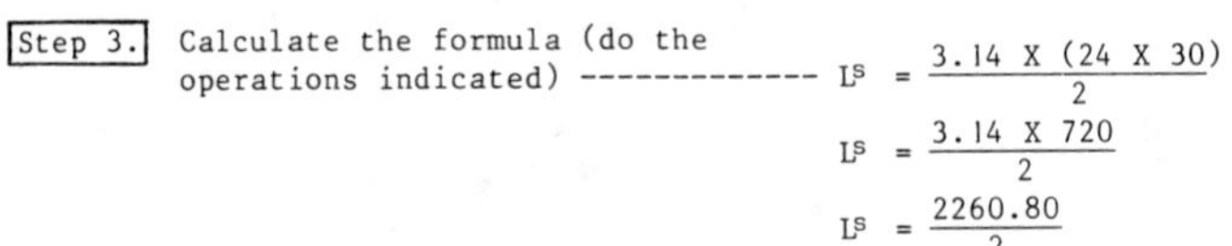

Step 3. Calculate the formula (do the operations indicated) ------------- $L^s = \frac{3.14 \times (24 \times 30)}{2}$

$L^s = \frac{3.14 \times 720}{2}$

$L^s = \frac{2260.80}{2}$

$L^s = 1130.40$ sq. in.

Step 4. Find the total surface area. $L^s = 1130.40 + 452.16$

Total surface area = 1582.56 sq. in.

Area of the Base

$A = \pi r^2$
$A = 3.14 \times 12^2$
$A = 3.14 \times 144$
$A = 452.16$ sq. in.

30"
24"

K8A ◂ Task No. LATERAL SURFACE AREA OF A FRUSTUM OF A CONE

To find the lateral surface area of a frustum of a cone, multiply pi (π) times the slant height times the sum of the large diameter plus the small diameter and divide by 2.

RULE: Lateral Surface area equals pi times the slant height times the sum of the large diameter (d) plus the small diameter (d') divided by 2.

FORMULA: $L^s = \pi s^h (d + d')/2$

Example Problem. Find the lateral surface area of the frustum of the cone on the right.

8"
10"
12"

Step 1. Use the formula ------------------- $L^s = \frac{\pi s^h (d + d')}{2}$

Step 2. Substitute the known values into the formula ------------------ $L^s = \frac{3.14 \times 10(12 + 8)}{2}$

Step 3. Calculate the formula (do the operations indicated) ------------- $L^s = \frac{3.14 \times 10 \times 20}{2}$

$L^s = \frac{3.14 \times 200}{2}$

$L^s = \frac{628}{2}$

$L^s = 314$ sq. in.

Step 4. Find the total surface area. $L^s = 314.00 + 113.04 + 50.24$

Total surface area = 477.28 sq. in.

Area of the Bases

$A = \pi r^2$	$A = \pi r^2$
$A = 3.14 \times 6^2$	$A = 3.14 \times 4^2$
$A = 3.14 \times 36$	$A = 3.14 \times 16$
$A = 113.04$ sq. in.	$A = 50.24$ sq. in.

LATERAL SURFACE AREA OF A PYRAMID — Task No. ▸ K9

To find the lateral surface area of a pyramid, multiply the perimeter of the base times the slant height and divide by 2.

RULE: Lateral surface area equals the perimeter of the base times the slant height divided by 2.

FORMULA: $L^s = ps^h/2$

14"
14"
17"

Example Problem. Find the lateral surface area of the pyramid on the right.

Step 1. Use the formula ------------------- $L^s = \frac{ps^h}{2}$

Step 2. Substitute the known values into the formula ------------------- $L^s = \frac{56 \times 17}{2}$
[base perimeter = 4 X 14 = 56]

Step 3. Calculate the formula (do the operations indicated) ------------- $L^s = \frac{952}{2}$

$L^s = 476$ sq. in.

Area of the Base

A = lw
A = 14 X 14
A = 196 sq. in.

Step 4. Find the total surface area. $L^s = 476 + 196$

Total surface area = 672 sq. in.

LATERAL SURFACE AREA OF A FRUSTUM OF A PYRAMID — Task No. ▸ K9A

To find the lateral surface area of a frustum of a pyramid, multiply the slant height times the sum of the perimeter of the lower base (b) plus the perimeter of the upper base (b') and divide by 2.

RULE: Lateral surface area equals the slant height times the sum of the lower base (b) plus the upper base (b') divided by 2.

FORMULA: $L^s = s^h(b + b')/2$

7"
7"
8"
10"
10"

Example Problem. Find the lateral surface area of the frustum of the pyramid on the right.

Step 1. Use the formula ------------------- $L^s = \frac{s^h(b + b')}{2}$

Step 2. Substitute the known values into the formula ------------------- $L^s = \frac{8(40 + 28)}{2}$
[base perimeters = 4 X 10 = 40 and 4 X 7 = 28]

Step 3. Calculate the formula (do the operations indicated) ------------- $L^s = \frac{8 \times 68}{2}$

$L^s = \frac{544}{2}$

$L^s = 272$ sq. in.

Area of the Bases

A = lw	A = lw
A = 10 X 10	A = 7 X 7
A = 100 sq. in.	A = 49 sq. in.

Step 4. Find the total surface area. $L^s = 272 + 100 + 49$

Total surface area = 421 sq. in.

K10 ◂ Task No. **SURFACE AREA OF A SPHERE**

To find the surface area of a sphere, multiply pi (π) times the diameter squared.

RULE: Surface area equals pi times the diameter squared.

FORMULA: $S = \pi d^2$

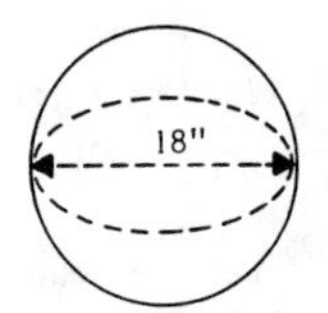

Example Problem. Find the surface area of the sphere on the right.

Step 1. Use the formula -------------- $S = \pi d^2$

Step 2. Substitute the known values into the formula ------------- $S = 3.14 \times 18^2$

Step 3. Calculate the formula (do the operations indicated) -------- $S = 3.14 \times (18 \times 18)$

$S = 3.14 \times 324$

$S = 1017.36$ sq. in.

Step 4. Convert to square feet. $S = 1017.36 \div 144 = 7.065$ sq. ft.

Or Use the Formula

$S = 4\pi r^2$

$S = 4 \times 3.14 \times 9^2$

$S = 4 \times 3.14 \times (9 \times 9)$

$S = 4 \times (3.14 \times 81)$

$S = 4 \times 254.34$

$S = 1017.36$ sq. in.

Note: For greater accuracy use 3.1416 for pi.

"This page intentionally blank."

REVIEW AND PRACTICAL APPLICATION OF BASIC GEOMETRY

This apply-it module is designed to put into perspective and to summarize all of the preceding plane geometry tasks. It also will serve as a review for other plane geometry concepts not covered or included as a task in this dictionary. It can be used as a quick reference and a study guide of the basics of plane geometry.

Area

The area of a geometric figure is the amount of surface it covers. The surface it covers is in the same plane. Square units are used in measuring area.

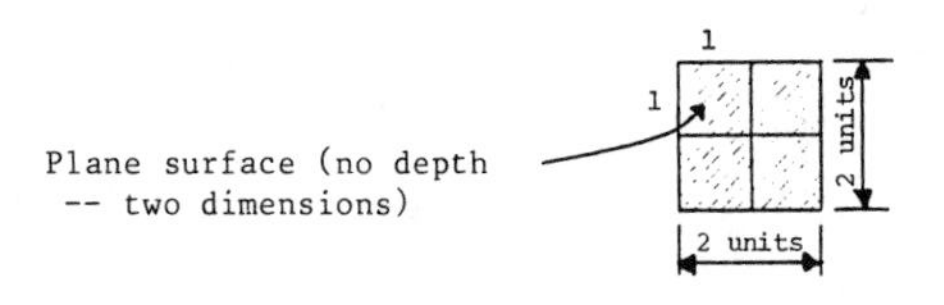

Note

There are 4 square units (each unit is 1 X 1) of area covered in the figure to the left.

Volume

Volume is the space taken up by a geometric figure that has three dimensions and is not in the same plane. Cubic units are used in measuring volume.

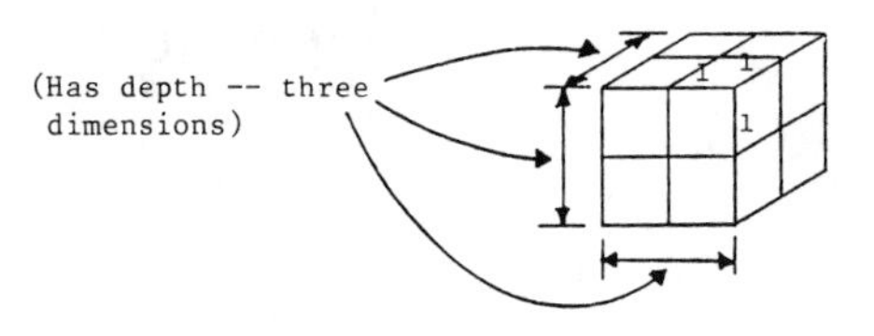

Note

There are 8 cubic units (each unit is 1 X 1 X 1) of the space (volume) taken up in the figure on the left.

Angles

An angle is the opening between two meeting lines or any opening between lines that cross each other.

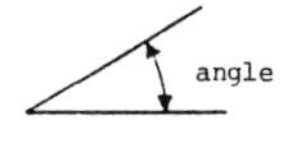

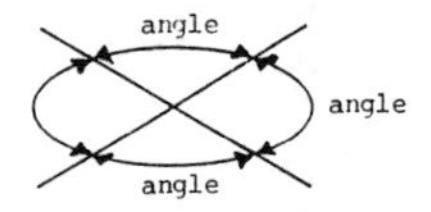

The sides of an angle are the lines that form the angle.

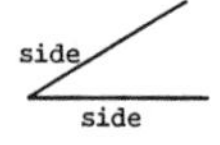

The vertex of an angle is the point where the sides meet.

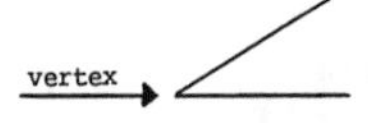

Degree is the unit of measure used to measure angles.

The sign of degree or degrees is °. Thus, 1° = 1 degree, 5° = 5 degrees.

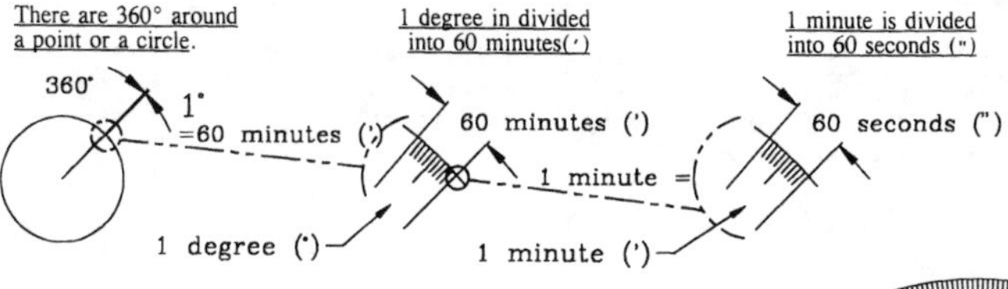

A protractor is used to measure angles in degrees.

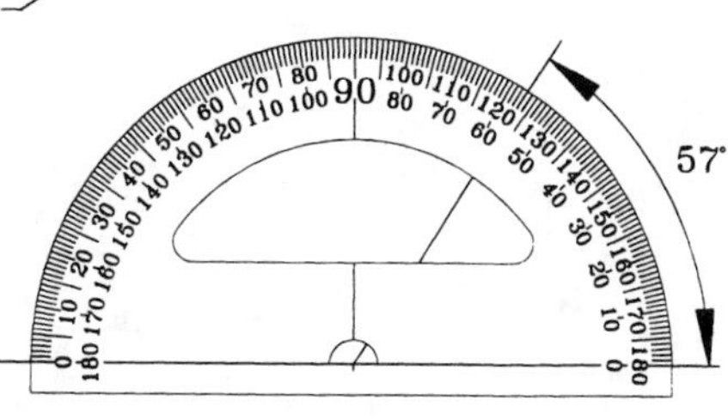

Protractor

A vernier protractor is used to measure angles in degrees and minutes. In this example, read 33 degrees 50 minutes. ⟶

[See Task No. Q2, pages 334 and 335 for further information.]

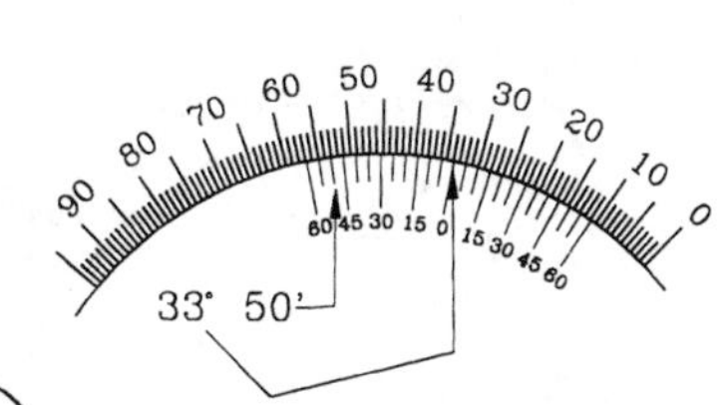

Types of angles

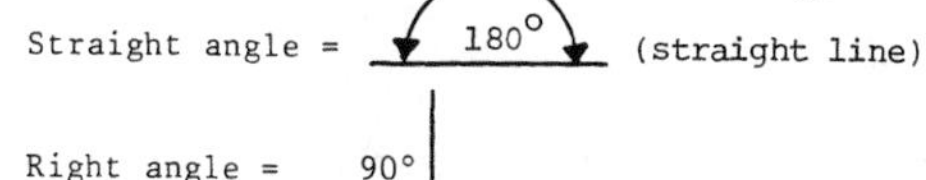

Right angle = 90°

Acute angle = 60° (any angle less than 90°)

Also called oblique angles

Obtuse angle = 150° (any angle more than 90°)

The complement of an angle is the angle, which, when added to another angle, equals 90°.

The complement of angle c (65°) is angle b (25°) because 65° + 25° = 90°.

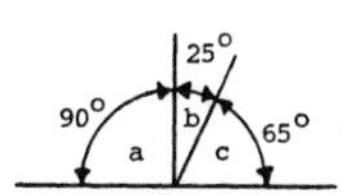

The supplement of an angle is the angle, which, when added to another angle, equals 180°.

The supplement of angle b (120°) is angle a (60°) because 120° + 60° = 180°.

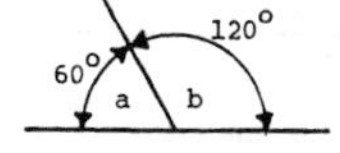

Polygons

Polygons are geometric figures that are made up of straight lines with closed sides that lie in a plane (flat surface).

Types of polygons: Parallelograms are quadrilaterals (figures with four sides) that have opposite sides that are parallel (lines the same distance away from each other at all points and which, therefore, will never meet).

Types of Parallelograms

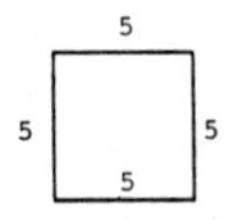

Square
Opposite sides are parallel. All sides are equal. All angles are 90°.

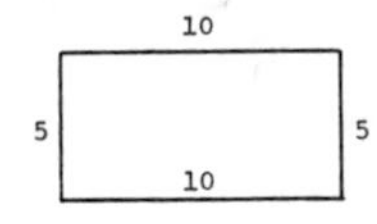

Rectangle
Opposite sides are parallel. Adjacent sides are not equal. All angles are 90°.

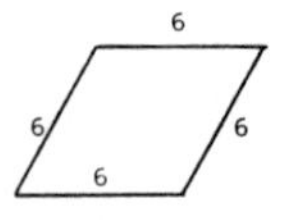

Rhombus
Opposite sides are parallel. All sides are equal. No right angles.

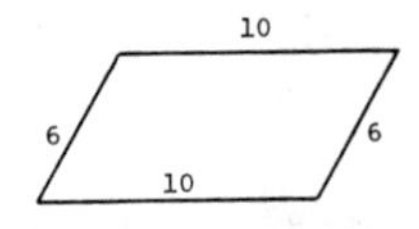

Rhomboid
Opposite sides are parallel. Adjacent sides are not equal. No right angles.

[To find the area of a square or rectangle see Task Number K2 .]

[To find the area of a rhombus or rhomboid see Task Number K3 .]

Trapezoids are quadrilaterals (figures with four sides that have only two sides that are parallel).

Examples of Trapezoids

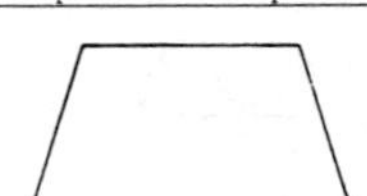

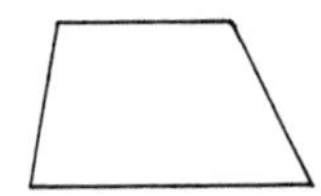

[To find the area of a trapezoid see Task Number K4 .]

Trapeziums are quadrilaterals (figures with four sides) that have no two sides parallel.

Examples of Trapeziums

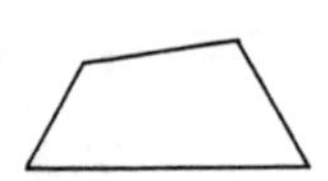

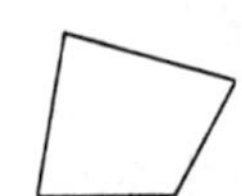

[To find the area of a trapezium see Task Number K4A , or use Task Number K4 area of a trapezoid, and Task Number K5 , area of a triangle, together.]

Triangles are polygons that have three sides. (All angles of a triangle add up to 180°.)

Types of Triangles

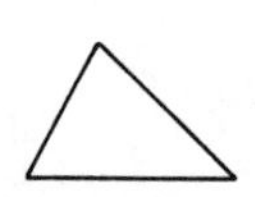

Scalene
No two sides are equal.

Isosceles
Two sides equal.

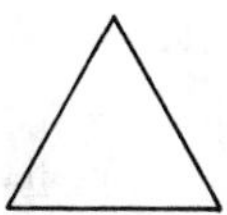

Equilateral
All sides equal

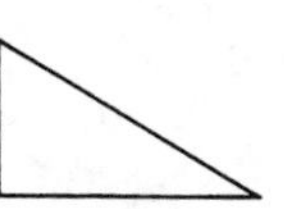

Right triangle
One angle is 90°.

Note: These triangles are also known as oblique because they have no angles that are 90°.

Oblique Triangles

Acute: A triangle with all angles less than 90°.

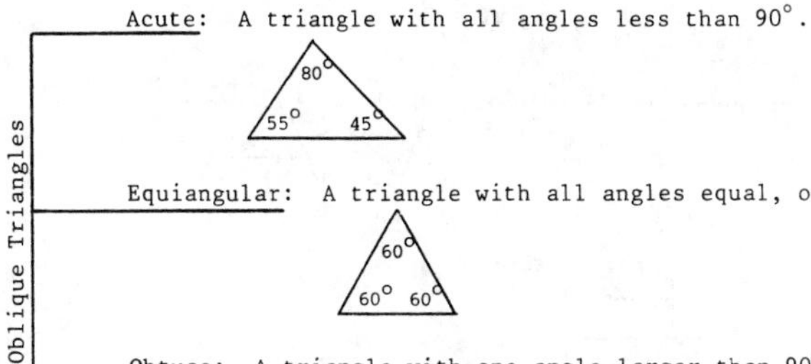

Equiangular: A triangle with all angles equal, or 60°.

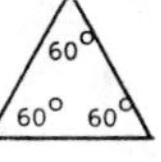

Obtuse: A triangle with one angle larger than 90°.

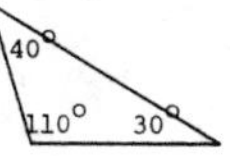

Right: A triangle with one right angle (90°) and two acute angles (less than 90°).

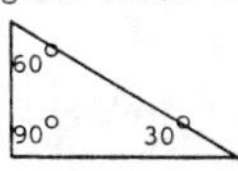

[To find the area of a triangle see Task Number K5 .]

The perimeter of any polygon is the distance around the figure (the sum of all the sides).

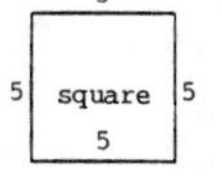

Perimeter equals 5 + 5 + 5 + 5 = 20

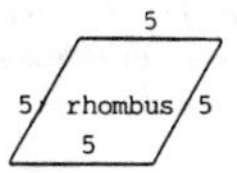

Perimeter equals 5 + 5 + 5 + 5 = 20

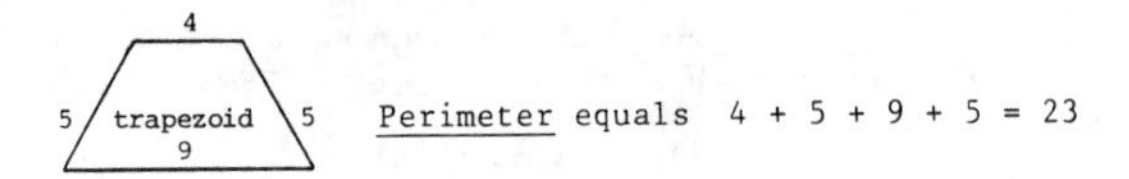

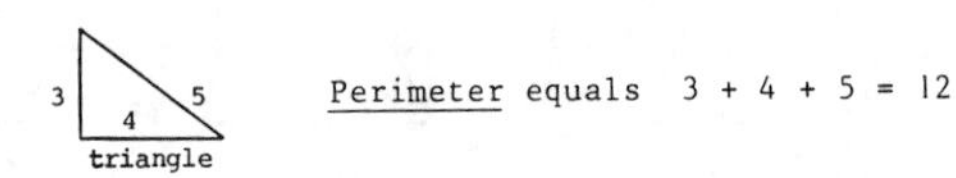

Parts of a Polygon

Vertex is a point where two lines meet.

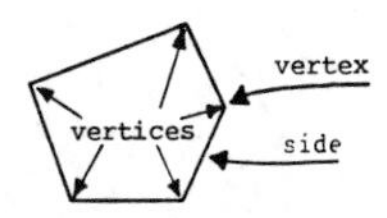

Vertices is the plural of vertex.

Diagonal is a line that joins two non-consecutive vertices.

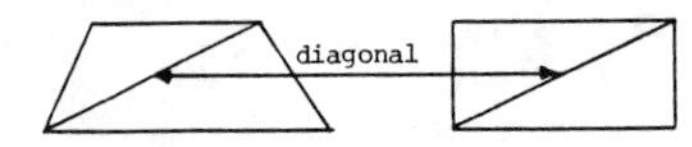

Generally, the base of a geometric figure is the side on which it seems to stand. A trapezoid, however, is considered to have both a top and a bottom base.

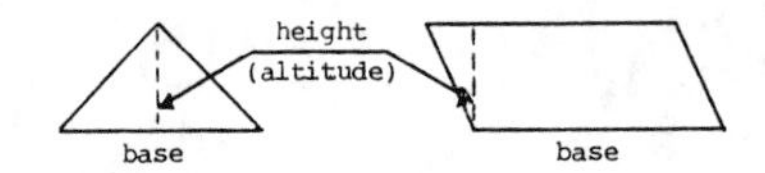

The height (altitude) of of a polygon is the length of the line drawn from the highest point to the base, forming a right angle with the base. The height is said to be perpendicular to the base.

A perpendicular is a line that is 90° from another line.

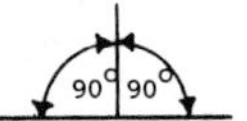

Circle and Circumference

Definitions: A circle is a closed plane curve all points of which are the same distance from a point inside the curve. That point is called its center.

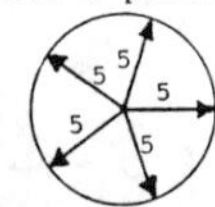

[There are 360° around a circle. See page 369 for more information.]

Definitions: (continued)

A semi circle is one-half of a circle.

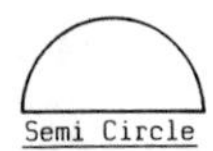

[There are 180° around a semi circle.]

The radius of a circle is a line drawn from the center of the circle to the curved line.

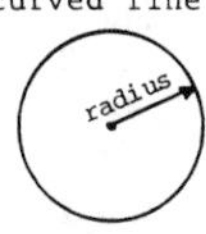

[To find the area of a circle, see Task Number K7, page 140.]

The diameter is a line drawn through the center of the circle, going completely across the circle. The diameter of a circle is twice the radius.

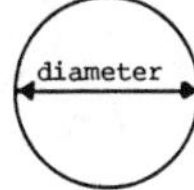

The circumference of a circle is the distance around the circle.

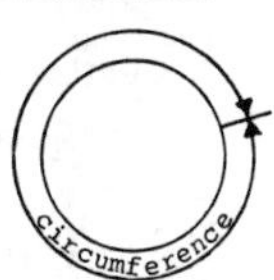

[To find the circumference of a circle, see Task Number K7A, page 141.]

Ellipse

Definition:

An ellipse is a closed plane curve with two focuses and shaped so that the sum of the distances from any point on the curve to the two focuses is always equal.

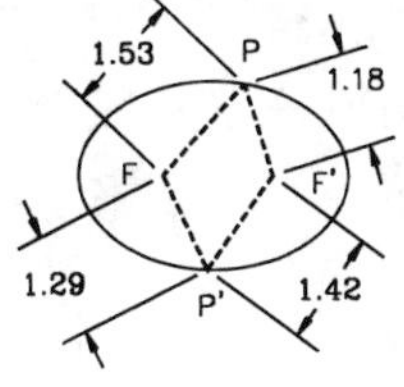

[To find the area of an ellipse, see Task Number K7E, page 145. To find the perimeter of an ellipse, see Task Number K7F, page 145. To lay out an ellipse, see Task Number O10, page 311.]

[PF + PF' = P'F + P'F']
[1.53 + 1.18 = 1.29 + 1.42]
[2.71 = 2.71]

Solid geometry is the part of mathematics which deals with three-dimensional figures.

Volume is the space taken up by a geometric figure that has three dimensions (not in one plane). Cubic units are used in measuring volume.

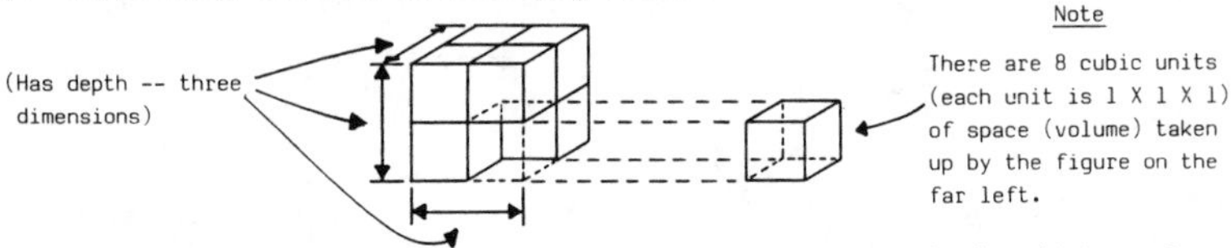

A cube is a geometric figure that is one unit in length, one unit in width, and one unit in height. All six sides are square and equal. Each is parallel to its opposite side and perpendicular to the other four sides.

In solid geometry, measurements are not of just one plane surface. For example, a cube or rectangular solid has three dimensions called the length, the width, and the height (or depth). To find the amount of space (called volume) encompassed within its dimensions, multiply the length times the width times the height. Therefore comes the rule, volume equals the length (l) times the width (w) times the height (h). From the rule comes the formula, $V = lwh$.

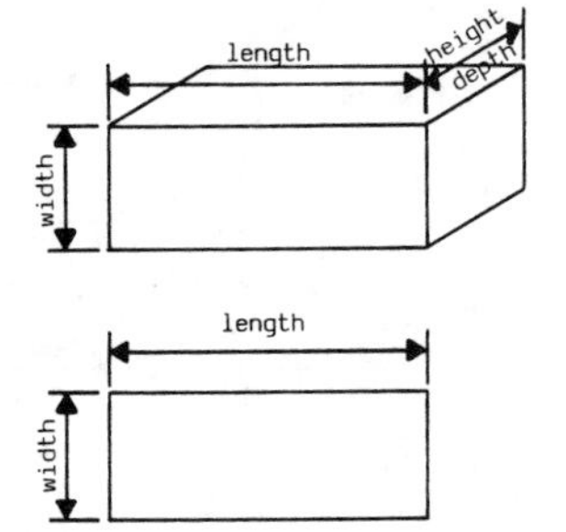

[In reviewing, remember that plane geometry is surface measurement (all measurements relate to one plane surface). For example, a square or rectangle has two dimensions called the length and the width. To find the amount of surface space (called area) encompassed within its dimensions, multiply the length times the width. Therefore comes the rule, area equals the length (l) times the width (w). From the rule comes the formula, $A = lw$.

A rectangular solid is a geometric figure that has six rectangular faces that meet to form right angles (90°). Cubic units are used to find the volume of rectangular solids.

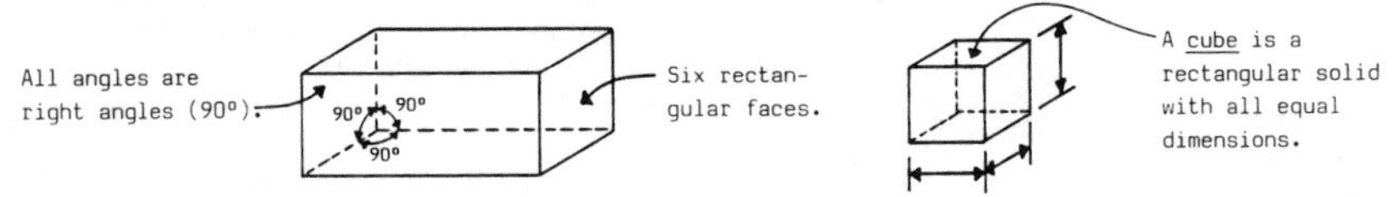

A prism is a solid geometric figure that has equal and parallel bases at each end.

A right prism is a solid in which the bases or ends are perpendicular (90°) to the sides. Prisms are named from their bases because all the sides are parallelograms.

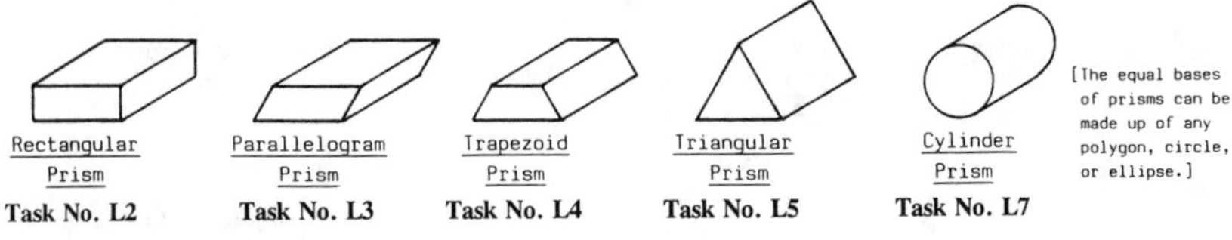

A cylinder is a right prism with a circular base.

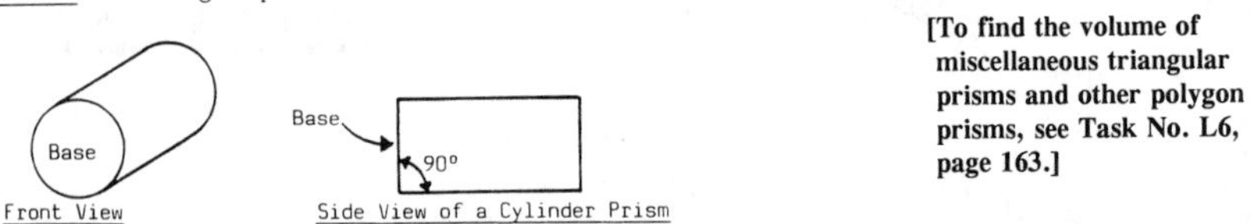

[To find the volume of miscellaneous triangular prisms and other polygon prisms, see Task No. L6, page 163.]

A pyramid is a geometric figure (solid) whose base is a polygon and whose sides are equal triangles that meet in a common point, or vertex.

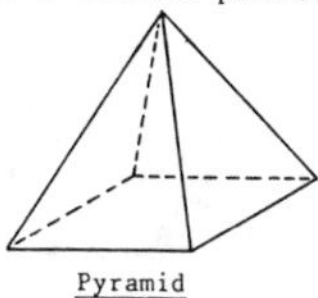

Pyramid

[To find the lateral surface area of a pyramid, see Task No. K9, page 147.]

[To find the volume of a pyramid, see Task No. L9, page 166.]

The frustum of a pyramid is the part of the pyramid left after a uniform top has been cut off. The top of a frustum of a pyramid is parallel to its base.

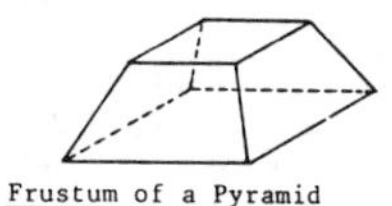

Frustum of a Pyramid

[To find the lateral surface area of a frustum of a pyramid, see Task No. K9A, page 147.]

[To find the volume of a frustum of a pyramid, see Task No. L9A, page 166.]

A cone is a geometric figure (solid) whose base is a circle and whose curved surface tapers uniformly to a common point, or vertex.

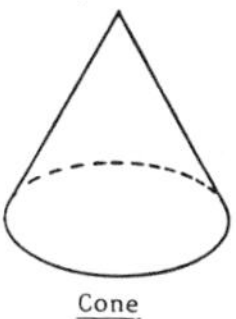

Cone

[To find the lateral surface area of a cone, see Task No. K8, page 146.]

[To find the volume of a cone, see Task No. L8, page 165.]

The frustum of a cone is the part of the cone left after a uniform top has been cut off. the top of a frustum of a cone is parallel to its base.

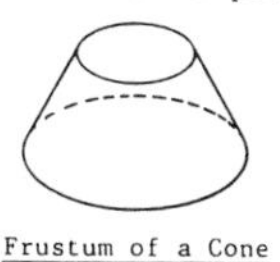

Frustum of a Cone

[To find the lateral surface area of a frustum of a cone, see Task No. K8A, page 146.]

[To find the volume of a frustum of a cone, see Task No. L8A, page 165.]

A sphere is a geometric figure (solid) whose surface at every point is equally distant from the center of the sphere. A sphere is a perfectly round ball.

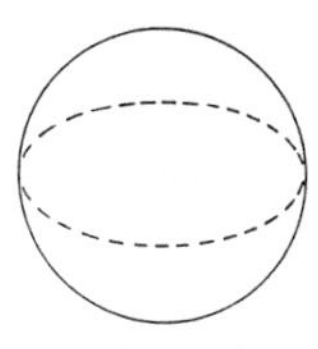

Sphere

[To find the surface area of a sphere, see Task No. K10, page 148.]

[To find the volume of a sphere, see Task No. L10, page 167.]

L2 ◄ Task No. VOLUME OF A CUBE OR RECTANGULAR PRISM

To find the volume of a cube or rectangular solid, multiply the length by the width by the height.

RULE: Volume equals the length times the width times the height.

FORMULA: V = lwh

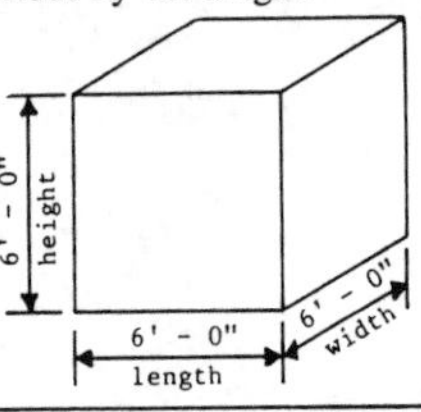

Example
Problem 1. Find the volume of the cube on the right.

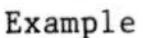

Step 1. Use the formula ------------→ V = lwh

Step 2. Substitute the known values into the formula ------------→ V = 6 X 6 X 6

Step 3. Calculate the formula (do the operations indicated) -------- V = 6 X (6 X 6)

V = 6 X 36

V = 216 cubic feet

Or use the formula:
V = side X side X side
$V = S^3$
V = 6 X 6 X 6
V = 216 cu. ft.

Step 4. Convert to cubic yards. Divide by 27 ----------------→ V = 8 cubic yards

$$27\overline{)216} = 8, \quad 216$$

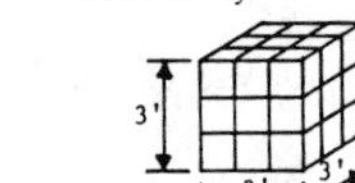

1 cu. yd. = 27 cu. ft.

Example
Problem 2. Find the volume of the rectangular prism on the right.

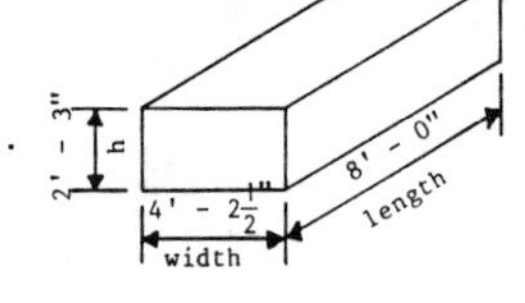

Step 1. Use the formula ------------→ V = lwh

Change to feet*
(solve with decimal fractions)

$4'2\frac{1}{2}" = 4$ and $\frac{2\frac{1}{2}}{12}$ or $\frac{2.5}{12}$ or $12\overline{)2.5} = .208$ = 4.208 ft.

$2'3" = 2\frac{3}{12}$ or $2\frac{1}{4}$ or 2.25 ft.

Step 2. Substitute the known values into the formula ------------→ V = 8 X 4.208 X 2.25

Step 3. Calculate the formula (do the operations indicated) -------→ V = 8 X (4.208 X 2.25)

V = 8 X 9.468

V = 75.744 cu. ft.

Step 4. Convert to cubic yards and to gallons. ------→ V = 2.805 cu. yd.

$$27\overline{)75.744} = 2.805$$

(1 cu. ft. = approx. 7.5 gallons)

```
   75.744
    X 7.5
   378720
  530208
 568.0800 gallons
```

***Note that the dimensions must be changed to the same units of measure before multiplying.**

VOLUME OF PARALLELOGRAM PRISMS — Task No. ▸ L3

To find the volume of a parallelogram prism, multiply the base by the height by the length.

RULE: Volume equals the base times the height times the length.

FORMULA: V = bhl

Example
Problem 1. Find the volume of the parallelogram on the right.

13" height 20" base 24" length

Step 1. Use the formula ------------▶ V = bhl

Step 2. Substitute the known values into the formula -----------▶ V = 20 X 13 X 24

Step 3. Calculate the formula (do the operations indicated) ------▶ V = (20 X 13) X 24

V = 260 X 24

V = 6240 cu. in.

Step 4. Convert to cubic feet ------▶ V = $1728\overline{)6240}$ = 3.61 cu. ft. and to gallons. $231\overline{)6240}$ = 27.01 gallons

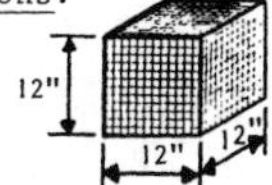

1 cu. ft. = 1728 cu. in.

1 gallon = 231 cu. in.

Example
Problem 2. Find the volume of the parallelogram on the right.

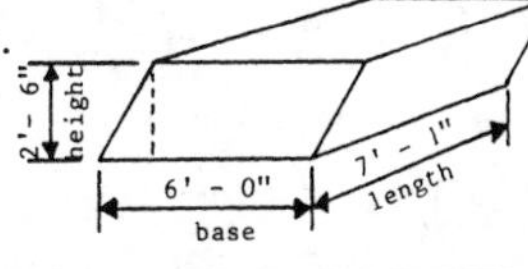

Step 1. Use the formula ------------▶ V = bhl

Step 2. Substitute the known values into the formula -----------▶ $V = 6 \times 2\frac{1}{2} \times 7\frac{1}{12}$

Change to feet*
(solve with mixed numbers)

$2'6" = \frac{6}{12}$ or $2\frac{1}{2}$ ft. and $7'1" = 7\frac{1}{12}$

Step 3. Calculate the formula (do the operations indicated) ------▶ $V = \frac{6}{1} \times \frac{5}{2} \times \frac{85}{12}$

$V = (\frac{\not 6^{\,3}}{1} \times \frac{5}{\not 2_{\,1}}) \times \frac{85}{12}$

$V = \frac{\not{15}^{\,5}}{1} \times \frac{85}{\not{12}_{\,4}} = \frac{425}{4}$ = 106.25 cu. ft.

Change to feet (solve with decimal fractions)
$2\frac{1}{2}$ = 2.5 ft. and $7\frac{1}{12}$ = 7.083 ft.
Step 1. V = bhl
Step 2. V = 6 X 2.5 X 7.083
Step 3. V = 6 X (2.5 X 7.083)
V = 6 X 17.7075
V = 106.245 cu. ft.

Step 4. Convert to cubic yards -----▶ V = $27\overline{)106.25}$ = 3.93 cu. yd.

```
   106.25
  X   7.5
   53125
  74375
  796.875 gallons
```

(27 cu. ft. = 1 cu. yd.)   (1 cu. ft. = 7.5 gallons)

***Note that the dimensions must be changed to the same units of measure before multiplying.**

L4	◂ Task No.	VOLUME OF A TRAPEZOID PRISM

To find the volume of a trapezoid prism, multiply ½ the sum of the bases by the height by the length.

RULE: Volume equals ½ the sum of the bases X the height X the length.

FORMULA: $V = \frac{1}{2}(b + b')hl$

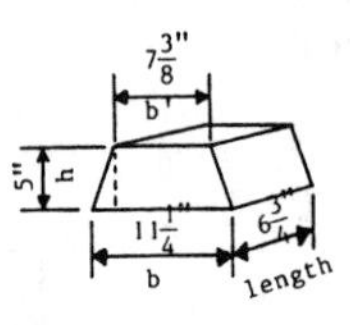

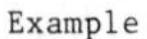
Example

Problem 1. Find the volume of the trapezoid prism on the right.

Step 1. Use the formula ---------▶ $V = \frac{1}{2}(b + b')hl$

Step 2. Substitute the known values into the formula ---------▶ $V = \frac{1}{2}(11\frac{1}{4} + 7\frac{3}{8}) \times 5 \times 6\frac{3}{4}$

Step 3. Calculate the formula (do the operations indicated) ---▶ $V = \frac{1}{2}(18\frac{5}{8}) \times (\frac{5}{1} \times \frac{27}{4})$

$$11\frac{1}{4} = \frac{2}{8}, \quad + 7\frac{3}{8} = \frac{3}{8}, \quad 18\frac{5}{8}$$

$$\frac{5}{1} \times \frac{27}{4} = \frac{135}{4} \qquad \frac{149}{8} \times \frac{135}{4} = \frac{20115}{32}$$

$$V = \frac{1}{2}(\frac{149}{8} \times \frac{135}{4})$$

$$V = \frac{1}{2} \times \frac{20115}{32} = \frac{20115}{64}$$

$$V = 64/\overline{20115.00000} = 314.29687 \text{ cu. in.}$$

Step 4. Convert to gallons. (1 gallon = 231 cu. in.)

$$231/\overline{314.30} = 1.36 \text{ gallons}$$

Solve with decimal fractions

$$V = \frac{1}{2}(b + b')hl$$

$$V = \frac{1}{2}(11.25 + 7.375) \times (5 \times 6.75)$$

$$V = \frac{1}{2}(18.625) \times (33.75)$$

$$V = \frac{1}{2}(18.625 \times 33.75)$$

$$V = \frac{1}{2} \times 628.59375$$

$$V = 2/\overline{628.59375} = 314.29687 \text{ cu. in.}$$

Example

Problem 2. Find the volume of the trapezoid prism on the right.

3'- 2"
b'
2'- 0"
height
4' - 7"
b
5'- 6"
length

Step 1. Use the formula ----------▶ $V = \frac{1}{2}(b + b')hl$

Change to feet*
(solve with mixed numbers)

$4'7" = 4\frac{7}{12}$ ft., $3'2" = 3\frac{2}{12}$ or $3\frac{1}{6}$ ft., and $5'6" = 5\frac{6}{12}$ or $5\frac{1}{2}$ ft.

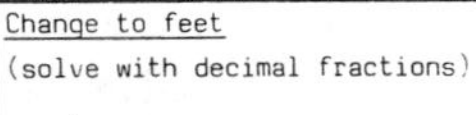
Change to feet
(solve with decimal fractions)

$$V = \frac{1}{2}(b + b')\,hl$$

$$V = \frac{1}{2}(4.583 + 3.166) \times (2 \times 5.5)$$

$$V = \frac{1}{2}(7.649) \times (11)$$

$$V = \frac{1}{2}(7.649 \times 11)$$

$$V = \frac{1}{2} \times 84.139$$

$$V = 2/\overline{84.139} = 42.069 \text{ cu. ft.}$$

Note: reduced accuracy due to rounding

Step 2. Substitute the known values into the formula ---------▶ $V = \frac{1}{2}(4\frac{7}{12} + 3\frac{1}{6}) \times 2 \times 5\frac{1}{2}$

Step 3. Calculate the formula (do the operations indicated) ----▶ $V = \frac{1}{2}(4\frac{7}{12} + 3\frac{1}{6}) \times (\frac{\not{2}}{1} \times \frac{11}{\not{2}})$

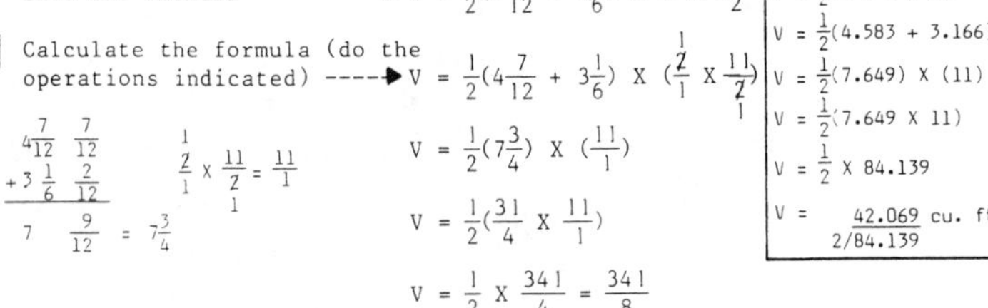

$$4\frac{7}{12} = \frac{7}{12}, \quad +3\frac{1}{6} = \frac{2}{12}, \quad 7\frac{9}{12} = 7\frac{3}{4}$$

$$\frac{\not{2}}{1} \times \frac{11}{\not{2}} = \frac{11}{1}$$

$$V = \frac{1}{2}(7\frac{3}{4}) \times (\frac{11}{1})$$

$$V = \frac{1}{2}(\frac{31}{4} \times \frac{11}{1})$$

$$V = \frac{1}{2} \times \frac{341}{4} = \frac{341}{8}$$

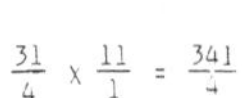

$$\frac{31}{4} \times \frac{11}{1} = \frac{341}{4}$$

$$V = 8/\overline{341} = 42.625 \text{ cu. ft.}$$

Step 4. Convert to cubic yards ---▶ $V = 27/\overline{42.625} = 1.578$ cu. yd. and to gallons.

```
   42.625 cu. ft.
  X  7.5
   213125
  298375
V = 319.6875 gallons
```

***Note that the dimensions must be changed to the same units of measure before multiplying.**

VOLUME OF A TRAPEZIUM PRISM — Task No. ▸ L4A

To find the volume of a trapezium prism, find the sum of the area of the triangle ($A = \frac{1}{2}bh$) plus the area of the trapezoid [$A = \frac{1}{2}(b + b')h$] times the length.

FORMULAS:

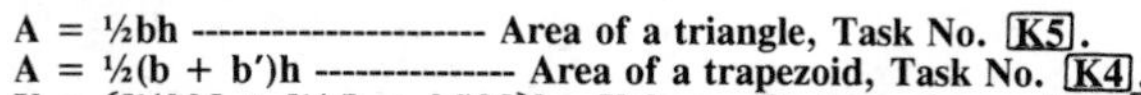

$A = \frac{1}{2}bh$ ---------------------- Area of a triangle, Task No. K5.
$A = \frac{1}{2}(b + b')h$ --------------- Area of a trapezoid, Task No. K4.
$V = \{[\frac{1}{2}bh] + [\frac{1}{2}(b + b')h]\}l$ -- Volume of a trapezium prism, L4A.

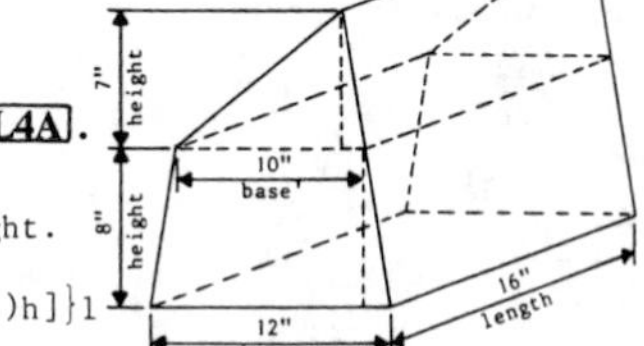

Example Problem. Find the volume of the trapezium prism on the right.

Step 1. Use the formula ----------- $V = \{[\frac{1}{2}bh] + [\frac{1}{2}(b + b')h]\}l$

Step 2. Substitute the known values into the formula ---------- $V = \{[\frac{1}{2} \times 10 \times 7] + [\frac{1}{2} \times (12 + 10) \times 8]\} \times 16$

Step 3. Calculate the formula (do the operations indicated) ----- $V = \{[\frac{1}{2} \times (10 \times 7)] + [\frac{1}{2} \times (12 + 10) \times 8]\} \times 16$

$V = \{[\frac{1}{2} \times 70] + [\frac{1}{2} \times 22 \times 8]\} \times 16$

$V = \{[35] + [\frac{1}{2} \times (22 \times 8)]\} \times 16$

$V = \{[35] + [\frac{1}{2} \times 176]\} \times 16$

$V = \{35 + 88\} \times 16$

$V = 123 \times 16$

$V = 1968$ cu. in. Answer

Or find the sum of the formulas separately, and then multiply that sum times the length.

Step 1. Use the formula --------------- $A = \frac{1}{2}bh$

Step 2. Substitute -------------------- $A = \frac{1}{2} \times 10 \times 7$

Step 3. Calculate --------------------- $A = \frac{1}{2} \times (10 \times 7)$

$A = \frac{1}{2} \times 70$

$A = 35$ sq. in.

Step 4. Use the formula --------------- $A = \frac{1}{2}(b + b')h$

Step 5. Substitute -------------------- $A = \frac{1}{2} \times (12 + 10) \times 8$

Step 6. Calculate --------------------- $A = \frac{1}{2} \times 22 \times 8$
[12 + 10 = 22]

$A = \frac{1}{2} \times (22 \times 8)$

$A = \frac{1}{2} \times 176$

$A = 88$ sq. in.

Step 7. Add Step 3 ------------------------ 35 sq. in.
and Step 6 ---------------------- + 88 sq. in.
123 sq. in.

Step 8. Multiply 123 times 16.

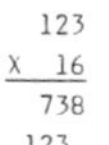

```
   123
 X  16
   738
  123
```

V = 1968 cu. in. Answer

L5 ◂ Task No. VOLUME OF A TRIANGULAR PRISM

To find the volume of a triangular prism, multiply ½ the base by the height by the length.

RULE: Volume equals ½ times the base times the height times the length.

FORMULA: V = ½bhl

Example Problem 1. Find the volume of the triangular prism on the right.

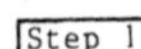

Step 1. Use the formula ----------→ $V = \frac{1}{2}bhl$

Step 2. Substitute the known values into the formula ----------→ $V = \frac{1}{2} \times 17 \times 13 \times 23\frac{1}{2}$

Step 3. Calculate the formula (do the operations indicated) ------→ $V = (\frac{1}{2} \times \frac{17}{1}) \times (13 \times 23.5)$

$V = \frac{17}{2} \times \frac{305.5}{1} = \frac{5193.5}{2}$

$V = 2596.75$ cu. in. (2/5193.5)

Step 4. Convert to cubic feet and to gallons.

$V = 1.50$ cu. ft. (1728/2596.75)

11.24 gallons (231/2596.75)

Example Problem 2. Find the volume of the triangular prism on the right.

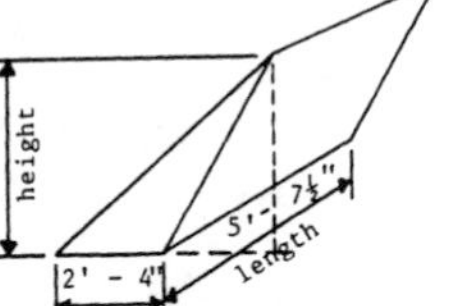

Step 1. Use the formula ----------→ $V = \frac{1}{2}bhl$

Change to feet*
(solve with decimal fractions)

$2'4" = 2\frac{4}{12}$ or $2\frac{1}{3}$ or 2.33 ft.

$5'7\frac{1}{2}" = 5\frac{7\frac{1}{2}}{12}$ or $5\frac{7.5}{12}$ or 5.625 ft.

Step 2. Substitute the known values into the formula ----------→ $V = \frac{1}{2} \times 2.33 \times 4 \times 5.625$

Step 3. Calculate the formula (do the operations indicated) ------→ $V = (\frac{1}{2} \times \frac{2.33}{1}) \times (4 \times 5.625)$

$V = \frac{2.33}{2} \times \frac{22.5}{1} = \frac{52.425}{2}$

$V = 26.2125$ cu. ft. (2/52.425)

Step 4. Convert to cubic yards and to gallons. ---→ $V = .9708$ or approx. 1 cu. yd. (27/26.2125)

Change to inches

$V = \frac{1}{2} \times 28 \times 48 \times 67.5$

$V = (\frac{1}{2} \times 28) \times (48 \times 67.5)$

$V = 14 \times 3240$

$V = 45360$ cu. in.

$V = 26.25$ cu. ft. (1728/45360.00)

```
   26.2125
 X     7.5
   1310625
  1834875
 196.59375 gal.
```

(1 cu. yd. = 27 cu. ft.)

(1 cu. ft. = 7.5 gal.)

***Note that the dimensions must be changed to the same units of measure before multiplying.**

To find the volume of the right prisms with polygon bases shown below, multiply the area of their bases times their length (height). Refer to each specific geometric figure (solid) for the proper formula to use.

[Note how the prisms are named from their base.]

Task No. K6B, page 134.

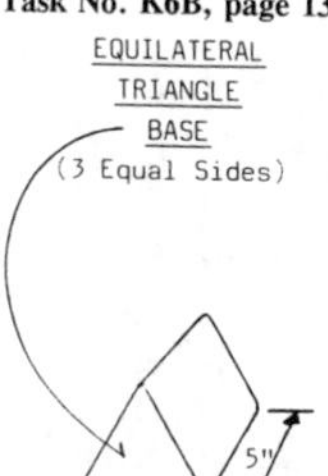

EQUILATERAL TRIANGULAR PRISM

Formula: V = $.433S^2L$
Substitute: V = .433 X 6^2 X 5
Calculate: V = .433 X (6 X 6) X 5
V = .433 X (36 X 5)
V = .433 X 180
V = 77.94 cu. in.

Task No. K6C, page 135.

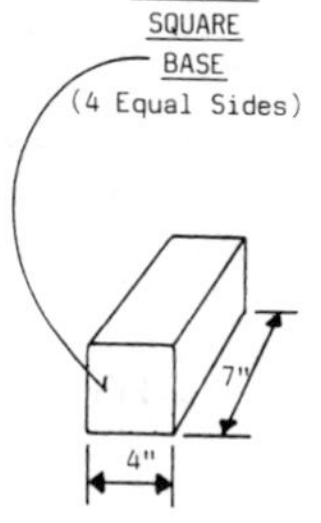

TETRAGON SQUARE PRISM

V = $1S^2L$
V = 1 X 4^2 X 7
V = 1 X (4 X 4) X 7
V = 1 X (16 X 7)
V = 1 X 112
V = 112 cu. in.

Task No. K6D, page 136.

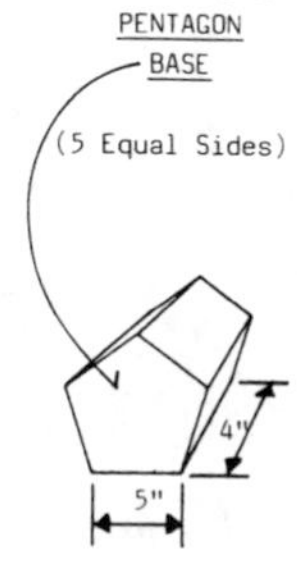

PENTAGON PRISM

V = $1.72S^2L$
V = 1.72 X 5^2 X 4
V = 1.72 X (5 X 5) X 4
V = 1.72 X (25 X 4)
V = 1.72 X 100
V = 172 cu. in.

Task No. K6E, page 137.

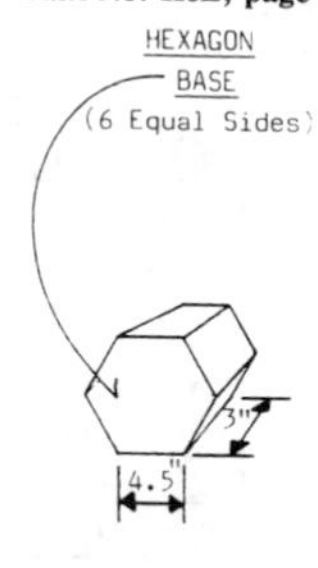

HEXAGON PRISM

Formula: V = $2.598S^2L$
Substitute: V = 2.598 X 4.5^2 X 3
Calculate: V = 2.598 X (4.5 X 4.5) X 3
V = 2.598 X (20.25 X 3)
V = 2.598 X 60.75
V = 157.828 cu. in.

Task No. K6F, page 138.

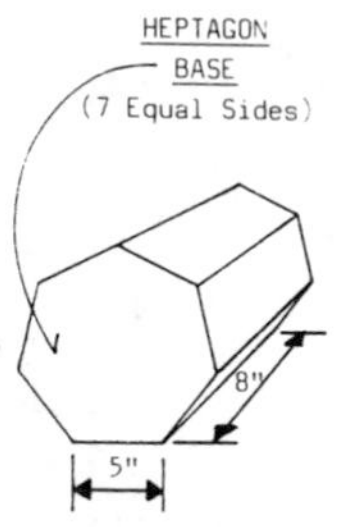

HEPTAGON PRISM

V = $3.633S^2L$
V = 3.633 X 5^2 X 8
V = 3.633 X (5 X 5) X 8
V = 3.633 X (25 X 8)
V = 3.633 X 200
V = 726.6 cu. in.

Task No. K6G, page 139.

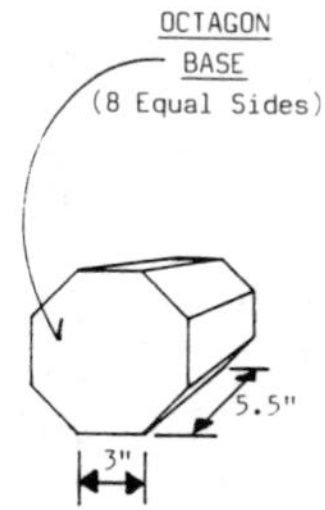

OCTAGON PRISM

V = $4.828S^2L$
V = 4.828 X 3^2 X 5.5
V = 4.828 X(3 X 3)X 5.5
V = 4.828 X (9 X 5.5)
V = 4.828 X 49.5
V = 238.986 cu. in.

L7	◂ Task No.	VOLUME OF A CYLINDER

To find the volume of a cylinder, multiply pi (π) by the radius squared by the height. Or, multiply ¼ pi by the diameter squared by the height.

RULES: Volume equals pi times the radius squared times the height. Or, Volume equals ¼ pi times the diameter squared times the height.

FORMULAS: $V = \pi r^2 h$ **Or** $V = .7854d^2h$

The reason for ¼ times pi.

$A = \pi r^2$

$A = 3.1416 \left(\frac{d}{2}\right)^2$

$A = \frac{3.1416d^2}{4}$

$A = .7854d^2$

($\frac{d}{2}$ is the same as the radius)

6" radius

11½" height

Example Problem 1. Find the volume of the cylinder on the right.

Step 1. Use the formula ----------- $V = \pi r^2 h$

Step 2. Substitute the known values into the formula ---------- $V = 3.1416 \times 6^2 \times 11\frac{1}{2}$

Step 3. Calculate the formula (do the operations indicated) ----- $V = 3.1416 \times 36 \times 11.5$
[6 X 6 = 36]

$V = 3.1416 \times (36 \times 11.5)$

$V = 3.1416 \times 414$

$V = 1300.6224$ cu. in.

Step 4. convert to gallons. $V = 5.6304$ gallons
[1 gallon = 231 cu. in.] 231/1300.6224

Finding the volume in terms of the diameter.

$V = .7854d^2h$

$V = .7854 \times 12^2 \times 11\frac{1}{2}$

$V = .7854 \times 144 \times 11.5$

$V = .7854 \times (144 \times 11.5)$

$V = .7854 \times 1656$

$V = 1300.6224$ cu. in.

Use one method to check the other method.

7'- 0" diameter

20' - 0" height

Example Problem 2. Find the volume of the cylinder on the right.

Step 1. Use the formula ----------- $V = .7854d^2h$

Step 2. Substitute the known values into the formula ---------- $V = .7854 \times 7^2 \times 20$

Step 3. Calculate the formula (do the operations indicated) ----- $V = .7854 \times 49 \times 20$
[7 X 7 = 49]

$V = .7854 \times (49 \times 20)$

$V = .7854 \times 980$

$V = 769.692$ cu. ft.

Finding the volume in terms of the radius.

$V = \pi r^2 h$

$V = 3.1416 \times 3.5^2 \times 20$

$V = 3.1416 \times 12.25 \times 20$

$V = 3.1416 \times (12.25 \times 20)$

$V = 3.1416 \times 245$

$V = 769.692$ cu. ft.

Use one method to check the other method.

Step 4. Convert to cubic yards and to gallons. $V = 28.507$ cu. yd.
27/769.692

[1 cubic foot = 7.48 gallons]

```
    769.692
 X     7.48
    6157536
   3078768
  5387844
V = 5757.29616 gallons
```

VOLUME OF A CONE — Task No. ▸ L8

To find the volume of a cone, multiply pi (π) times the radius squared times the height divided by 3.

RULE: Volume equals pi times the radius squared times the height divided by 3.

FORMULA: $V = \pi r^2h/3$

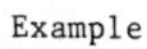

Example
Problem. Find the volume of the cone on the right.

(Cone diagram: height 42", diameter 36")

Step 1. Use the formula ------------------- $V = \frac{\pi r^2 h}{3}$

Step 2. Substitute the known values into the formula ------------------ $V = \frac{3.14 \times 18^2 \times 42}{3}$

Step 3. Calculate the formula (do the operations indicated) ------------- $V = \frac{3.14 \times (18 \times 18) \times 42}{3}$

[Note for greater accuracy use 3.1416 for pi.]

$V = \frac{3.14 \times (324 \times 42)}{3}$

$V = \frac{3.14 \times 13608}{3}$

$V = \frac{42729.12}{3}$

$V = 14243.04$ cu. in.

Step 4. Convert to cubic feet and to gallons. $V =$ 1728/14243 = 8.24 cu. ft.   231/14243 = 61.66 gallons

VOLUME OF A FRUSTUM OF A CONE — Task No. ▸ L8A

To find the volume of a frustum of a cone, add the area of the two bases plus the square root of the product of the area of the two bases times 1/3 of the height.

RULE: Volume equals the sum of the area of the two bases plus the square root of the product of the area of the two bases times 1/3 of the height.

FORMULA: $V = (b + b' + \sqrt{bb'})h/3$

Example
Problem. Find the volume of the frustum of the cone on the right.

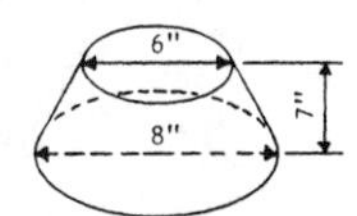

Step 1. Use the formula ----------- $V = (b + b' + \sqrt{bb'})\frac{h}{3}$

Step 2. Substitute the known values into the formula ---------- $V = [(3.14 \times 4 \times 4) + (3.14 \times 3 \times 3) + \sqrt{(3.14 \times 4 \times 4) \times (3.14 \times 3 \times 3)}] \times \frac{7}{3}$

[Area of the bases equals pi times r²]

Step 3. Calculate the formula (do the operations indicated) $V = (50.24 + 28.26 + \sqrt{50.24 \times 28.26}) \times \frac{7}{3}$

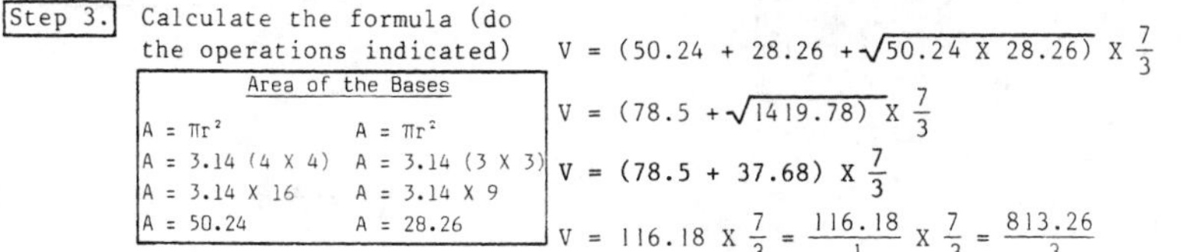

Area of the Bases	
$A = \pi r^2$	$A = \pi r^2$
$A = 3.14\ (4 \times 4)$	$A = 3.14\ (3 \times 3)$
$A = 3.14 \times 16$	$A = 3.14 \times 9$
$A = 50.24$	$A = 28.26$

$V = (78.5 + \sqrt{1419.78}) \times \frac{7}{3}$

$V = (78.5 + 37.68) \times \frac{7}{3}$

$V = 116.18 \times \frac{7}{3} = \frac{116.18}{1} \times \frac{7}{3} = \frac{813.26}{3}$

$V = 271.09$ cu. in.

Step 4. Convert to gallons. $V =$ 231/271.09 = 1.17 gallons.

L9 ◂ Task No. VOLUME OF A PYRAMID

To find the volume of a pyramid, multiply the area of the base times the height and divide by 3.

RULE: Volume equals the area of the base times the height divided by 3.

FORMULA: V = bh/3

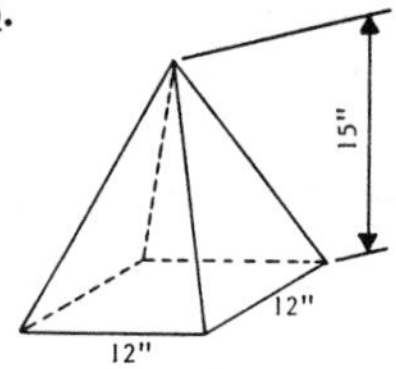

Example Problem. Find the volume of the pyramid on the right.

(Pyramid: height 15", base 12" × 12")

Step 1. Use the formula ------------------- $V = \frac{bh}{3}$

Step 2. Substitute the known values into the formula ------------------- $V = \frac{(12 \times 12) \times 15}{3}$

[The base of a pyramid is square, therefore, 12 X 12 = the area of the base.]

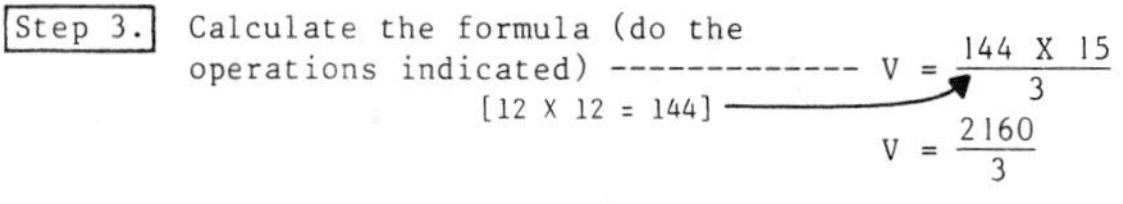

Step 3. Calculate the formula (do the operations indicated) ------------- $V = \frac{144 \times 15}{3}$

[12 X 12 = 144]

$V = \frac{2160}{3}$

V = 720 cu. in.

Step 4. Convert to gallons. V = $231\overline{)720}$ = 3.11 gallons

L9A ◂ Task No. VOLUME OF A FRUSTUM OF A PYRAMID

To find the volume of a frustum of a pyramid, add the area of the two bases plus the square root of the product of the area of the two bases times 1/3 of the height.

RULE: Volume equals the sum of the area of the two bases plus the square root of the product of the area of the two bases times 1/3 of the height.

FORMULA: $V = (b + b' + \sqrt{b'b})h/3$

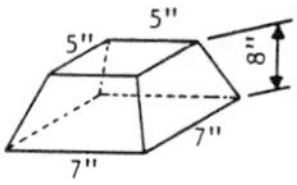

Example Problem. Find the volume of the frustum of the pyramid on the right.

(Frustum: top 5" × 5", bottom 7" × 7", height 8")

Step 1. Use the formula ------------------- $V = (b + b' + \sqrt{bb'})\frac{h}{3}$

Step 2. Substitute the known values into the formula ------------------- $V = (49 + 25 + \sqrt{49 \times 25})\frac{8}{3}$

Area of the Bases	
A = lw	A = lw
A = 7 X 7	A = 5 X 5
A = 49	A = 25

Step 3. Calculate the formula (do the operations indicated) ------------- $V = (74 + \sqrt{1225})\frac{8}{3}$

[49 + 25 = 74]

[49 X 25 = 1225] $V = (74 + 35)\frac{8}{3}$

$V = 109 \times \frac{8}{3} = \frac{109}{1} \times \frac{8}{3} = \frac{872}{3} = 290.66$

V = 290.66 cu. in.

Step 4. Convert to cubic feet and to gallons. V = $1728\overline{)290.66}$ = .168 cu. ft. $231\overline{)290.66}$ = 1.258 gallons

VOLUME OF A SPHERE

Task No. ▸ L10

To find the volume of a sphere, multiply pi (π) times the diameter cubed and divided by 6.

RULE: Volume equals pi times the diameter cubed divided by 6.

FORMULA: $V = \pi d^3/6$

Example Problem. Find the volume of the sphere on the right.

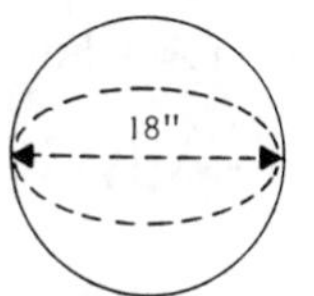

Step 1. Use the formula -------------- $V = \frac{\pi d^3}{6}$

Step 2. Substitute the known values into the formula ------------- $V = \frac{3.14 \times 18^3}{6}$

Step 3. Calculate the formula (do the operations indicated) -------- $V = \frac{3.14 \times (18 \times 18 \times 18)}{6}$

$V = \frac{3.14 \times [18 \times (18 \times 18)]}{6}$

$V = \frac{3.14 \times (18 \times 324)}{6}$

$V = \frac{3.14 \times 5832}{6}$

$V = \frac{18312.48}{6}$

V = 3052.08 cu. in.

Or Use the Formula

$V = \frac{4\pi r^3}{3}$

$V = \frac{4 \times 3.14 \times 9^3}{3}$

$V = \frac{4 \times 3.14 \times [9 \times (9 \times 9)]}{3}$

$V = \frac{4 \times 3.14 \times (9 \times 81)}{3}$

$V = \frac{4 \times (3.14 \times 729)}{3}$

$V = \frac{4 \times 2289.06}{3}$

$V = \frac{9156.24}{3}$

V = 3052.08 cu.in.

Step 4. Convert to cubic feet and to gallons.

V = 1728/3052.08 = 1.76 cu. ft.   231/3052.08 = 13.21 gallons

Note: For greater accuracy use 3.1416 for pi.

REVIEW AND PRACTICAL APPLICATION OF SOLID GEOMETRY

Volume is the space taken up by a geometric figure that has three dimensions and is not in the same plane. Cubic units are used in measuring volume.

How to Calculate Cubic Inches: (Note a cubic inch is a cube 1 inch by 1 inch by 1 inch.)

How many cubic inches are in a rectangular solid 3 inches by 5 inches by 6 inches?

Use the Formula: V = lwh

Substitute known values: V = 3 X 5 X 6

Calculate: V = (3 X 5) X 6

V = 15 X 6

V = 90 Cubic inches

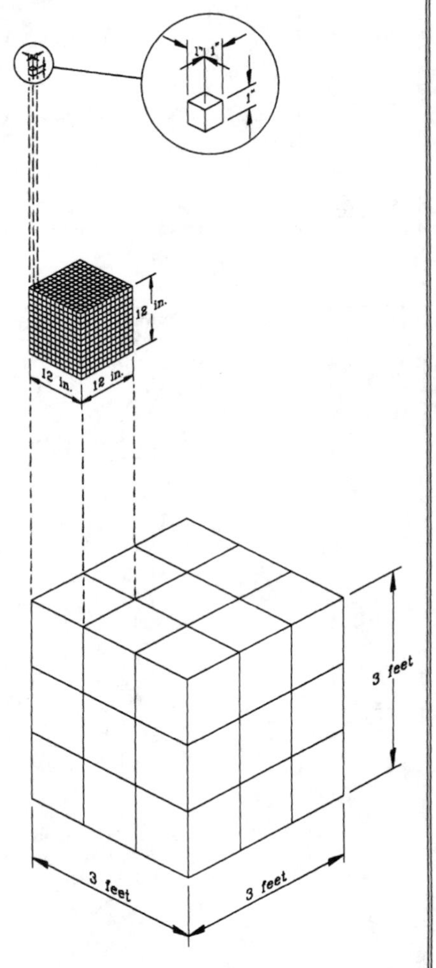

How to Calculate Cubic Feet:
(Note a cubic foot consists of 1728 cu. in.)

How many cubic feet are in a rectangular solid 16 inches by 20 inches by 35 inches?

Use the formula: V = lwh
Substitute known values: V = 16 X 20 X 35
Calculate: V = (16 X 20) X 28
V = 320 X 28
V = 8960 cu. inches

Convert to cubic feet divide by 1728 cu. in.
V = 8960 ÷ 1728
V = 5.185 cu. ft.

How to Calculate Cubic Yards:
(Note a cubic yard consists of 27 cu. ft.)

How many cubic yards are in a rectangular solid 12 ft. by 30 feet by 4 inches?

Use the formula: V = lwh
Substitute known values: V = 12 X 30 X .33
Calculate: V = (12 X 30) X .33
V = 360 X .33
V = 118.8 cubic feet

Convert to cubic yards divide by 27 cubic feet.
V = 118.8 ÷ 27
V = 4.4 cubic yards

How to Calculate Board Feet: (Note a board foot is 1 inch by 1 foot by 1 foot.)

How many board feet are in 24 - 2 X 4's X 8 feet.

Use the formula: Board Feet = Number of pieces X thickness (in.) X width (in.) X length (ft.) ÷ 12

Substitute known values: $BF = \frac{24 \times 2 \times 4 \times 8}{12}$

Calculate: $BF = \frac{(24 \times 2) \times (4 \times 8)}{12}$

$BF = \frac{48 \times 32}{12}$

$BF = \frac{1536}{12}$

BF = 128 board feet.

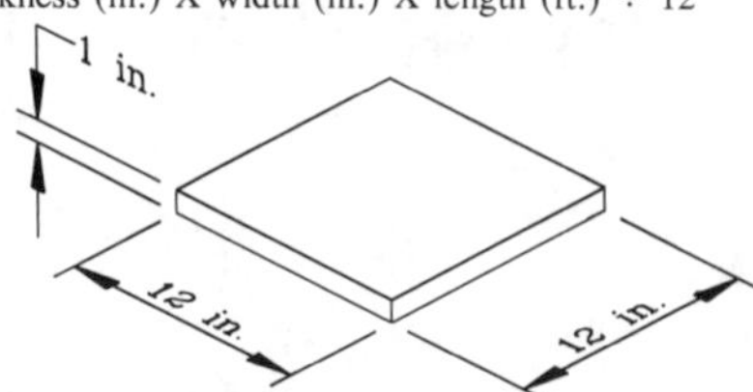

How to Calculate Full Cord Measure:
(Note a full cord of wood consists of 128 cu. ft.)
[4 X 4 X 8 = 128 cu. ft.]

To calculate wood measure in terms of cords, the width is 4 feet, the height is 4 feet and the length is 8 feet. Example: A pile of logs 4 feet wide by 4 feet high by 12 feet long would be 1 1/2 full cords of wood.

[4 X 4 X 12 = 192 cu. ft. And 192 ÷ 128 = 1 1/2 full cords.]

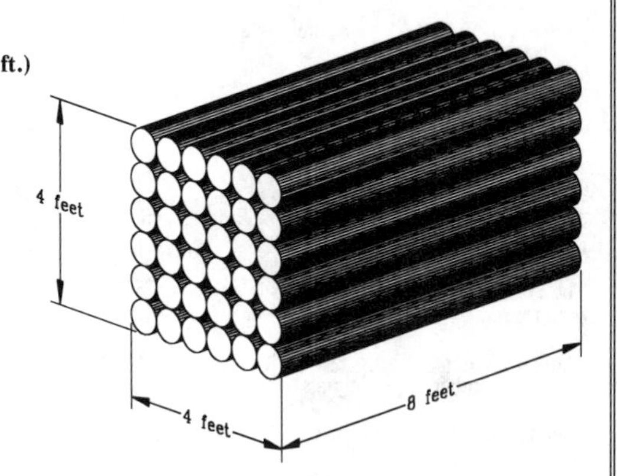

How to Calculate Face Cord Measure:
(Note a face cord of wood consists of 42 2/3 cu. ft.)
[4' X 8' X 1 1/3' (16") = 42 2/3 cu. ft.]

To calculate wood measure in terms of face cords, the width is 8 feet, the height is 4 feet and the length is 16 inches (1 1/3 feet). Example: 1 full cord of logs 4 feet wide by 4 feet high by 8 feet long would be 128 cu. ft. or 3 face cords of wood.

[1 cord = 128 cu. ft. and 1 face cord = 42 2/3 cu. ft. Therefore, 128 ÷ 42 2/3 = 3 face cords.]

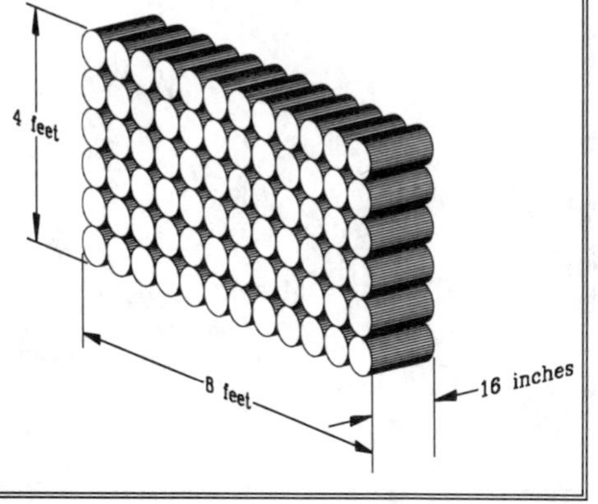

M1	◄ Task No.	BASIC TRIGONOMETRY

Trigonometry means triangular measurement. It is a branch of mathematics which studies the relationships between sides and angles of a triangle. Its principles can be utilized in the trades to find the measurements of the sides or angles of any triangle related problem, which can not be readily computed from the principles discussed previously in basic mathematics or geometry.

REVIEW OF **ANGLES** AND **TRIANGLES**:

Types of Angles:

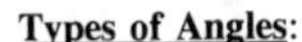

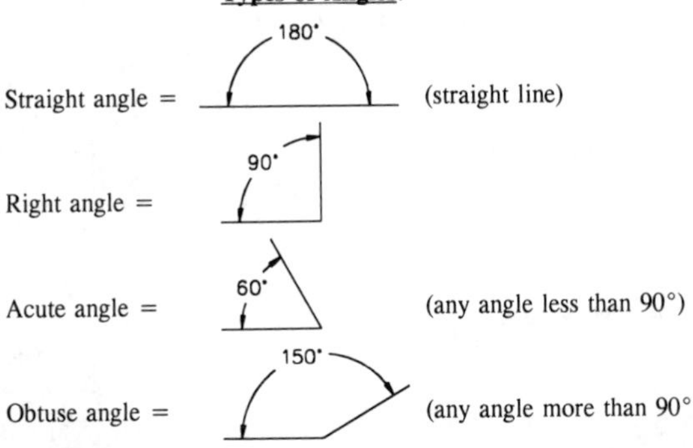

The **complement of an angle** is the angle, which, when added to another angle, equals 90°.

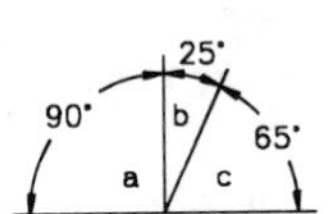

The complement of angle c (65°) is angle b (25°) because 65° + 25° = 90°.

The **supplement of an angle** is the angle, which, when added to another angle, equals 180°.

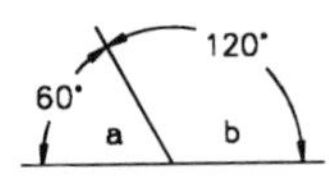

The supplement of angle b (120°) is angle a (60°) because 120° + 60° = 180°.

Types of Triangles:

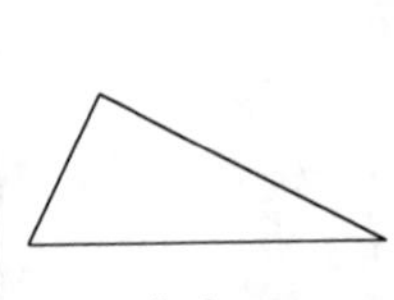

Scalene
No two sides are equal.
(No angles are equal.)

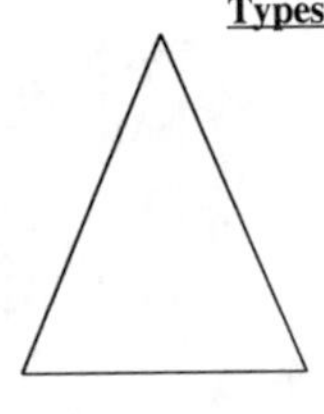

Isosceles
Two sides are equal.
(Two angles are equal.)

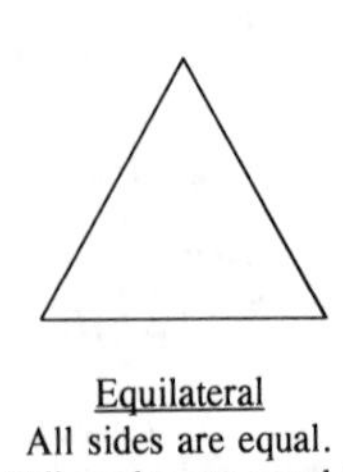

Equilateral
All sides are equal.
(All angles are equal.)

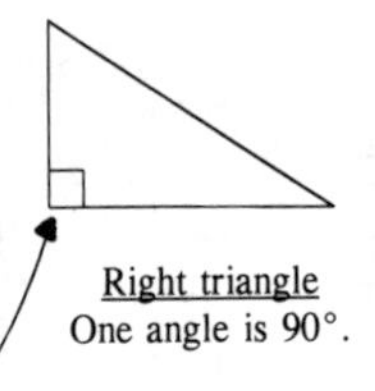

Right triangle
One angle is 90°.

[A little square drawn inside a triangle, as shown, indicates 90° or a right triangle.]

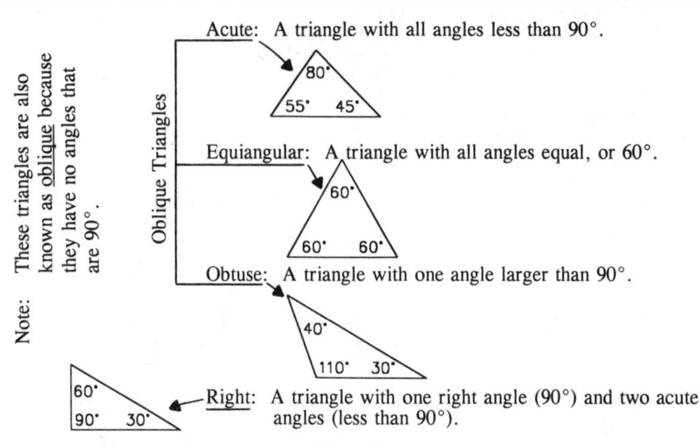

[RULE: **The sum of the angles inside a triangle always add up to 180°.** Notice in all the types of triangles shown above the sum of the angles equals 180°.]

The **basic trigonometric relationships** are as described below:

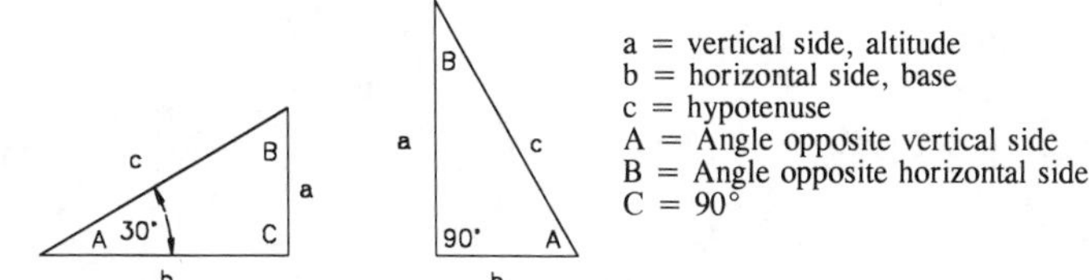

a = vertical side, altitude
b = horizontal side, base
c = hypotenuse
A = Angle opposite vertical side
B = Angle opposite horizontal side
C = 90°

[Note the proper way to label the 3 sides and the 3 angles of a right triangle is; A is the angle opposite the altitude a, B is the angle opposite the base b, and C is the angle opposite the hypotenuse c.]

Basic Functions Of Angles

sine A = $\frac{a}{c}$	cosecant A = $\frac{c}{a}$
cosine A = $\frac{b}{c}$	secant A = $\frac{c}{b}$
tangent A = $\frac{a}{b}$	cotangent A = $\frac{b}{a}$

There exists for any given **angle A** of any given right triangle ABC as shown above, specific values which are known as the **sine**, **cosine** and **tangent functions**, of the angle. The cosecant, secant, and cotangent are the reciprocal of these functions. They are also the less used of the functions. The basic functions of angles are shown above. These functions or ratios will never change because they are always proportional to each other. The diagrams on the next page will help illustrate this concept and correlates with the problems used throughout this module on trigonometry.

[Note how the three functions on the right are the reciprocal of the three functions on the left. When calculating trigonometry problems and using the trigonometry tables (sample shown on the next page), you can plainly see how one is the reciprocal of the other.]

STANDARD ABBREVIATIONS

SINE =	SIN	HYPOTENUSE =	HYP.	See page 179.
COSINE =	COS	SIDE OPPOSITE =	OPP.	
TANGENT =	TAN	SIDE ADJACENT =	ADJ.	
COTANGENT =	COT			
SECANT =	SEC			
COSECANT =	CSC			

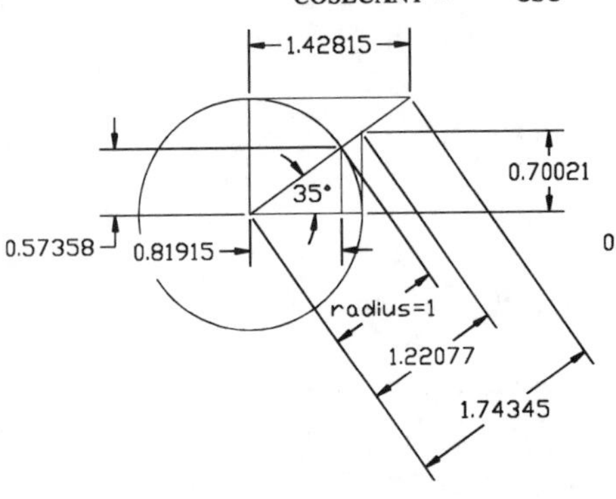

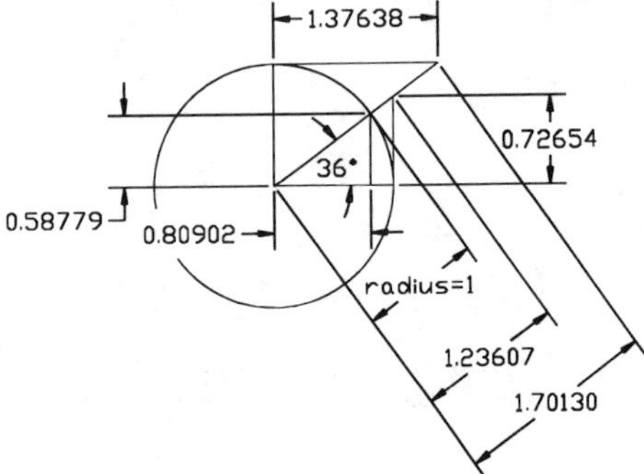

35°

M	Sine	Cosine	Tan.	Cotan.	Secant	Cosec.	M
0	.57358	.81915	.70021	1.4281	1.2208	1.7434	60
1	.57381	.81898	.70064	.4273	.2210	.7427	59
2	.57405	.81882	.70107	.4264	.2213	.7420	58
3	.57429	.81865	.70151	.4255	.2215	.7413	57

36°

M	Sine	Cosine	Tan.	Cotan.	Secant	Cosec.	M
0	.58778	.80902	.72654	1.3764	1.2361	1.7013	60
1	.58802	.80885	.72699	.3755	.2363	.7006	59
2	.58825	.80867	.72743	.3747	.2366	.6999	58
3	.58849	.80850	.72788	.3738	.2368	.6993	57

From the drawings above, notice how trig table calculations are based on the radius being a unit of 1. The radius is always the same, but as the angle changes so do the other function dimensions proportionally. Refer to each drawing above and the excerpt of its corresponding trig table below it, to see how this is true.

From plane geometry, the relationships of a 30° - 60° - 90°, a 60° - 30° - 90°, and a 45° - 45° - 90° right triangles are familiar as illustrated below.

Example 1.

In a 30° - 60° right triangle, the side opposite the 30° angle is equal to $\frac{1}{2}$ of the hypotenuse.

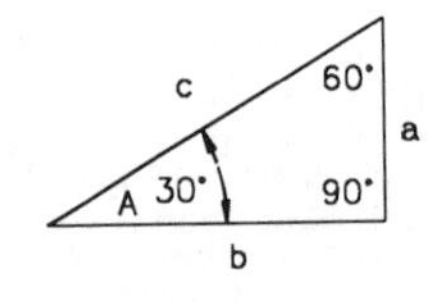

Minutes → **30°** ← Minutes

M	Sine	Cosine	Tan.	Cotan.	Secant	Cosec.	M
0	.50000	.86603	.57735	1.7320	1.1547	2.0000	60
1	.50025	.86588	.57774	.7309	.1549	1.9990	59
2	.50050	.86573	.57813	.7297	.1551	.9980	58
3	.50075	.86559	.57851	.7286	.1553	.9970	57
4	.50101	.86544	.57890	.7274	.1555	.9960	56

Relating the above to trig, it can be stated $\frac{a}{c}$ = sine A or in the above example sine A = $\frac{1}{2}$ = 0.500

Example 2.

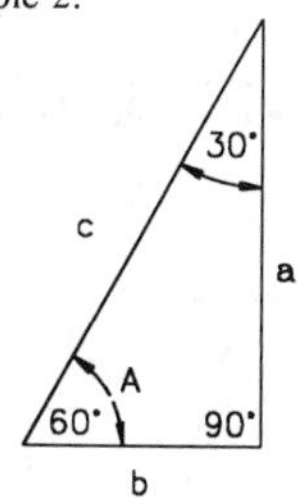

In a 60° - 30° - 90° right triangle $b = \frac{1}{2}$ of c since it is the side opposite the 30° angle.

55	.49874	.86675	.57541	1.7379	1.1537	2.0050	5
56	.49899	.86661	.57580	.7367	.1539	.0040	4
57	.49924	.86646	.57619	.7355	.1541	.0030	3
58	.49950	.86632	.57657	.7344	.1543	.0020	2
59	.49975	.86617	.57696	.7332	.1545	.0010	1
60	.50000	.86603	.57735	1.7320	1.1547	2.0000	0
M	Cosine	Sine	Cotan.	Tan.	Cosec.	Secant	M

60°

Relating the above to trig, it can be stated cosine A = $\frac{b}{c} = \frac{1}{2} = 0.500$

Example 3.

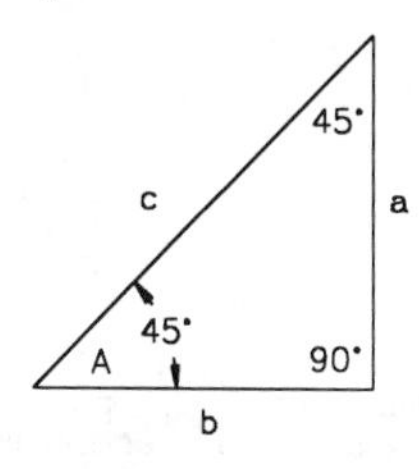

In a 45° - 45° - 90° right triangle, which is the same as an isosceles triangle, the sides a and b are equal.

56	.70628	.70793	.99767	.0023	.4126	.4159	4
57	.70649	.70772	.99826	.0017	.4130	.4154	3
58	.70669	.70752	.99884	.0012	.4134	.4150	2
59	.70690	.70731	.99942	.0006	.4138	.4146	1
60	.70711	.70711	1.0000	1.0000	1.4142	1.4142	0
M	Cosine	Sine	Cotan.	Tan.	Cosec.	Secant	M

45°

Relating the above to trig, it can be stated tangent A or Tangent 45° = $\frac{a}{b}$ = 1.00 (because a = b)

Finding the value of the <u>sine</u>, <u>cosine</u>, and <u>tangent functions</u> for any triangle other than those illustrated above deals with studies and calculations which are beyond the intended scope of this book, however **the values of all the trigonometric functions, for angles from 0° to 90°, can be solved by using the trigonometry tables on pages 220-264**. See the example shown below.

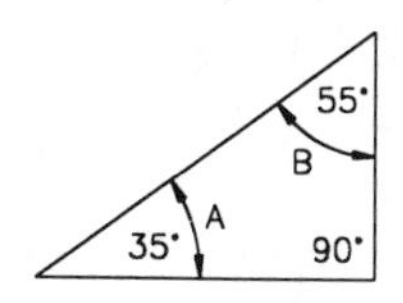

35°

M	Sine	Cosine	Tan.	Cotan.	Secant	Cosec.	M
0	.57358	.81915	.70021	1.4281	1.2208	1.7434	60

60	.57358	.81915	.70021	1.4281	1.2208	1.7434	0
M	Cosine	Sine	Cotan.	Tan.	Cosec.	Secant	M

55°

Note that angle A, 35° is the compliment of angle B, 55°. This brings to light another important relationship between ratios of the sides and the reciprocal function which are the functions of the complimentary angles. The trig tables above should help illustrate that the functions shown below are true.

Sine A = Cosine B
Cosine B = Sine A
Tangent A = Cotangent B

Cosecant A = Secant B
Secant B = Cosecant A
Cotangent A = Tangent B

Hopefully, the above examples will help you understand the relationships of the ratios of the sides and angles of triangles for using the basic trigonometry formulas and solving the trig problems you will encounter.

M1 ◂ **Task No.**

SOLVING BASIC RIGHT TRIANGLE TRIGONOMETRY PROBLEMS:

From previous discussion of the principles of mathematics, algebra, and geometry, and now with the basic principles of trigonometry, **the following right triangle related measurements can be calculated for any applicable problem which may be encountered in the world of work.**

1. To find the **Value of angle A** (angle opposite the vertical side), see the formulas listed below.

Formula 1A. **Sine A** $= \frac{a}{c}$ See page 186, **(a and c are known)**

Formula 2A. **Cosine A** $= \frac{b}{c}$ See page 188, **(b and c are known)**

Formula 3A. **Tangent A** $= \frac{a}{b}$ See page 190, **(a and b are known)**

2. To find the **Value of angle B** (angle opposite the horizontal side), see the formulas listed below.

Formula 1B. **Cosine B** $= \frac{a}{c}$ See page 186, **(a and c are known)**

Formula 2B. **Sine B** $= \frac{b}{c}$ See page 188, **(b and c are known)**

Formula 3B. **Cotangent B** $= \frac{a}{b}$ See page 190, **(a and b are known)**

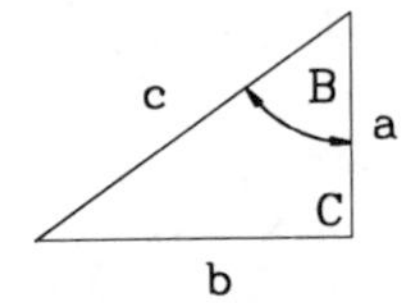

3. The **Value of angle C** (opposite the hypotenuse) will always be 90° when solving right triangle problems. However in calculations for oblique triangles, the value of C will be more or less than 90°. **Formulas for solving basic oblique triangle problems are listed in Task No. M4, pages**

4. To find the **Value of the vertical side (a)**, see the formulas listed below.

Formula 3A. **Tangent A** $= \frac{a}{b}$ See page 192, **(A and b are known)**

Formula 3B. **Cotangent B** $= \frac{a}{b}$ See page 192, **(B and b are known)**

Formula 1A. **Sine A** $= \frac{a}{c}$ See page 194, **(A and c are known)**

Formula 1B. **Cosine B** $= \frac{a}{c}$ See page 194, **(B and c are known)**

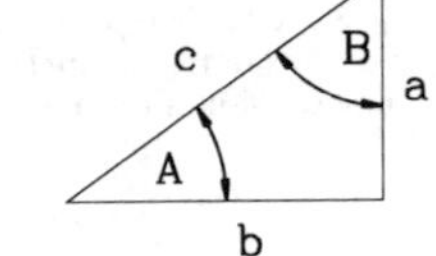

$a^2 = c^2 - b^2$ See page 204 & 132, **(b and c are known)**

5. To find the **Value of the horizontal side (b)**, see the formulas listed below.

Formula 3A. **Tangent A** $= \frac{a}{b}$ See page 196, **(A and a are known)**

Formula 3B. **Cotangent B** $= \frac{a}{b}$ See page 196, **(B and a are known)**

Formula 2A. **Cosine A** $= \frac{b}{c}$ See page 198, **(A and c are known)**

Formula 2B. **Sine B** $= \frac{b}{c}$ See page 198, **(B and c are known)**

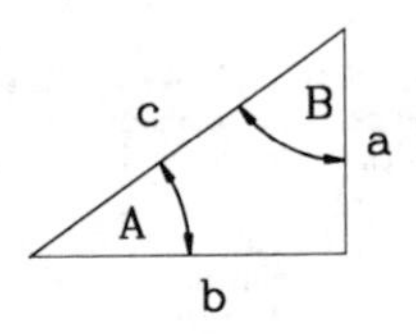

$b^2 = c^2 - a^2$ See page 204 & 132, **(a and c are known)**

6. To find the **Value of the hypotenuse (c)**, see the formulas listed below.

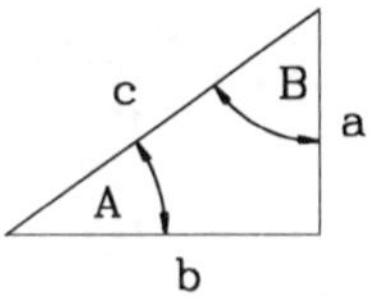

Formula 1A. **Sine A** $= \frac{a}{c}$ See page 200, **(A and a are known)**

Formula 1B **Cosine B** $= \frac{a}{c}$ See page 200, **(B and a are known)**

Formula 2A. **Cosine A** $= \frac{b}{c}$ See page 202, **(A and b are known)**

Formula 2B. **Sine B** $= \frac{b}{c}$ See page 202, **(B and b are known)**

$a^2 + b^2 = c^2$ See page 204 & 132, **(a and b are known)**

For more information on the $a^2 + b^2 = c^2$, see The Rule of Pythagoras, Task No. K5A, page 132.

HOW TO MANIPULATE BASIC TRIG FORMULAS USING ALGEBRA:

Algebra has been used to arrange formulas so that any side or angle can be found as long as two values are known, angles or sides.

Example 1. Consider the **FORMULA: Sine A** $= \frac{a}{c}$. Assume **A** and **c** are known and **a** needs to be solved.

Step 1. Set up as follows. **Sine A** $= \frac{a}{c}$

Step 2. The line of a fraction indicates division, therefore multiply both sides by **c** and cancel. **c x Sine A** $= \frac{a}{c} \times c$

New FORMULA: c x Sine A = a

Example 2. Solve for **c** in the formula **Sine A** $= \frac{a}{c}$.

Step 1. Set up as follows. **Sine A** $= \frac{a}{c}$

Step 2. Multiply both sides by **c** and cancel. **c x Sine A** $= \frac{a}{c} \times c$

Step 3. To get **c** by itself, divide both sides by **A** and cancel. $\frac{c \times \text{Sin A}}{\text{Sin A}} = \frac{a}{\text{Sin A}}$

New FORMULA: $c = \frac{a}{\text{Sin A}}$

ANGULAR MEASUREMENT:

There are 360° around a point or a circle.

1 degree in divided into 60 minutes(')

1 minute is divided into 60 seconds (")

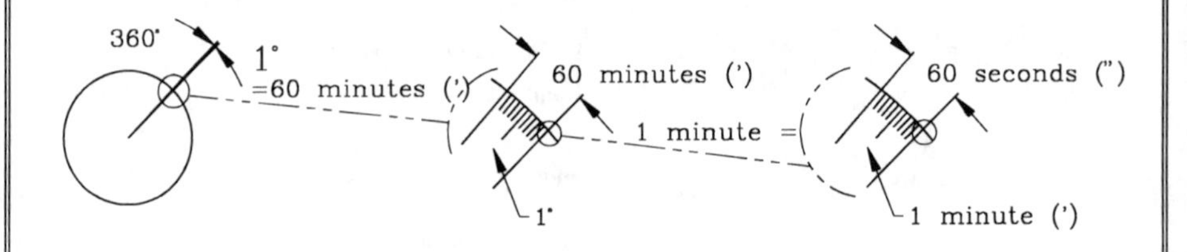

A vernier protractor is used to measure angles in degrees and minutes.

(In this example, read 33 degrees 50 minutes.)

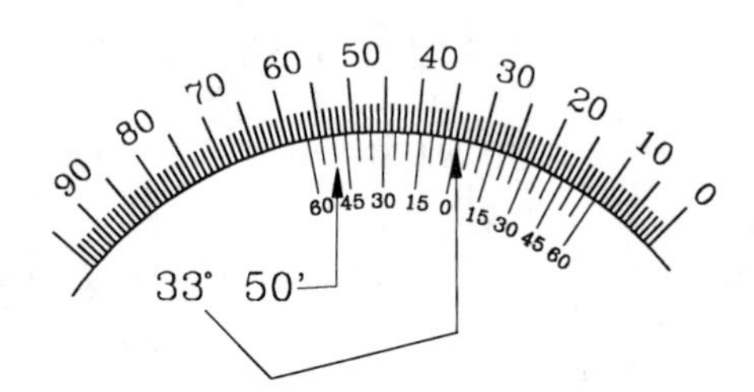

A protractor is used to measure angles in whole degrees.

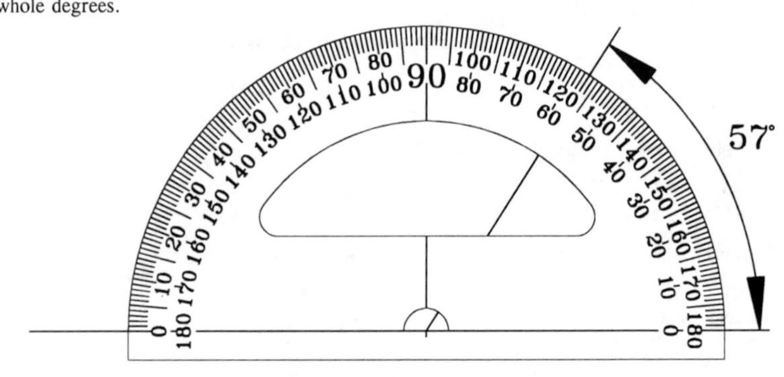

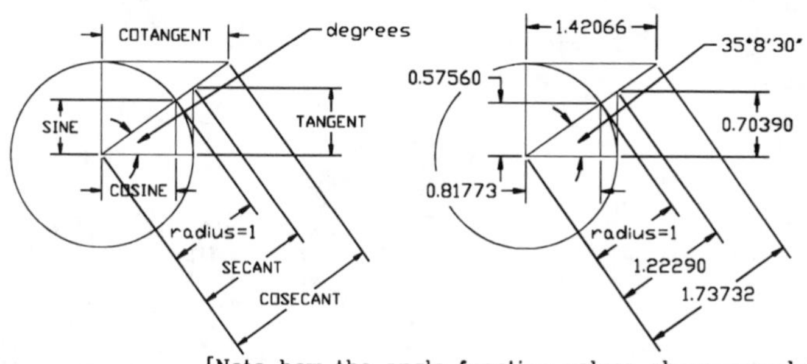

[Note how the angle function values above correlate with the interpolated angle values shown below.]

INTERPOLATION:

Many times when calculating trig problems that involve finding **Angle A** or **Angle B** as shown below, the **Sine, Cosine, Tangent,** and other trigonometric calculations will not coincide with an angle listed in the trig tables. When this happens, calculate the difference between the two values, make a fraction out of the difference between the two values, and change the converted decimal value to seconds of a degree as shown below. This process is known as **interpolation**.

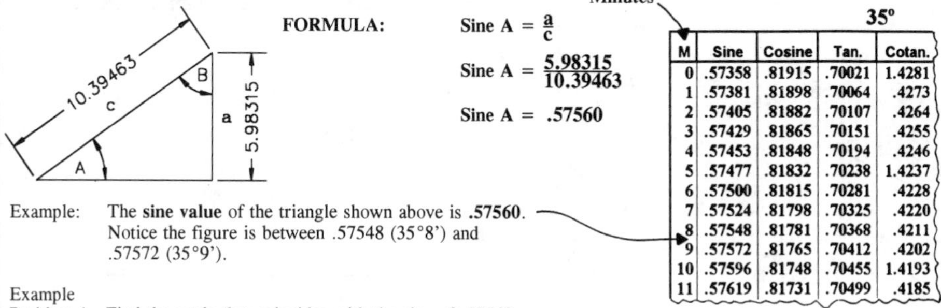

FORMULA:

$$\text{Sine A} = \frac{a}{c}$$

$$\text{Sine A} = \frac{5.98315}{10.39463}$$

$$\text{Sine A} = .57560$$

35°

M	Sine	Cosine	Tan.	Cotan.
0	.57358	.81915	.70021	1.4281
1	.57381	.81898	.70064	.4273
2	.57405	.81882	.70107	.4264
3	.57429	.81865	.70151	.4255
4	.57453	.81848	.70194	.4246
5	.57477	.81832	.70238	1.4237
6	.57500	.81815	.70281	.4228
7	.57524	.81798	.70325	.4220
8	.57548	.81781	.70368	.4211
9	.57572	.81765	.70412	.4202
10	.57596	.81748	.70455	1.4193
11	.57619	.81731	.70499	.4185

Example: The **sine value** of the triangle shown above is **.57560**. Notice the figure is between .57548 (35°8') and .57572 (35°9').

Example Problem 1. Find the angle that coincides with the sine of .57560.

Step 1. Write down the sine value that is greater than .57560
Write down the sine value that is less than .57560
Subtract to get the total difference ------------▸

$$\begin{array}{r} .57572 \\ -\underline{.57548} \\ .00024 \end{array} \quad \textbf{Total difference}$$

Step 2. Subtract the sine value that is greater than .57560 from the calculated sine value, .57560.
The result is the partial difference ------------▸

$$\begin{array}{r} .57572 \\ -\underline{.57560} \\ .00012 \end{array} \quad \textbf{Partial difference}$$

Step 3. Make a fraction out of the difference between the total difference 24 (denominator) and the partial difference 12 (numerator) as shown.

$$24 \left\{ \begin{array}{l} 57572 \\ 57560 \\ 57548 \end{array} \right\} \frac{12}{24} = \frac{1}{2}$$

Step 4. Convert the fraction to a decimal and multiply that quotient times 60 seconds (1 degree minute).

$1 \div 2 = .5 \times 60 = 30''$

Answer 35°8'30"

Read as 35 degrees 8 minutes 30 seconds

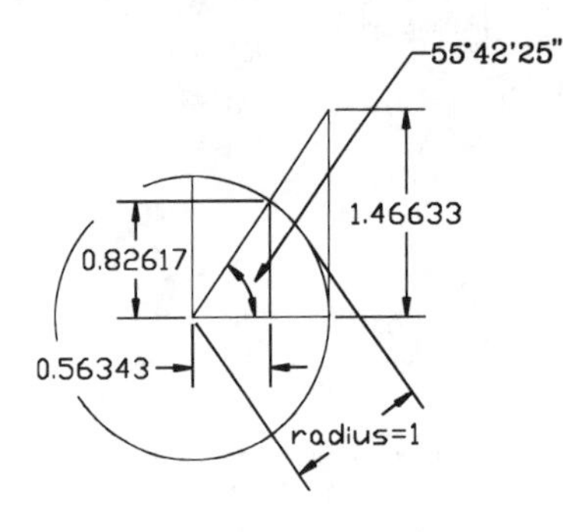

[Note how the angle function values above correlate with the interpolated angle values in the trig tables and the angle values shown below.]

M	Cosine	Sine	Cotan.	Tan.	Cosec.	Secant	M
14	.56256	.82675	.68045	.4696	.2095	.7776	46
15	.56280	.82659	.68087	1.4687	1.2098	1.7768	45
16	.56304	.82643	.68130	.4678	.2100	.7760	44
17	.56328	.82626	.68173	.4669	.2103	.7753	43
18	.56353	.82610	.68215	.4659	.2105	.7745	42
19	.56377	.82593	.68258	.4650	.2107	.7738	41
20	.56401	.82577	.68301	1.4641	1.2110	1.7730	40
21	.56425	.82561	.68343	.4632	.2112	.7723	39
56	.57262	.81982	.69847	.4317	.2198	.7463	4
57	.57286	.81965	.69891	.4308	.2200	.7456	3
58	.57310	.81948	.69934	.4299	.2203	.7449	2
59	.57334	.81932	.69977	.4290	.2205	.7442	1
60	.57358	.81915	.70021	1.4281	1.2208	1.7434	0

55°

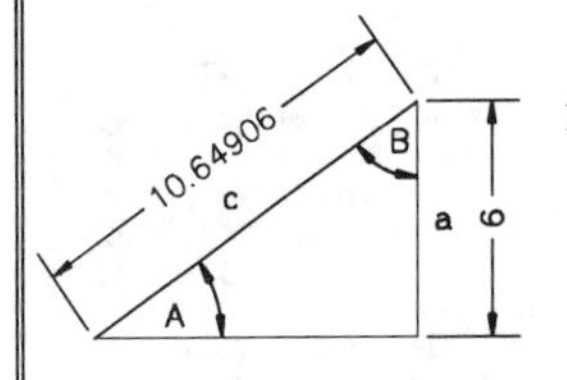

FORMULA:

$$\text{Cosine B} = \frac{a}{c}$$

$$\text{Cosine B} = \frac{6}{10.64906}$$

$$\text{Cosine B} = .56343$$

Example: The **cosine value** of the triangle shown above is **.56343**. Notice the figure is between .56328 (55°43') and .56353 (55°42').

Example Problem 2. Find the angle that coincides with the cosine of .56343.

Step 1. Write down the cosine value that is greater than .56343 — **.56353**
Write down the cosine value that is less than .56343 — **- .56328**
Subtract to get the total difference — **.00025 Total difference**

Step 2. Subtract the cosine value that is greater than .56343 from the calculated cosine value, .56343. — **.56353** **- .56343**
The result is the partial difference — **.00010 Partial difference**

Step 3. Make a fraction out of the difference between the total difference 25 (denominator) and the partial difference 12 (numerator) as shown.

$$25\left\{\begin{matrix}56353\\56343\\56328\end{matrix}\right. \quad \frac{10}{25} = \frac{2}{5}$$

Step 4. Convert the fraction to a decimal and multiply that quotient times 60 seconds (1 degree minute).

5 ÷ 2 = .4 X 60 = **24"**

Answer 55°42'24"

Read as 55 degrees 42 minutes 24 seconds

[FOR AN ALTERNATE METHOD TO FIND THE ANGLES AND SIDES OF RIGHT TRIANGLES, SEE TRICK OF THE TRADES FORMULA: SOH-CAH-TOA BELOW.]

One of the biggest advantages of using this formula versus the traditional method on the preceding pages is that you don't have to label the angles or the sides of the triangle, and it is much easier to memorize all the formulas. This book is set up so you can choose either this method or the traditional method. The example problems will have the traditional method on the left page and the SOH-CAH-TOA method on the right page for your convenience. Use the method(s) that work best for you.

<u>STANDARD ABBREVIATIONS</u>: HYPOTENUSE = HYP.
SIDE OPPOSITE = OPP.
SIDE ADJACENT = ADJ.

FORMULA: { **SOH** / **CAH** (Pronounced "SO-CA-TOE-A") / **TOA** }

SOH = (Sine = $\frac{\text{side opp.}}{\text{hyp.}}$)

CAH = (Cosine = $\frac{\text{side adj.}}{\text{hyp.}}$)

TOA = (Tangent = $\frac{\text{side opp.}}{\text{side adj.}}$)

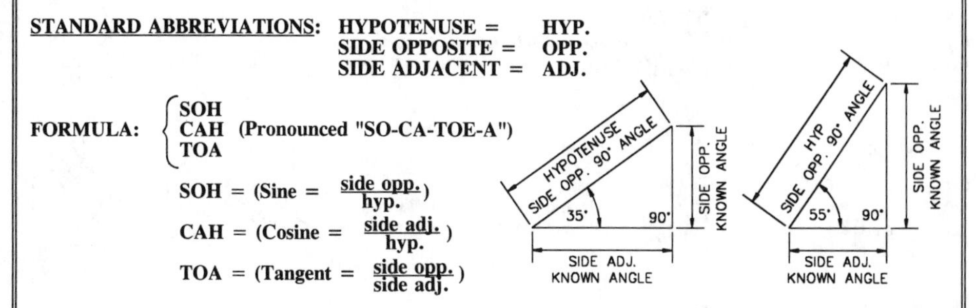

Algebra has been used to arrange formulas so that any <u>side</u> or <u>angle</u> can be found as long as two values are known, angles or sides.

SOH = (Sine = $\frac{\text{side opp.}}{\text{hyp.}}$) = (Sine X hyp. = side opp.) = (hyp. = $\frac{\text{side opp.}}{\text{Sine}}$)

CAH = (Cosine = $\frac{\text{side adj.}}{\text{hyp.}}$) = (Cosine X hyp. = side adj.) = (hyp. = $\frac{\text{side adj.}}{\text{Cosine}}$)

TOA = (Tangent = $\frac{\text{side opp.}}{\text{side adj.}}$) = (Tangent X side adj. = side opp.) = (side adj. = $\frac{\text{side opp.}}{\text{Tangent}}$)

<u>SOLVING RIGHT TRIANGLE TRIG PROBLEMS USING SOH-CAH-TOA:</u>

1. **Find angle opposite the vertical side,**

Formula: SOH Sine = $\frac{\text{side opp.}}{\text{hyp.}}$ See page 187
(side opp. and hyp. are known)

Formula: CAH Cosine = $\frac{\text{side adj.}}{\text{hyp.}}$ **(side adj. and hyp. are known)**

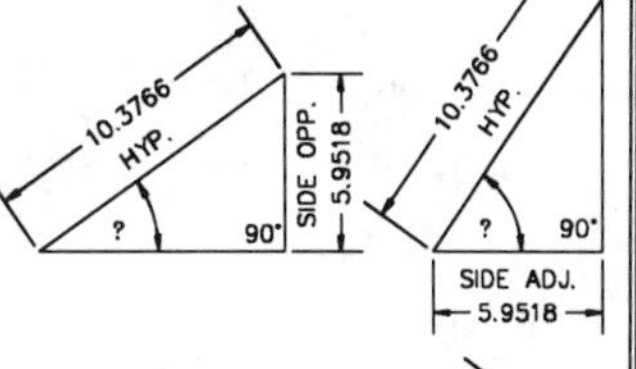

2. **Find angle opposite the vertical side,**

Formula: CAH Cosine = $\frac{\text{side adj.}}{\text{hyp.}}$ See page 189
(side adj. and hyp. are known)

Formula: SOH Sine = $\frac{\text{side opp.}}{\text{hyp.}}$ **(side opp. and hyp. are known)**

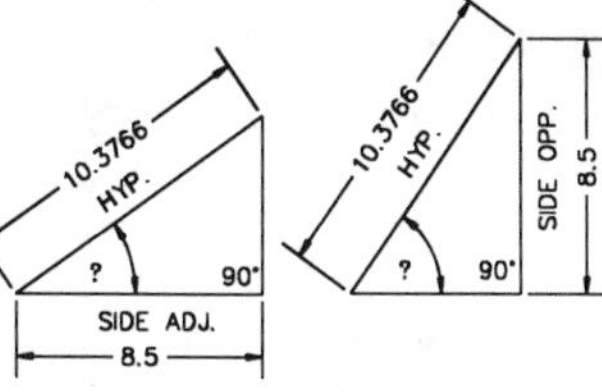

 Trick of the Trade: SOH-CAH-TOA continued

3. **Find angle opposite the vertical side,**

Formula: TOA Tangent = $\frac{\text{side opp.}}{\text{side adj.}}$ See page 191 **(side adj. and side opp. are known)**

Formula: TOA Tangent = $\frac{\text{side opp.}}{\text{side adj.}}$ **(side opp. and side adj. are known)**

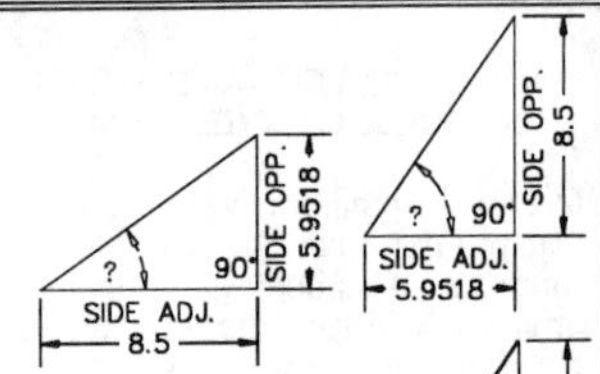

4. **Find opposite side and adjacent side,**

Formula: TOA Tangent = $\frac{\text{side opp.}}{\text{side adj.}}$ See page 193 **(angle and side adj. are known)**

Formula: TOA Tangent = $\frac{\text{side opp.}}{\text{side adj.}}$ **(angle and side opposite are known)**

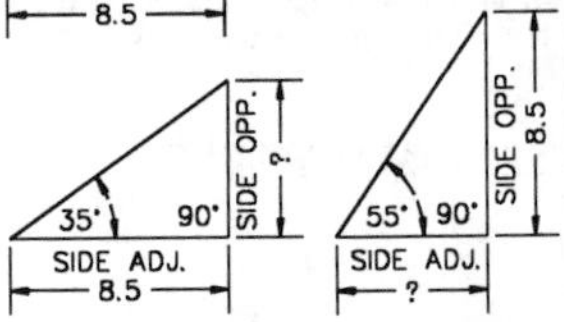

5. **Find opposite side and adjacent side,**

Formula: SOH Sine = $\frac{\text{side opp.}}{\text{hyp.}}$ See page 195 **(angle and hypotenuse are known)**

Formula: CAH Cosine = $\frac{\text{side adj.}}{\text{hyp.}}$ **(angle and hypotenuse are known)**

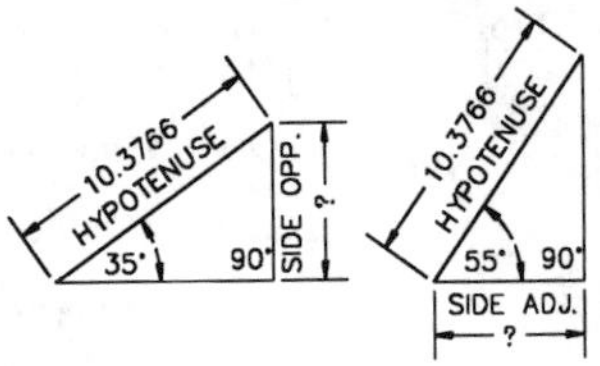

6. **Find adjacent side and opposite side,**

Formula: TOA Tangent = $\frac{\text{side opp.}}{\text{side adj.}}$ See page 197 **(angle and side opposite are known)**

Formula: TOA Tangent = $\frac{\text{side opp.}}{\text{side adj.}}$ **(angle and side adjacent are known)**

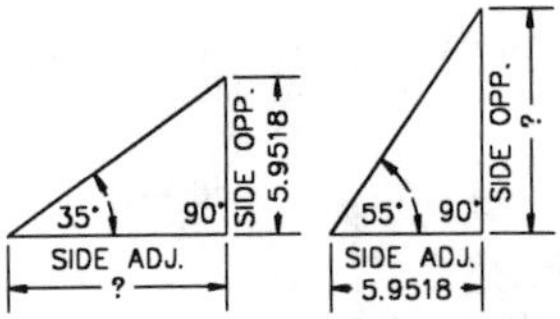

7. **Find adjacent side and opposite side,**

Formula: TOA Tangent = $\frac{\text{side opp.}}{\text{side adj.}}$ See page 199 **(angle and hypotenuse are known)**

Formula: TOA Tangent = $\frac{\text{side opp.}}{\text{side adj.}}$ **(angle and hypotenuse are known)**

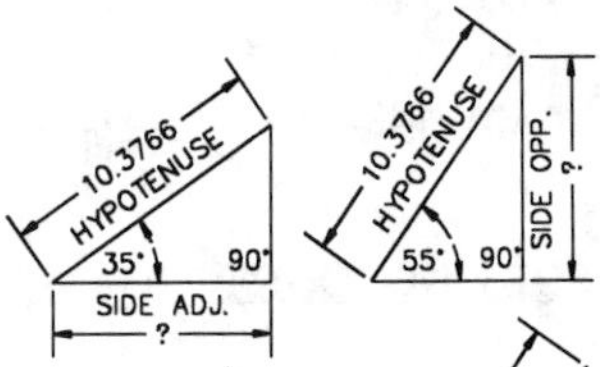

8. **Find hypotenuse,**

Formula: SOH Sine = $\frac{\text{side opp.}}{\text{hyp.}}$ See page 201 **(angle and side opposite are known)**

Formula: CAH Cosine = $\frac{\text{side adj.}}{\text{hyp.}}$ **(angle and side adjacent are known)**

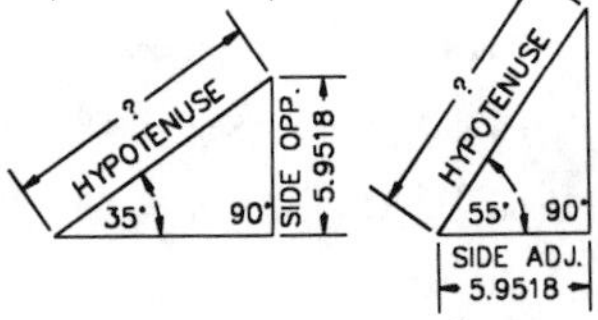

9. **Find hypotenuse,**

Formula: CAH Cosine = $\frac{\text{side adj.}}{\text{hyp.}}$ See page 203 **(angle and side adjacent are known)**

Formula: SOH Sine = $\frac{\text{side opp.}}{\text{hyp.}}$ **(angle and side opposite are known)**

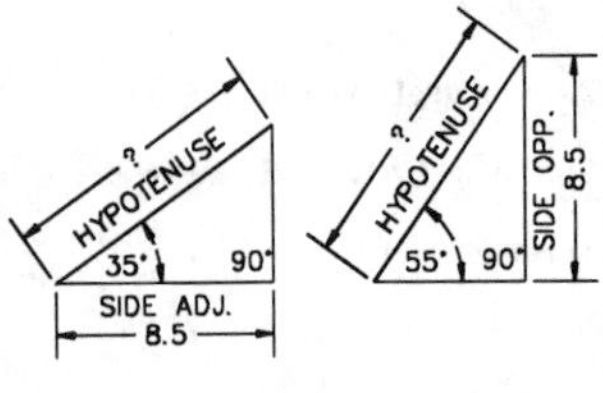

10. **Find the length of a side when angle C and two sides are known.**

Formula: $a^2 + b^2 = c^2$ See page 204. Or see the Rule of Pythagoras, Task Number K5A, page 132.

How to Use the Scientific Calculator to Find Trig Functions and Angles:

Note: Make sure the Calculator is in the DEG mode.
The display should show:

Texas Instruments
TI-36X Solar
DEG 0.
Display

SCIENTIFIC

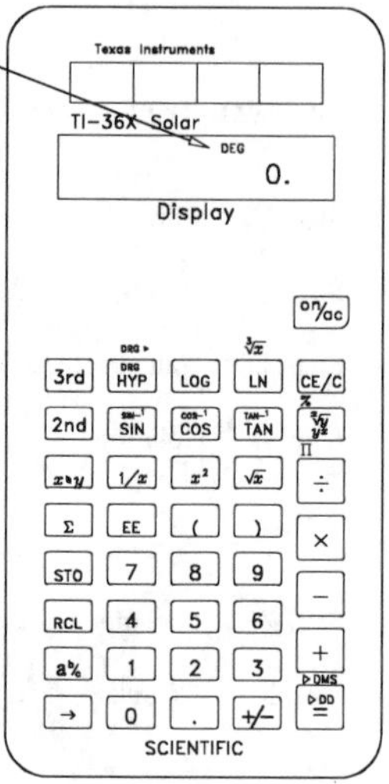

How to Calculate the Functions of Angles:

1. Calculate the Sine, Cosine, and Tangent of angles.

Example 1. Find the SIN of 45° angle.

Select: [4] [5] [SIN]

The display should show: DEG 0.707106781 Display

Example 2. Find the COS of 52° angle.

Select: [5] [2] [COS]

The display should show: DEG 0.615661475 Display

Example 3. Find the TAN of 15° angle.

Select: [1] [5] [TAN]

The display should show: DEG 0.267949192 Display

Calculate the Angles from the Functions of Angles:

2. Calculate the Angle from the Sine, Cosine, and Tangent angle functions.

Example 1. Find the Angle from the SIN. 0.707106781 (Carry Decimal 3–5 Places).

Select: [.] [7] [0] [7] [1] [2nd] [SIN]

The display should show: 45° DEG 44.99945053 Display

Example 2. Find the Angle from the COS. 0.615661475 (Carry Decimal 3–5 Places).

Select: [.] [6] [1] [5] [6] [2nd] [COS]

The display should show: 52° DEG 52.0044697 Display

Example 3. Find the Angle from the TAN. 0.267949192 (Carry Decimal 3–5 Places).

Select: [.] [2] [6] [7] [9] [2nd] [TAN]

The display should show: 15° DEG 14.99737025 Display

[Note how the Functions of Angles and the Angles themselves relate to each other. This can also be verified in the Trig Tables, pages 220 – 264.]

[Note: Most scientific calculators operate similar to the one above.]

TO SOLVE BASIC TRIG PROBLEMS WITH A SCIENTIFIC CALCULATOR, SEE THE NEXT PAGE.

PRACTICAL PROBLEMS USING BASIC TRIG AND THE SCIENTIFIC CALCULATOR:

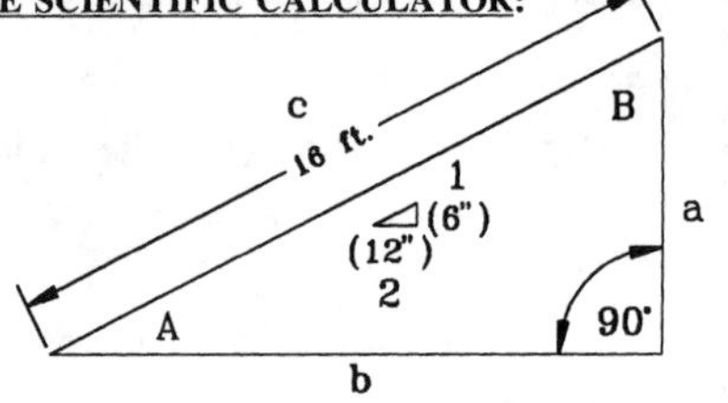

Example Problem 1. Given a utility shed with a 16 ft. length roof with a 2 to 1 slope as per sketch.

Find: a) Vertical height **a**
b) Angle **A**

First, solve by using trigonometry and the trig tables:

Find angle A.
- Step 1. Formula: **Tan A** $= \frac{a}{b}$
- Step 2. Substitute: **Tan A** $= \frac{1}{2}$
- Step 3. Calculate: **Tan A** $= 1 \div 2$ or $2\overline{)1.0} = .5$
- Step 4. Solution: **Angle A** = 26°34' (from trig tables, angle whose tan = 0.5 = 26°34')

Find side a.
- Step 1. Formula: **Sin A** $= \frac{a}{c}$
- Step 2. Substitute: **.4472** $= \frac{a}{16}$ (Sin A = .4472 from trig tables)
- Step 3. Calculate: **a** = 16 X .4472 (derived from multiplying both sides by 16)
- Step 4. Solution: **a = 7.1558' or 7'1 7/8"** (.1558' X 12 = 1.869")

Second, solve by using trig and the scientific calculator (Radio Shack EC-4039):

[NOTE: The procedure for calculating trig problems with a calculator could vary from calculator to calculator. Therefore, read the instructions with the owner's manual before calculating.]

Find Angle A.
- Step 1. Enter 1 (Display 1.) ← **From Step 2 above**
- Step 2. Enter ÷ (Display 1. ÷)
- Step 3. Enter 2 (Display 2.)
- Step 4. Enter = (Display 0.5)
- Step 5. Enter shift (Display [S] --- 0.5) note shift key used to obtain 2nd function of Tan Key which is indicated as Tan^{-1}, (or Arc Tan, or Angle whose Tangent is a certain value).
- Step 6. Enter Tan (Display Tan^{-1} --- 0.5)
- Step 7. Enter (Display 26.56505°) = Angle A

Find side a.
- Step 1. Enter Sin (Display Sin -- 0)
- Step 2. Enter 26.565 (Display Sin --- 26.565)
- Step 3. Enter = (Display 0.4472--) = Sin A
- Step 4. Enter X (Display 0.4472--X)
- Step 5. Enter 16 (Display 16.)
- Step 6. Enter = (Display 7.1554--) = 7'1 7/8"

Check by using the Rule of Pythagoras

$a^2 + b^2 = c^2$
$b^2 = c^2 - a^2$ (from algebra)
$b^2 = 16^2 - 7.1554^2$
$b^2 = 256 - 51.1997$
$b^2 = 204.8002$
$b = \sqrt{204.8002}$
b = 14.3108 or 14' 3 3/4"
$(7.155)^2 + (14.3108)^2 = (16)^2$
51.1997 + 204.8002 = 256
256 = 256

Task No. ▸ M1

Example
Problem 2. Given the vertical and slope distance with slope angle for 1/2" plywood gusset.

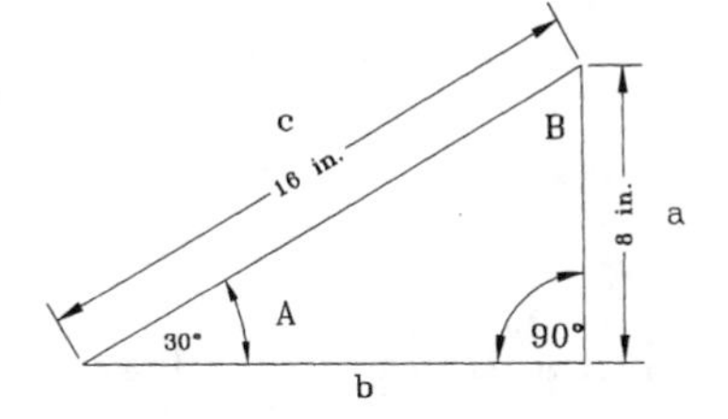

Find: a) Width of plywood cut needed (side **b**)
b) Angle **B**

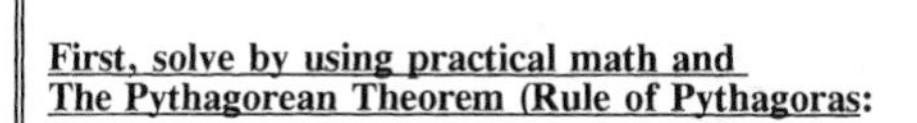

First, solve by using practical math and The Pythagorean Theorem (Rule of Pythagoras:

Find side **b**

Step 1.	Formula:	$\mathbf{a^2 + b^2 = c^2}$
Step 2.	Substitute:	$\mathbf{8^2 + b^2 = 16^2}$
Step 3.	Calculate:	$\mathbf{b^2 = 16^2 - 8^2}$ (from basic algebra; 8^2 is transposed to other side of equation with sign changed from + to -)
Step 4.	Calculate:	$\mathbf{b^2 = 256 - 64 = 192}$
Step 5.	Solution:	$\mathbf{b = \sqrt{192} = 13.856 = 13\ 7/8}$" (nearest 1/16")

Find angle **B**

Step 1.	Calculate:	**90°- 30°** (3 angles must total 180°)
Step 2.	Solution:	**Angle B = 60°**

Second, solve by using trigonometry and the trig tables:

Find side **b**.

Step 1.	Formula:	$\mathbf{Cosine\ A = \frac{b}{c}}$ (b = c X Cosine A from algebra)
Step 2.	Substitute:	**b = 16 X Cosine 30°** (multiply both sides by 16)
Step 3.	Calculate:	**b = 16 X .86603** (from trig tables)
Step 4.	Solution:	**b = 13.856** (= 13 7/8")

Third, solve by using trig and the scientific calculator (Radio Shack EC-4039):

Find side **b**.

Step 1.	**b = 16 X cos 30°** (from Step 2 above)
Step 2.	Enter cos (Display cos -- 0.)
Step 3.	Enter 30° (Display cos -- 30.)
Step 4.	Enter = (Display 0.8660---)
Step 5.	Enter X (Display 0.8660 --- X)
Step 6.	Enter 16 (Display 16.)
Step 7.	Enter = (Display 13.8564---) = 13 7/8"

Fourth, other methods:

Step 1. Mark 8" on short leg of framing square (see sketch)

Step 2. Hold 0" with tape at 8" mark.

Step 3. Move tape (while holding 0 at 8" mark) until 16" intersects long side of square.

Step 4. Reading at point of intersection should be 13 7/8".

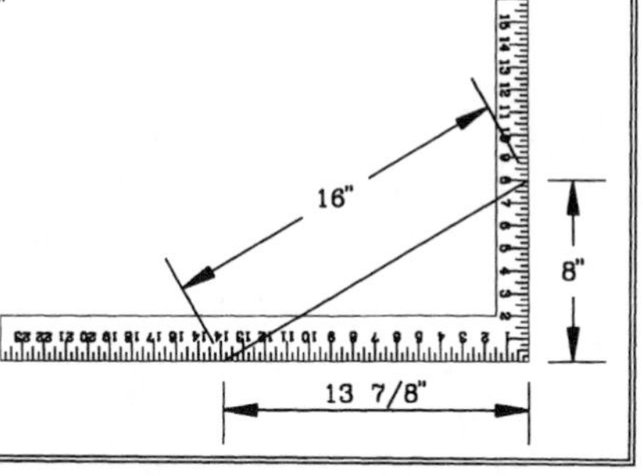

Example Problem 3. Given a garage roof that has a 6 foot rise and a 18 foot run as per sketch.

Find: a) Slope distance (side **c**)
b) Angle **A**

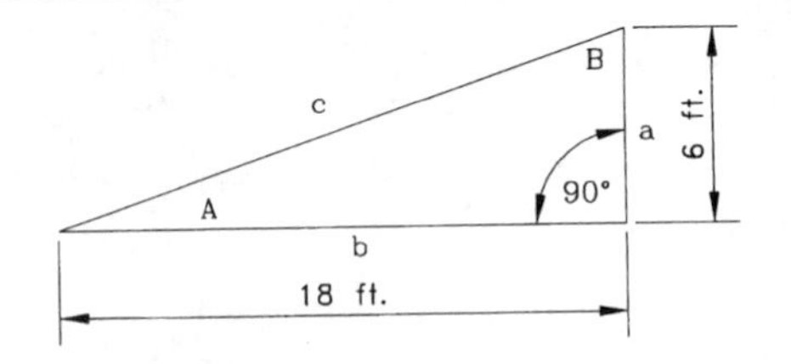

First, solve by using practical math and The Pythagorean Theorem (Rule of Pythagoras):

Find side **c**

Step 1.	Formula:	$c^2 = a^2 + b^2$
Step 2.	Substitute:	$c^2 = 6^2 + 18^2$
Step 3.	Calculate:	$c^2 = 36 + 324$
Step 4.	Calculate:	$c^2 = 360$
Step 5.	Solution:	$c = \sqrt{360}$ = **18.9736 = 18' 11 11/16"** (18' + .937 X 12")

[NOTE: **The procedure for calculating trig problems with a calculator could vary from calculator to calculator. Therefore, read the instructions with the owner's manual before calculating.]**

Second, solve by using practical math and the scientific calculator (Radio Shack EC-4039):

Find side **c**.

Step 1. Enter 6 (Display 6.) from Step 2 above.
Step 2. Enter X^2 (Display 36)
Step 3. Enter Shift (S 36.)
Step 4. Enter Min (Display 36.) (Min is "Memory in")
Step 5. Enter 18 (Display 18)
Step 6. Enter X^2 (Display 324)
Step 7. Enter + (Display 324 +)
Step 8. Enter MR (Display 36) (MR is "Memory Recall")
Step 9. Enter = (Display 360)
Step 10. Enter √ (Display √ 360)
Step 11. Enter = (18.97366') = 18' 11 11/16" (.97366 X 12" = 11.683")

Third, solve by using trigonometry and the trig tables:

Find angle **A**.

Step 1**A**.	Formula:	**Tan A** $= \frac{a}{b}$
Step 2**A**.	Substitute:	**Tan A** $= \frac{6}{18}$
Step 3**A**.	Calculate:	**Tan A = 6 ÷ 18 or 6/18 = 0.3333**
Step 4**A**.	Solution:	**Angle A = 18°26'**(from trig tables, Angle whose tan = 0.3333)

Find side **c**

Step 1**c**.	Formula:	**Sin A** $= \frac{a}{c}$
Step 2**c**.	Substitute:	**.3162** $= \frac{6}{c}$ (Sin A = .3162 from trig tables)
Step 3**c**.	Manipulate:	**c X .3162** $= \frac{6}{c}$ **X c**
Step 4**c**.	Calculate:	**c = 6 ÷ .3162**
Step 5**c**.	Solution:	Side **c = 18.975'* = 18' 11 11/16"** (18' + .975 X 12") *difference due to rounding off in tables.

Fourth, solve by using trig and the scientific calculator (Radio Shack EC-4039):

Find
Angle **A**

Step 1. From Step 2A, enter 6 (Display 6.)
Step 2. Enter ÷ (Display 6 ÷)
Step 3. Enter 18 (Display 18.)
Step 4. Enter = (Display 0.3333) = Tan A
Step 5. Enter shift (Display [S]0.33333--)
Step 6. Enter Tan^{-1} (Display Tan^{-1} 0.33333--)
Step 7. Enter = (Display 18.43494--) = Angle A

Find
side **c**

Step 1. From Step 4c, enter 6 (Display --- 6)
Step 2. Enter ÷ (Display --- 6 ÷)
Step 3. Enter .3162 (Display --- 0.3162)
Step 4. Enter (Display 18.9753 ---) = side c
= 18' - 11 11/16"

Fifth, tricks of the trade/other methods:

Step 1. Mark off 18" on long side (blade) of framing square (from 90° angle).

Step 2. Mark off 6" on short side (tongue) of framing square (from 90° angle).

Step 3. Check hypotenuse (should be 18.975")

Step 4. Multiply by 12 to get correct feet and inches (18.975 X 12 = 227.7" = 18' 11 11/16")

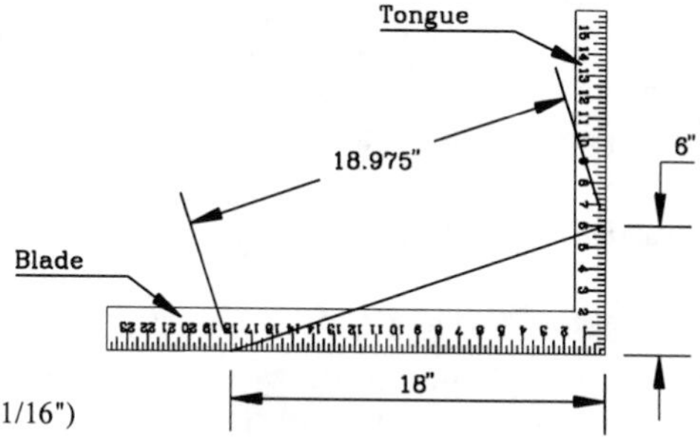

[FIND ANGLE A OR ANGLE B WHEN SIDE a AND SIDE c ARE KNOWN]

Use Formula 1A: Sine A = $\frac{a}{c}$

Example 1. **Find Angle A** (when a and c are known)

Given:

FORMULA: **Sine A = $\frac{a}{c}$**

Substitute: Sine A = $\frac{5.9518}{10.3766}$

Calculate: Sine A = .57358

Angle A = 35°

Use Formula 1B: Cosine B = $\frac{a}{c}$

Example 2. **Find Angle B** (when a and c are known)

Cosine B = $\frac{a}{c}$

Cosine B = $\frac{5.9518}{10.3766}$

Cosine B = .57358

Angle B = 55°

[TRIG TABLES]

35°

M	Sine	Cosine	Tan.	Cotan.	Secant	Cosec.	M
0	.57358	.81915	.70021	1.4281	1.2208	1.7434	60
1	.57381	.81898	.70064	.4273	.2210	.7427	59
2	.57405	.81882	.70107	.4264	.2213	.7420	58
3	.57429	.81865	.70151	.4255	.2215	.7413	57
4	.57453	.81848	.70194	.4246	.2218	.7405	56

56	.57262	.81982	.69847	.4317	.2198	.7463	4
57	.57286	.81965	.69891	.4308	.2200	.7456	3
58	.57310	.81948	.69934	.4299	.2203	.7449	2
59	.57334	.81932	.69977	.4290	.2205	.7442	1
60	.57358	.81915	.70021	1.4281	1.2208	1.7434	0
M	Cosine	Sine	Cotan.	Tan.	Cosec.	Secant	M

55°

Note: Sine A = $\frac{a}{c}$ or

Cosecant A = $\frac{c}{a}$

Therefore: Cosecant A = $\frac{10.3766}{5.9518}$

Cosecant A = 1.7434

Angle A = 35°

Note: Cosine B = $\frac{a}{c}$ or

Secant B = $\frac{c}{a}$

Therefore: Secant B = $\frac{10.3766}{5.9518}$

Secant B = 1.7434

Angle B = 55°

RECIPROCAL FORMULAS

[FIND ANGLE WHEN SIDE OPPOSITE OR SIDE ADJACENT, AND HYPOTENUSE ARE KNOWN]

Example 1. **Find Angle Shown.** (when side opp. and hyp. are known)

Given:

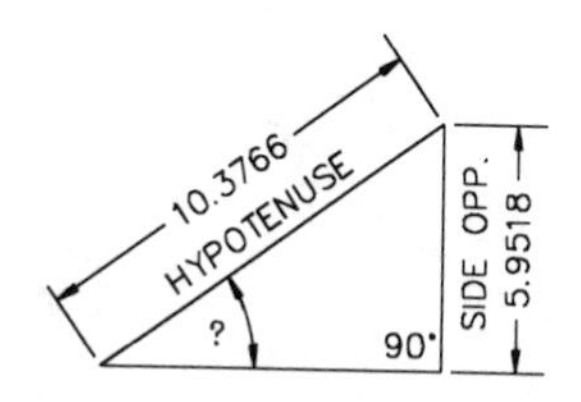

Example 2. **Find Angle Shown.** (when side adj. and hyp. are known)

10.3766
HYPOTENUSE
?
SIDE ADJ
5.9518

Select Correct Formula:

- **SOH (Sine = $\frac{\text{side opp.}}{\text{hyp.}}$)**
- **CAH (Cosine = $\frac{\text{side adj.}}{\text{hyp.}}$)**
- **TOA (Tangent = $\frac{\text{side opp.}}{\text{side adj.}}$)**

	Example 1	Example 2
FORMULA:	**Sine = $\frac{\text{side opp.}}{\text{hyp.}}$**	**Cosine = $\frac{\text{side adj.}}{\text{hyp.}}$**
Substitute:	Sine = $\frac{5.9518}{10.3766}$	Cosine = $\frac{5.9518}{10.3766}$
Calculate:	Sine = .57358	Cosine = .57358
	[Refer to trig. tables for sine .57358]	[Refer to the trig. tables for the cosine .57358]
	Angle = 35°	**Angle = 55°**

NOTE: One of the biggest advantages of using this formula versus the traditional method on the preceding page, is that you don't have to label the angles or the sides of the triangle, and it is much easier to memorize all the formulas. Use the method(s) that work best for you.

Calculator Solution (Find Angle A):

Step 1. Enter 5.9518 (Display 5.9518)
Step 2. Enter ÷ (Display 5.9518 ÷)
Step 3. Enter 10.3766 (Display 10.3766)
Step 4. Enter = (Display 0.57357---) = Sin A
Step 5. Enter Shift (Display [S] 0.57357---)
Step 6. Enter *Sin^{-1} (Display Sin^{-1} 0.57357---)
Step 7. Enter = (Display 35.000---) = Angle A
*Second function of "Sin" key

(Find Angle B):

Enter 5.9518 (Display 5.9518)
Enter ÷ (Display 5.9518 ÷)
Enter 10.3766 (Display 10.3766)
Enter = (Display 0.5735---) = Cos B
Enter Shift (Display [S] 0.5735---)
Enter Cos^{-1} (Display Cos^{-1} 0.5735---)
Enter = (Display 54.9998---) = 55° = Angle B

NOTE: Make sure the calculator is in the degree mode (instructions with owner's manual for EC-4039 Scientific Calculator). Display will show "DEG" at top of center display area.

NOTE: **The procedure for calculating trig problems with a calculator would vary from calculator to calculator. Therefore, read the instructions with the owner's manual before calculating.**

[FIND ANGLE A OR ANGLE B WHEN SIDE b AND SIDE c ARE KNOWN]

Use Formula 2A: Cosine A = $\frac{b}{c}$

Example 1. **Find Angle A** (when b and c are known)

Given:

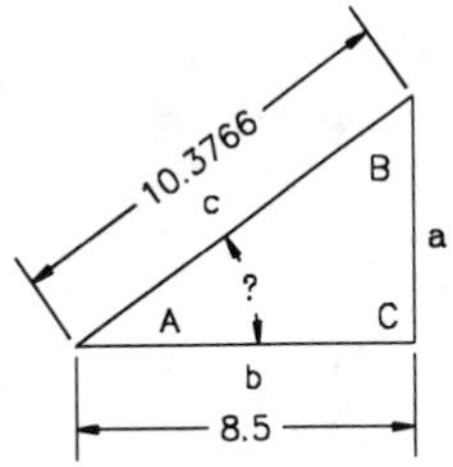

FORMULA: Cosine A = $\frac{b}{c}$

Substitute: Cosine A = $\frac{8.5}{10.3766}$

Calculate: Cosine A = .81915

Angle A = 35°

Use Formula 2B: Sine B = $\frac{b}{c}$

Example 2. **Find Angle B** (when b and c are known)

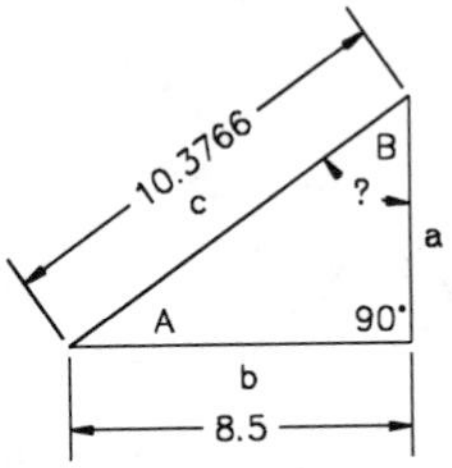

Sine B = $\frac{b}{c}$

Sine B = $\frac{8.5}{10.3766}$

Sine B = .81915

Angle B = 55°

[TRIG TABLES]

35°

M	Sine	Cosine	Tan.	Cotan.	Secant	Cosec.	M
0	.57358	.81915	.70021	1.4281	1.2208	1.7434	60
1	.57381	.81898	.70064	.4273	.2210	.7427	59
2	.57405	.81882	.70107	.4264	.2213	.7420	58
3	.57429	.81865	.70151	.4255	.2215	.7413	57
4	.57453	.81848	.70194	.4246	.2218	.7405	56

56	.57262	.81982	.69847	.4317	.2198	.7463	4
57	.57286	.81965	.69891	.4308	.2200	.7456	3
58	.57310	.81948	.69934	.4299	.2203	.7449	2
59	.57334	.81932	.69977	.4290	.2205	.7442	1
60	.57358	.81915	.70021	1.4281	1.2208	1.7434	0
M	Cosine	Sine	Cotan.	Tan.	Cosec.	Secant	M

55°

Note: Cosine A = $\frac{b}{c}$ <u>or</u>

Secant A = $\frac{c}{b}$

Therefore: Secant A = $\frac{10.3766}{8.5}$

Secant A = 1.2208

Angle A = 35°

Note: Sine B = $\frac{b}{c}$ <u>or</u>

Cosecant B = $\frac{c}{b}$

Therefore: Cosecant B = $\frac{10.3766}{8.5}$

Cosecant B = 1.2208

Angle B = 55°

RECIPROCAL FORMULAS

[FIND ANGLE WHEN SIDE ADJACENT OR SIDE OPPOSITE, AND HYPOTENUSE ARE KNOWN]

Example 1. **Find Angle Shown.** (when side adj. and hyp. are known)

Given:

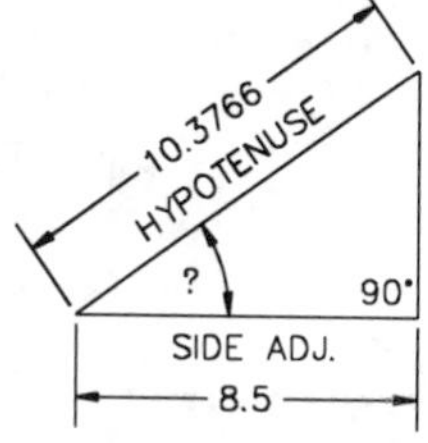

Select Correct Formula:

- **SOH (Sine = $\frac{\text{side opp.}}{\text{hyp.}}$)**
- **CAH (Cosine = $\frac{\text{side adj.}}{\text{hyp.}}$)**
- **TOA (Tangent = $\frac{\text{side opp.}}{\text{side adj.}}$)**

FORMULA: **Cosine = $\frac{\text{side adj.}}{\text{hyp.}}$**

Substitute: Cosine = $\frac{8.5}{10.3766}$

Calculate: Cosine = .81915

[Refer to trig. tables for cosine .81915]

Angle = 35°

Example 2. **Find Angle Shown.** (when side opp. and hyp. are known)

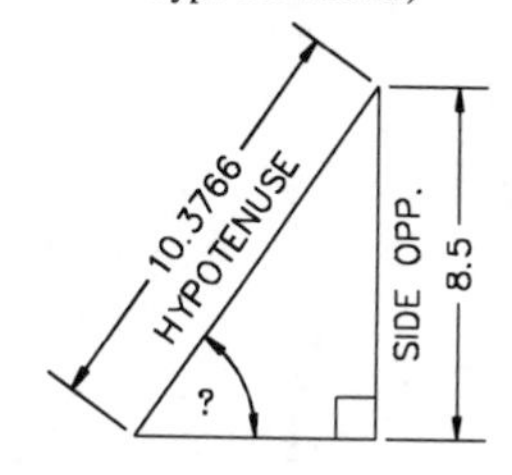

Sine = $\frac{\text{side opp.}}{\text{hyp.}}$

Sine = $\frac{8.5}{10.3766}$

Sine = .81915

[Refer to the trig. tables for the sine .81915]

Angle = 55°

NOTE: One of the biggest advantages of using this formula versus the traditional method on the preceding page, is that you don't have to label the angles or the sides of the triangle, and it is much easier to memorize all the formulas. Use the method(s) that work best for you.

M2	◂ Task No.	TRIGONOMETRY AND RIGHT TRIANGLES

[FIND ANGLE A OR ANGLE B WHEN SIDE a AND SIDE b ARE KNOWN]

Use Formula 3A: Tangent A = $\frac{a}{b}$

Example 1. **Find Angle A**
(when a and b are known)

Given:

FORMULA: Tangent A = $\frac{a}{b}$

Substitute: Tangent A = $\frac{5.9518}{8.5}$

Calculate: Tangent A = .70021

Angle A = 35°

Use Formula 3B: Cotangent B = $\frac{a}{b}$

Example 2. **Find Angle B**
(when a and b are known)

Cotangent B = $\frac{a}{b}$

Cotangent B = $\frac{5.9518}{8.5}$

Cotangent B = .70021

Angle B = 55°

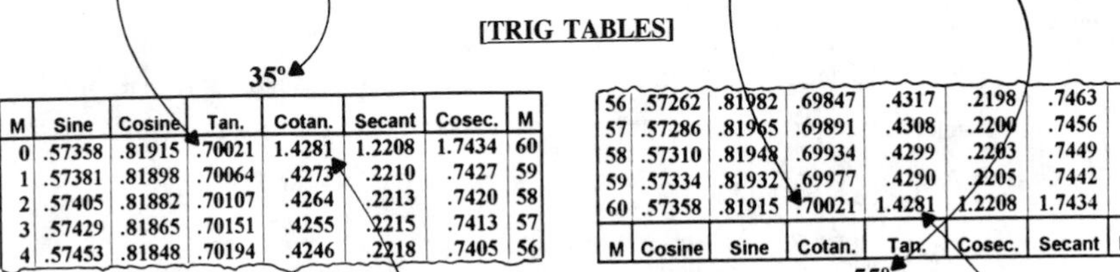

[TRIG TABLES]

35°

M	Sine	Cosine	Tan.	Cotan.	Secant	Cosec.	M
0	.57358	.81915	.70021	1.4281	1.2208	1.7434	60
1	.57381	.81898	.70064	.4273	.2210	.7427	59
2	.57405	.81882	.70107	.4264	.2213	.7420	58
3	.57429	.81865	.70151	.4255	.2215	.7413	57
4	.57453	.81848	.70194	.4246	.2218	.7405	56

56	.57262	.81982	.69847	.4317	.2198	.7463	4
57	.57286	.81965	.69891	.4308	.2200	.7456	3
58	.57310	.81948	.69934	.4299	.2203	.7449	2
59	.57334	.81932	.69977	.4290	.2205	.7442	1
60	.57358	.81915	.70021	1.4281	1.2208	1.7434	0
M	Cosine	Sine	Cotan.	Tan.	Cosec.	Secant	M

55°

Note: Tangent A = $\frac{a}{b}$ <u>or</u>

Cotangent A = $\frac{b}{a}$

Therefore: Cotangent A = $\frac{8.5}{5.9518}$

Cotangent A = 1.4281

Angle A = 35°

Note: Cotangent B = $\frac{a}{b}$ <u>or</u>

Tangent B = $\frac{b}{a}$

Therefore: Tangent B = $\frac{8.5}{5.9518}$

Tangent B = 1.4281

Angle B = 55°

RECIPROCAL FORMULAS

[FIND ANGLE WHEN SIDE ADJACENT AND SIDE OPPOSITE ARE KNOWN]

Example 1. **Find Angle Shown.** (when side adj. and and side opp. are known)

Given:

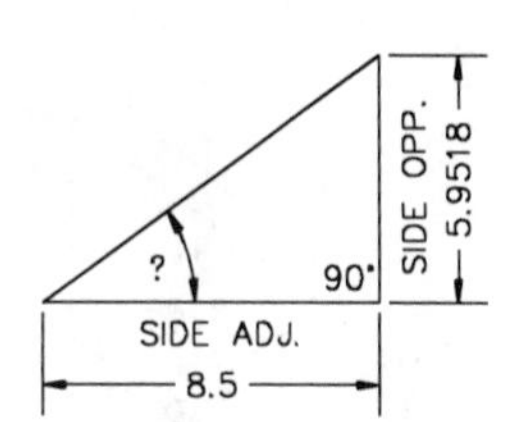

Select Correct Formula:

- **SOH** (Sine $= \frac{\text{side opp.}}{\text{hyp.}}$)
- **CAH** (Cosine $= \frac{\text{side adj.}}{\text{hyp.}}$)
- **TOA** (Tangent $= \frac{\text{side opp.}}{\text{side adj.}}$)

FORMULA: **Tangent** $= \frac{\text{side opp.}}{\text{side adj.}}$

Substitute: Tangent $= \frac{5.9518}{8.5}$

Calculate: Tangent = .70021

[Refer to trig. tables for tangent .70021]

Angle = 35°

Example 2. **Find Angle Shown.** (when side opp. and side adj. are known)

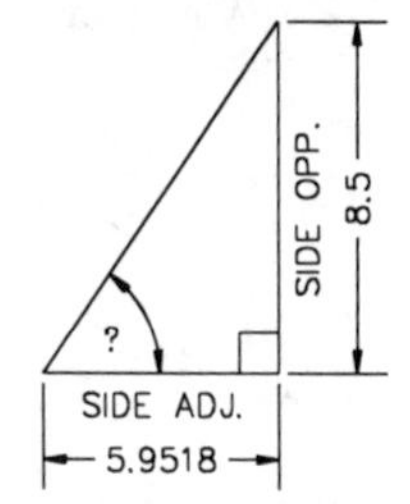

Tangent $= \frac{\text{side opp.}}{\text{side adj.}}$

Tangent $= \frac{8.5}{5.9518}$

Tangent = 1.4281

[Refer to the trig. tables for the tangent 1.4281]

Angle = 55°

NOTE: One of the biggest advantages of using this formula versus the traditional method on the preceding page, is that you don't have to label the angles or the sides of the triangle, and it is much easier to memorize all the formulas. Use the method(s) that work best for you.

[FIND LENGTH OF a WHEN ANGLE A OR ANGLE B, AND SIDE b ARE KNOWN]

Use Formula 3A: Tangent A = $\frac{a}{b}$

Example 1. **Find length of a.**
(when A and b are known)

Given:

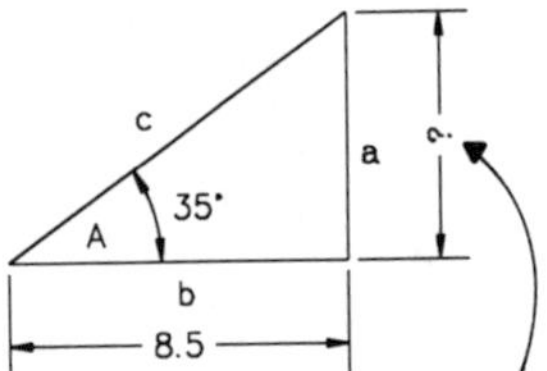

FORMULA: Tangent A = $\frac{a}{b}$

b X Tan A = a

Substitute: 8.5 X .70021 = a

Calculate: **5.9518 = a**

Use Formula 3B: Cotangent B = $\frac{a}{b}$

Example 2. **Find Length of a.**
(when B and b are known)

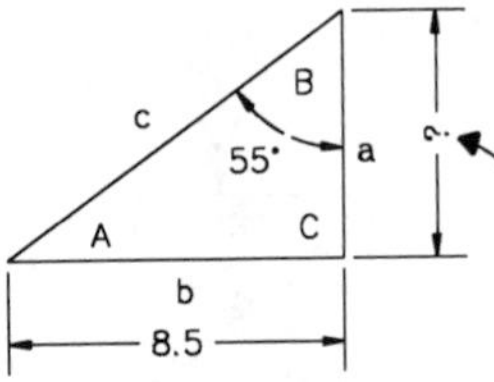

Cotangent B = $\frac{a}{b}$

b X Cot B = a

8.5 X .70021 = a

5.9518 = a

[TRIG TABLES]

35°

M	Sine	Cosine	Tan.	Cotan.	Secant	Cosec.	M
0	.57358	.81915	.70021	1.4281	1.2208	1.7434	60
1	.57381	.81898	.70064	.4273	.2210	.7427	59
2	.57405	.81882	.70107	.4264	.2213	.7420	58
3	.57429	.81865	.70151	.4255	.2215	.7413	57
4	.57453	.81848	.70194	.4246	.2218	.7405	56

56	.57262	.81982	.69847	.4317	.2198	.7463	4
57	.57286	.81965	.69891	.4308	.2200	.7456	3
58	.57310	.81948	.69934	.4299	.2203	.7449	2
59	.57334	.81932	.69977	.4290	.2205	.7442	1
60	.57358	.81915	.70021	1.4281	1.2208	1.7434	0
M	Cosine	Sine	Cotan.	Tan.	Cosec.	Secant	M

55°

RECIPROCAL FORMULAS

Note: Tangent A = $\frac{a}{b}$ or

Cotangent A = $\frac{b}{a}$

Therefore: Cotangent A = $\frac{b}{a}$

a X Cot A = b

$a = \frac{b}{\text{Cot A}}$

$a = \frac{8.5}{1.4281}$

a = 5.9519

Note: Cotangent B = $\frac{a}{b}$ or

Tangent B = $\frac{b}{a}$

Therefore: Tangent B = $\frac{b}{a}$

a X Tan B = b

$a = \frac{b}{\text{Tan B}}$

$a = \frac{8.5}{1.4281}$

a = 5.9519

[<u>FIND A SIDE</u> WHEN AN ANGLE AND ANOTHER SIDE ARE KNOWN]

Example 1. **Find side opposite.** (when an angle and side adj. are known)

Given:

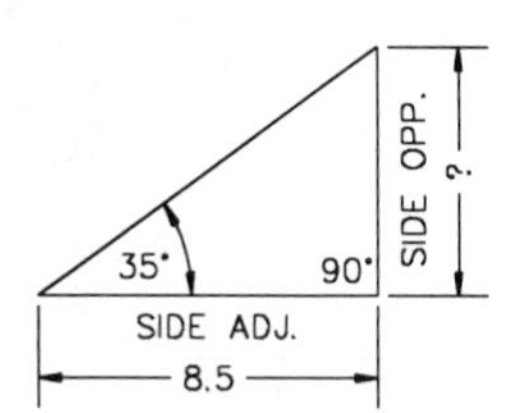

Example 2. **Find side adjacent.** (when an angle and side opp. are known)

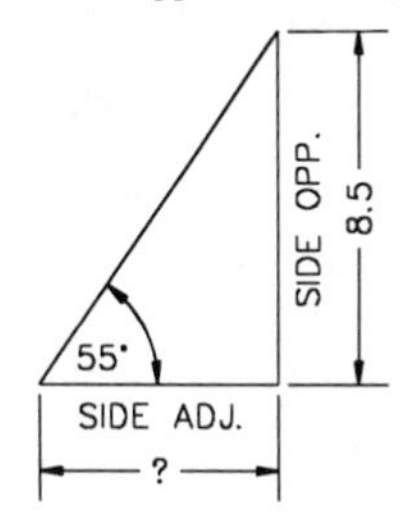

Select Correct Formula:

- **SOH (Sine = $\frac{\text{side opp.}}{\text{hyp.}}$)**
- **CAH (Cosine = $\frac{\text{side adj.}}{\text{hyp.}}$)**
- **TOA (Tangent = $\frac{\text{side opp.}}{\text{side adj.}}$)**

Example 1:

FORMULA: **Tangent = $\frac{\text{side opp.}}{\text{side adj.}}$**

side adj. X Tangent = side opp.

Substitute: 8.5 X .70021 = side opp.

Calculate: **5.9518 = side opp.**

Example 2:

Tangent = $\frac{\text{side opp.}}{\text{side adj.}}$

side adj. X Tangent = side opp.

side adj. = $\frac{\text{side opp.}}{\text{Tangent}}$

side adj. = $\frac{8.5}{1.4281}$

side adj. = 5.9519

NOTE: One of the biggest advantages of using this formula versus the traditional method on the preceding page, is that you don't have to label the angles or the sides of the triangle, and it is much easier to memorize all the formulas. Use the method(s) that work best for you.

M2	◂ Task No.	TRIGONOMETRY AND RIGHT TRIANGLES

[FIND LENGTH OF a WHEN ANGLE A OR ANGLE B, AND SIDE c ARE KNOWN]

Use Formula 1A: Sine A = $\frac{a}{c}$

Example 1. **Find length of a.** (when A and c are known)

Given:

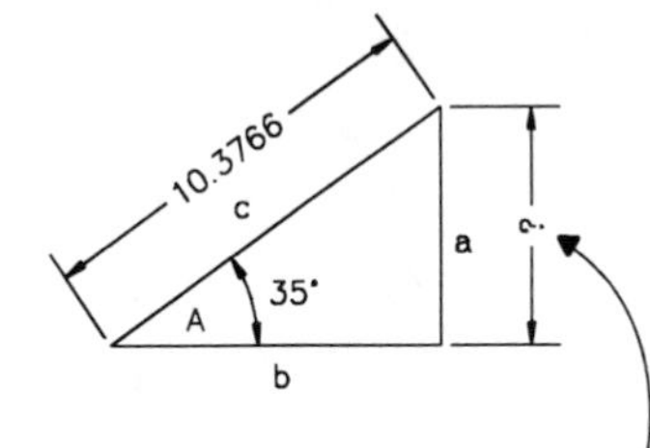

FORMULA: **Sine A = $\frac{a}{c}$**

c X Sin A = a

Substitute: 10.3766 X .57358 = a

Calculate: **5.9518 = a**

Use Formula 1B: Cosine B = $\frac{a}{c}$

Example 2. **Find Length of a.** (when B and c are known)

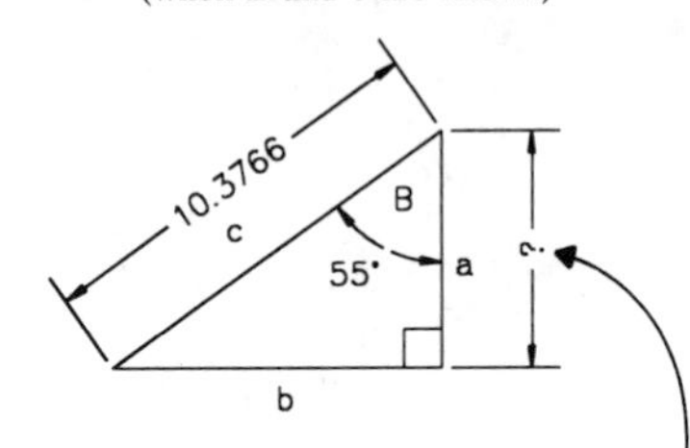

Cosine B = $\frac{a}{c}$

c X Cos B = a

10.3766 X .57358 = a

5.9518 = a

[TRIG TABLES]

35°

M	Sine	Cosine	Tan.	Cotan.	Secant	Cosec.	M
0	.57358	.81915	.70021	1.4281	1.2208	1.7434	60
1	.57381	.81898	.70064	.4273	.2210	.7427	59
2	.57405	.81882	.70107	.4264	.2213	.7420	58
3	.57429	.81865	.70151	.4255	.2215	.7413	57
4	.57453	.81848	.70194	.4246	.2218	.7405	56

56	.57262	.81982	.69847	.4317	.2198	.7463	4
57	.57286	.81965	.69891	.4308	.2200	.7456	3
58	.57310	.81948	.69934	.4299	.2203	.7449	2
59	.57334	.81932	.69977	.4290	.2205	.7442	1
60	.57358	.81915	.70021	1.4281	1.2208	1.7434	0
M	Cosine	Sine	Cotan.	Tan.	Cosec.	Secant	M

55°

RECIPROCAL FORMULAS

Note: Sine A = $\frac{a}{c}$ or

Cosecant A = $\frac{c}{a}$

Therefore: Cosecant A = $\frac{c}{a}$

a X Csc A = c

$a = \frac{c}{\text{Csc A}}$

$a = \frac{10.3766}{1.7434}$

a = 5.9519

Note: Cosine B = $\frac{a}{c}$ or

Secant B = $\frac{c}{a}$

Therefore: Secant B = $\frac{c}{a}$

a X Sec B = c

$a = \frac{c}{\text{Sec B}}$

$a = \frac{1.7434}{10.3766}$

a = 5.9519

(Trick of the Trades Formula: SOH-CAH-TOA) Task No. ▸ M2

[FIND A SIDE WHEN AN ANGLE AND THE HYPOTENUSE ARE KNOWN]

Example 1. **Find side opposite.** (when an angle and hyp. are known)

Example 2. **Find side adjacent.** (when an angle and hyp. are known)

Given:

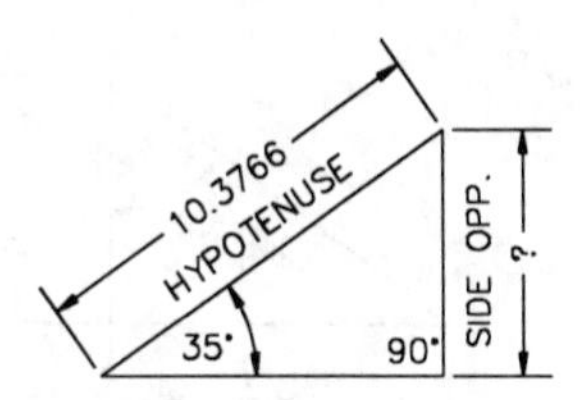

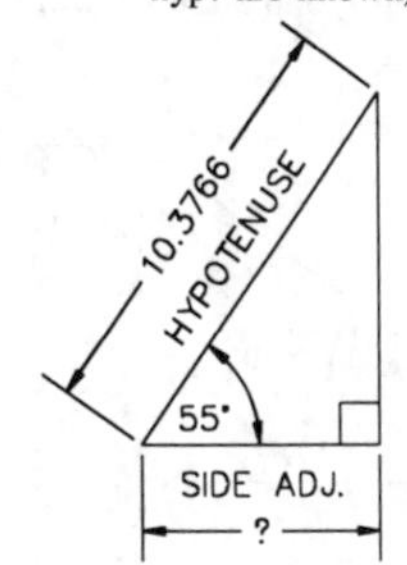

Select Correct Formula:

SOH (Sine = $\frac{\text{side opp.}}{\text{hyp.}}$)

CAH (Cosine = $\frac{\text{side adj.}}{\text{hyp.}}$)

TOA (Tangent = $\frac{\text{side opp.}}{\text{side adj.}}$)

FORMULA:

Example 1: **Sine = $\frac{\text{side opp.}}{\text{hyp.}}$**

hyp. X Sine = side opp.*

Example 2: **Cosine = $\frac{\text{side adj.}}{\text{hyp.}}$**

hyp. X Cosine = side adj.*

Substitute:

Example 1: 10.3766 X .57358 = side opp.

Example 2: 10.3766 X .57358 = side adj.

Calculate:

Example 1: **5.9518 = side opp.**

Example 2: **side adj. = 5.9518**

NOTE: One of the biggest advantages of using this formula versus the traditional method on the preceding page, is that you don't have to label the angles or the sides of the triangle, and it is much easier to memorize all the formulas. Use the method(s) that work best for you.

*Obtained by multiplying each side of equation by "hyp." (refer to examples 1 and 2 on page 175 .

[<u>FIND LENGTH OF b</u> WHEN ANGLE A OR ANGLE B, AND SIDE a ARE KNOWN]

Use Formula 3A: Tangent A = $\frac{a}{b}$

Example 1. **<u>Find length of b</u>.**
(when A and a are known)

Given:

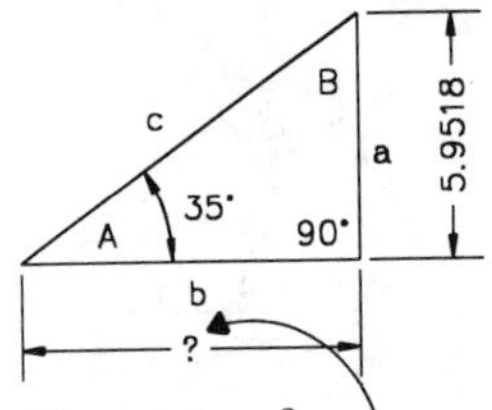

FORMULA: **Tangent A = $\frac{a}{b}$**

$b \times \text{Tan } A = a$

$b = \dfrac{a}{\text{Tan } A}$

Substitute: $b = \dfrac{5.9518}{.70021}$

Calculate: **b = 8.5**

Use Formula 3B: Cotangent B = $\frac{a}{b}$

Example 2. **<u>Find Length of b</u>.**
(when B and a are known)

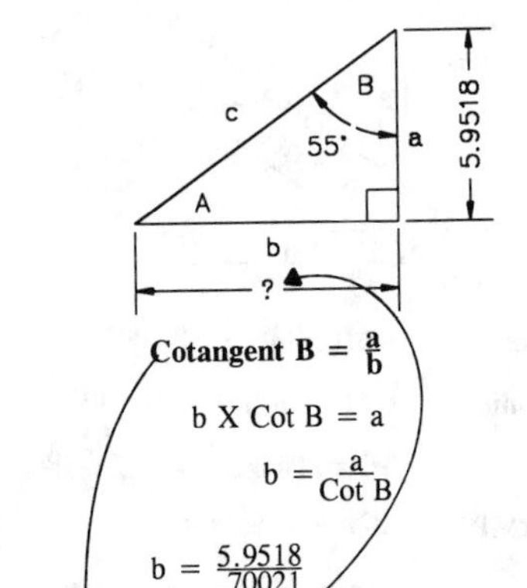

Cotangent B = $\frac{a}{b}$

$b \times \text{Cot } B = a$

$b = \dfrac{a}{\text{Cot } B}$

$b = \dfrac{5.9518}{.70021}$

b = 8.5

<u>[TRIG TABLES]</u>

35°

M	Sine	Cosine	Tan.	Cotan.	Secant	Cosec.	M
0	.57358	.81915	.70021	1.4281	1.2208	1.7434	60
1	.57381	.81898	.70064	.4273	.2210	.7427	59
2	.57405	.81882	.70107	.4264	.2213	.7420	58
3	.57429	.81865	.70151	.4255	.2215	.7413	57
4	.57453	.81848	.70194	.4246	.2218	.7405	56

M	Cosine	Sine	Cotan.	Tan.	Cosec.	Secant	M
56	.57262	.81982	.69847	.4317	.2198	.7463	4
57	.57286	.81965	.69891	.4308	.2200	.7456	3
58	.57310	.81948	.69934	.4299	.2203	.7449	2
59	.57334	.81932	.69977	.4290	.2205	.7442	1
60	.57358	.81915	.70021	1.4281	1.2208	1.7434	0

55°

RECIPROCAL FORMULAS

Note: Tangent A = $\frac{a}{b}$ <u>or</u>

Cotangent A = $\frac{b}{a}$

Therefore: Cotangent A = $\frac{b}{a}$

$a \times \text{Cot } A = b$

$5.9518 \times 1.4281 = b$

8.5 = b

Note: Cotangent B = $\frac{a}{b}$ <u>or</u>

Tangent B = $\frac{b}{a}$

Therefore: Tangent B = $\frac{b}{a}$

$a \times \text{Tan } B = b$

$5.9518 \times 1.4281 = b$

8.5 = b

(Trick of the Trades Formula: SOH-CAH-TOA) Task No. ▸ M2

[FIND A SIDE WHEN AN ANGLE AND ANOTHER SIDE ARE KNOWN]

Example 1. **Find adjacent side.** (when an angle and side opp. are known)

Given:

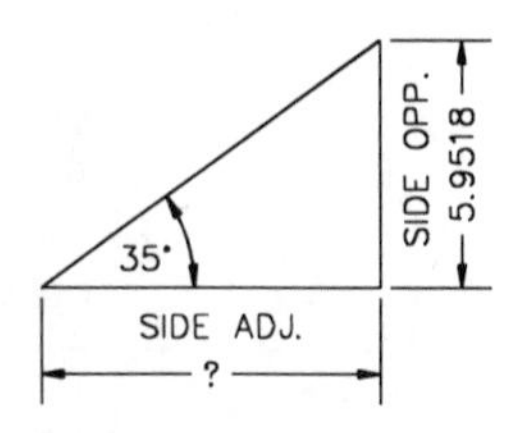

Select Correct Formula:

SOH $(\text{Sine} = \frac{\text{side opp.}}{\text{hyp.}})$

CAH $(\text{Cosine} = \frac{\text{side adj.}}{\text{hyp.}})$

TOA $(\text{Tangent} = \frac{\text{side opp.}}{\text{side adj.}})$

FORMULA: $\text{Tangent} = \frac{\text{side opp.}}{\text{side adj.}}$

side adj. X Tan. = side opp.*

$\text{side adj.} = \frac{\text{side opp.}}{\text{Tangent}}$

Substitute: $\text{side adj.} = \frac{5.9518}{.70021}$

Calculate: **side adj. = 8.5**

Example 2. **Find opposite side.** (when an angle and side adj. are known)

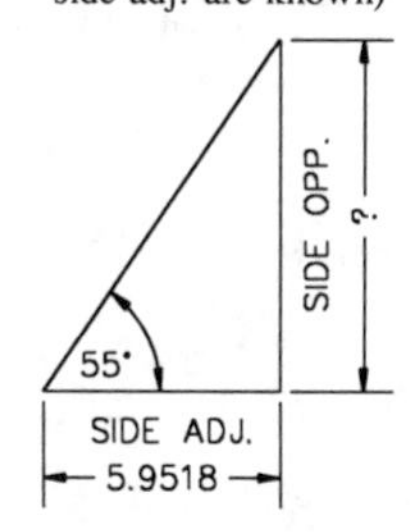

$\text{Tangent} = \frac{\text{side opp.}}{\text{side adj.}}$

side adj. X Tan. = side opp.*

5.9518 X 1.4281 = side opp.

8.499765 or 8.5 = side opp.

NOTE: One of the biggest advantages of using this formula versus the traditional method on the preceding page, is that you don't have to label the angles or the sides of the triangle, and it is much easier to memorize all the formulas. Use the method(s) that work best for you.

*Obtained by multiplying each side of equation by "side adj." (refer to examples 1 and 2 on page 175.

[FIND LENGTH OF b WHEN ANGLE A OR ANGLE B, AND SIDE c ARE KNOWN]

Use Formula 2A: Cosine A = $\frac{b}{c}$

Example 1. **Find length of b.** (when A and c are known)

Given:

10.3766 c B a 35° A 90° b ?

FORMULA: **Cosine A = $\frac{b}{c}$**

c X Cos A = b

Substitute: 10.3766 X .81915 = b

Calculate: **8.5 = b**

Use Formula 2B: Sine B = $\frac{b}{c}$

Example 2. **Find Length of b.** (when B and c are known)

10.3766 c B 55° a A b ?

Sine B = $\frac{b}{c}$

c X Sin B = b

10.3766 X .81915 = b

8.5 = b

[TRIG TABLES]

35°

M	Sine	Cosine	Tan.	Cotan.	Secant	Cosec.	M
0	.57358	.81915	.70021	1.4281	1.2208	1.7434	60
1	.57381	.81898	.70064	.4273	.2210	.7427	59
2	.57405	.81882	.70107	.4264	.2213	.7420	58
3	.57429	.81865	.70151	.4255	.2215	.7413	57
4	.57453	.81848	.70194	.4246	.2218	.7405	56

56	.57262	.81982	.69847	.4317	.2198	.7463	4
57	.57286	.81965	.69891	.4308	.2200	.7456	3
58	.57310	.81948	.69934	.4299	.2203	.7449	2
59	.57334	.81932	.69977	.4290	.2205	.7442	1
60	.57358	.81915	.70021	1.4281	1.2208	1.7434	0
M	**Cosine**	**Sine**	**Cotan.**	**Tan.**	**Cosec.**	**Secant**	**M**

55°

RECIPROCAL FORMULAS

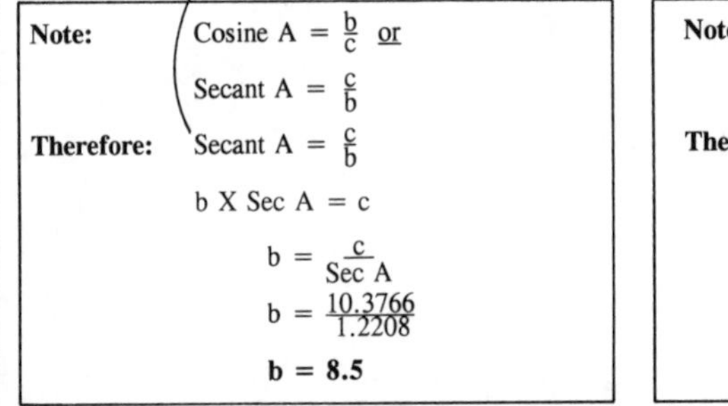

Note: Cosine A = $\frac{b}{c}$ or

Secant A = $\frac{c}{b}$

Therefore: Secant A = $\frac{c}{b}$

b X Sec A = c

$b = \frac{c}{\text{Sec A}}$

$b = \frac{10.3766}{1.2208}$

b = 8.5

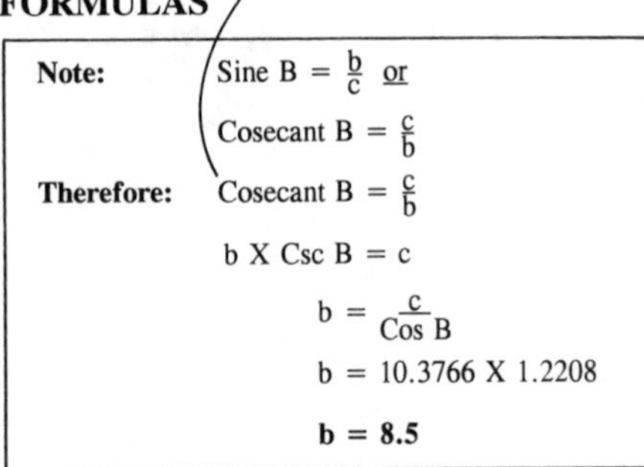

Note: Sine B = $\frac{b}{c}$ or

Cosecant B = $\frac{c}{b}$

Therefore: Cosecant B = $\frac{c}{b}$

b X Csc B = c

$b = \frac{c}{\text{Cos B}}$

b = 10.3766 X 1.2208

b = 8.5

[FIND A SIDE WHEN AN ANGLE AND THE HYPOTENUSE ARE KNOWN]

Example 1. **Find side adjacent.** (when an angle and hyp. are known)

Given:

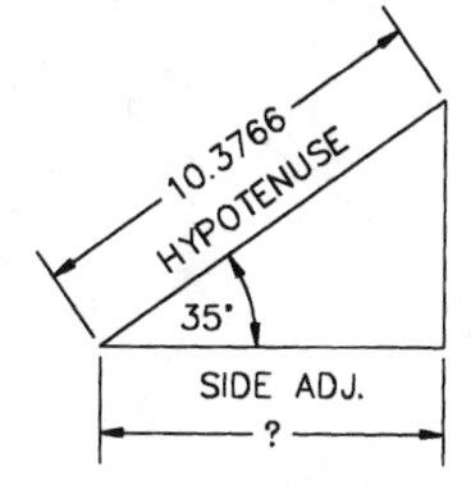

Example 2. **Find side opposite.** (when an angle and hyp. are known)

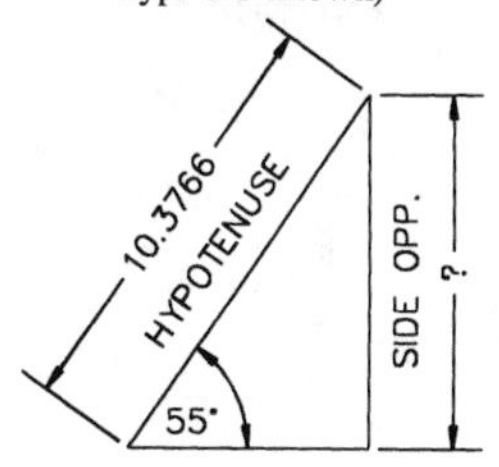

Select Correct Formula:

SOH (Sine = $\frac{\text{side opp.}}{\text{hyp.}}$)

CAH (Cosine = $\frac{\text{side adj.}}{\text{hyp.}}$)

TOA (Tangent = $\frac{\text{side opp.}}{\text{side adj.}}$)

	Example 1	Example 2
FORMULA:	**Cosine = $\frac{\text{side adj.}}{\text{hyp.}}$**	**Sine = $\frac{\text{side opp.}}{\text{hyp.}}$**
	Hyp. X Cos.. = side adj.*	Hyp. X Sin. = side opp.*
Substitute:	10.3766 X .81915 = side adj.	10.3766 X .81915 = side opp.
Calculate:	**8.49999 = 8.5**	**8.49999 or 8.5 = side opp.**

NOTE: One of the biggest advantages of using this formula versus the traditional method on the preceding page, is that you don't have to label the angles or the sides of the triangle, and it is much easier to memorize all the formulas. Use the method(s) that work best for you.

*Obtained by manipulating the formula. See example 1 on page 175.

M2	◂ Task No.	TRIGONOMETRY AND RIGHT TRIANGLES

[FIND LENGTH OF c WHEN ANGLE A OR ANGLE B, AND SIDE a ARE KNOWN]

Use Formula 1A: Sine A = $\frac{a}{c}$

Example 1. **Find length of c.** (when A and a are known)

Given:

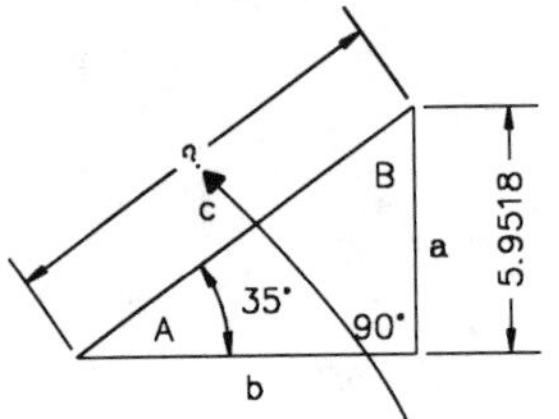

FORMULA: Sine A = $\frac{a}{c}$

c X Sin A = a

$c = \frac{a}{\text{Sin A}}$

Substitute: $c = \frac{5.9518}{.57358}$

Calculate: **c = 10.3766**

Use Formula 1B: Cosine B = $\frac{a}{c}$

Example 2. **Find Length of c.** (when B and a are known)

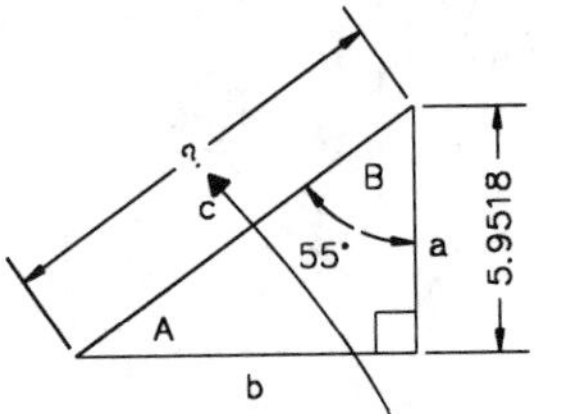

Cosine B = $\frac{a}{c}$

c X Cos B = a

$c = \frac{b}{\text{Cos B}}$

$c = \frac{5.9518}{.57358}$

c = 10.3766

[TRIG TABLES]

35°

M	Sine	Cosine	Tan.	Cotan.	Secant	Cosec.	M
0	.57358	.81915	.70021	1.4281	1.2208	1.7434	60
1	.57381	.81898	.70064	.4273	.2210	.7427	59
2	.57405	.81882	.70107	.4264	.2213	.7420	58
3	.57429	.81865	.70151	.4255	.2215	.7413	57
4	.57453	.81848	.70194	.4246	.2218	.7405	56

M	Cosine	Sine	Cotan.	Tan.	Cosec.	Secant	M
56	.57262	.81982	.69847	.4317	.2198	.7463	4
57	.57286	.81965	.69891	.4308	.2200	.7456	3
58	.57310	.81948	.69934	.4299	.2203	.7449	2
59	.57334	.81932	.69977	.4290	.2205	.7442	1
60	.57358	.81915	.70021	1.4281	1.2208	1.7434	0

55°

RECIPROCAL FORMULAS

Note: Sine A = $\frac{a}{c}$ or

Cosecant A = $\frac{c}{a}$

Therefore: Cosecant A = $\frac{c}{a}$

a X Csc A = c

5.9518 X 1.7434 = c

10.3766 = c

Note: Sine B = $\frac{a}{c}$ or

Secant B = $\frac{c}{a}$

Therefore: Secant B = $\frac{c}{a}$

a X Sec B = c

5.9518 X 1.7434 = c

10.3766 = c

(Trick of the Trades Formula: SOH-CAH-TOA) Task No. ▸ M2

[FIND HYPOTENUSE WHEN AN ANGLE AND A SIDE ARE KNOWN]

Example 1. **Find The hypotenuse.** (when an angle and side opp. are known

Given:

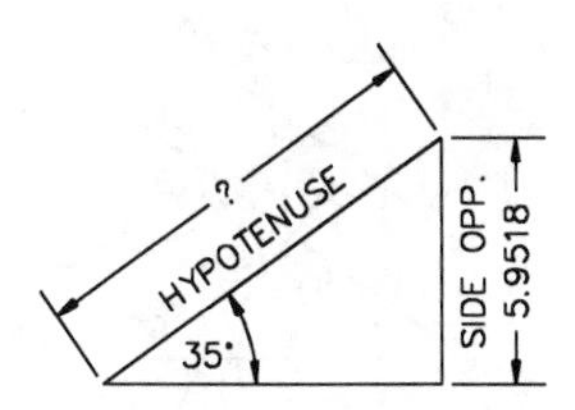

Example 2. **Find the hypotenuse.** (when an angle and side adj. are known)

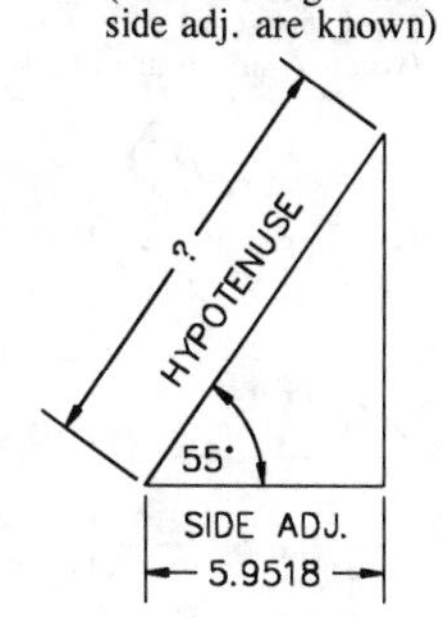

Select Correct Formula:

SOH $\left(\text{Sine} = \frac{\text{side opp.}}{\text{hyp.}}\right)$

CAH $\left(\text{Cosine} = \frac{\text{side adj.}}{\text{hyp.}}\right)$

TOA $\left(\text{Tangent} = \frac{\text{side opp.}}{\text{side adj.}}\right)$

Example 1:

FORMULA: $\text{Sine} = \frac{\text{side opp.}}{\text{hyp.}}$

hyp. X Sine = side opp.*

$\text{hyp.} = \frac{\text{side opp.}}{\text{Sine}}$

Substitute: $\text{hyp.} = \frac{5.9518}{.57358}$

Calculate: **hyp. = 10.3766**

Example 2:

$\text{Cosine} = \frac{\text{side adj.}}{\text{hyp.}}$

hyp. X Cosine = side adj.*

$\text{hyp.} = \frac{\text{side adj.}}{\text{Cosine}}$

$\text{hyp.} = \frac{5.9518}{.57358}$

hyp. = 10.3766

NOTE: One of the biggest advantages of using this formula versus the traditional method on the preceding page, is that you don't have to label the angles or the sides of the triangle, and it is much easier to memorize all the formulas. Use the method(s) that work best for you.

*Obtained by multiplying each side of equation by "hyp." (refer to examples 1 and 2 on page 175 .

M2 ◄ Task No. TRIGONOMETRY AND RIGHT TRIANGLES

[FIND LENGTH OF c WHEN ANGLE A OR ANGLE B, AND SIDE b ARE KNOWN]

Use Formula 2A: Cosine A = $\frac{b}{c}$

Example 1. **Find length of c.**
(when A and b are known)

Given:

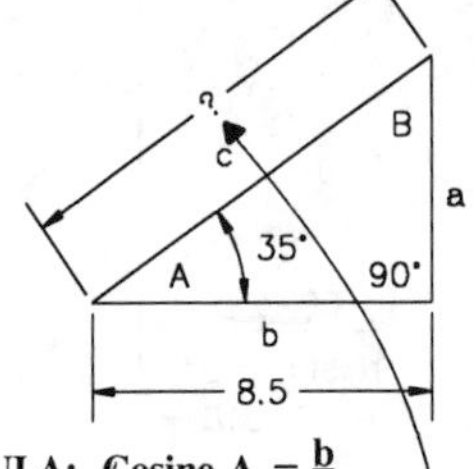

FORMULA: **Cosine A = $\frac{b}{c}$**

c X Cos A = b

$c = \frac{b}{\text{Cos A}}$

Substitute: $c = \frac{8.5}{.81915}$

Calculate: **c = 10.3766**

Use Formula 2B: Sine B = $\frac{b}{c}$

Example 2. **Find Length of c.**
(when B and b are known)

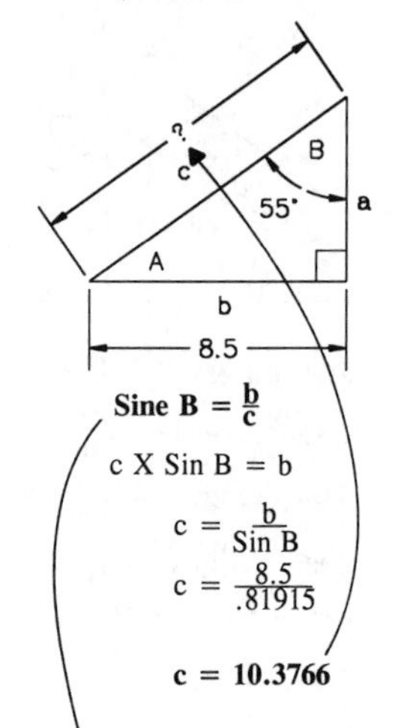

Sine B = $\frac{b}{c}$

c X Sin B = b

$c = \frac{b}{\text{Sin B}}$

$c = \frac{8.5}{.81915}$

c = 10.3766

[TRIG TABLES]

35°

M	Sine	Cosine	Tan.	Cotan.	Secant	Cosec.	M
0	.57358	.81915	.70021	1.4281	1.2208	1.7434	60
1	.57381	.81898	.70064	.4273	.2210	.7427	59
2	.57405	.81882	.70107	.4264	.2213	.7420	58
3	.57429	.81865	.70151	.4255	.2215	.7413	57
4	.57453	.81848	.70194	.4246	.2218	.7405	56

M	Cosine	Sine	Cotan.	Tan.	Cosec.	Secant	M
56	.57262	.81982	.69847	.4317	.2198	.7463	4
57	.57286	.81965	.69891	.4308	.2200	.7456	3
58	.57310	.81948	.69934	.4299	.2203	.7449	2
59	.57334	.81932	.69977	.4290	.2205	.7442	1
60	.57358	.81915	.70021	1.4281	1.2208	1.7434	0

55°

RECIPROCAL FORMULAS

Note: Cosine A = $\frac{b}{c}$ <u>or</u>

Secant A = $\frac{c}{b}$

Therefore: Secant A = $\frac{c}{b}$

b X Sec A = c

8.5 X 1.2208 = c

10.3766 = c

Note: Sine B = $\frac{b}{c}$ <u>or</u>

Cosecant B = $\frac{c}{b}$

Therefore: Cosecant B = $\frac{c}{b}$

b X Csc B = c

8.5 X 1.2208 = c

10.3766 = c

[FIND HYPOTENUSE WHEN AN ANGLE AND A SIDE ARE KNOWN]

Example 1. **Find The hypotenuse.** (when an angle and side adj. are known)

Given:

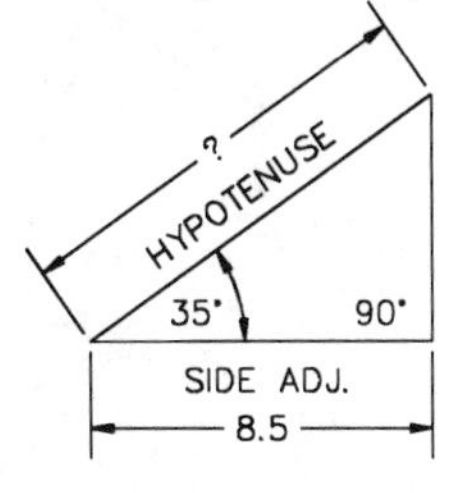

Select Correct Formula:

SOH (Sine = $\frac{\text{side opp.}}{\text{hyp.}}$)

CAH (Cosine = $\frac{\text{side adj.}}{\text{hyp.}}$)

TOA (Tangent = $\frac{\text{side opp.}}{\text{side adj.}}$)

FORMULA: **Cosine = $\frac{\text{side adj.}}{\text{hyp.}}$**

hyp. X Cosine = side adj.*

hyp. = $\frac{\text{side adj.}}{\text{Cosine}}$

Substitute: hyp. = $\frac{8.5}{.81915}$

Calculate: **hyp. = 10.3766**

Example 2. **Find the hypotenuse.** (when an angle and side opp. are known)

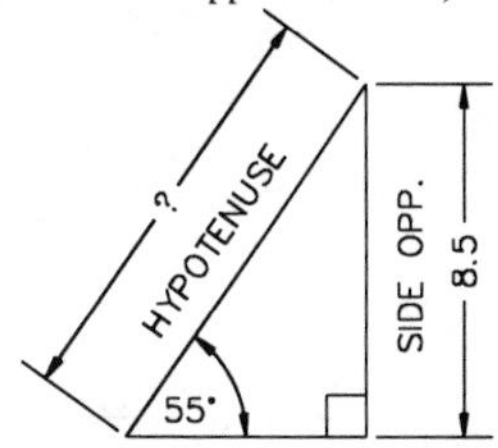

Sine = $\frac{\text{side opp.}}{\text{hyp.}}$

hyp. X Sine = side opp.*

hyp. = $\frac{\text{side opp.}}{\text{Sine}}$

hyp. = $\frac{8.5}{.81915}$

hyp. = 10.3766

NOTE: One of the biggest advantages of using this formula versus the traditional method on the preceding page, is that you don't have to label the angles or the sides of the triangle, and it is much easier to memorize all the formulas. Use the method(s) that work best for you.

*Obtained by multiplying each side of equation by "hyp." (refer to examples 1 and 2 on page 175 .

[FIND LENGTH OF A SIDE WHEN ANGLE C AND TWO SIDES ARE KNOWN]
"THE RULE OF PYTHAGORAS", TASK NUMBER K5A

Example 1. **Find length (c).**
(when C, a, and b are known)

Given:

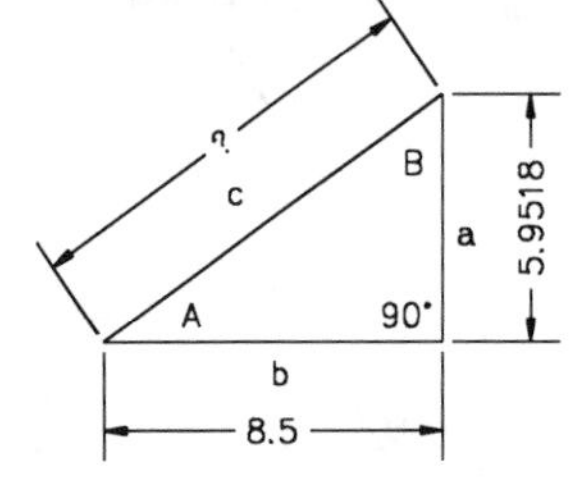

FORMULA: $\mathbf{a^2 + b^2 = c^2}$

Substitute: $(5.9518)^2 + (8.5)^2 = c^2$

Calculate: $35.4239 + 72.25 = c^2$

$107.6739 = c^2$

$\sqrt{107.6739} = c^2$

$\mathbf{10.3766 = c}$

Example 2. **Find length of (a).**
(when C, b, and c are known)

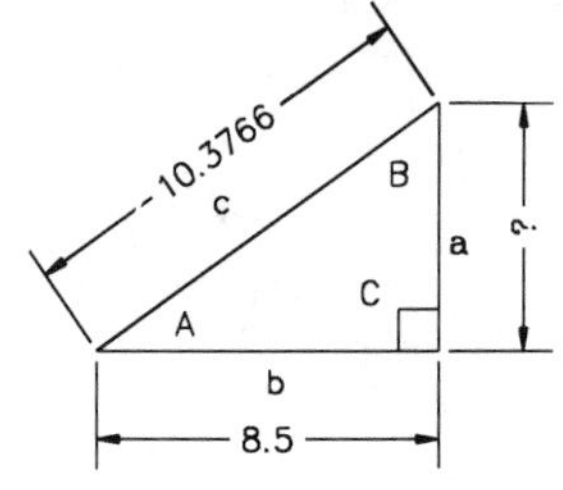

$\mathbf{a^2 + b^2 = c^2}$
$a^2 = c^2 - b^2$

$a^2 = (10.3766)^2 - (8.5)^2$

$a^2 = 107.6738 - 72.25$

$a^2 = 35.4238$

$a^2 = \sqrt{35.4238}$

$\mathbf{a = 5.9518}$

Example Problem 3. **Find length of (b).**
(when C, a, and c are known)

Given:

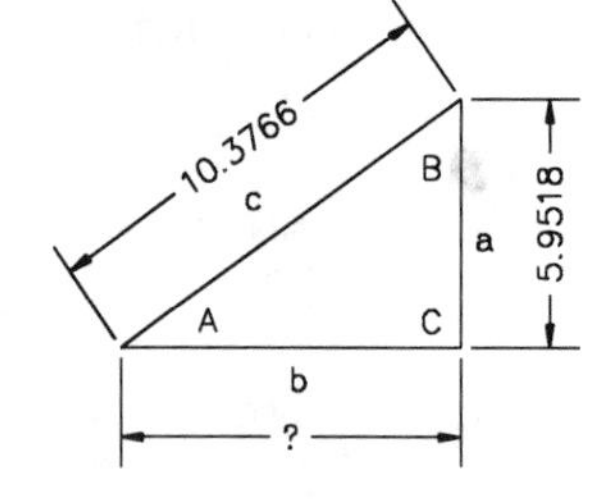

FORMULA: $\mathbf{a^2 + b^2 = c^2}$
$b^2 = c^2 - a^2$

Substitute: $b^2 = (10.3766)^2 - (5.9518)^2$

$b^2 = 107.6738 - 35.4239$

$b^2 = 72.2499$

$b = \sqrt{72.2449}$

$\mathbf{b = 8.4999 \text{ or } 8.5}$

Unique Application of the Rule of Pythagoras

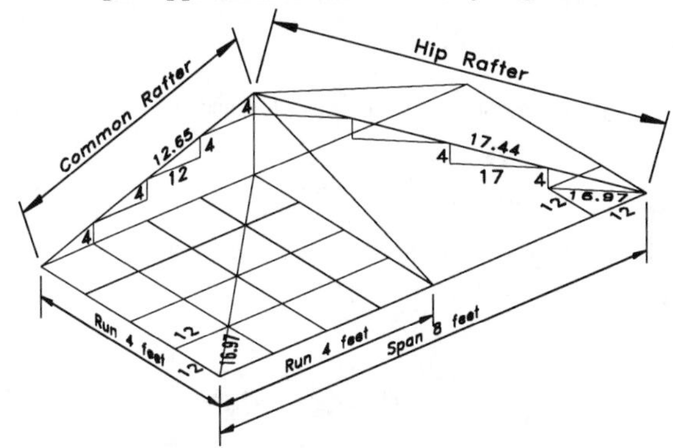

Solve Rafter Problems by using the "Rule of Pythagoras":

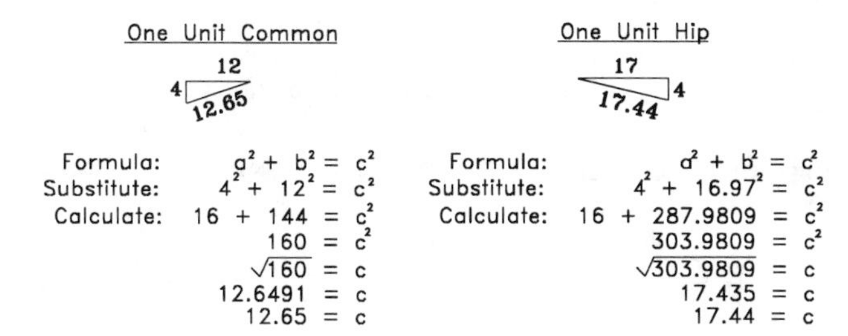

Formula: $a^2 + b^2 = c^2$
Substitute: $4^2 + 12^2 = c^2$
Calculate: $16 + 144 = c^2$

$160 = c^2$

$\sqrt{160} = c$

$12.6491 = c$

$12.65 = c$

Formula: $a^2 + b^2 = c^2$
Substitute: $4^2 + 16.97^2 = c^2$
Calculate: $16 + 287.9809 = c^2$

$303.9809 = c^2$

$\sqrt{303.9809} = c$

$17.435 = c$

$17.44 = c$

Solve Rafter Problems by using the "Framing Square":

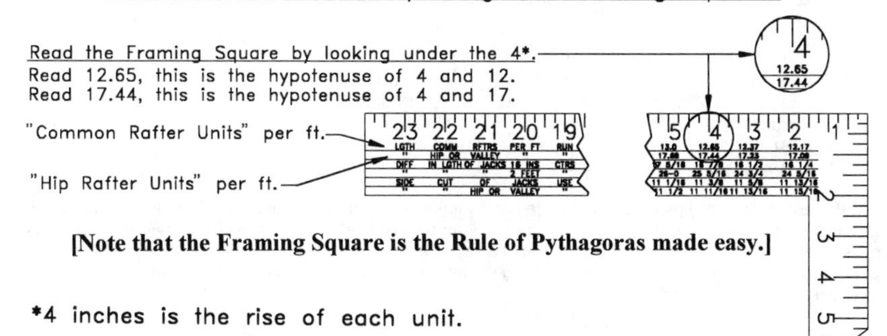

[Note that the Framing Square is the Rule of Pythagoras made easy.]

*4 inches is the rise of each unit.

How to Read the Framing Square, plus much more is in the Work Book!

[TO SOLVE BASIC OBLIQUE TRIANGLE TRIG PROBLEMS, SEE THE FOLLOWING PAGES.]

REVIEW OF **OBLIQUE TRIANGLES:**

Oblique triangles are triangles that have no 90° angles.

Note: These triangles are also known as oblique because they have no angles that are 90°.

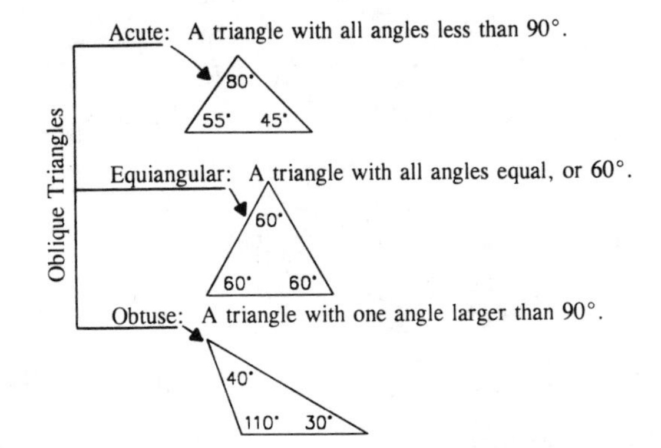

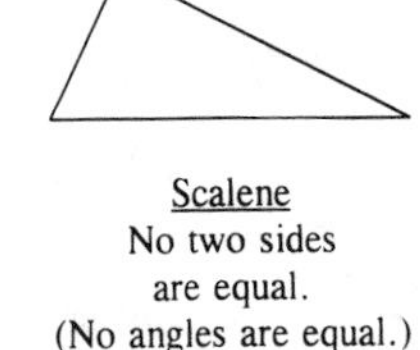

Scalene
No two sides are equal.
(No angles are equal.)

[RULE: The sum of the angles inside a triangle always add up to 180°. Notice in all the types of triangles shown above the sum of the angles equals 180°.]

The **basic trigonometric relationships** (as applied to oblique triangles) are as described below:

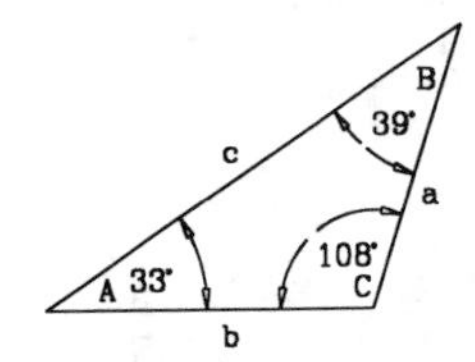

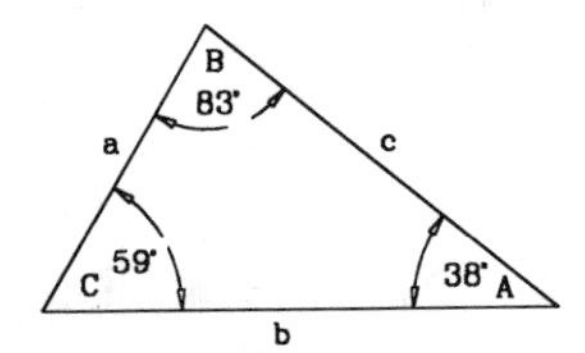

a = altitude (side closest to vertical)
b = horizontal side (base or side the triangle seems to rest on)
c = hypotenuse (side that appears to be adjacent to the altitude)
A = Angle opposite the altitude
B = Angle opposite the base
C = Angle opposite the hypotenuse

[Note the proper way to label the 3 sides and the 3 angles of an oblique triangle is; A is the angle opposite the altitude a (in the case of oblique triangles there are no perpendicular sides, therefore, choose the side that is closest to vertical), B is the angle opposite the base b (the side the triangle seems to rest on), and C is the angle opposite the hypotenuse c (in the case of oblique triangles the side that appears to be adjacent to the altitude since there are no right angles).

Basic Functions Of Angles

sine A $= \frac{a}{c}$	cosecant A $= \frac{c}{a}$
cosine A $= \frac{b}{c}$	secant A $= \frac{c}{b}$
tangent A $= \frac{a}{b}$	cotangent A $= \frac{b}{a}$

SOLVING BASIC OBLIQUE TRIANGLE PROBLEMS USING TRIGONOMETRY FORMULAS:

From previous discussion of the principles of mathematics, algebra, and geometry, the basic principles of right triangle trigonometry, and now with a few more principles of oblique triangles and with the trig function relationships to the quadrants of a circle, basic oblique triangle related measurements can be calculated for the oblique triangles shown on the next page.

Trig Functions in Relationship to the Quadrants of a Circle:

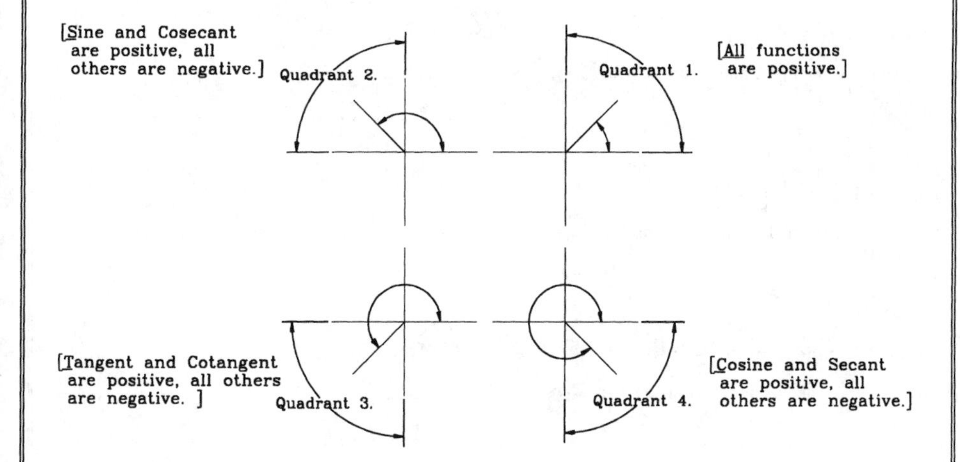

Trick of the Trades:

An easy way to memorize the above function relationships is to use the following sentence.

All students take calc. All represents all the functions in a right triangle which are positive; the s in students represents the sine and its reciprocal the cosecant are positive, whereas the other functions are negative; the t in take represents the tangent and its reciprocal the cotangent are positive, whereas the other functions are negative; the c in calc represents the cosine and its reciprocal the secant are positive, whereas the other functions are negative.

M3 ◂ Task No. **REVIEW OF OBLIQUE TRIANGLES (continued)**

Solving Basic Oblique Triangle Trigonometry Problems:

Formula 1: $a = \frac{b \times \sin A}{\sin B}$

Find side a when b, A, and B are known. See page 210.

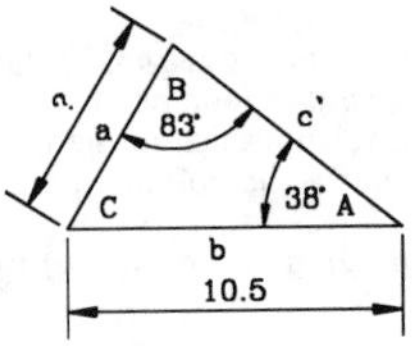

Formula 2: $b = \frac{a \times \sin B}{\sin A}$

Find side b when a, A, and B are known. See page 210.

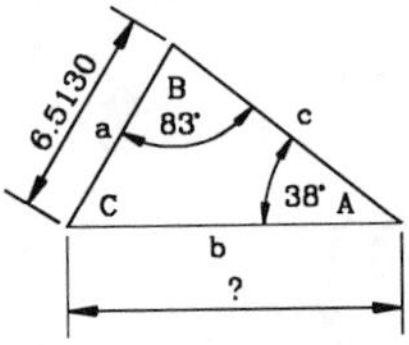

Formula 3: $c = \frac{b \times \sin C}{\sin B}$

Find side c when b, B, and C are known. See page 211.

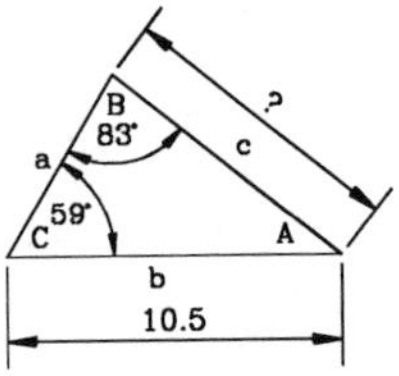

Formula 4: $\text{Sine } A = \frac{a \times \sin B}{b}$

Find angle A when a, b, and B are known. See page 211.

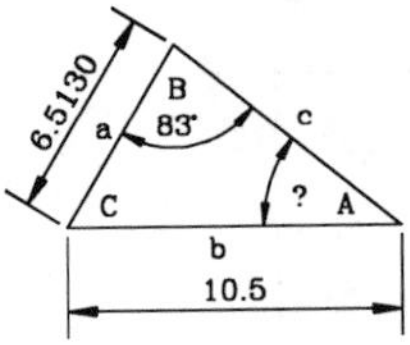

Formula 5: $\text{Sine } B = \frac{b \times \sin A}{a}$

Find angle B when a, b, and A are known. See page 212.

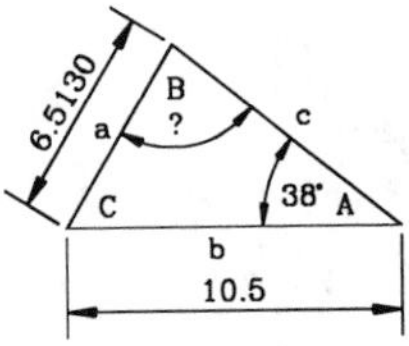

Formula 6: $\text{Sine } C = \frac{c \times \sin B}{b}$

Find angle C when b, c, and B are known. See page 212.

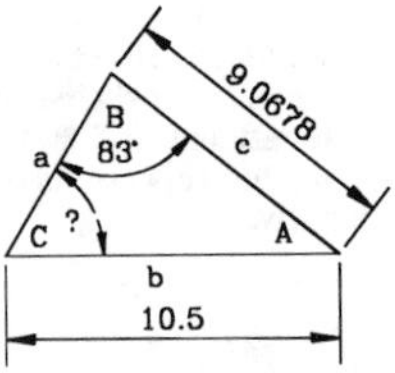

Formula 7: $b = \frac{c \times \sin B}{\sin C}$

Find side b when c, B, and C are known. See page 213.

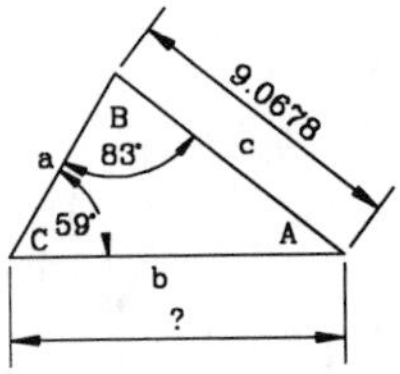

Formula 8: Tangent B = $\dfrac{b \times \sin C}{a - (b \times \cos C)}$

Find tangent B when a, b, and C are known.
See page 213.

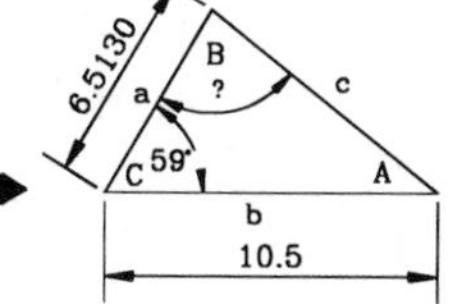

Formula 9: Cosine B = $\dfrac{a^2 + c^2 - b^2}{2ac}$

Find angle B when sides a, b, and c are known.
See page 214.

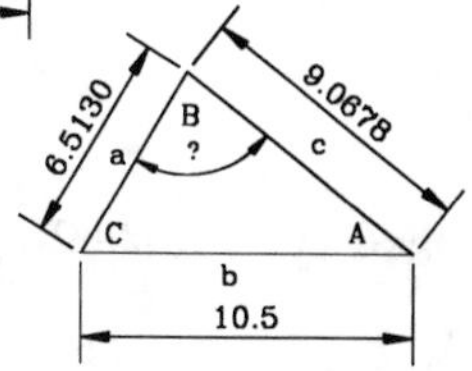

Formula 10: Cotangent A = $\dfrac{b \times \csc C}{a} - \cot C$

Find angle A when a, b, and C are known.
See page 214.

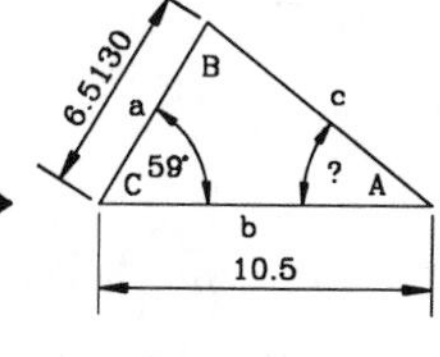

Formula 11: $c = a \times \sin C \times \csc A$ **Or** $c = \dfrac{a \times \sin C}{\sin A}$

Find side c when a, A, and C are known.
See page 215.

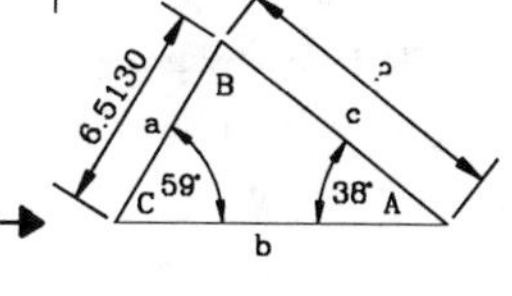

Formula 12: Sine A = $\dfrac{a \times \sin C}{c}$

Find angle A when a, c, and C are known.
See page 216.

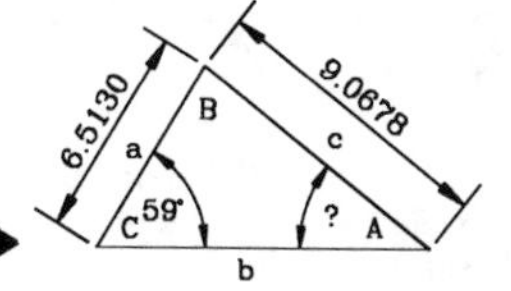

Formula 13: Area = $\dfrac{a \times b \times \sin C}{2}$

Find the area of an oblique triangle when a, b, and C are known.
See page 216.

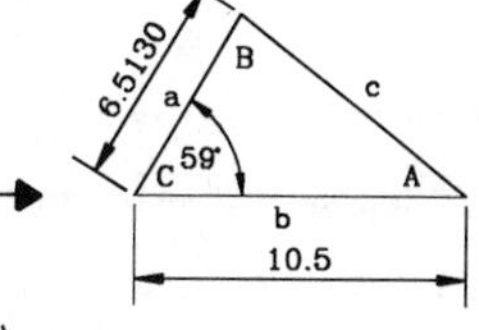

Formula 14: The Ambiguous Case

Find side b^1 and b when a, c, and A are known.
See page 217.

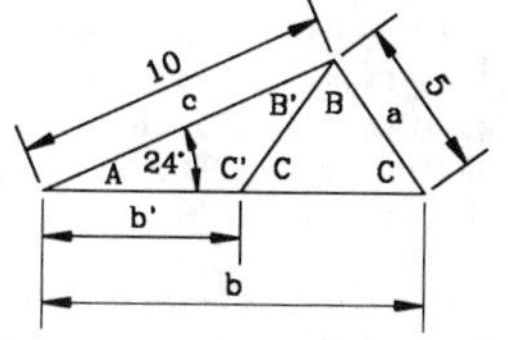

[FIND SIDE a WHEN SIDE b AND ANGLES A AND B ARE KNOW]

Example 1. Find side a when b, A, and B are known.

Given:

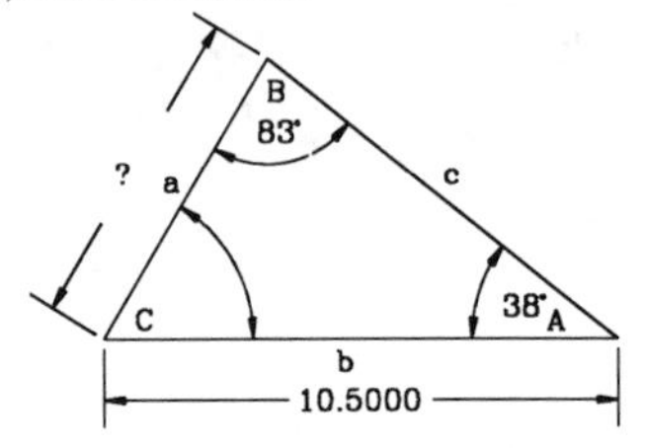

FORMULA: $a = \frac{b \times \sin A}{\sin B}$

Substitute: $a = \frac{10.5 \times .61566}{.99255}$

Calculate: **a = 6.5130**

[FIND SIDE b WHEN SIDE a AND ANGLES A AND B ARE KNOW]

Example 2. Find side b when a, A, and B are known.

Given:

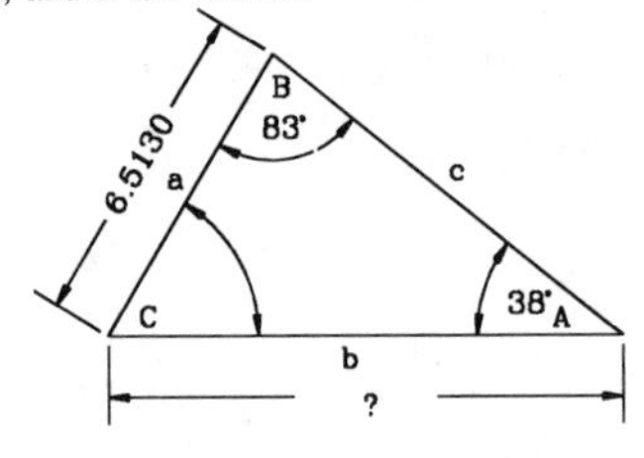

FORMULA: $b = \frac{a \times \sin B}{\sin A}$

Substitute: $b = \frac{6.5130 \times .99255}{.61566}$

Calculate: **b = 10.5**

Calculator Solution (Find side a):

Step 1. Enter Sin (Display Sin --- 0)
Step 2. Enter 83 (Display Sin --- 83)
Step 3. Enter = (Display 0.9925---)
Step 4. Enter Shift (Display [s] 0.9925---)
Step 5. Enter Min (Display 0.9925---m)
Step 6. Enter Sin (Display Sin 0.9925---)
Step 7. Enter 38 (Display Sin--- 38)
Step 8. Enter = (Display 0.6156---)
Step 9. Enter X (Display 0.6156---)
Step 10. Enter 10.5 (Display 10.5)
Step 11. Enter = (Display 6.4644---)
Step 12. Enter ÷ (Display 6.4644--- ÷)
Step 13. Enter MR (Display 0.9925---m)
Step 14. Enter = (Display = 6.5129---)
Side **a** = 6.5129

NOTE: **The procedure for calculating trig problems with a calculator would vary from calculator to calculator. Therefore, read the instructions with the owner's manual before calculating.**

[FIND SIDE c WHEN SIDE b AND ANGLES B AND C ARE KNOWN]

Example 3. Find side c when b, B, and C are known.

Given:

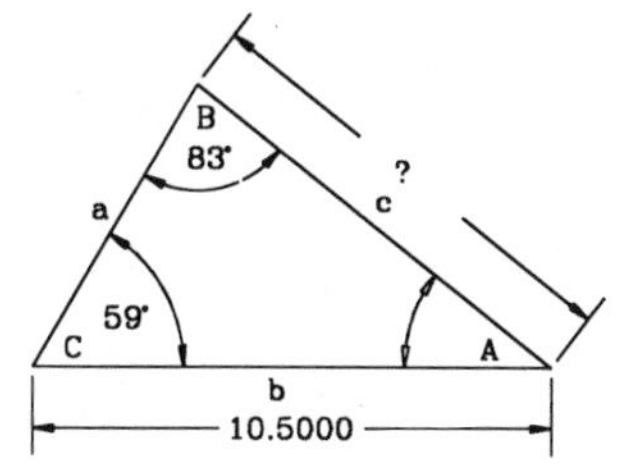

FORMULA: $c = \frac{b \times \sin C}{\sin B}$

Substitute: $c = \frac{10.5 \times .85717}{.99255}$

Calculate: **c = 9.0678**

[FIND ANGLE A WHEN SIDES a AND b, AND ANGLE B ARE KNOWN]

Example 4. Find angle A when a, b, and B are known.

Given:

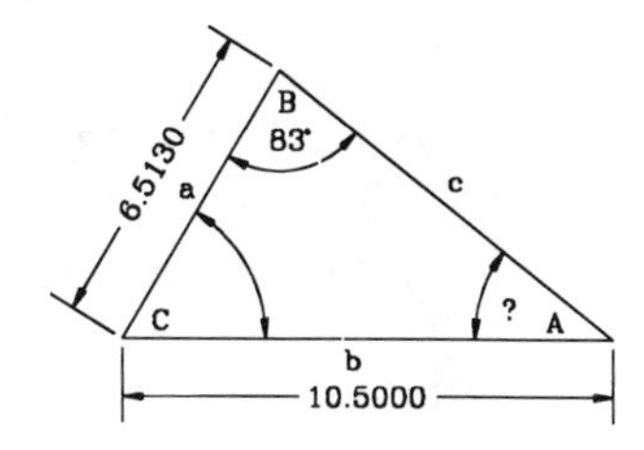

FORMULA: $\text{Sine } A = \frac{a \times \sin B}{b}$

Substitute: $\text{Sine } A = \frac{6.5130 \times .99255}{10.5}$

Calculate: Sine A = .6156 or 38°

Angle A = 38°

CAUTION: **In some cases, when working with oblique triangles, the answer will not calculate correctly. This will depend on the function of the angle, the degree of the angle, and which quadrant it lies in.** **See Task No. M3, page 207, for further information.**

[FIND ANGLE B WHEN SIDES a AND b, AND ANGLE A ARE KNOWN]

Example 5. Find angle B when a, b, and A are known.

Given:

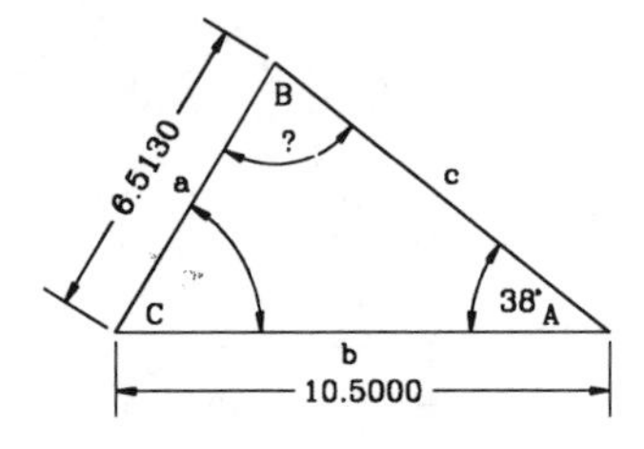

FORMULA: $\text{Sine B} = \dfrac{\text{b X sin A}}{\text{a}}$

Substitute: $\text{Sine B} = \dfrac{10.5 \text{ X } .61566}{6.5130}$

Calculate: Sine B = .99254 or 83°

Angle B = 83°

[FIND ANGLE C WHEN SIDES b AND c, AND ANGLE B ARE KNOW]

Example 6. Find angle C when b, c, and B are known.

Given:

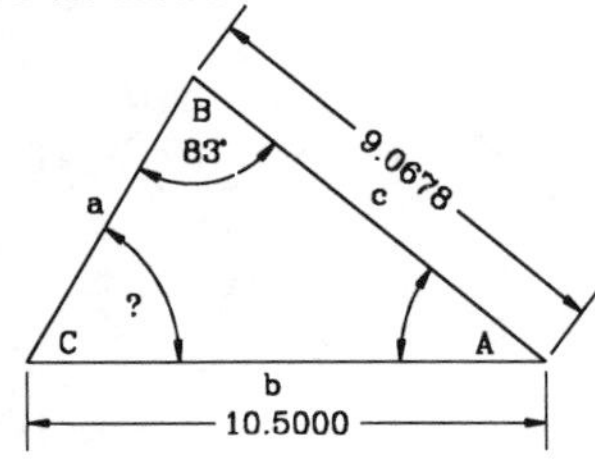

FORMULA: $\text{Sine C} = \dfrac{\text{c X sin B}}{\text{b}}$

Substitute: $\text{Sine C} = \dfrac{9.0678 \text{ X } .99254}{10.5}$

Calculate: Sine C = .857157 or 59°

Angle C = 59°

Task No. ▸ M4

[FIND SIDE b WHEN SIDE c AND ANGLES B AND C ARE KNOWN]

Example 7. Find side b when c, B, and C are known.

Given:

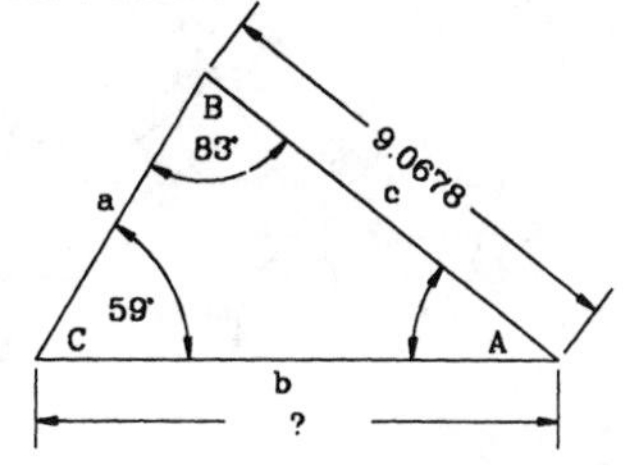

FORMULA: $b = \dfrac{c \text{ X } \sin B}{\sin C}$

Substitute: $b = \dfrac{9.0678 \text{ X } .99254}{.857157}$

Calculate: **b = 10.5**

[FIND ANGLE B WHEN SIDES a AND b AND ANGLE C ARE KNOWN]

Example 8. Find tan B when a, b, and C are known.

Given:

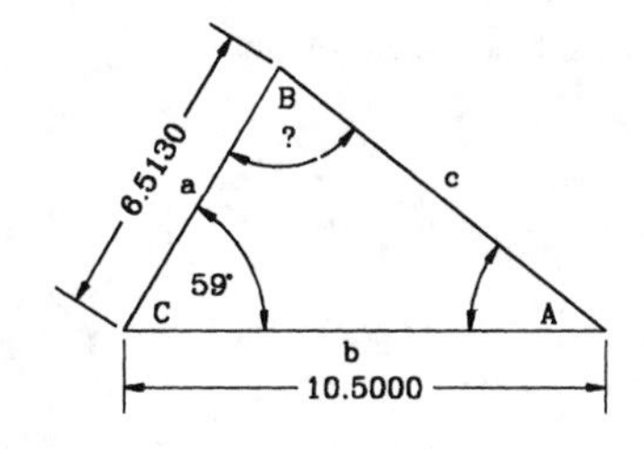

FORMULA: $\text{Tangent } B = \dfrac{b \text{ X } \sin C}{a - (b \text{ X } \cos C)}$

Substitute: $\text{Tangent } B = \dfrac{10.5 \text{ X } .857157}{6.5130 - (10.5 \text{ X } .51504)}$

Calculate: $\text{Tangent } B = \dfrac{9.0001485}{6.5130 - (10.5 \text{ X } .51504)}$

$\text{Tangent } B = \dfrac{9.0001485}{1.10508}$

Tangent B = 8.14434 or 83°

Angle B = 83°

CAUTION: **In some cases, when working with oblique triangles, the answer will not calculate correctly. This will depend on the function of the angle, the degree of the angle, and which quadrant it lies in.** **See Task No. M3, page 207, for further information.**

[FIND ANGLE B WHEN SIDES a, b, AND c ARE KNOWN]

Example 9. Find angle B when sides a, b, and c are known.

Given:

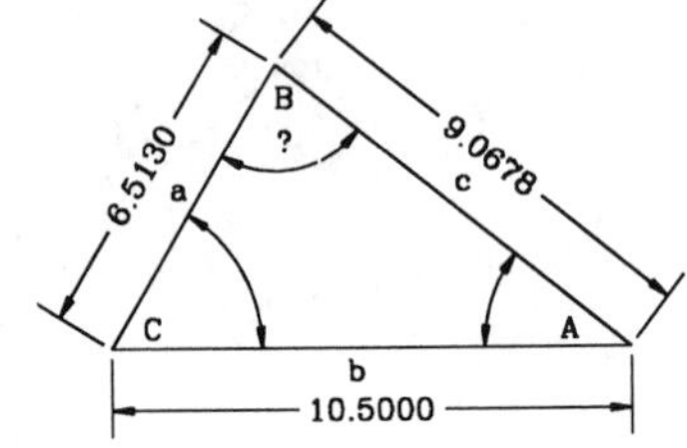

FORMULA: **Cosine B =** $\frac{a^2 + c^2 - b^2}{2ac}$

Substitute: Cosine B = $\frac{6.513^2 + 9.0678^2 - 10.5^2}{(2 \times 6.513) \times 9.0678}$

Calculate: Cosine B = $\frac{(42.419169 + 82.2224996) - 110.25}{13.026 \times 9.0678}$

Cosine B = $\frac{124.64166 - 110.25}{118.11716}$

Cosine B = .1218422 or 83°
Angle B = 83°

[FIND ANGLE A WHEN SIDES a AND b, AND ANGLE C ARE KNOWN]

Example 10. Find angle A when a, b, and C are known.

Given:

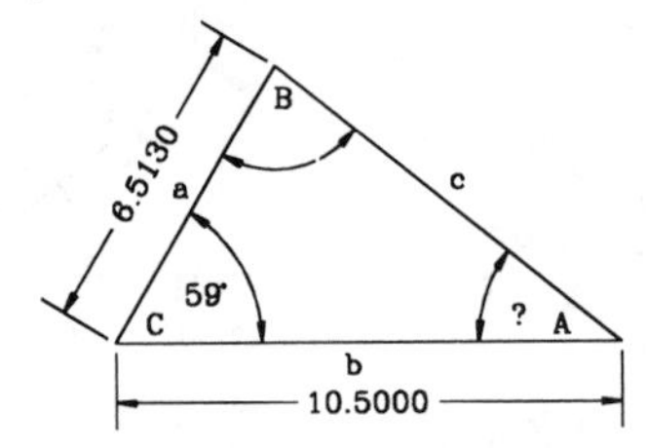

FORMULA: **Cotangent A =** $\frac{b \times \csc C}{a} - \cot C$

Substitute: Cotangent A = $\frac{10.5 \times 1.1666}{6.5130} - .60086$

Calculate: Cotangent A = $\frac{12.2493}{6.5130} - .60086$

Cotangent A = 1.8807461 - .60086

Cotangent A = 1.279886 or 38°
Angle A = 38°

[FIND SIDE c WHEN SIDE a AND ANGLES A AND C ARE KNOWN]

Example 11. Find side c when a, A, and C are known.

Given:

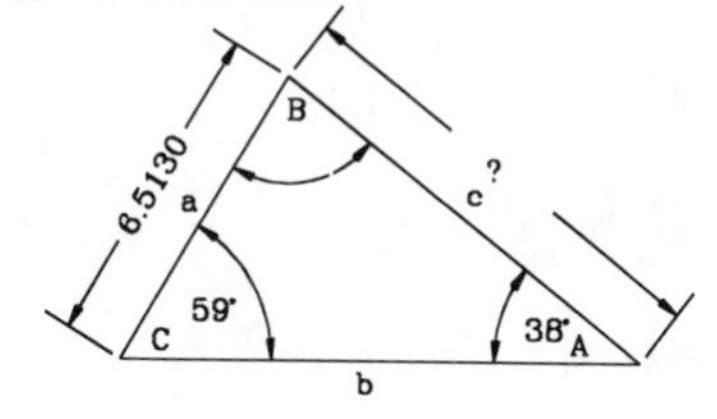

FORMULA: **c = a X sin C X csc A**

Substitute: c = 6.5130 X .857157 X 1.6243

Calculate: c = (6.513 X .857157) X 1.6243

c = 5.5826635 X 1.6243

c = 9.06792

[OR]

FORMULA: $c = \frac{a \text{ X } \sin C}{\sin A}$

Substitute: $c = \frac{6.5130 \text{ X } .857157}{.61566}$

Calculate: $c = \frac{5.5826635}{.61566}$

c = 9.0677703

CAUTION: **In some cases, when working with oblique triangles, the answer will not calculate correctly. This will depend on the function of the angle, the degree of the angle, and which quadrant it lies in.** **See Task No. M3, page 207, for further information.**

[FIND ANGLE A WHEN SIDES a AND c, AND ANGLE C ARE KNOWN]

Example 12. Find angle A when a, c, and C are known. **Given:**

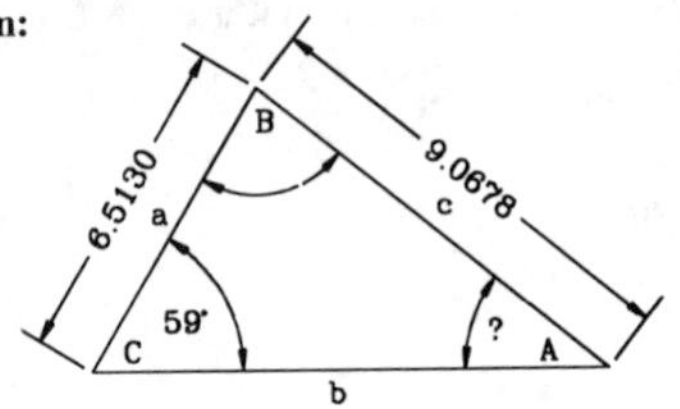

FORMULA: **Sine A =** $\frac{a \times \sin C}{c}$

Substitute: Sine A = $\frac{6.5130 \times .857157}{9.0678}$

Calculate: Sine A = $\frac{5.9826635}{9.0678}$

Sine A = .6156579 or 38°

Angle A = 38°

[FIND THE AREA OF AN OBLIQUE TRIANGLE]

Example 13. Find the area of an oblique triangle.

Given:

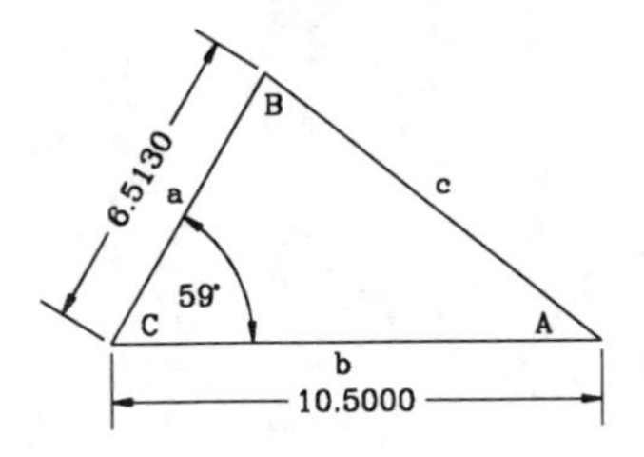

FORMULA: **Area =** $\frac{a \times b \times \sin C}{2}$

Substitute: Area = $\frac{6.5130 \times 10.5 \times .857157}{2}$

Calculate: Area = $\frac{(6.5130 \times 10.5) \times .857157}{2}$

Area = $\frac{68.3865 \times .857157}{2}$

Area = $\frac{58.617967}{2}$

Area = 29.31 square units.

[Find the area of any triangle when only the sides are known.]

Hero's Formula:

Formula: $A = \sqrt{s(s-a)(s-b)(s-c)}$

$s = \frac{1}{2}(a + b + c)$

Example: Find the area of the triangle shown below.

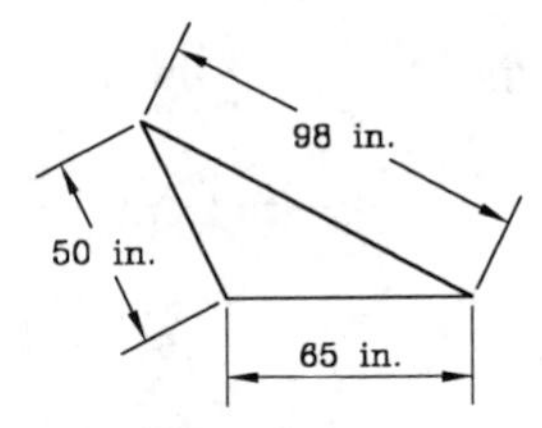

Step 1. Solve for s.

FORMULA: $s = \frac{1}{2}(a + b + c)$

Substitute: s = .5(50" + 65" + 98")

Calculate: s = .5(213)

s = 106.5

Step 2. Solve for A.

FORMULA: $\sqrt{s(s-a)(s-b)(s-c)}$

Substitute: $\sqrt{106.5(106.5 - 50)(106.5 - 65)(106.5 - 98)}$

Calculate: $\sqrt{2122585}$

Answer: **1456.9 sq. in.**

THE AMBIGUOUS CASE — Task No. ▸ M4

[FIND SIDE b^1 AND SIDE b WHEN a, c, AND A ARE KNOWN]

Example 14. Find b^1 and b when a, c, and A are known.

Given:

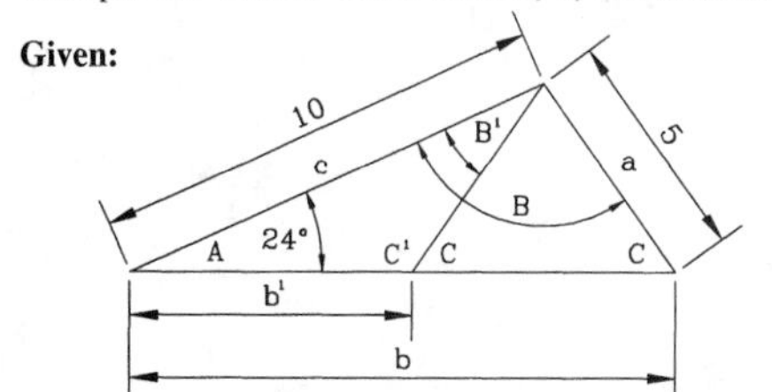

BASIC FORMULA:

$$\frac{a}{\text{Sin A}} = \frac{b}{\text{Sin B}} = \frac{c}{\text{Sin C}}$$

Or

$$\frac{\text{Sin A}}{a} = \frac{\text{Sin B}}{b} = \frac{\text{Sin C}}{c}$$

First, find angle C^1 (so angle B^1 can be determined):

Find angle $\mathbf{C^1}$

Step 1. Formula: $\text{Sin C} = \frac{c \text{ X Sin A}}{a}$ [$\frac{\text{Sin C}}{c} = \frac{\text{Sin A}}{a}$ **from Law of Sines and algebra]**

Step 2. Substitute: $\text{Sin C} = \frac{10 \text{ X } .40673 \text{ (Sin 24°)}}{5}$

Step 3. Calculate: **Angle C = 54.4367°** (From trig tables or calculator)

Step 4. Solution: **Angle C^1** = 180° - 54.4367° = **125.5633°**

Second, find angle B^1 (so side b^1 can be found using the Law of Sines):

Find angle $\mathbf{B^1}$

Step 1. $B^1 = 180° - (C^1 + A)$

Step 2. $B^1 = 180° - (125.5633° + 24°)$

Step 3. $B^1 = \mathbf{30.4367°}$

Third, find side b^1:

Find side $\mathbf{b^1}$

Step 1. Formula: $b^1 = \frac{a \text{ X Sin } B^1}{\text{Sin A}}$ [$\frac{b^1}{\text{Sin } B^1} = \frac{a}{\text{Sin A}}$ **from Law of Sines and algebra]**

Step 2. Substitute: $b^1 = \frac{5 \text{ X } .50658 \text{ (Sin 30.4367°)}}{.40673 \text{ (Sin 24°)}}$

Step 3. Calculate: $b^1 = \frac{2.5329}{.40673}$

Step 4. Solution: $b^1 = \mathbf{6.2274}$

Fourth, find side b:

Find side **b**

Step 1. Find B

B = 180° - (A + C)
B = 180° - (24° + 54.4367°)
B = 180° - (78.4367°)
B = 101.5633°

Step 2. Formula: $b = \frac{a \text{ X Sin B}}{\text{Sin A}}$ [$\frac{b}{\text{Sin B}} = \frac{a}{\text{Sin A}}$ **from Law of Sines and algebra]**

Step 3. Substitute: $b = \frac{5 \text{ X } .97970 \text{ (Sin 101.5633°)}}{.40673 \text{ (Sin 24°)}}$

Step 4. Calculate: $b = \frac{4.8985}{.40673}$

Step 5. Solution: **b = 12.0436**

[TO FIND THE ABOVE SOLUTIONS USING A CALCULATOR, SEE THE NEXT PAGE]

M4	◂ Task No. OBLIQUE TRIANGLES - AMBIGUOUS CASE (continued)

[FIND SIDE b^1 AND SIDE b WHEN a, c, AND A ARE KNOWN]

Example 14a. Find b^1 and b when a, c, and A are known by using a scientific calculator (Radio Shack EC-4039).

Given:

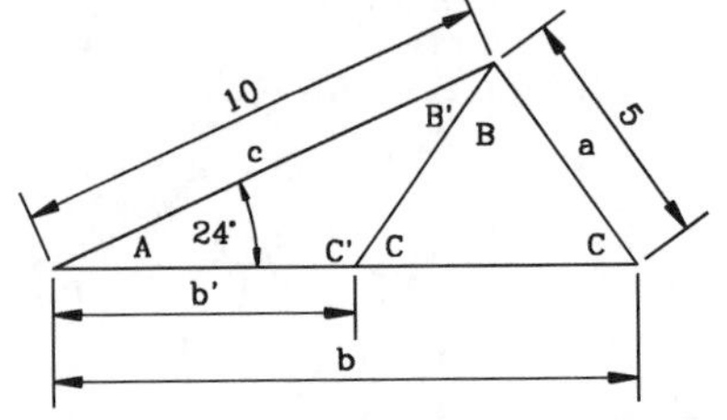

<u>First, find angle C^1 (so angle B^1 can be determined):</u>

Find angle $\mathbf{C^1}$

Step 1. Enter Sin (Display Sin 0)
Step 2. Enter 24 (Display Sin 24)
Step 3. Enter = (Display .4067---)
Step 4. Enter X (Display .4067---X)
Step 5. Enter 10 (Display 10)
Step 6. Enter = (Display 4.067---)
Step 7. Enter ÷ (Display 4.0673--÷)
Step 8. Enter 5 (Display 5)
Step 9. Enter = (Display .81347---)
Step 10. Enter Shift (Display [S] 0.81347---)
Step 11. Enter Sin (Display Sin^{-1} 0.81347---)
Step 12. Enter = (Display 54.4366---)
Step 13. Enter - (Display 54.4366 -)
Step 14. Enter 180 (Display 180)
Step 15. Enter + (Display -125.5633---)
Step 16. Enter +/- (Display **125.5633**) = **Angle C^1**

<u>Second, find angle B^1 (so side b^1 can be found using the Law of Sines):</u>

Find angle $\mathbf{B^1}$

Step 1. Enter + (Display 125.5633 +)
Step 2. Enter 24 (Display 24)
Step 3. Enter = (Display 149.5633)
Step 4. Enter - (Display 149.5633 -)
Step 5. Enter 180 (Display 180)
Step 6. Enter + (Display -30.4367)
Step 7. Enter ± (Display **30.4367**) = **Angle B^1**

<u>Third, find side b^1:</u>

Find side $\mathbf{b^1}$

Step 1. Enter Sin (Display Sin 0)
Step 2. Enter 30.4367 (Display Sin 30.4367)
Step 3. Enter = (Display 0.5065---)
Step 4. Enter X (Display 0.5065---)
Step 5. Enter 5 (Display 5)
Step 6. Enter = (Display 2.5329---)
Step 7. Enter Shift (Display [S] 2.5329---)
Step 8. Enter Min (Display Sin 2.5329---)
Step 9. Enter Sin (Display 24---)

Step 10. Enter = (Display 0.4067---)
Step 11. Enter MR (Display 2.53293)
Step 12. Enter ÷ (Display 2.53293---÷)
Step 13. Enter 4.067 (Display **6.2280**---) = **Side b**[1]

Fourth, find side b:

Find side **b**

First find Angle **B**

Step 1. Enter 180 (Display 180)
Step 2. Enter - (Display 180 -)
Step 3. Enter 24 (Display 24)
Step 4. Enter - (Display 156 -)
Step 5. Enter 54.4367 (Display 54.4367)
Step 6. Enter + (Display 101.5633 +)
Step 7. Enter = (Display **101.5633**) = **Angle B**

Second find side **b**

Step 8. Enter Sin (Display Sin 0)
Step 9. Enter 24 (Display Sin 24)
Step 10. Enter = (Display 0.4067---)
Step 11. Enter Shift (Display [S] 0.4067---)
Step 12. Enter Min (Display 0.4067---)
Step 13. Enter Sin (Display Sin 0.4067---)
Step 14. Enter 101.5633 (Display Sin 101.5633---)
Step 15. Enter = (Display 0.97970---)
Step 16. Enter X (Display 0.97970---X)
Step 17. Enter 5 (Display 5)
Step 18. Enter = (Display 4.8985---)
Step 19. Enter ÷ (Display 4.8985---÷)
Step 20. Enter MR (Display 0.4067---)
Step 21. Enter = (Display **12.0434**---) = **side b**

NOTE: **The procedure for calculating trig problems with a calculator would vary from calculator to calculator. Therefore, read the instructions with the owner's manual before calculating.**

0°

M	Sine	Cosine	Tangent	Cotan.	Secant	Cosec.	M
0	.00000	1.0000	.00000	Infinite	1.0000	Infinite	60
1	.00029	.0000	.00029	3437.7	.0000	3437.7	59
2	.00058	.0000	.00058	1718.9	.0000	1718.9	58
3	.00087	.0000	.00087	1145.9	.0000	1145.9	57
4	.00116	.0000	.00116	859.44	.0000	859.44	56
5	.00145	1.0000	.00145	687.55	1.0000	687.55	55
6	.00174	.0000	.00174	572.96	.0000	572.96	54
7	.00204	.0000	.00204	491.11	.0000	491.11	53
8	.00233	.0000	.00233	429.72	.0000	429.72	52
9	.00262	.0000	.00262	381.97	.0000	381.97	51
10	.00291	.99999	.00291	343.77	1.0000	343.77	50
11	.00320	.99999	.00320	312.52	.0000	312.52	49
12	.00349	.99999	.00349	286.48	.0000	286.48	48
13	.00378	.99999	.00378	264.44	.0000	264.44	47
14	.00407	.99999	.00407	245.55	.0000	245.55	46
15	.00436	.99999	.00436	229.18	1.0000	229.18	45
16	.00465	.99999	.00465	214.86	.0000	214.86	44
17	.00494	.99999	.00494	202.22	.0000	202.22	43
18	.00524	.99999	.00524	190.98	.0000	190.99	42
19	.00553	.99998	.00553	180.93	.0000	180.93	41
20	.00582	.99998	.00582	171.88	1.0000	171.89	40
21	.00611	.99998	.00611	163.70	.0000	163.70	39
22	.00640	.99998	.00640	156.26	.0000	156.26	38
23	.00669	.99998	.00669	149.46	.0000	149.47	37
24	.00698	.99997	.00698	143.24	.0000	143.24	36
25	.00727	.99997	.00727	137.51	1.0000	137.51	35
26	.00756	.99997	.00756	132.22	.0000	132.22	34
27	.00785	.99997	.00785	127.32	.0000	127.32	33
28	.00814	.99997	.00814	122.77	.0000	122.78	32
29	.00843	.99996	.00844	118.54	.0000	118.54	31
30	.00873	.99996	.00873	114.59	1.0000	114.59	30
31	.00902	.99996	.00902	110.89	.0000	110.90	29
32	.00931	.99996	.00931	107.43	.0000	107.43	28
33	.00960	.99995	.00960	104.17	.0000	104.17	27
34	.00989	.99995	.00989	101.11	.0000	101.11	26
35	.01018	.99995	.01018	98.218	1.0000	98.223	25
36	.01047	.99994	.01047	95.489	.0000	95.495	24
37	.01076	.99994	.01076	92.908	.0000	92.914	23
38	.01105	.99994	.01105	90.463	.0001	90.469	22
39	.01134	.99993	.01134	88.143	.0001	88.149	21
40	.01163	.99993	.01164	85.940	1.0001	85.946	20
41	.01193	.99993	.01193	83.843	.0001	83.849	19
42	.01222	.99992	.01222	81.847	.0001	81.853	18
43	.01251	.99992	.01251	79.943	.0001	79.950	17
44	.01280	.99992	.01280	78.126	.0001	78.133	16
45	.01309	.99991	.01309	76.390	1.0001	76.396	15
46	.01338	.99991	.01338	74.729	.0001	74.736	14
47	.01367	.99991	.01367	73.139	.0001	73.146	13
48	.01396	.99990	.01396	71.615	.0001	71.622	12
49	.01425	.99990	.01425	70.153	.0001	70.160	11
50	.01454	.99989	.01454	68.750	1.0001	68.757	10
51	.01483	.99989	.01484	67.402	.0001	67.409	9
52	.01512	.99988	.01513	66.105	.0001	66.113	8
53	.01542	.99988	.01542	64.858	.0001	64.866	7
54	.01571	.99988	.01571	63.657	.0001	63.664	6
55	.01600	.99987	.01600	62.499	1.0001	62.507	5
56	.01629	.99987	.01629	61.383	.0001	61.391	4
57	.01658	.99987	.01658	60.306	.0001	60.314	3
58	.01687	.99986	.01687	59.266	.0001	59.274	2
59	.01716	.99985	.01716	58.261	.0001	58.270	1
60	.01745	.99985	.01745	57.290	1.0001	57.299	0
M	Cosine	Sine	Cotan.	Tangent	Cosec.	Secant	M

89°

1°

M	Sine	Cosine	Tangent	Cotan.	Secant	Cosec.	M
0	.01745	.99985	.01745	57.290	1.0001	57.299	60
1	.01774	.99984	.01775	56.350	.0001	56.359	59
2	.01803	.99984	.01804	55.441	.0001	55.450	58
3	.01832	.99983	.01833	54.561	.0002	54.570	57
4	.01861	.99983	.01862	53.708	.0002	53.718	56
5	.01891	.99982	.01891	52.882	1.0002	52.891	55
6	.01920	.99981	.01920	52.081	.0002	52.090	54
7	.01949	.99981	.01949	51.303	.0002	51.313	53
8	.01978	.99980	.01978	50.548	.0002	50.558	52
9	.02007	.99980	.02007	49.816	.0002	49.826	51
10	.02036	.99979	.02036	49.104	1.0002	49.114	50
11	.02065	.99979	.02066	48.412	.0002	48.422	49
12	.02094	.99978	.02095	47.739	.0002	47.750	48
13	.02123	.99977	.02124	47.085	.0002	47.096	47
14	.02152	.99977	.02153	46.449	.0002	46.460	46
15	.02181	.99976	.02182	45.829	1.0002	45.840	45
16	.02210	.99975	.02211	45.226	.0002	45.237	44
17	.02240	.99975	.02240	44.638	.0002	44.650	43
18	.02269	.99974	.02269	44.066	.0002	44.077	42
19	.02298	.99974	.02298	43.508	.0003	43.520	41
20	.02326	.99973	.02327	42.964	1.0003	42.976	40
21	.02356	.99972	.02357	42.433	.0003	42.445	39
22	.02385	.99971	.02386	41.916	.0003	41.928	38
23	.02414	.99971	.02415	41.410	.0003	41.423	37
24	.02443	.99970	.02444	40.917	.0003	40.930	36
25	.02472	.99969	.02473	40.436	1.0003	40.448	35
26	.02501	.99969	.02502	39.965	.0003	39.978	34
27	.02530	.99968	.02531	39.506	.0003	39.518	33
28	.02559	.99967	.02560	39.057	.0003	39.069	32
29	.02589	.99966	.02589	38.618	.0003	38.631	31
30	.02618	.99966	.02618	38.188	1.0003	38.201	30
31	.02647	.99965	.02648	37.769	.0003	37.782	29
32	.02676	.99964	.02677	37.358	.0003	37.371	28
33	.02705	.99963	.02706	36.956	.0004	36.969	27
34	.02734	.99963	.02735	36.563	.0004	36.576	26
35	.02763	.99962	.02764	36.177	1.0004	36.191	25
36	.02792	.99961	.02793	35.800	.0004	35.814	24
37	.02821	.99960	.02822	35.431	.0004	35.445	23
38	.02850	.99959	.02851	35.069	.0004	35.084	22
39	.02879	.99958	.02880	34.715	.0004	34.729	21
40	.02908	.99958	.02910	34.368	1.0004	34.382	20
41	.02937	.99957	.02939	34.027	.0004	34.042	19
42	.02967	.99956	.02968	33.693	.0004	33.708	18
43	.02996	.99955	.02997	33.366	.0004	33.381	17
44	.03025	.99954	.03026	33.045	.0004	33.060	16
45	.03054	.99953	.03055	32.730	1.0005	32.745	15
46	.03083	.99952	.03084	32.421	.0005	32.437	14
47	.03112	.99951	.03113	32.118	.0005	32.134	13
48	.03141	.99951	.03143	31.820	.0005	31.836	12
49	.03170	.99950	.03172	31.528	.0005	31.544	11
50	.03199	.99949	.03201	31.241	1.0005	31.257	10
51	.03228	.99948	.03230	30.960	.0005	30.976	9
52	.03257	.99947	.03259	30.683	.0005	30.699	8
53	.03286	.99946	.03288	30.411	.0005	30.428	7
54	.03315	.99945	.03317	30.145	.0005	30.161	6
55	.03344	.99944	.03346	29.882	1.0005	29.899	5
56	.03374	.99943	.03375	29.624	.0006	29.641	4
57	.03403	.99942	.03405	29.371	.0006	29.388	3
58	.03432	.99941	.03434	29.122	.0006	29.139	2
59	.03461	.99940	.03463	28.877	.0006	28.894	1
60	.03490	.99939	.03492	28.636	1.0006	28.654	0
M	Cosine	Sine	Cotan.	Tangent	Cosec.	Secant	M

88°

2°

M	Sine	Cosine	Tangent	Cotan.	Secant	Cosec.	M
0	.03490	.99939	.03492	28.636	1.0006	28.654	60
1	.03519	.99938	.03521	28.399	.0006	28.417	59
2	.03548	.99937	.03550	28.166	.0006	28.184	58
3	.03577	.99936	.03579	27.937	.0006	27.955	57
4	.03606	.99935	.03608	27.712	.0006	27.730	56
5	.03635	.99934	.03638	27.490	1.0007	27.508	55
6	.03664	.99933	.03667	27.271	.0007	27.290	54
7	.03693	.99932	.03696	27.056	.0007	27.075	53
8	.03722	.99931	.03725	26.845	.0007	26.864	52
9	.03751	.99930	.03754	26.637	.0007	26.655	51
10	.03781	.99928	.03783	26.432	1.0007	26.450	50
11	.03810	.99927	.03812	26.230	.0007	26.249	49
12	.03839	.99926	.03842	26.031	.0007	26.050	48
13	.03868	.99925	.03871	25.835	.0007	25.854	47
14	.03897	.99924	.03900	25.642	.0008	25.661	46
15	.03926	.99923	.03929	25.452	1.0008	25.471	45
16	.03955	.99922	.03958	25.264	.0008	25.284	44
17	.03984	.99921	.03987	25.080	.0008	25.100	43
18	.04013	.99919	.04016	24.898	.0008	24.918	42
19	.04042	.99918	.04045	24.718	.0008	24.739	41
20	.04071	.99917	.04075	24.542	1.0008	24.562	40
21	.04100	.99916	.04104	24.367	.0008	24.388	39
22	.04129	.99915	.04133	24.196	.0008	24.216	38
23	.04158	.99913	.04162	24.026	.0009	24.047	37
24	.04187	.99912	.04191	23.859	.0009	23.880	36
25	.04217	.99911	.04220	23.694	1.0009	23.716	35
26	.04246	.99910	.04249	23.532	.0009	23.553	34
27	.04275	.99908	.04279	23.372	.0009	23.393	33
28	.04304	.99907	.04308	23.214	.0009	23.235	32
29	.04333	.99906	.04337	23.058	.0009	23.079	31
30	.04362	.99905	.04366	22.904	1.0009	22.925	30
31	.04391	.99903	.04395	22.752	.0010	22.774	29
32	.04420	.99902	.04424	22.602	.0010	22.624	28
33	.04449	.99901	.04453	22.454	.0010	22.476	27
34	.04478	.99900	.04483	22.308	.0010	22.330	26
35	.04507	.99898	.04512	22.164	1.0010	22.186	25
36	.04536	.99897	.04541	22.022	.0010	22.044	24
37	.04565	.99896	.04570	21.881	.0010	21.904	23
38	.04594	.99894	.04599	21.742	.0010	21.765	22
39	.04623	.99893	.04628	21.606	.0011	21.629	21
40	.04652	.99892	.04657	21.470	1.0011	21.494	20
41	.04681	.99890	.04687	21.337	.0011	21.360	19
42	.04711	.99889	.04716	21.205	.0011	21.228	18
43	.04740	.99888	.04745	21.075	.0011	21.098	17
44	.04769	.99886	.04774	20.946	.0011	20.970	16
45	.04798	.99885	.04803	20.819	1.0011	20.843	15
46	.04827	.99883	.04832	20.693	.0012	20.717	14
47	.04856	.99882	.04862	20.569	.0012	20.593	13
48	.04885	.99881	.04891	20.446	.0012	20.471	12
49	.04914	.99879	.04920	20.325	.0012	20.350	11
50	.04943	.99878	.04949	20.205	1.0012	20.230	10
51	.04972	.99876	.04978	20.087	.0012	20.112	9
52	.05001	.99875	.05007	19.970	.0012	19.995	8
53	.05030	.99873	.05037	19.854	.0013	19.880	7
54	.05059	.99872	.05066	19.740	.0013	19.766	6
55	.05088	.99870	.05095	19.627	1.0013	19.653	5
56	.05117	.99869	.05124	19.515	.0013	19.541	4
57	.05146	.99867	.05153	19.405	.0013	19.431	3
58	.05175	.99866	.05182	19.296	.0013	19.322	2
59	.05204	.99864	.05212	19.188	.0013	19.214	1
60	.05234	.99863	.05241	19.081	1.0014	19.107	0
M	Cosine	Sine	Cotan.	Tangent	Cosec.	Secant	M

87°

3°

M	Sine	Cosine	Tangent	Cotan.	Secant	Cosec.	M
0	.05234	.99863	.05241	19.081	1.0014	19.107	60
1	.05263	.99861	.05270	18.975	.0014	19.002	59
2	.05292	.99860	.05299	18.871	.0014	18.897	58
3	.05321	.99858	.05328	18.768	.0014	18.794	57
4	.05350	.99857	.05357	18.665	.0014	18.692	56
5	.05379	.99855	.05387	18.564	1.0014	18.591	55
6	.05408	.99854	.05416	18.464	.0015	18.491	54
7	.05437	.99852	.05445	18.365	.0015	18.393	53
8	.05466	.99850	.05474	18.268	.0015	18.295	52
9	.05495	.99849	.05503	18.171	.0015	18.198	51
10	.05524	.99847	.05532	18.075	1.0015	18.103	50
11	.05553	.99846	.05562	17.980	.0015	18.008	49
12	.05582	.99844	.05591	17.886	.0016	17.914	48
13	.05611	.99842	.05620	17.793	.0016	17.821	47
14	.05640	.99841	.05649	17.701	.0016	17.730	46
15	.05669	.99839	.05678	17.610	1.0016	17.639	45
16	.05698	.99837	.05707	17.520	.0016	17.549	44
17	.05727	.99836	.05737	17.431	.0016	17.460	43
18	.05756	.99834	.05766	17.343	.0017	17.372	42
19	.05785	.99832	.05795	17.256	.0017	17.285	41
20	.05814	.99831	.05824	17.169	1.0017	17.198	40
21	.05843	.99829	.05853	17.084	.0017	17.113	39
22	.05872	.99827	.05883	16.999	.0017	17.028	38
23	.05902	.99826	.05912	16.915	.0017	16.944	37
24	.05931	.99824	.05941	16.832	.0018	16.861	36
25	.05960	.99822	.05970	16.750	1.0018	16.779	35
26	.05989	.99820	.05999	16.668	.0018	16.698	34
27	.06018	.99819	.06029	16.587	.0018	16.617	33
28	.06047	.99817	.06058	16.507	.0018	16.538	32
29	.06076	.99815	.06087	16.428	.0018	16.459	31
30	.06105	.99813	.06116	16.350	1.0019	16.380	30
31	.06134	.99812	.06145	16.272	.0019	16.303	29
32	.06163	.99810	.06175	16.195	.0019	16.226	28
33	.06192	.99808	.06204	16.119	.0019	16.150	27
34	.06221	.99806	.06233	16.043	.0019	16.075	26
35	.06250	.99804	.06262	15.969	1.0019	16.000	25
36	.06279	.99803	.06291	15.894	.0020	15.926	24
37	.06308	.99801	.06321	15.821	.0020	15.853	23
38	.06337	.99799	.06350	15.748	.0020	15.780	22
39	.06366	.99797	.06379	15.676	.0020	15.708	21
40	.06395	.99795	.06408	15.605	1.0020	15.637	20
41	.06424	.99793	.06437	15.534	.0021	15.566	19
42	.06453	.99791	.06467	15.464	.0021	15.496	18
43	.06482	.99790	.06496	15.394	.0021	15.427	17
44	.06511	.99788	.06525	15.325	.0021	15.358	16
45	.06540	.99786	.06554	15.257	1.0021	15.290	15
46	.06569	.99784	.06583	15.189	.0022	15.222	14
47	.06598	.99782	.06613	15.122	.0022	15.155	13
48	.06627	.99780	.06642	15.056	.0022	15.089	12
49	.06656	.99778	.06671	14.990	.0022	15.023	11
50	.06685	.99776	.06700	14.924	1.0022	14.958	10
51	.06714	.99774	.06730	14.860	.0023	14.893	9
52	.06743	.99772	.06759	14.795	.0023	14.829	8
53	.06772	.99770	.06788	14.732	.0023	14.765	7
54	.06801	.99768	.06817	14.668	.0023	14.702	6
55	.06830	.99766	.06846	14.606	1.0023	14.640	5
56	.06859	.99764	.06876	14.544	.0024	14.578	4
57	.06888	.99762	.06905	14.482	.0024	14.517	3
58	.06918	.99760	.06934	14.421	.0024	14.456	2
59	.06947	.99758	.06963	14.361	.0024	14.395	1
60	.06976	.99756	.06993	14.301	1.0024	14.335	0
M	Cosine	Sine	Cotan.	Tangent	Cosec.	Secant	M

86°

4°

M	Sine	Cosine	Tangent	Cotan.	Secant	Cosec.	M
0	.06976	.99756	.06993	14.301	1.0024	14.335	60
1	.07005	.99754	.07022	14.241	.0025	14.276	59
2	.07034	.99752	.07051	14.182	.0025	14.217	58
3	.07063	.99750	.07080	14.123	.0025	14.159	57
4	.07092	.99748	.07110	14.065	.0025	14.101	56
5	.07121	.99746	.07139	14.008	1.0025	14.043	55
6	.07150	.99744	.07168	13.951	.0026	13.986	54
7	.07179	.99742	.07197	13.894	.0026	13.930	53
8	.07208	.99740	.07226	13.838	.0026	13.874	52
9	.07237	.99738	.07256	13.782	.0026	13.818	51
10	.07266	.99736	.07285	13.727	1.0026	13.763	50
11	.07295	.99733	.07314	13.672	.0027	13.708	49
12	.07324	.99731	.07343	13.617	.0027	13.654	48
13	.07353	.99729	.07373	13.533	.0027	13.600	47
14	.07382	.99727	.07402	13.510	.0027	13.547	46
15	.07411	.99725	.07431	13.457	1.0027	13.494	45
16	.07440	.99723	.07460	13.404	.0028	13.441	44
17	.07469	.99721	.07490	13.351	.0028	13.389	43
18	.07498	.99718	.07519	13.299	.0028	13.337	42
19	.07527	.99716	.07548	13.248	.0028	13.286	41
20	.07556	.99714	.07577	13.197	1.0029	13.235	40
21	.07585	.99712	.07607	13.146	.0029	13.184	39
22	.07614	.99710	.07636	13.096	.0029	13.134	38
23	.07643	.99707	.07665	13.046	.0029	13.084	37
24	.07672	.99705	.07694	12.996	.0029	13.034	36
25	.07701	.99703	.07724	12.947	1.0030	12.985	35
26	.07730	.99701	.07753	12.898	.0030	12.937	34
27	.07759	.99698	.07782	12.849	.0030	12.888	33
28	.07788	.99696	.07812	12.801	.0030	12.840	32
29	.07817	.99694	.07841	12.754	.0031	12.793	31
30	.07846	.99692	.07870	12.706	1.0031	12.745	30
31	.07875	.99689	.07899	12.659	.0031	12.698	29
32	.07904	.99687	.07929	12.612	.0031	12.652	28
33	.07933	.99685	.07958	12.566	.0032	12.606	27
34	.07962	.99682	.07987	12.520	.0032	12.560	26
35	.07991	.99680	.08016	12.474	1.0032	12.514	25
36	.08020	.99678	.08046	12.429	.0032	12.469	24
37	.08049	.99675	.08075	12.384	.0032	12.424	23
38	.08078	.99673	.08104	12.339	.0033	12.379	22
39	.08107	.99671	.08134	12.295	.0033	12.335	21
40	.08136	.99668	.08163	12.250	1.0033	12.291	20
41	.08165	.99666	.08192	12.207	.0033	12.248	19
42	.08194	.99664	.08221	12.163	.0034	12.204	18
43	.08223	.99661	.08251	12.120	.0034	12.161	17
44	.08252	.99659	.08280	12.077	.0034	12.118	16
45	.08281	.99656	.08309	12.035	1.0034	12.076	15
46	.08310	.99654	.08339	11.992	.0035	12.034	14
47	.08339	.99652	.08368	11.950	.0035	11.992	13
48	.08368	.99649	.08397	11.909	.0035	11.950	12
49	.08397	.99647	.08426	11.867	.0035	11.909	11
50	.08426	.99644	.08456	11.826	1.0036	11.868	10
51	.08455	.99642	.08485	11.785	.0036	11.828	9
52	.08484	.99639	.08514	11.745	.0036	11.787	8
53	.08513	.99637	.08544	11.704	.0036	11.747	7
54	.08542	.99634	.08573	11.664	.0037	11.707	6
55	.08571	.99632	.08602	11.625	1.0037	11.668	5
56	.08600	.99629	.08632	11.585	.0037	11.628	4
57	.08629	.99627	.08661	11.546	.0037	11.589	3
58	.08658	.99624	.08690	11.507	.0038	11.550	2
59	.08687	.99622	.08719	11.468	.0038	11.512	1
60	.08715	.99619	.08749	11.430	1.0038	11.474	0
M	Cosine	Sine	Cotan.	Tangent	Cosec.	Secant	M

85°

5°

M	Sine	Cosine	Tangent	Cotan.	Secant	Cosec.	M
0	.08715	.99619	.08749	11.430	1.0038	11.474	60
1	.08744	.99617	.08778	11.392	.0038	11.436	59
2	.08773	.99614	.08807	11.354	.0039	11.398	58
3	.08802	.99612	.08837	11.316	.0039	11.360	57
4	.08831	.99609	.08866	11.279	.0039	11.323	56
5	.08860	.99607	.08895	11.242	1.0039	11.286	55
6	.08889	.99604	.08925	11.205	.0040	11.249	54
7	.08918	.99601	.08954	11.168	.0040	11.213	53
8	.08947	.99599	.08983	11.132	.0040	11.176	52
9	.08976	.99596	.09013	11.095	.0040	11.140	51
10	.09005	.99594	.09042	11.059	1.0041	11.104	50
11	.09034	.99591	.09071	11.024	.0041	11.069	49
12	.09063	.99588	.09101	10.988	.0041	11.033	48
13	.09092	.99586	.09130	10.953	.0041	10.998	47
14	.09121	.99583	.09159	10.918	.0042	10.963	46
15	.09150	.99580	.09189	10.883	1.0042	10.929	45
16	.09179	.99578	.09218	10.848	.0042	10.894	44
17	.09208	.99575	.09247	10.814	.0043	10.860	43
18	.09237	.99572	.09277	10.780	.0043	10.826	42
19	.09266	.99570	.09306	10.746	.0043	10.792	41
20	.09295	.99567	.09335	10.712	1.0043	10.758	40
21	.09324	.99564	.09365	10.678	.0044	10.725	39
22	.09353	.99562	.09394	10.645	.0044	10.692	38
23	.09382	.99559	.09423	10.612	.0044	10.659	37
24	.09411	.99556	.09453	10.579	.0044	10.626	36
25	.09440	.99553	.09482	10.546	1.0045	10.593	35
26	.09469	.99551	.09511	10.514	.0045	10.561	34
27	.09498	.99548	.09541	10.481	.0045	10.529	33
28	.09527	.99545	.09570	10.449	.0046	10.497	32
29	.09556	.99542	.09599	10.417	.0046	10.465	31
30	.09584	.99540	.09629	10.385	1.0046	10.433	30
31	.09613	.99537	.09658	10.354	.0046	10.402	29
32	.09642	.99534	.09688	10.322	.0047	10.371	28
33	.09671	.99531	.09717	10.291	.0047	10.340	27
34	.09700	.99528	.09746	10.260	.0047	10.309	26
35	.09729	.99525	.09776	10.229	1.0048	10.278	25
36	.09758	.99523	.09805	10.199	.0048	10.248	24
37	.09787	.99520	.09834	10.168	.0048	10.217	23
38	.09816	.99517	.09864	10.138	.0048	10.187	22
39	.09845	.99514	.09893	10.108	.0049	10.157	21
40	.09874	.99511	.09922	10.078	1.0049	10.127	20
41	.09903	.99508	.09952	10.048	.0049	10.098	19
42	.09932	.99505	.09981	10.019	.0050	10.068	18
43	.09961	.99503	.10011	9.9893	.0050	10.039	17
44	.09990	.99500	.10040	9.9601	.0050	10.010	16
45	.10019	.99497	.10069	9.9310	1.0050	9.9812	15
46	.10048	.99494	.10099	9.9021	.0051	9.9525	14
47	.10077	.99491	.10128	9.8734	.0051	9.9239	13
48	.10106	.99488	.10158	9.8448	.0051	9.8955	12
49	.10134	.99485	.10187	9.8164	.0052	9.8672	11
50	.10163	.99482	.10216	9.7882	1.0052	9.8391	10
51	.10192	.99479	.10246	9.7601	.0052	9.8112	9
52	.10221	.99476	.10275	9.7322	.0053	9.7834	8
53	.10250	.99473	.10305	9.7044	.0053	9.7558	7
54	.10279	.99470	.10334	9.6768	.0053	9.7283	6
55	.10308	.99467	.10363	9.6493	1.0053	9.7010	5
56	.10337	.99464	.10393	9.6220	.0054	9.6739	4
57	.10366	.99461	.10422	9.5949	.0054	9.6469	3
58	.10395	.99458	.10452	9.5679	.0054	9.6200	2
59	.10424	.99455	.10481	9.5411	.0055	9.5933	1
60	.10453	.99452	.10510	9.5144	1.0055	9.5668	0
M	Cosine	Sine	Cotan.	Tangent	Cosec.	Secant	M

84°

6°

M	Sine	Cosine	Tangent	Cotan.	Secant	Cosec.	M
0	.10453	.99452	.10510	9.5144	1.0055	9.5668	60
1	.10482	.99449	.10540	.4878	.0055	.5404	59
2	.10511	.99446	.10569	.4614	.0056	.5141	58
3	.10540	.99443	.10599	.4351	.0056	.4880	57
4	.10568	.99440	.10628	.4090	.0056	.4620	56
5	.10597	.99437	.10657	9.3831	1.0057	9.4362	55
6	.10626	.99434	.10687	.3572	.0057	.4105	54
7	.10655	.99431	.10716	.3315	.0057	.3850	53
8	.10684	.99428	.10746	.3060	.0057	.3596	52
9	.10713	.99424	.10775	.2806	.0058	.3343	51
10	.10742	.99421	.10805	9.2553	1.0058	9.3092	50
11	.10771	.99418	.10834	.2302	.0058	.2842	49
12	.10800	.99415	.10863	.2051	.0059	.2593	48
13	.10829	.99412	.10893	.1803	.0059	.2346	47
14	.10858	.99409	.10922	.1555	.0059	.2100	46
15	.10887	.99406	.10952	9.1309	1.0060	9.1855	45
16	.10916	.99402	.10981	.1064	.0060	.1612	44
17	.10944	.99399	.11011	.0821	.0060	.1370	43
18	.10973	.99396	.11040	.0579	.0061	.1129	42
19	.11002	.99393	.11069	.0338	.0061	.0890	41
20	.11031	.99390	.11099	9.0098	1.0061	9.0651	40
21	.11060	.99386	.11128	8.9860	.0062	.0414	39
22	.11089	.99383	.11158	.9623	.0062	.0179	38
23	.11118	.99380	.11187	.9387	.0062	8.9944	37
24	.11147	.99377	.11217	.9152	.0063	.9711	36
25	.11176	.99373	.11246	8.8918	1.0063	8.9479	35
26	.11205	.99370	.11276	.8686	.0063	.9248	34
27	.11234	.99367	.11305	.8455	.0064	.9018	33
28	.11262	.99364	.11335	.8225	.0064	.8790	32
29	.11291	.99360	.11364	.7996	.0064	.8563	31
30	.11320	.99357	.11393	8.7769	1.0065	8.8337	30
31	.11349	.99354	.11423	.7542	.0065	.8112	29
32	.11378	.99350	.11452	.7317	.0065	.7888	28
33	.11407	.99347	.11482	.7093	.0066	.7665	27
34	.11436	.99344	.11511	.6870	.0066	.7444	26
35	.11465	.99341	.11541	8.6648	1.0066	8.7223	25
36	.11494	.99337	.11570	.6427	.0067	.7004	24
37	.11523	.99334	.11600	.6208	.0067	.6786	23
38	.11551	.99330	.11629	.5989	.0067	.6569	22
39	.11580	.99327	.11659	.5772	.0068	.6353	21
40	.11609	.99324	.11688	8.5555	1.0068	8.6138	20
41	.11638	.99320	.11718	.5340	.0068	.5924	19
42	.11667	.99317	.11747	.5126	.0069	.5711	18
43	.11696	.99314	.11777	.4913	.0069	.5499	17
44	.11725	.99310	.11806	.4701	.0069	.5289	16
45	.11754	.99307	.11836	8.4489	1.0070	8.5079	15
46	.11783	.99303	.11865	.4279	.0070	.4871	14
47	.11811	.99300	.11895	.4070	.0070	.4663	13
48	.11840	.99296	.11924	.3862	.0071	.4457	12
49	.11869	.99293	.11954	.3655	.0071	.4251	11
50	.11898	.99290	.11983	8.3449	1.0071	8.4046	10
51	.11927	.99286	.12013	.3244	.0072	.3843	9
52	.11956	.99283	.12042	.3040	.0072	.3640	8
53	.11985	.99279	.12072	.2837	.0073	.3439	7
54	.12014	.99276	.12101	.2635	.0073	.3238	6
55	.12042	.99272	.12131	8.2434	1.0073	8.3039	5
56	.12071	.99269	.12160	.2234	.0074	.2840	4
57	.12100	.99265	.12190	.2035	.0074	.2642	3
58	.12129	.99262	.12219	.1837	.0074	.2446	2
59	.12158	.99258	.12249	.1640	.0075	.2250	1
60	.12187	.99255	.12278	8.1443	1.0075	8.2055	0
M	Cosine	Sine	Cotan.	Tangent	Cosec.	Secant	M

83°

7°

M	Sine	Cosine	Tangent	Cotan.	Secant	Cosec.	M
0	.12187	.99255	.12278	8.1443	1.0075	8.2055	60
1	.12216	.99251	.12308	.1248	.0075	.1861	59
2	.12245	.99247	.12337	.1053	.0076	.1668	58
3	.12273	.99244	.12367	.0860	.0076	.1476	57
4	.12302	.99240	.12396	.0667	.0076	.1285	56
5	.12331	.99237	.12426	8.0476	1.0077	8.1094	55
6	.12360	.99233	.12456	.0285	.0077	.0905	54
7	.12389	.99229	.12485	.0095	.0078	.0717	53
8	.12418	.99226	.12515	7.9906	.0078	.0529	52
9	.12447	.99222	.12544	.9717	.0078	.0342	51
10	.12476	.99219	.12574	7.9530	1.0079	8.0156	50
11	.12504	.99215	.12603	.9344	.0079	7.9971	49
12	.12533	.99211	.12633	.9158	.0079	.9787	48
13	.12562	.99208	.12662	.8973	.0080	.9604	47
14	.12591	.99204	.12692	.8789	.0080	.9421	46
15	.12620	.99200	.12722	7.8606	1.0080	7.9240	45
16	.12649	.99197	.12751	.8424	.0081	.9059	44
17	.12678	.99193	.12781	.8243	.0081	.8879	43
18	.12706	.99189	.12810	.8062	.0082	.8700	42
19	.12735	.99186	.12840	.7882	.0082	.8522	41
20	.12764	.99182	.12869	7.7703	1.0082	7.8344	40
21	.12793	.99178	.12899	.7525	.0083	.8168	39
22	.12822	.99174	.12928	.7348	.0083	.7992	38
23	.12851	.99171	.12958	.7171	.0084	.7817	37
24	.12879	.99167	.12988	.6996	.0084	.7642	36
25	.12908	.99163	.13017	7.6821	1.0084	7.7469	35
26	.12937	.99160	.13047	.6646	.0085	.7296	34
27	.12966	.99156	.13076	.6473	.0085	.7124	33
28	.12995	.99152	.13106	.6300	.0085	.6953	32
29	.13024	.99148	.13136	.6129	.0086	.6783	31
30	.13053	.99144	.13165	7.5957	1.0086	7.6613	30
31	.13081	.99141	.13195	.5787	.0087	.6444	29
32	.13110	.99137	.13224	.5617	.0087	.6276	28
33	.13139	.99133	.13254	.5449	.0087	.6108	27
34	.13168	.99129	.13284	.5280	.0088	.5942	26
35	.13197	.99125	.13313	7.5113	1.0088	7.5776	25
36	.13226	.99121	.13343	.4946	.0089	.5611	24
37	.13254	.99118	.13372	.4780	.0089	.5446	23
38	.13283	.99114	.13402	.4615	.0089	.5282	22
39	.13312	.99110	.13432	.4451	.0090	.5119	21
40	.13341	.99106	.13461	7.4287	1.0090	7.4957	20
41	.13370	.99102	.13491	.4124	.0090	.4795	19
42	.13399	.99098	.13520	.3961	.0091	.4634	18
43	.13427	.99094	.13550	.3800	.0091	.4474	17
44	.13456	.99090	.13580	.3639	.0092	.4315	16
45	.13485	.99086	.13609	7.3479	1.0092	7.4156	15
46	.13514	.99083	.13639	.3319	.0092	.3998	14
47	.13543	.99079	.13669	.3160	.0093	.3840	13
48	.13571	.99075	.13698	.3002	.0093	.3683	12
49	.13600	.99071	.13728	.2844	.0094	.3527	11
50	.13629	.99067	.13757	7.2687	1.0094	7.3372	10
51	.13658	.99063	.13787	.2531	.0094	.3217	9
52	.13687	.99059	.13817	.2375	.0095	.3063	8
53	.13716	.99055	.13846	.2220	.0095	.2909	7
54	.13744	.99051	.13876	.2066	.0096	.2757	6
55	.13773	.99047	.13906	7.1912	1.0096	7.2604	5
56	.13802	.99043	.13935	.1759	.0097	.2453	4
57	.13831	.99039	.13965	.1607	.0097	.2302	3
58	.13860	.99035	.13995	.1455	.0097	.2152	2
59	.13888	.99031	.14024	.1304	.0098	.2002	1
60	.13917	.99027	.14054	7.1154	1.0098	7.1853	0
M	**Cosine**	**Sine**	**Cotan.**	**Tangent**	**Cosec.**	**Secant**	**M**

82°

8°

M	Sine	Cosine	Tangent	Cotan.	Secant	Cosec.	M
0	.13917	.99027	.14054	7.1154	1.0098	7.1853	60
1	.13946	.99023	.14084	.1004	.0099	.1704	59
2	.13975	.99019	.14113	.0854	.0099	.1557	58
3	.14004	.99015	.14143	.0706	.0099	.1409	57
4	.14032	.99010	.14173	.0558	.0100	.1263	56
5	.14061	.99006	.14202	7.0410	1.0100	7.1117	55
6	.14090	.99002	.14232	.0264	.0101	.0972	54
7	.14119	.98998	.14262	.0117	.0101	.0827	53
8	.14148	.98994	.14291	6.9972	.0102	.0683	52
9	.14176	.98990	.14321	.9827	.0102	.0539	51
10	.14205	.98986	.14351	6.9682	1.0102	7.0396	50
11	.14234	.98982	.14380	.9538	.0103	.0254	49
12	.14263	.98978	.14410	.9395	.0103	.0112	48
13	.14292	.98973	.14440	.9252	.0104	6.9971	47
14	.14320	.98969	.14470	.9110	.0104	.9830	46
15	.14349	.98965	.14499	6.8969	1.0104	6.9690	45
16	.14378	.98961	.14529	.8828	.0105	.9550	44
17	.14407	.98957	.14559	.8687	.0105	.9411	43
18	.14436	.98952	.14588	.8547	.0106	.9273	42
19	.14464	.98948	.14618	.8408	.0106	.9135	41
20	.14493	.98944	.14648	6.8269	1.0107	6.8998	40
21	.14522	.98940	.14677	.8131	.0107	.8861	39
22	.14551	.98936	.14707	.7993	.0107	.8725	38
23	.14579	.98931	.14737	.7856	.0108	.8589	37
24	.14608	.98927	.14767	.7720	.0108	.8454	36
25	.14637	.98923	.14796	6.7584	1.0109	6.8320	35
26	.14666	.98919	.14826	.7448	.0109	.8185	34
27	.14695	.98914	.14856	.7313	.0110	.8052	33
28	.14723	.98910	.14886	.7179	.0110	.7919	32
29	.14752	.98906	.14915	.7045	.0111	.7787	31
30	.14781	.98901	.14945	6.6911	1.0111	6.7655	30
31	.14810	.98897	.14975	.6779	.0111	.7523	29
32	.14838	.98893	.15004	.6646	.0112	.7392	28
33	.14867	.98889	.15034	.6514	.0112	.7262	27
34	.14896	.98884	.15064	.6383	.0113	.7132	26
35	.14925	.98880	.15094	6.6252	1.0113	6.7003	25
36	.14953	.98876	.15123	.6122	.0114	.6874	24
37	.14982	.98871	.15153	.5992	.0114	.6745	23
38	.15011	.98867	.15183	.5863	.0115	.6617	22
39	.15040	.98862	.15213	.5734	.0115	.6490	21
40	.15068	.98858	.15243	6.5605	1.0115	6.6363	20
41	.15097	.98854	.15272	.5478	.0116	.6237	19
42	.15126	.98849	.15302	.5350	.0116	.6111	18
43	.15155	.98845	.15332	.5223	.0117	.5985	17
44	.15183	.98840	.15362	.5097	.0117	.5860	16
45	.15212	.98836	.15391	6.4971	1.0118	6.5736	15
46	.15241	.98832	.15421	.4845	.0118	.5612	14
47	.15270	.98827	.15451	.4720	.0119	.5488	13
48	.15298	.98823	.15481	.4596	.0119	.5365	12
49	.15328	.98818	.15511	.4472	.0119	.5243	11
50	.15356	.98814	.15540	6.4348	1.0120	6.5121	10
51	.15385	.98809	.15570	.4225	.0120	.4999	9
52	.15413	.98805	.15600	.4103	.0121	.4878	8
53	.15442	.98800	.15630	.3980	.0121	.4757	7
54	.15471	.98796	.15659	.3859	.0122	.4637	6
55	.15500	.98791	.15689	6.3737	1.0122	6.4517	5
56	.15528	.98787	.15719	.3616	.0123	.4398	4
57	.15557	.98782	.15749	.3496	.0123	.4279	3
58	.15586	.98778	.15779	.3376	.0124	.4160	2
59	.15615	.98773	.15809	.3257	.0124	.4042	1
60	.15643	.98769	.15838	6.3137	1.0125	6.3924	0
M	Cosine	Sine	Cotan.	Tangent	Cosec.	Secant	M

81°

9°

M	Sine	Cosine	Tangent	Cotan.	Secant	Cosec.	M
0	.15643	.98769	.15838	6.3137	1.0125	6.3924	60
1	.15672	.98764	.15868	.3019	.0125	.3807	59
2	.15701	.98760	.15898	.2901	.0125	.3690	58
3	.15730	.98755	.15928	.2783	.0126	.3574	57
4	.15758	.98750	.15958	.2665	.0126	.3458	56
5	.15787	.98746	.15987	6.2548	1.0127	6.3343	55
6	.15816	.98741	.16017	.2432	.0127	.3228	54
7	.15844	.98737	.16047	.2316	.0128	.3113	53
8	.15873	.98732	.16077	.2200	.0128	.2999	52
9	.15902	.98727	.16107	.2085	.0129	.2885	51
10	.15931	.98723	.16137	6.1970	1.0129	6.2772	50
11	.15959	.98718	.16167	.1856	.0130	.2659	49
12	.15988	.98714	.16196	.1742	.0130	.2546	48
13	.16017	.98709	.16226	.1628	.0131	.2434	47
14	.16045	.98704	.16256	.1515	.0131	.2322	46
15	.16074	.98700	.16286	6.1402	1.0132	6.2211	45
16	.16103	.98695	.16316	.1290	.0132	.2100	44
17	.16132	.98690	.16346	.1178	.0133	.1990	43
18	.16160	.98685	.16376	.1066	.0133	.1880	42
19	.16189	.98681	.16405	.0955	.0134	.1770	41
20	.16218	.98676	.16435	6.0844	1.0134	6.1661	40
21	.16246	.98671	.16465	.0734	.0135	.1552	39
22	.16275	.98667	.16495	.0624	.0135	.1443	38
23	.16304	.98662	.16525	.0514	.0136	.1335	37
24	.16333	.98657	.16555	.0405	.0136	.1227	36
25	.16361	.98652	.16585	6.0296	1.0136	6.1120	35
26	.16390	.98648	.16615	.0188	.0137	.1013	34
27	.16419	.98643	.16644	.0080	.0137	.0906	33
28	.16447	.98638	.16674	5.9972	.0138	.0800	32
29	.16476	.98633	.16704	.9865	.0138	.0694	31
30	.16505	.98628	.16734	5.9758	1.0139	6.0588	30
31	.16533	.98624	.16764	.9651	.0139	.0483	29
32	.16562	.98619	.16794	.9545	.0140	.0379	28
33	.16591	.98614	.16824	.9439	.0140	.0274	27
34	.16619	.98609	.16854	.9333	.0141	.0170	26
35	.16648	.98604	.16884	5.9228	1.0141	6.0066	25
36	.16677	.98600	.16914	.9123	.0142	5.9963	24
37	.16705	.98595	.16944	.9019	.0142	.9860	23
38	.16734	.98590	.16973	.8915	.0143	.9758	22
39	.16763	.98585	.17003	.8811	.0143	.9655	21
40	.16791	.98580	.17033	5.8708	1.0144	5.9554	20
41	.16820	.98575	.17063	.8605	.0144	.9452	19
42	.16849	.98570	.17093	.8502	.0145	.9351	18
43	.16878	.98565	.17123	.8400	.0145	.9250	17
44	.16906	.98560	.17153	.8298	.0146	.9150	16
45	.16935	.98556	.17183	5.8196	1.0146	5.9049	15
46	.16964	.98551	.17213	.8095	.0147	.8950	14
47	.16992	.98546	.17243	.7994	.0147	.8850	13
48	.17021	.98541	.17273	.7894	.0148	.8751	12
49	.17050	.98536	.17303	.7794	.0148	.8652	11
50	.17078	.98531	.17333	5.7694	1.0149	5.8554	10
51	.17107	.98526	.17363	.7594	.0150	.8456	9
52	.17136	.98521	.17393	.7495	.0150	.8358	8
53	.17164	.98516	.17423	.7396	.0151	.8261	7
54	.17193	.98511	.17453	.7297	.0151	.8163	6
55	.17221	.98506	.17483	5.7199	1.0152	5.8067	5
56	.17250	.98501	.17513	.7101	.0152	.7970	4
57	.17279	.98496	.17543	.7004	.0153	.7874	3
58	.17307	.98491	.17573	.6906	.0153	.7778	2
59	.17336	.98486	.17603	.6809	.0154	.7683	1
60	.17365	.98481	.17633	5.6713	1.0154	5.7588	0
M	Cosine	Sine	Cotan.	Tangent	Cosec.	Secant	M

80°

10°

M	Sine	Cosine	Tangent	Cotan.	Secant	Cosec.	M
0	.17365	.98481	.17633	5.6713	1.0154	5.7588	60
1	.17393	.98476	.17663	.6616	.0155	.7493	59
2	.17422	.98471	.17693	.6520	.0155	.7398	58
3	.17451	.98465	.17723	.6425	.0156	.7304	57
4	.17479	.98460	.17753	.6329	.0156	.7210	56
5	.17508	.98455	.17783	5.6234	1.0157	5.7117	55
6	.17537	.98450	.17813	.6140	.0157	.7023	54
7	.17565	.98445	.17843	.6045	.0158	.6930	53
8	.17594	.98440	.17873	.5951	.0158	.6838	52
9	.17622	.98435	.17903	.5857	.0159	.6745	51
10	.17651	.98430	.17933	5.5764	1.0159	5.6653	50
11	.17680	.98425	.17963	.5670	.0160	.6561	49
12	.17708	.98419	.17993	.5578	.0160	.6470	48
13	.17737	.98414	.18023	.5485	.0161	.6379	47
14	.17766	.98409	.18053	.5393	.0162	.6288	46
15	.17794	.98404	.18083	5.5301	1.0162	5.6197	45
16	.17823	.98399	.18113	.5209	.0163	.6107	44
17	.17852	.98394	.18143	.5117	.0163	.6017	43
18	.17880	.98388	.18173	.5026	.0164	.5928	42
19	.17909	.98383	.18203	.4936	.0164	.5838	41
20	.17937	.98378	.18233	5.4845	1.0165	5.5749	40
21	.17966	.98373	.18263	.4755	.0165	.5660	39
22	.17995	.98368	.18293	.4665	.0166	.5572	38
23	.18023	.98362	.18323	.4575	.0166	.5484	37
24	.18052	.98357	.18353	.4486	.0167	.5396	36
25	.18080	.98352	.18383	5.4396	1.0167	5.5308	35
26	.18109	.98347	.18413	.4308	.0168	.5221	34
27	.18138	.98341	.18444	.4219	.0169	.5134	33
28	.18166	.98336	.18474	.4131	.0169	.5047	32
29	.18195	.98331	.18504	.4043	.0170	.4960	31
30	.18223	.98325	.18534	5.3955	1.0170	5.4874	30
31	.18252	.98320	.18564	.3868	.0171	.4788	29
32	.18281	.98315	.18594	.3780	.0171	.4702	28
33	.18309	.98309	.18624	.3694	.0172	.4617	27
34	.18338	.98304	.18654	.3607	.0172	.4532	26
35	.18366	.98299	.18684	5.3521	1.0173	5.4447	25
36	.18395	.98293	.18714	.3434	.0174	.4362	24
37	.18424	.98288	.18745	.3349	.0174	.4278	23
38	.18452	.98283	.18775	.3263	.0175	.4194	22
39	.18481	.98277	.18805	.3178	.0175	.4110	21
40	.18509	.98272	.18835	5.3093	1.0176	5.4026	20
41	.18538	.98267	.18865	.3008	.0176	.3943	19
42	.18567	.98261	.18895	.2923	.0177	.3860	18
43	.18595	.98256	.18925	.2839	.0177	.3777	17
44	.18624	.98250	.18955	.2755	.0178	.3695	16
45	.18652	.98245	.18985	5.2671	1.0179	5.3612	15
46	.18681	.98240	.19016	.2588	.0179	.3530	14
47	.18709	.98234	.19046	.2505	.0180	.3449	13
48	.18738	.98229	.19076	.2422	.0180	.3367	12
49	.18767	.98223	.19106	.2339	.0181	.3286	11
50	.18795	.98218	.19136	5.2257	1.0181	5.3205	10
51	.18824	.98212	.19166	.2174	.0182	.3124	9
52	.18852	.98207	.19197	.2092	.0182	.3044	8
53	.18881	.98201	.19227	.2011	.0183	.2963	7
54	.18909	.98196	.19257	.1929	.0184	.2883	6
55	.18938	.98190	.19287	5.1848	1.0184	5.2803	5
56	.18967	.98185	.19317	.1767	.0185	.2724	4
57	.18995	.98179	.19347	.1686	.0185	.2645	3
58	.19024	.98174	.19378	.1606	.0186	.2566	2
59	.19052	.98168	.19408	.1525	.0186	.2487	1
60	.19081	.98163	.19438	5.1445	1.0187	5.2408	0
M	Cosine	Sine	Cotan.	Tangent	Cosec.	Secant	M

79°

11°

M	Sine	Cosine	Tangent	Cotan.	Secant	Cosec.	M
0	.19081	.98163	.19438	5.1445	1.0187	5.2408	60
1	.19109	.98157	.19468	.1366	.0188	.2330	59
2	.19138	.98152	.19498	.1286	.0188	.2252	58
3	.19166	.98146	.19529	.1207	.0189	.2174	57
4	.19195	.98140	.19559	.1128	.0189	.2097	56
5	.19224	.98135	.19589	5.1049	1.0190	5.2019	55
6	.19252	.98129	.19619	.0970	.0191	.1942	54
7	.19281	.98124	.19649	.0892	.0191	.1865	53
8	.19309	.98118	.19680	.0814	.0192	.1788	52
9	.19338	.98112	.19710	.0736	.0192	.1712	51
10	.19366	.98107	.19740	5.0658	1.0193	5.1636	50
11	.19395	.98101	.19770	.0581	.0193	.1560	49
12	.19423	.98095	.19800	.0504	.0194	.1484	48
13	.19452	.98090	.19831	.0427	.0195	.1409	47
14	.19480	.98084	.19861	.0350	.0195	.1333	46
15	.19509	.98078	.19891	5.0273	1.0196	5.1258	45
16	.19537	.98073	.19921	.0197	.0196	.1183	44
17	.19566	.98067	.19952	.0121	.0197	.1109	43
18	.19595	.98061	.19982	.0045	.0198	.1034	42
19	.19623	.98056	.20012	4.9969	.0198	.0960	41
20	.19652	.98050	.20042	4.9894	1.0199	5.0886	40
21	.19680	.98044	.20073	.9819	.0199	.0812	39
22	.19709	.98039	.20103	.9744	.0200	.0739	38
23	.19737	.98033	.20133	.9669	.0201	.0666	37
24	.19766	.98027	.20163	.9594	.0201	.0593	36
25	.19794	.98021	.20194	4.9520	1.0202	5.0520	35
26	.19823	.98016	.20224	.9446	.0202	.0447	34
27	.19851	.98010	.20254	.9372	.0203	.0375	33
28	.19880	.98004	.20285	.9298	.0204	.0302	32
29	.19908	.97998	.20315	.9225	.0204	.0230	31
30	.19937	.97992	.20345	4.9151	1.0205	5.0158	30
31	.19965	.97987	.20375	.9078	.0205	.0087	29
32	.19994	.97981	.20406	.9006	.0206	.0015	28
33	.20022	.97975	.20436	.8933	.0207	4.9944	27
34	.20051	.97969	.20466	.8860	.0207	.9873	26
35	.20079	.97963	.20497	4.8788	1.0208	4.9802	25
36	.20108	.97957	.20527	.8716	.0208	.9732	24
37	.20136	.97952	.20557	.8644	.0209	.9661	23
38	.20165	.97946	.20588	.8573	.0210	.9591	22
39	.20193	.97940	.20618	.8501	.0210	.9521	21
40	.20222	.97934	.20648	4.8430	1.0211	4.9452	20
41	.20250	.97928	.20679	.8359	.0211	.9382	19
42	.20279	.97922	.20709	.8288	.0212	.9313	18
43	.20307	.97916	.20739	.8217	.0213	.9243	17
44	.20336	.97910	.20770	.8147	.0213	.9175	16
45	.20364	.97904	.20800	4.8077	1.0214	4.9106	15
46	.20393	.97899	.20830	.8007	.0215	.9037	14
47	.20421	.97893	.20861	.7937	.0215	.8969	13
48	.20450	.97887	.20891	.7867	.0216	.8901	12
49	.20478	.97881	.20921	.7798	.0216	.8833	11
50	.20506	.97875	.20952	4.7728	1.0217	4.8765	10
51	.20535	.97869	.20982	.7659	.0218	.8697	9
52	.20563	.97863	.21012	.7591	.0218	.8630	8
53	.20592	.97857	.21043	.7522	.0219	.8563	7
54	.20620	.97851	.21073	.7453	.0220	.8496	6
55	.20649	.97845	.21104	4.7385	1.0220	4.8429	5
56	.20677	.97839	.21134	.7317	.0221	.8362	4
57	.20706	.97833	.21164	.7249	.0221	.8296	3
58	.20734	.97827	.21195	.7181	.0222	.8229	2
59	.20763	.97821	.21225	.7114	.0223	.8163	1
60	.20791	.97815	.21256	4.7046	1.0223	4.8097	0
M	Cosine	Sine	Cotan.	Tangent	Cosec.	Secant	M

78°

12°

M	Sine	Cosine	Tangent	Cotan.	Secant	Cosec.	M
0	.20791	.97815	.21256	4.7046	1.0223	4.8097	60
1	.20820	.97809	.21286	.6979	.0224	.8032	59
2	.20848	.97803	.21316	.6912	.0225	.7966	58
3	.20876	.97797	.21347	.6845	.0225	.7901	57
4	.20905	.97790	.21377	.6778	.0226	.7835	56
5	.20933	.97784	.21408	4.6712	1.0226	4.7770	55
6	.20962	.97778	.21438	.6646	.0227	.7706	54
7	.20990	.97772	.21468	.6580	.0228	.7641	53
8	.21019	.97766	.21499	.6514	.0228	.7576	52
9	.21047	.97760	.21529	.6448	.0229	.7512	51
10	.21076	.97754	.21560	4.6382	1.0230	4.7448	50
11	.21104	.97748	.21590	.6317	.0230	.7384	49
12	.21132	.97741	.21621	.6252	.0231	.7320	48
13	.21161	.97735	.21651	.6187	.0232	.7257	47
14	.21189	.97729	.21682	.6122	.0232	.7193	46
15	.21218	.97723	.21712	4.6057	1.0233	4.7130	45
16	.21246	.97717	.21742	.5993	.0234	.7067	44
17	.21275	.97711	.21773	.5928	.0234	.7004	43
18	.21303	.97704	.21803	.5864	.0235	.6942	42
19	.21331	.97698	.21834	.5800	.0235	.6879	41
20	.21360	.97692	.21864	4.5736	1.0236	4.6817	40
21	.21388	.97686	.21895	.5673	.0237	.6754	39
22	.21417	.97680	.21925	.5609	.0237	.6692	38
23	.21445	.97673	.21956	.5546	.0238	.6631	37
24	.21473	.97667	.21986	.5483	.0239	.6569	36
25	.21502	.97661	.22017	4.5420	1.0239	4.6507	35
26	.21530	.97655	.22047	.5357	.0240	.6446	34
27	.21559	.97648	.22078	.5294	.0241	.6385	33
28	.21587	.97642	.22108	.5232	.0241	.6324	32
29	.21615	.97636	.22139	.5169	.0242	.6263	31
30	.21644	.97630	.22169	4.5107	1.0243	4.6201	30
31	.21672	.97623	.22200	.5045	.0243	.6142	29
32	.21701	.97617	.22230	.4983	.0244	.6081	28
33	.21729	.97611	.22261	.4921	.0245	.6021	27
34	.21757	.97604	.22291	.4860	.0245	.5961	26
35	.21786	.97598	.22322	4.4799	1.0246	4.5901	25
36	.21814	.97592	.22353	.4737	.0247	.5841	24
37	.21843	.97585	.22383	.4676	.0247	.5782	23
38	.21871	.97579	.22414	.4615	.0248	.5722	22
39	.21899	.97573	.22444	.4555	.0249	.5663	21
40	.21928	.97566	.22475	4.4494	1.0249	4.5604	20
41	.21956	.97560	.22505	.4434	.0250	.5545	19
42	.21985	.97553	.22536	.4373	.0251	.5486	18
43	.22013	.97547	.22566	.4313	.0251	.5428	17
44	.22041	.97541	.22597	.4253	.0252	.5369	16
45	.22070	.97534	.22628	4.4194	1.0253	4.5311	15
46	.22098	.97528	.22658	.4134	.0253	.5253	14
47	.22126	.97521	.22689	.4074	.0254	.5195	13
48	.22155	.97515	.22719	.4015	.0255	.5137	12
49	.22183	.97508	.22750	.3956	.0255	.5079	11
50	.22211	.97502	.22781	4.3897	1.0256	4.5021	10
51	.22240	.97495	.22811	.3838	.0257	.4964	9
52	.22268	.97489	.22842	.3779	.0257	.4907	8
53	.22297	.97483	.22872	.3721	.0258	.4850	7
54	.22325	.97476	.22903	.3662	.0259	.4793	6
55	.22353	.97470	.22934	4.3604	1.0260	4.4736	5
56	.22382	.97463	.22964	.3546	.0260	.4679	4
57	.22410	.97457	.22995	.3488	.0261	.4623	3
58	.22438	.97450	.23025	.3430	.0262	.4566	2
59	.22467	.97443	.23056	.3372	.0262	.4510	1
60	.22495	.97437	.23087	4.3315	1.0263	4.4454	0
M	Cosine	Sine	Cotan.	Tangent	Cosec.	Secant	M

77°

13°

M	Sine	Cosine	Tangent	Cotan.	Secant	Cosec.	M
0	.22495	.97437	.23087	4.3315	1.0263	4.4454	60
1	.22523	.97430	.23117	.3257	.0264	.4398	59
2	.22552	.97424	.23148	.3200	.0264	.4342	58
3	.22580	.97417	.23179	.3143	.0265	.4287	57
4	.22608	.97411	.23209	.3086	.0266	.4231	56
5	.22637	.97404	.23240	4.3029	1.0266	4.4176	55
6	.22665	.97398	.23270	.2972	.0267	.4121	54
7	.22693	.97391	.23301	.2916	.0268	.4065	53
8	.22722	.97384	.23332	.2859	.0268	.4011	52
9	.22750	.97378	.23363	.2803	.0269	.3956	51
10	.22778	.97371	.23393	4.2747	1.0270	4.3901	50
11	.22807	.97364	.23424	.2691	.0271	.3847	49
12	.22835	.97358	.23455	.2635	.0271	.3792	48
13	.22863	.97351	.23485	.2579	.0272	.3738	47
14	.22892	.97344	.23516	.2524	.0273	.3684	46
15	.22920	.97338	.23547	4.2468	1.0273	4.3630	45
16	.22948	.97331	.23577	.2413	.0274	.3576	44
17	.22977	.97324	.23608	.2358	.0275	.3522	43
18	.23005	.97318	.23639	.2303	.0276	.3469	42
19	.23033	.97311	.23670	.2248	.0276	.3415	41
20	.23061	.97304	.23700	4.2193	1.0277	4.3362	40
21	.23090	.97298	.23731	.2139	.0278	.3309	39
22	.23118	.97291	.23762	.2084	.0278	.3256	38
23	.23146	.97284	.23793	.2030	.0279	.3203	37
24	.23175	.97277	.23823	.1976	.0280	.3150	36
25	.23203	.97271	.23854	4.1921	1.0280	4.3098	35
26	.23231	.97264	.23885	.1867	.0281	.3045	34
27	.23260	.97257	.23916	.1814	.0282	.2993	33
28	.23288	.97250	.23946	.1760	.0283	.2941	32
29	.23316	.97244	.23977	.1706	.0283	.2888	31
30	.23344	.97237	.24008	4.1653	1.0284	4.2836	30
31	.23373	.97230	.24039	.1600	.0285	.2785	29
32	.23401	.97223	.24069	.1546	.0285	.2733	28
33	.23429	.97216	.24100	.1493	.0286	.2681	27
34	.23458	.97210	.24131	.1440	.0287	.2630	26
35	.23486	.97203	.24162	4.1388	1.0288	4.2579	25
36	.23514	.97196	.24192	.1335	.0288	.2527	24
37	.23542	.97189	.24223	.1282	.0289	.2476	23
38	.23571	.97182	.24254	.1230	.0290	.2425	22
39	.23599	.97175	.24285	.1178	.0291	.2375	21
40	.23627	.97169	.24316	4.1126	1.0291	4.2324	20
41	.23655	.97162	.24346	.1073	.0292	.2273	19
42	.23684	.97155	.24377	.1022	.0293	.2223	18
43	.23712	.97148	.24408	.0970	.0293	.2173	17
44	.23740	.97141	.24439	.0918	.0294	.2122	16
45	.23768	.97134	.24470	4.0867	1.0295	4.2072	15
46	.23797	.97127	.24501	.0815	.0296	.2022	14
47	.23825	.97120	.24531	.0764	.0296	.1972	13
48	.23853	.97113	.24562	.0713	.0297	.1923	12
49	.23881	.97106	.24593	.0662	.0298	.1873	11
50	.23910	.97099	.24624	4.0611	1.0299	4.1824	10
51	.23938	.97092	.24655	.0560	.0299	.1774	9
52	.23966	.97086	.24686	.0509	.0300	.1725	8
53	.23994	.97079	.24717	.0458	.0301	.1676	7
54	.24023	.97072	.24747	.0408	.0302	.1627	6
55	.24051	.97065	.24778	4.0358	1.0302	4.1578	5
56	.24079	.97058	.24809	.0370	.0303	.1529	4
57	.24107	.97051	.24840	.0257	.0304	.1481	3
58	.24136	.97044	.24871	.0207	.0305	.1432	2
59	.24164	.97037	.24902	.0157	.0305	.1384	1
60	.24192	.97029	.24933	4.0108	1.0306	4.1336	0
M	Cosine	Sine	Cotan.	Tangent	Cosec.	Secant	M

76°

14°

M	Sine	Cosine	Tangent	Cotan.	Secant	Cosec.	M
0	.24192	.97029	.24933	4.0108	1.0306	4.1336	60
1	.24220	.97022	.24964	.0058	.0307	.1287	59
2	.24249	.97015	.24995	.0009	.0308	.1239	58
3	.24277	.97008	.25025	3.9959	.0308	.1191	57
4	.24305	.97001	.25056	.9910	.0309	.1144	56
5	.24333	.96994	.25087	3.9861	1.0310	4.1096	55
6	.24361	.96987	.25118	.9812	.0311	.1048	54
7	.24390	.96980	.25149	.9763	.0311	.1001	53
8	.24418	.96973	.25180	.9714	.0312	.0953	52
9	.24446	.96966	.25211	.9665	.0313	.0906	51
10	.24474	.96959	.25242	3.9616	1.0314	4.0859	50
11	.24502	.96952	.25273	.9568	.0314	.0812	49
12	.24531	.96944	.25304	.9520	.0315	.0765	48
13	.24559	.96937	.25335	.9471	.0316	.0718	47
14	.24587	.96930	.25366	.9423	.0317	.0672	46
15	.24615	.96923	.25397	3.9375	1.0317	4.0625	45
16	.24643	.96916	.25428	.9327	.0318	.0579	44
17	.24672	.96909	.25459	.9279	.0319	.0532	43
18	.24700	.96901	.25490	.9231	.0320	.0486	42
19	.24728	.96894	.25521	.9184	.0320	.0440	41
20	.24756	.96887	.25552	3.9136	1.0321	4.0394	40
21	.24784	.96880	.25583	.9089	.0322	.0348	39
22	.24813	.96873	.25614	.9042	.0323	.0302	38
23	.24841	.96865	.25645	.8994	.0323	.0256	37
24	.24869	.96858	.25676	.8947	.0324	.0211	36
25	.24897	.96851	.25707	3.8900	1.0325	4.0165	35
26	.24925	.96844	.25738	.8853	.0326	.0120	34
27	.24953	.96836	.25769	.8807	.0327	.0074	33
28	.24982	.96829	.25800	.8760	.0327	.0029	32
29	.25010	.96822	.25831	.8713	.0328	3.9984	31
30	.25038	.96815	.25862	3.8667	1.0329	3.9939	30
31	.25066	.96807	.25893	.8621	.0330	.9894	29
32	.25094	.96800	.25924	.8574	.0330	.9850	28
33	.25122	.96793	.25955	.8528	.0331	.9805	27
34	.25151	.96785	.25986	.8482	.0332	.9760	26
35	.25179	.96778	.26017	3.8436	1.0333	3.9716	25
36	.25207	.96771	.26048	.8390	.0334	.9672	24
37	.25235	.96763	.26079	.8345	.0334	.9627	23
38	.25263	.96756	.26110	.8299	.0335	.9583	22
39	.25291	.96749	.26141	.8254	.0336	.9539	21
40	.25319	.96741	.26172	3.8208	1.0337	3.9495	20
41	.25348	.96734	.26203	.8163	.0338	.9451	19
42	.25376	.96727	.26234	.8118	.0338	.9408	18
43	.25404	.96719	.26266	.8073	.0339	.9364	17
44	.25432	.96712	.26297	.8027	.0340	.9320	16
45	.25460	.96704	.26328	3.7983	1.0341	3.9277	15
46	.25488	.96697	.26359	.7938	.0341	.9234	14
47	.25516	.96690	.26390	.7893	.0342	.9190	13
48	.25544	.96682	.26421	.7848	.0343	.9147	12
49	.25573	.96675	.26452	.7804	.0344	.9104	11
50	.25601	.96667	.26483	3.7759	1.0345	3.9061	10
51	.25629	.96660	.26514	.7715	.0345	.9018	9
52	.25657	.96652	.26546	.7671	.0346	.8976	8
53	.25685	.96645	.26577	.7627	.0347	.8933	7
54	.25713	.96638	.26608	.7583	.0348	.8890	6
55	.25741	.96630	.26639	3.7539	1.0349	3.8848	5
56	.25769	.96623	.26670	.7495	.0349	.8805	4
57	.25798	.96615	.26701	.7451	.0350	.8763	3
58	.25826	.96608	.26732	.7407	.0351	.8721	2
59	.25854	.96600	.26764	.7364	.0352	.8679	1
60	.25882	.96592	.26795	3.7320	1.0353	3.8637	0
M	Cosine	Sine	Cotan.	Tangent	Cosec.	Secant	M

75°

15°

M	Sine	Cosine	Tangent	Cotan.	Secant	Cosec.	M
0	.25882	.96592	.26795	3.7320	1.0353	3.8637	60
1	.25910	.96585	.26826	.7277	.0353	.8595	59
2	.25938	.96577	.26857	.7234	.0354	.8553	58
3	.25966	.96570	.26888	.7191	.0355	.8512	57
4	.25994	.96562	.26920	.7147	.0356	.8470	56
5	.26022	.96555	.26951	3.7104	1.0357	3.8428	55
6	.26050	.96547	.26982	.7062	.0358	.8387	54
7	.26078	.96540	.27013	.7019	.0358	.8346	53
8	.26107	.96532	.27044	.6976	.0359	.8304	52
9	.26135	.96524	.27076	.6933	.0360	.8263	51
10	.26163	.96517	.27107	3.6891	1.0361	3.8222	50
11	.26191	.96509	.27138	.6848	.0362	.8181	49
12	.26219	.96502	.27169	.6806	.0362	.8140	48
13	.26247	.96494	.27201	.6764	.0363	.8100	47
14	.26275	.96486	.27232	.6722	.0364	.8059	46
15	.26303	.96479	.27263	3.6679	1.0365	3.8018	45
16	.26331	.96471	.27294	.6637	.0366	.7978	44
17	.26359	.96463	.27326	.6596	.0367	.7937	43
18	.26387	.96456	.27357	.6554	.0367	.7897	42
19	.26415	.96448	.27388	.6512	.0368	.7857	41
20	.26443	.96440	.27419	3.6470	1.0369	3.7816	40
21	.26471	.96433	.27451	.6429	.0370	.7776	39
22	.26499	.96425	.27482	.6387	.0371	.7736	38
23	.26527	.96417	.27513	.6346	.0371	.7697	37
24	.26556	.96409	.27544	.6305	.0372	.7657	36
25	.26584	.96402	.27576	3.6263	1.0373	3.7617	35
26	.26612	.96394	.27607	.6222	.0374	.7577	34
27	.26640	.96386	.27638	.6181	.0375	.7538	33
28	.26668	.96378	.27670	.6140	.0376	.7498	32
29	.26696	.96371	.27701	.6100	.0376	.7459	31
30	.26724	.96363	.27732	3.6059	1.0377	3.7420	30
31	.26752	.96355	.27764	.6018	.0378	.7380	29
32	.26780	.96347	.27795	.5977	.0379	.7341	28
33	.26808	.96340	.27826	.5937	.0380	.7302	27
34	.26836	.96332	.27858	.5896	.0381	.7263	26
35	.26864	.96324	.27889	3.5856	1.0382	3.7224	25
36	.26892	.96316	.27920	.5816	.0382	.7186	24
37	.26920	.96308	.27952	.5776	.0383	.7147	23
38	.26948	.96301	.27983	.5736	.0384	.7108	22
39	.26976	.96293	.28014	.5696	.0385	.7070	21
40	.27004	.96285	.28046	3.5656	1.0386	3.7031	20
41	.27032	.96277	.28077	.5616	.0387	.6993	19
42	.27060	.96269	.28109	.5576	.0387	.6955	18
43	.27088	.96261	.28140	.5536	.0388	.6917	17
44	.27116	.96253	.28171	.5497	.0389	.6878	16
45	.27144	.96245	.28203	3.5457	1.0390	3.6840	15
46	.27172	.96238	.28234	.5418	.0391	.6802	14
47	.27200	.96230	.28266	.5378	.0392	.6765	13
48	.27228	.96222	.28297	.5339	.0393	.6727	12
49	.27256	.96214	.28328	.5300	.0393	.6689	11
50	.27284	.96206	.28360	3.5261	1.0394	3.6651	10
51	.27312	.96198	.28391	.5222	.0395	.6614	9
52	.27340	.96190	.28423	.5183	.0396	.6576	8
53	.27368	.96182	.28454	.5144	.0397	.6539	7
54	.27396	.96174	.28486	.5105	.0398	.6502	6
55	.27424	.96166	.28517	3.5066	1.0399	3.6464	5
56	.27452	.96158	.28549	.5028	.0399	.6427	4
57	.27480	.96150	.28580	.4989	.0400	.6390	3
58	.27508	.96142	.28611	.4951	.0401	.6353	2
59	.27536	.96134	.28643	.4912	.0402	.6316	1
60	.27564	.96126	.28674	3.4874	1.0403	3.6279	0
M	Cosine	Sine	Cotan.	Tangent	Cosec.	Secant	M

74°

16°

M	Sine	Cosine	Tangent	Cotan.	Secant	Cosec.	M
0	.27564	.96126	.28674	3.4874	1.0403	3.6279	60
1	.27592	.96118	.28706	.4836	.0404	.6243	59
2	.27620	.96110	.28737	.4798	.0405	.6206	58
3	.27648	.96102	.28769	4760	.0406	.6169	57
4	.27675	.96094	.28800	.4722	.0406	.6133	56
5	.27703	.96086	.28832	3.4684	1.0407	3.6096	55
6	.27731	.96078	.28863	.4646	.0408	.6060	54
7	.27759	.96070	.28895	.4608	.0409	.6024	53
8	.27787	.96062	.28926	.4570	.0410	.5987	52
9	.27815	.96054	.28958	.4533	.0411	.5951	51
10	.27843	.96045	.28990	3.4495	1.0412	3.5915	50
11	.27871	.96037	.29021	.4458	.0413	.5879	49
12	.27899	.96029	.29053	.4420	.0413	.5843	48
13	.27927	.96021	.29084	.4383	.0414	.5807	47
14	.27955	.96013	.29116	.4346	.0415	.5772	46
15	.27983	.96005	.29147	3.4308	1.0416	3.5736	45
16	.28011	.95997	.29179	.4271	.0417	.5700	44
17	.28039	.95989	.29210	.4234	.0418	.5665	43
18	.28067	.95980	.29242	.4197	.0419	.5629	42
19	.28094	.95972	.29274	.4160	.0420	.5594	41
20	.28122	.95964	.29305	3.4124	1.0420	3.5559	40
21	.28150	.95956	.29337	.4087	.0421	.5523	39
22	.28178	.95948	.29368	.4050	.0422	.5488	38
23	.28206	.95940	.29400	.4014	.0423	.5453	37
24	.28234	.95931	.29432	.3977	.0424	.5418	36
25	.28262	.95923	.29463	3.3941	1.0425	3.5383	35
26	.28290	.95915	.29495	.3904	.0426	.5348	34
27	.28318	.95907	.29526	.3868	.0427	.5313	33
28	.28346	.95898	.29558	.3832	.0428	.5279	32
29	.28374	.95890	.29590	.3795	.0428	.5244	31
30	.28401	.95882	.29621	3.3759	1.0429	3.5209	30
31	.28429	.95874	.29653	.3723	.0430	.5175	29
32	.28457	.95865	.29685	.3687	.0431	.5140	28
33	.28485	.95857	.29716	.3651	.0432	.5106	27
34	.28513	.95849	.29748	.3616	.0433	.5072	26
35	.28541	.95840	.29780	3.3580	1.0434	3.5037	25
36	.28569	.95832	.29811	.3544	.0435	.5003	24
37	.28597	.95824	.29843	.3509	.0436	.4969	23
38	.28624	.95816	.29875	.3473	.0437	.4935	22
39	.28652	.95807	.29906	.3438	.0438	.4901	21
40	.28680	.95799	.29938	3.3402	1.0438	3.4867	20
41	.28708	.95791	.29970	.3367	.0439	.4833	19
42	.28736	.95782	.30001	.3332	.0440	.4799	18
43	.28764	.95774	.30033	.3296	.0441	.4766	17
44	.28792	.95765	.30065	.3261	.0442	.4732	16
45	.28820	.95757	.30096	3.3226	1.0443	3.4698	15
46	.28847	.95749	.30128	.3191	.0444	.4665	14
47	.28875	.95740	.30160	.3156	.0445	.4632	13
48	.28903	.95732	.30192	.3121	.0446	.4598	12
49	.28931	.95723	.30223	.3087	.0447	.4565	11
50	.28959	.95715	.30255	3.3052	1.0448	3.4532	10
51	.28987	.95707	.30287	.3017	.0448	.4498	9
52	.29014	.95698	.30319	3.2983	.0449	.4465	8
53	.29042	.95690	.30350	.2948	.0450	.4432	7
54	.29070	.95681	.30382	.2914	.0451	.4399	6
55	.29098	.95673	.30414	3.2879	1.0452	3.4366	5
56	.29126	.95664	.30446	.2845	.0453	.4334	4
57	.29154	.95656	.30478	.2811	.0454	.4301	3
58	.29181	.95647	.30509	.2777	.0455	.4268	2
59	.29209	.95639	.30541	.2742	.0456	.4236	1
60	.29237	.95630	.30573	3.2708	1.0457	3.4203	0
M	Cosine	Sine	Cotan.	Tangent	Cosec.	Secant	M

73°

17°

M	Sine	Cosine	Tangent	Cotan.	Secant	Cosec.	M
0	.29237	.95630	.30573	3.2708	1.0457	3.4203	60
1	.29265	.95622	.30605	.2674	.0458	.4170	59
2	.29293	.95613	.30637	.2640	.0459	.4138	58
3	.29321	.95605	.30668	2607	.0460	.4106	57
4	.29348	.95596	.30700	.2573	.0461	.4073	56
5	.29376	.95588	.30732	3.2539	1.0461	3.4041	55
6	.29404	.95579	.30764	.2505	.0462	.4009	54
7	.29432	.95571	.30796	.2472	.0463	.3977	53
8	.29460	.95562	.30828	.2438	.0464	.3945	52
9	.29487	.95554	.30859	.2405	.0465	.3913	51
10	.29515	.95545	.30891	3.2371	1.0466	3.3881	50
11	.29543	.95536	.30923	.2338	.0467	.3849	49
12	.29571	.95528	.30955	.2305	.0468	.3817	48
13	.29598	.95519	.30987	.2271	.0469	.3785	47
14	.29626	.95511	.31019	.2238	.0470	.3754	46
15	.29654	.95502	.31051	3.2205	1.0471	3.3722	45
16	.29682	.95493	.31083	.2172	.0472	.3690	44
17	.29710	.95485	.31115	.2139	.0473	.3659	43
18	.29737	.95476	.31146	.2106	.0474	.3627	42
19	.29765	.95467	.31178	.2073	.0475	.3596	41
20	.29793	.95459	.31210	3.2041	1.0476	3.3565	40
21	.29821	.95450	.31242	.2008	.0477	.3534	39
22	.29848	.95441	.31274	.1975	.0478	.3502	38
23	.29876	.95433	.31306	.1942	.0478	.3471	37
24	.29904	.95424	.31338	.1910	.0479	.3440	36
25	.29932	.95415	.31370	3.1877	1.0480	3.3409	35
26	.29959	.95407	.31402	.1845	.0481	.3378	34
27	.29987	.95398	.31434	.1813	.0482	.3347	33
28	.30015	.95389	.31466	.1780	.0483	.3316	32
29	.30043	.95380	.31498	.1748	.0484	.3286	31
30	.30070	.95372	.31530	3.1716	1.0485	3.3255	30
31	.30098	.95363	.31562	.1684	.0486	.3224	29
32	.30126	.95354	.31594	.1652	.0487	.3194	28
33	.30154	.95345	.31626	.1620	.0488	.3163	27
34	.30181	.95337	.31658	.1588	.0489	.3133	26
35	.30209	.95328	.31690	3.1556	1.0490	3.3102	25
36	.30237	.95319	.31722	.1524	.0491	.3072	24
37	.30265	.95310	.31754	.1492	.0492	.3042	23
38	.30292	.95301	.31786	.1460	.0493	.3011	22
39	.30320	.95293	.31818	.1429	.0494	.2981	21
40	.30348	.95284	.31850	3.1397	1.0495	3.2951	20
41	.30375	.95275	.31882	.1366	.0496	.2921	19
42	.30403	.95266	.31914	.1334	.0497	.2891	18
43	.30431	.95257	.31946	.1303	.0498	.2861	17
44	.30459	.95248	.31978	.1271	.0499	.2831	16
45	.30486	.95239	.32010	3.1240	1.0500	3.2801	15
46	.30514	.95231	.32042	.1209	.0501	.2772	14
47	.30542	.95222	.32074	.1177	.0502	.2742	13
48	.30569	.95213	.32106	.1146	.0503	.2712	12
49	.30597	.95204	.32138	.1115	.0504	.2683	11
50	.30625	.95195	.32171	3.1084	1.0505	3.2653	10
51	.30653	.95186	.32203	.1053	.0506	.2624	9
52	.30680	.95177	.32235	.1022	.0507	.2594	8
53	.30708	.95168	.32267	.0991	.0508	.2565	7
54	.30736	.95159	.32299	.0960	.0509	.2535	6
55	.30763	.95150	.32331	3.0930	1.0510	3.2506	5
56	.30791	.95141	.32363	.0899	.0511	.2477	4
57	.30819	.95132	.32395	.0868	.0512	.2448	3
58	.30846	.95124	.32428	.0838	.0513	.2419	2
59	.30874	.95115	.32460	.0807	.0514	.2390	1
60	.30902	.95106	.32492	3.0777	1.0515	3.2361	0
M	Cosine	Sine	Cotan.	Tangent	Cosec.	Secant	M

72°

18°

M	Sine	Cosine	Tangent	Cotan.	Secant	Cosec.	M
0	.30902	.95106	.32492	3.0777	1.0515	3.2361	60
1	.30929	.95097	.32524	.0746	.0516	.2332	59
2	.30957	.95088	.32556	.0716	.0517	.2303	58
3	.30985	.95079	.32588	0686	.0518	.2274	57
4	.31012	.95070	.32621	.0655	.0519	.2245	56
5	.31040	.95061	.32653	3.0625	1.0520	3.2216	55
6	.31068	.95051	.32685	.0595	.0521	.2188	54
7	.31095	.95042	.32717	.0565	.0522	.2159	53
8	.31123	.95033	.32749	.0535	.0523	.2131	52
9	.31150	.95024	.32782	.0505	.0524	.2102	51
10	.31178	.95015	.32814	3.0475	1.0525	3.2074	50
11	.31206	.95006	.32846	.0445	.0526	.2045	49
12	.31233	.94997	.32878	.0415	.0527	.2017	48
13	.31261	.94988	.32910	.0385	.0528	.1989	47
14	.31289	.94979	.32943	.0356	.0529	.1960	46
15	.31316	.94970	.32975	3.0326	1.0530	3.1932	45
16	.31344	.94961	.33007	.0296	.0531	.1904	44
17	.31372	.94952	.33039	.0267	.0532	.1876	43
18	.31399	.94942	.33072	.0237	.0533	.1848	42
19	.31427	.94933	.33104	.0208	.0534	.1820	41
20	.31454	.94924	.33136	3.0178	1.0535	3.1792	40
21	.31482	.94915	.33169	.0149	.0536	.1764	39
22	.31510	.94906	.33201	.0120	.0537	.1736	38
23	.31537	.94897	.33233	.0090	.0538	.1708	37
24	.31565	.94888	.33265	.0061	.0539	.1681	36
25	.31592	.94878	.33298	3.0032	1.0540	3.1653	35
26	.31620	.94869	.33330	.0003	.0541	.1625	34
27	.31648	.94860	.33362	2.9974	.0542	.1598	33
28	.31675	.94851	.33395	.9945	.0543	.1570	32
29	.31703	.94841	.33427	.9916	.0544	.1543	31
30	.31730	.94832	.33459	2.9887	1.0545	3.1515	30
31	.31758	.94823	.33492	.9858	.0546	.1488	29
32	.31786	.94814	.33524	.9829	.0547	.1461	28
33	.31813	.94805	.33557	.9800	.0548	.1433	27
34	.31841	.94795	.33589	.9772	.0549	.1406	26
35	.31868	.94786	.33621	2.9743	1.0550	3.1379	25
36	.31896	.94777	.33654	.9714	.0551	.1352	24
37	.31923	.94767	.33686	.9686	.0552	.1325	23
38	.31951	.94758	.33718	.9657	.0553	.1298	22
39	.31978	.94749	.33751	.9629	.0554	.1271	21
40	.32006	.94740	.33783	2.9600	1.0555	3.1244	20
41	.32034	.94730	.33816	.9572	.0556	.1217	19
42	.32061	.94721	.33848	.9544	.0557	.1190	18
43	.32089	.94712	.33880	.9515	.0558	.1163	17
44	.32116	.94702	.33913	.9487	.0559	.1137	16
45	.32144	.94693	.33945	2.9459	1.0560	3.1110	15
46	.32171	.94684	.33978	.9431	.0561	.1083	14
47	.32199	.94674	.34010	.9403	.0562	.1057	13
48	.32226	.94665	.34043	.9375	.0563	.1030	12
49	.32254	.94655	.34075	.9347	.0565	.1004	11
50	.32282	.94646	.34108	2.9319	1.0566	3.0977	10
51	.32309	.94637	.34140	.9291	.0567	.0951	9
52	.32337	.94627	.34173	.9263	.0568	.0925	8
53	.32364	.94618	.34205	.9235	.0569	.0898	7
54	.32392	.94608	.34238	.9208	.0570	.0872	6
55	.32419	.94599	.34270	2.9180	1.0571	3.0846	5
56	.32447	.94590	.34303	.9152	.0572	.0820	4
57	.32474	.94580	.34335	.9125	.0573	.0793	3
58	.32502	.94571	.34368	.9097	.0574	.0767	2
59	.32529	.94561	.34400	.9069	.0575	.0741	1
60	.32557	.94552	.34433	2.9042	1.0576	3.0715	0
M	Cosine	Sine	Cotan.	Tangent	Cosec.	Secant	M

71°

19°

M	Sine	Cosine	Tangent	Cotan.	Secant	Cosec.	M
0	.32557	.94552	.34433	2.9042	1.0576	3.0715	60
1	.32584	.94542	.34465	.9015	.0577	.0690	59
2	.32612	.94533	.34498	.8987	.0578	.0664	58
3	.32639	.94523	.34530	8960	.0579	.0638	57
4	.32667	.94514	.34563	.8933	.0580	.0612	56
5	.32694	.94504	.34595	2.8905	1.0581	3.0586	55
6	.32722	.94495	.34628	.8878	.0582	.0561	54
7	.32749	.94485	.34661	.8851	.0584	.0535	53
8	.32777	.94476	.34693	.8824	.0585	.0509	52
9	.32804	.94466	.34726	.8797	.0586	.0484	51
10	.32832	.94457	.34758	2.8770	1.0587	3.0458	50
11	.32859	.94447	.34791	.8743	.0588	.0433	49
12	.32887	.94438	.34824	.8716	.0589	.0407	48
13	.32914	.94428	.34856	.8689	.0590	.0382	47
14	.32942	.94418	.34889	.8662	.0591	.0357	46
15	.32969	.94409	.34921	2.8636	1.0592	3.0331	45
16	.32996	.94399	.34954	.8609	.0593	.0306	44
17	.33024	.94390	.34987	.8582	.0594	.0281	43
18	.33051	.94380	.35019	.8555	.0595	.0256	42
19	.33079	.94370	.35052	.8529	.0596	.0231	41
20	.33106	.94361	.35085	2.8502	1.0598	3.0206	40
21	.33134	.94351	.35117	.8476	.0599	.0181	39
22	.33161	.94341	.35150	.8449	.0600	.0156	38
23	.33189	.94332	.35183	.8423	.0601	.0131	37
24	.33216	.94322	.35215	.8396	.0602	.0106	36
25	.33243	.94313	.35248	2.8370	1.0603	3.0081	35
26	.33271	.94303	.35281	.8344	.0604	.0056	34
27	.33298	.94293	.35314	.8318	.0605	.0031	33
28	.33326	.94283	.35346	.8291	.0606	.0007	32
29	.33353	.94274	.35379	.8265	.0607	2.9982	31
30	.33381	.94264	.35412	2.8239	1.0608	2.9957	30
31	.33408	.94254	.35445	.8213	.0609	.9933	29
32	.33435	.94245	.35477	.8187	.0611	.9908	28
33	.33463	.94235	.35510	.8161	.0612	.9884	27
34	.33490	.94225	.35543	.8185	.0613	.9859	26
35	.33518	.94215	.35576	2.8109	1.0614	2.9835	25
36	.33545	.94206	.35608	.8083	.0615	.9810	24
37	.33572	.94196	.35641	.8057	.0616	.9786	23
38	.33600	.94186	.35674	.8032	.0617	.9762	22
39	.33627	.94176	.35707	.8006	.0618	.9738	21
40	.33655	.94167	.35739	2.7980	1.0619	2.9713	20
41	.33682	.94157	.35772	.7954	.0620	.9689	19
42	.33709	.94147	.35805	.7929	.0622	.9665	18
43	.33737	.94137	.35838	.7903	.0623	.9641	17
44	.33764	.94127	.35871	.7878	.0624	.9617	16
45	.33792	.94118	.35904	2.7852	1.0625	2.9593	15
46	.33819	.94108	.35936	.7827	.0626	.9569	14
47	.33846	.94098	.35969	.7801	.0627	.9545	13
48	.33874	.94088	.36002	.7776	.0628	.9521	12
49	.33901	.94078	.36035	.7751	.0629	.9497	11
50	.33928	.94068	.36068	2.7725	1.0630	2.9474	10
51	.33956	.94058	.36101	.7700	.0632	.9450	9
52	.33983	.94049	.36134	.7675	.0633	.9426	8
53	.34011	.94039	.36167	.7650	.0634	.9402	7
54	.34038	.94029	.36199	.7625	.0635	.9379	6
55	.34065	.94019	.36232	2.7600	1.0636	2.9355	5
56	.34093	.94009	.36265	.7575	.0637	.9332	4
57	.34120	.93999	.36298	.7550	.0638	.9308	3
58	.34147	.93989	.36331	.7525	.0639	.9285	2
59	.34175	.93979	.36364	.7500	.0641	.9261	1
60	.34202	.93969	.36397	2.7475	1.0642	2.9238	0
M	Cosine	Sine	Cotan.	Tangent	Cosec.	Secant	M

70°

20°

M	Sine	Cosine	Tangent	Cotan.	Secant	Cosec.	M
0	.34202	.93969	.36397	2.7475	1.0642	2.9238	60
1	.34229	.93959	.36430	.7450	.0643	.9215	59
2	.34257	.93949	.36463	.7425	.0644	.9191	58
3	.34284	.93939	.36496	7400	.0645	.9168	57
4	.34311	.93929	.36529	.7376	.0646	.9145	56
5	.34339	.93919	.36562	2.7351	1.0647	2.9122	55
6	.34366	.93909	.36595	.7326	.0648	.9098	54
7	.34393	.93899	.36628	.7302	.0650	.9075	53
8	.34421	.93889	.36661	.7277	.0651	.9052	52
9	.34448	.93879	.36694	.7252	.0652	.9029	51
10	.34475	.93869	.36727	2.7228	1.0653	2.9006	50
11	.34502	.93859	.36760	.7204	.0654	.8983	49
12	.34530	.93849	.36793	.7179	.0655	.8960	48
13	.34557	.93839	.36826	.7155	.0656	.8937	47
14	.34584	.93829	.36859	.7130	.0658	.8915	46
15	.34612	.93819	.36892	2.7106	1.0659	2.8892	45
16	.34639	.93809	.36925	.7082	.0660	.8869	44
17	.34666	.93799	.36958	.7058	.0661	.8846	43
18	.34693	.93789	.36991	.7033	.0662	.8824	42
19	.34721	.93779	.37024	.7009	.0663	.8801	41
20	.34748	.93769	.37057	2.6985	1.0664	2.8778	40
21	.34775	.93758	.37090	.6961	.0666	.8756	39
22	.34803	.93748	.37123	.6937	.0667	.8733	38
23	.34830	.93738	.37156	.6913	.0668	.8711	37
24	.34857	.93728	.37190	.6889	.0669	.8688	36
25	.34884	.93718	.37223	2.6865	1.0670	2.8666	35
26	.34912	.93708	.37256	.6841	.0671	.8644	34
27	.34939	.93698	.37289	.6817	.0673	.8621	33
28	.34966	.93687	.37322	.6794	.0674	.8599	32
29	.34993	.93677	.37355	.6770	.0675	8577	31
30	.35021	.93667	.37388	2.6746	1.0676	2.8554	30
31	.35048	.93657	.37422	.6722	.0677	.8532	29
32	.35075	.93647	.37455	.6699	.0678	.8510	28
33	.35102	.93637	.37488	.6675	.0679	.8488	27
34	.35130	.93626	.37521	.6652	.0681	.8466	26
35	.35157	.93616	.37554	2.6628	1.0682	2.8444	25
36	.35184	.93606	.37587	.6604	.0683	.8422	24
37	.35211	.93596	.37621	.6581	.0684	.8400	23
38	.35239	.93585	.37654	.6558	.0685	.8378	22
39	.35266	.93575	.37687	.6534	.0686	.8356	21
40	.35293	.93565	.37720	2.6511	1.0688	2.8334	20
41	.35320	.93555	.37754	.6487	.0689	.8312	19
42	.35347	.93544	.37787	.6464	.0690	.8290	18
43	.35375	.93534	.37820	.6441	.0691	.8269	17
44	.35402	.93524	.37853	.6418	.0692	.8247	16
45	.35429	.93513	.37887	2.6394	1.0694	2.8225	15
46	.35456	.93503	.37920	.6371	.0695	.8204	14
47	.35483	.93493	.37953	.6348	.0696	.8182	13
48	.35511	.93482	.37986	.6325	.0697	.8160	12
49	.35538	.93472	.38020	.6302	.0698	.8139	11
50	.35565	.93462	.38053	2.6279	1.0699	2.8117	10
51	.35592	.93451	.38086	.6256	.0701	.8096	9
52	.35619	.93441	.38120	.6233	.0702	.8074	8
53	.35647	.93431	.38153	.6210	.0703	.8053	7
54	.35674	.93420	.38186	.6187	.0704	.8032	6
55	.35701	.93410	.38220	2.6164	1.0705	2.8010	5
56	.35728	.93400	.38253	.6142	.0707	.7989	4
57	.35755	.93389	.38286	.6119	.0708	.7968	3
58	.35782	.93379	.38320	.6096	.0709	.7947	2
59	.35810	.93368	.38353	.6073	.0710	7925	1
60	.35837	.93358	.38386	2.6051	1.0711	2.7904	0
M	Cosine	Sine	Cotan.	Tangent	Cosec.	Secant	M

69°

21°

M	Sine	Cosine	Tangent	Cotan.	Secant	Cosec.	M
0	.35837	.93358	.38386	2.6051	1.0711	2.7904	60
1	.35864	.93348	.38420	.6028	.0713	.7883	59
2	.35891	.93337	.38453	.6006	.0714	.7862	58
3	.35918	.93327	.38486	5983	.0715	.7841	57
4	.35945	.93316	.38520	.5960	.0716	.7820	56
5	.35972	.93306	.38553	2.5938	1.0717	2.7799	55
6	.36000	.93295	.38587	.5916	.0719	.7778	54
7	.36027	.93285	.38620	.5893	.0720	.7757	53
8	.36054	.93274	.38654	.5871	.0721	.7736	52
9	.36081	.93264	.38687	.5848	.0722	.7715	51
10	.36108	.93253	.38720	2.5826	1.0723	2.7694	50
11	.36135	.93243	.38754	.5804	.0725	.7674	49
12	.36162	.93232	.38787	.5781	.0726	.7653	48
13	.36189	.93222	.38821	.5759	.0727	.7632	47
14	.36217	.93211	.38854	.5737	.0728	.7611	46
15	.36244	.93201	.38888	2.5715	1.0729	2.7591	45
16	.36271	.93190	.38921	.5693	.0731	.7570	44
17	.36298	.93180	.38955	.5671	.0732	.7550	43
18	.36325	.93169	.38988	.5649	.0733	.7529	42
19	.36352	.93158	.39022	.5627	.0734	.7509	41
20	.36379	.93148	.39055	2.5605	1.0736	2.7488	40
21	.36406	.93137	.39089	.5583	.0737	.7468	39
22	.36433	.93127	.39122	.5561	.0738	.7447	38
23	.36460	.93116	.39156	.5539	.0739	.7427	37
24	.36488	.93105	.39189	.5517	.0740	.7406	36
25	.36515	.93095	.39223	2.5495	1.0742	2.7386	35
26	.36542	.93084	.39257	.5473	.0743	.7366	34
27	.36569	.93074	.39290	.5451	.0744	.7346	33
28	.36596	.93063	.39324	.5430	.0745	.7325	32
29	.36623	.93052	.39357	.5408	.0747	7305	31
30	.36650	.93042	.39391	2.5386	1.0748	2.7285	30
31	.36677	.93031	.39425	.5365	.0749	.7265	29
32	.36704	.93020	.39458	.5343	.0750	.7245	28
33	.36731	.93010	.39492	.5322	.0751	.7225	27
34	.36758	.92999	.39525	.5300	.0753	.7205	26
35	.36785	.92988	.39559	2.5278	1.0754	2.7185	25
36	.36812	.92978	.39593	.5257	.0755	.7165	24
37	.36839	.92967	.39626	.5236	.0756	.7145	23
38	.36866	.92956	.39660	.5214	.0758	.7125	22
39	.36893	.92945	.39694	.5193	.0759	.7105	21
40	.36921	.92935	.39727	2.5171	1.0760	2.7085	20
41	.36948	.92924	.39761	.5150	.0761	.7065	19
42	.36975	.92913	.39795	.5129	.0763	.7045	18
43	.37002	.92902	.39828	.5108	.0764	.7026	17
44	.37029	.92892	.39862	.5086	.0765	.7006	16
45	.37056	.92881	.39896	2.5065	1.0766	2.6986	15
46	.37083	.92870	.39930	.5044	.0768	.6967	14
47	.37110	.92859	.39963	.5023	.0769	.6947	13
48	.37137	.92848	.39997	.5002	.0770	.6927	12
49	.37164	.92838	.40031	.4981	.0771	.6908	11
50	.37191	.92827	.40065	2.4960	1.0773	2.6888	10
51	.37218	.92816	.40098	.4939	.0774	.6869	9
52	.37245	.92805	.40132	.4918	.0775	.6849	8
53	.37272	.92794	.40166	.4897	.0776	.6830	7
54	.37299	.92784	.40200	.4876	.0778	.6810	6
55	.37326	.92773	.40233	2.4855	1.0779	2.6791	5
56	.37353	.92762	.40267	.4834	.0780	.6772	4
57	.37380	.92751	.40301	.4813	.0781	.6752	3
58	.37407	.92740	.40335	.4792	.0783	.6733	2
59	.37434	.92729	.40369	.4772	.0784	6714	1
60	.37461	.92718	.40403	2.4751	1.0785	2.6695	0
M	Cosine	Sine	Cotan.	Tangent	Cosec.	Secant	M

68°

22°

M	Sine	Cosine	Tangent	Cotan.	Secant	Cosec.	M
0	.37461	.92718	.40403	2.4751	1.0785	2.6695	60
1	.37488	.92707	.40436	.4730	.0787	.6675	59
2	.37514	.92696	.40470	.4709	.0788	.6656	58
3	.37541	.92686	.40504	.4689	.0789	.6637	57
4	.37568	.92675	.40538	.4668	.0790	.6618	56
5	.37595	.92664	.40572	2.4647	1.0792	2.6599	55
6	.37622	.92653	.40606	.4627	.0793	.6580	54
7	.37649	.92642	.40640	.4606	.0794	.6561	53
8	.37676	.92631	.40673	.4586	.0795	.6542	52
9	.37703	.92620	.40707	.4565	.0797	.6523	51
10	.37730	.92609	.40741	2.4545	1.0798	2.6504	50
11	.37757	.92598	.40775	.4525	.0799	.6485	49
12	.37784	.92587	.40809	.4504	.0801	.6466	48
13	.37811	.92576	.40843	.4484	.0802	.6447	47
14	.37838	.92565	.40877	.4463	.0803	.6428	46
15	.37865	.92554	.40911	2.4443	1.0804	2.6410	45
16	.37892	.92543	.40945	.4423	.0806	.6391	44
17	.37919	.92532	.40979	.4403	.0807	.6372	43
18	.37946	.92521	.41013	.4382	.0808	.6353	42
19	.37972	.92510	.41047	.4362	.0810	.6335	41
20	.37999	.92499	.41081	2.4342	1.0811	2.6316	40
21	.38026	.92488	.41115	.4322	.0812	.6297	39
22	.38053	.92477	.41149	.4302	.0813	.6279	38
23	.38080	.92466	.41183	.4282	.0815	.6260	37
24	.38107	.92455	.41217	.4262	.0816	.6242	36
25	.38134	.92443	.41251	2.4242	1.0817	2.6223	35
26	.38161	.92432	.41285	.4222	.0819	.6205	34
27	.38188	.92421	.41319	.4202	.0820	.6186	33
28	.38214	.92410	.41353	.4182	.0821	.6168	32
29	.38241	.92399	.41387	.4162	.0823	6150	31
30	.38268	.92388	.41421	2.4142	1.0824	2.6131	30
31	.38295	.92377	.41455	.4122	.0825	.6113	29
32	.38322	.92366	.41489	.4102	.0826	.6095	28
33	.38349	.92354	.41524	.4083	.0828	.6076	27
34	.38376	.92343	.41558	.4063	.0829	.6058	26
35	.38403	.92332	.41592	2.4043	1.0830	2.6040	25
36	.38429	.92321	.41626	.4023	.0832	.6022	24
37	.38456	.92310	.41660	.4004	.0833	.6003	23
38	.38483	.92299	.41694	.3984	.0834	.5985	22
39	.38510	.92287	.41728	.3964	.0836	.5967	21
40	.38537	.92276	.41762	2.3945	1.0837	2.5949	20
41	.38564	.92265	.41797	.3925	.0838	.5931	19
42	.38591	.92254	.41831	.3906	.0840	.5913	18
43	.38617	.92242	.41865	.3886	.0841	.5895	17
44	.38644	.92231	.41899	.3867	.0842	.5877	16
45	.38671	.92220	.41933	2.3847	1.0844	2.5859	15
46	.38698	.92209	.41968	.3828	.0845	.5841	14
47	.38725	.92197	.42002	.3808	.0846	.5823	13
48	.38751	.92186	.42036	.3789	.0847	.5805	12
49	.38778	.92175	.42070	.3770	.0849	.5787	11
50	.38805	.92164	.42105	2.3750	1.0850	2.5770	10
51	.38832	.92152	.42139	.3731	.0851	.5752	9
52	.38859	.92141	.42173	.3712	.0853	.5734	8
53	.38886	.92130	.42207	.3692	.0854	.5716	7
54	.38912	.92118	.42242	.3673	.0855	.5699	6
55	.38939	.92107	.42276	2.3654	1.0857	2.5681	5
56	.38966	.92096	.42310	.3635	.0858	.5663	4
57	.38993	.92084	.42344	.3616	.0859	.5646	3
58	.39019	.92073	.42379	.3597	.0861	.5628	2
59	.39046	.92062	.42413	.3577	.0862	5610	1
60	.39073	.92050	.42447	2.3558	1.0864	2.5593	0
M	Cosine	Sine	Cotan.	Tangent	Cosec.	Secant	M

67°

23°

M	Sine	Cosine	Tangent	Cotan.	Secant	Cosec.	M
0	.39073	.92050	.42447	2.3558	1.0864	2.5593	60
1	.39100	.92039	.42482	.3539	.0865	.5575	59
2	.39126	.92028	.42516	.3520	.0866	.5558	58
3	.39153	.92016	.42550	3501	.0868	.5540	57
4	.39180	.92005	.42585	.3482	.0869	.5523	56
5	.39207	.91993	.42619	2.3463	1.0870	2.5506	55
6	.39234	.91982	.42654	.3445	.0872	.5488	54
7	.39260	.91971	.42688	.3426	.0873	.5471	53
8	.39287	.91959	.42722	.3407	.0874	.5453	52
9	.39314	.91948	.42757	.3388	.0876	.5436	51
10	.39341	.91936	.42791	2.3369	1.0877	2.5419	50
11	.39367	.91925	.42826	.3350	.0878	.5402	49
12	.39394	.91913	.42860	.3332	.0880	.5384	48
13	.39421	.91902	.42894	.3313	.0881	.5367	47
14	.39448	.91891	.42929	.3294	.0882	.5350	46
15	.39474	.91879	.42963	2.3276	1.0884	2.5333	45
16	.39501	.91868	.42998	.3257	.0885	.5316	44
17	.39528	.91856	.43032	.3238	.0886	.5299	43
18	.39554	.91845	.43067	.3220	.0888	.5281	42
19	.39581	.91833	.43101	.3201	.0889	.5264	41
20	.39608	.91822	.43136	2.3183	1.0891	2.5247	40
21	.39635	.91810	.43170	.3164	.0892	.5230	39
22	.39661	.91798	.43205	.3145	.0893	.5213	38
23	.39688	.91787	.43239	.3127	.0895	.5196	37
24	.39715	.91775	.43274	.3109	.0896	.5179	36
25	.39741	.91764	.43308	2.3090	1.0897	2.5163	35
26	.39768	.91752	.43343	.3072	.0899	.5146	34
27	.39795	.91741	.43377	.3053	.0900	.5129	33
28	.39821	.91729	.43412	.3035	.0902	.5112	32
29	.39848	.91718	.43447	.3017	.0903	5095	31
30	.39875	.91706	.43481	2.2998	1.0904	2.5078	30
31	.39901	.91694	.43516	.2980	.0906	.5062	29
32	.39928	.91683	.43550	.2962	.0907	.5045	28
33	.39955	.91671	.43585	.2944	.0908	.5028	27
34	.39981	.91659	.43620	.2925	.0910	.5011	26
35	.40008	.91648	.43654	2.2907	1.0911	2.4995	25
36	.40035	.91636	.43689	.2889	.0913	.4978	24
37	.40061	.91625	.43723	.2871	.0914	.4961	23
38	.40088	.91613	.43758	.2853	.0915	.4945	22
39	.40115	.91601	.43793	.2835	.0917	.4928	21
40	.40141	.91590	.43827	2.2817	1.0918	2.4912	20
41	.40168	.91578	.43862	.2799	.0920	.4895	19
42	.40195	.91566	.43897	.2781	.0921	.4879	18
43	.40221	.91554	.43932	.2763	.0922	.4862	17
44	.40248	.91543	.43966	.2745	.0924	.4846	16
45	.40275	.91531	.44001	2.2727	1.0925	2.4829	15
46	.40301	.91519	.44036	.2709	.0927	.4813	14
47	.40328	.91508	.44070	.2691	.0928	.4797	13
48	.40354	.91496	.44105	.2673	.0929	.4780	12
49	.40381	.91484	.44140	.2655	.0931	.4764	11
50	.40408	.91472	.44175	2.2637	1.0932	2.4748	10
51	.40434	.91461	.44209	.2619	.0934	.4731	9
52	.40461	.91449	.44244	.2602	.0935	.4715	8
53	.40487	.91437	.44279	.2584	.0936	.4699	7
54	.40514	.91425	.44314	.2566	.0938	.4683	6
55	.40541	.91414	.44349	2.2548	1.0939	2.4666	5
56	.40567	.91402	.44383	.2531	.0941	.4650	4
57	.40594	.91390	.44418	.2513	.0942	.4634	3
58	.40620	.91378	.44453	.2495	.0943	.4618	2
59	.40647	.91366	.44488	.2478	.0945	4602	1
60	.40674	.91354	.44523	2.2460	1.0946	2.4586	0
M	Cosine	Sine	Cotan.	Tangent	Cosec.	Secant	M

66°

24°

M	Sine	Cosine	Tangent	Cotan.	Secant	Cosec.	M
0	.40674	.91354	.44523	2.2460	1.0946	2.4586	60
1	.40700	.91343	.44558	.2443	.0948	.4570	59
2	.40727	.91331	.44593	.2425	.0949	.4554	58
3	.40753	.91319	.44627	.2408	.0951	.4538	57
4	.40780	.91307	.44662	.2390	.0952	.4522	56
5	.40806	.91295	.44697	2.2373	1.0953	2.4506	55
6	.40833	.91283	.44732	.2355	.0955	.4490	54
7	.40860	.91271	.44767	.2338	.0956	.4474	53
8	.40886	.91260	.44802	.2320	.0958	.4458	52
9	.40913	.91248	.44837	.2303	.0959	.4442	51
10	.40939	.91236	.44872	2.2286	1.0961	2.4426	50
11	.40966	.91224	.44907	.2268	.0962	.4411	49
12	.40992	.91212	.44942	.2251	.0963	.4395	48
13	.41019	.91200	.44977	.2234	.0965	.4379	47
14	.41045	.91188	.45012	.2216	.0966	.4363	46
15	.41072	.91176	.45047	2.2199	1.0968	2.4347	45
16	.41098	.91164	.45082	.2182	.0969	.4332	44
17	.41125	.91152	.45117	.2165	.0971	.4316	43
18	.41151	.91140	.45152	.2147	.0972	.4300	42
19	.41178	.91128	.45187	.2130	.0973	.4285	41
20	.41204	.91116	.45222	2.2113	1.0975	2.4269	40
21	.41231	.91104	.45257	.2096	.0976	.4254	39
22	.41257	.91092	.45292	.2079	.0978	.4238	38
23	.41284	.91080	.45327	.2062	.0979	.4222	37
24	.41310	.91068	.45362	.2045	.0981	.4207	36
25	.41337	.91056	.45397	2.2028	1.0982	2.4191	35
26	.41363	.91044	.45432	.2011	.0984	.4176	34
27	.41390	.91032	.45467	.1994	.0985	.4160	33
28	.41416	.91020	.45502	.1977	.0986	.4145	32
29	.41443	.91008	.45537	.1960	.0988	.4130	31
30	.41469	.90996	.45573	2.1943	1.0989	2.4114	30
31	.41496	.90984	.45608	.1926	.0991	.4099	29
32	.41522	.90972	.45643	.1909	.0992	.4083	28
33	.41549	.90960	.45678	.1892	.0994	.4068	27
34	.41575	.90948	.45713	.1875	.0995	.4053	26
35	.41602	.90936	.45748	2.1859	1.0997	2.4037	25
36	.41628	.90924	.45783	.1842	.0998	.4022	24
37	.41654	.90911	.45819	.1825	.1000	.4007	23
38	.41681	.90899	.45854	.1808	.1001	.3992	22
39	.41707	.90887	.45889	.1792	.1003	.3976	21
40	.41734	.90875	.45924	2.1775	1.1004	2.3961	20
41	.41760	.90863	.45960	.1758	.1005	.3946	19
42	.41787	.90851	.45995	.1741	.1007	.3931	18
43	.41813	.90839	.46030	.1725	.1008	.3916	17
44	.41838	.90826	.46065	.1708	.1010	.3901	16
45	.41866	.90814	.46101	2.1692	1.1011	2.3886	15
46	.41892	.90802	.46136	.1675	.1013	.3871	14
47	.41919	.90790	.46171	.1658	.1014	.3856	13
48	.41945	.90778	.46206	.1642	.1016	.3841	12
49	.41972	.90765	.46242	.1625	.1017	.3826	11
50	.41998	.90753	.46277	2.1609	1.1019	2.3811	10
51	.42024	.90741	.46312	.1592	.1020	.3796	9
52	.42051	.90729	.46348	.1576	.1022	.3781	8
53	.42077	.90717	.46383	.1559	.1023	.3766	7
54	.42103	.90704	.46418	.1543	.1025	.3751	6
55	.42130	.90692	.46454	2.1527	1.1026	2.3736	5
56	.42156	.90680	.46489	.1510	.1028	.3721	4
57	.42183	.90668	.46524	.1494	.1029	.3706	3
58	.42209	.90655	.46560	.1478	.1031	.3691	2
59	.42235	.90643	.46595	.1461	.1032	.3677	1
60	.42262	.90631	.46631	2.1445	1.1034	2.3662	0
M	**Cosine**	**Sine**	**Cotan.**	**Tangent**	**Cosec.**	**Secant**	**M**

65°

25°

M	Sine	Cosine	Tangent	Cotan.	Secant	Cosec.	M
0	.42262	.90631	.46631	2.1445	1.1034	2.3662	60
1	.42288	.90618	.46666	.1429	.1035	.3647	59
2	.42314	.90606	.46702	.1412	.1037	.3632	58
3	.42341	.90594	.46737	1396	.1038	.3618	57
4	.42367	.90581	.46772	.1380	.1040	.3603	56
5	.42394	.90569	.46808	2.1364	1.1041	2.3588	55
6	.42420	.90557	.46843	.1348	.1043	.3574	54
7	.42446	.90544	.46879	.1331	.1044	.3559	53
8	.42473	.90532	.46914	.1315	.1046	.3544	52
9	.42499	.90520	.46950	.1299	.1047	.3530	51
10	.42525	.90507	.46985	2.1283	1.1049	2.3515	50
11	.42552	.90495	.47021	.1267	.1050	.3501	49
12	.42578	.90483	.47056	.1251	.1052	.3486	48
13	.42604	.90470	.47092	.1235	.1053	.3472	47
14	.42630	.90458	.47127	.1219	.1055	.3457	46
15	.42657	.90445	.47163	2.1203	1.1056	2.3443	45
16	.42683	.90433	.47199	.1187	.1058	.3428	44
17	.42709	.90421	.47234	.1171	.1059	.3414	43
18	.42736	.90408	.47270	.1155	.1061	.3399	42
19	.42762	.90396	.47305	.1139	.1062	.3385	41
20	.42788	.90383	.47341	2.1123	1.1064	2.3371	40
21	.42815	.90371	.47376	.1107	.1065	.3356	39
22	.42841	.90358	.47412	.1092	.1067	.3342	38
23	.42867	.90346	.47448	.1076	.1068	.3328	37
24	.42893	.90333	.47483	.1060	.1070	.3313	36
25	.42920	.90321	.47519	2.1044	1.1072	2.3299	35
26	.42946	.90308	.47555	.1028	.1073	.3285	34
27	.42972	.90296	.47590	.1013	.1075	.3271	33
28	.42998	.90283	.47626	.0997	.1076	.3256	32
29	.43025	.90271	.47662	.0981	.1078	3242	31
30	.43051	.90258	.47697	2.0965	1.1079	2.3228	30
31	.43077	.90246	.47733	.0950	.1081	.3214	29
32	.43104	.90233	.47769	.0934	.1082	.3200	28
33	.43130	.90221	.47805	.0918	.1084	.3186	27
34	.43156	.90208	.47840	.0903	.1085	.3172	26
35	.43182	.90196	.47876	2.0887	1.1087	2.3158	25
36	.43208	.90183	.47912	.0872	.1088	.3143	24
37	.43235	.90171	.47948	.0856	.1090	.3129	23
38	.43261	.90158	.47983	.0840	.1092	.3115	22
39	.43287	.90145	.48019	.0825	.1093	.3101	21
40	.43313	.90133	.48055	2.0809	1.1095	2.3087	20
41	.43340	.90120	.48091	.0794	.1096	.3073	19
42	.43366	.90108	.48127	.0778	.1098	.3059	18
43	.43392	.90095	.48162	.0763	.1099	.3046	17
44	.43418	.90082	.48198	.0747	.1101	.3032	16
45	.43444	.90070	.48234	2.0732	1.1102	2.3018	15
46	.43471	.90057	.48270	.0717	.1104	.3004	14
47	.43497	.90044	.48306	.0701	.1106	.2990	13
48	.43523	.90032	.48342	.0686	.1107	.2976	12
49	.43549	.90019	.48378	.0671	.1109	.2962	11
50	.43575	.90006	.48414	2.0655	1.1110	2.2949	10
51	.43602	.89994	.48449	.0640	.1112	.2935	9
52	.43628	.89981	.48485	.0625	.1113	.2921	8
53	.43654	.89968	.48521	.0609	.1115	.2907	7
54	.43680	.89956	.48557	.0594	.1116	.2894	6
55	.43706	.89943	.48593	2.0579	1.1118	2.2880	5
56	.43732	.89930	.48629	.0564	.1120	.2866	4
57	.43759	.89918	.48665	.0548	.1121	.2853	3
58	.43785	.89905	.48701	.0533	.1123	.2839	2
59	.43811	.89892	.48737	.0518	.1124	.2825	1
60	.43837	.89879	.48773	2.0503	1.1126	2.2812	0
M	Cosine	Sine	Cotan.	Tangent	Cosec.	Secant	M

64°

26°

M	Sine	Cosine	Tangent	Cotan.	Secant	Cosec.	M
0	.43837	.89879	.48773	2.0503	1.1126	2.2812	60
1	.43863	.89867	.48809	.0488	.1127	.2798	59
2	.43889	.89854	.48845	.0473	.1129	.2784	58
3	.43915	.89841	.48881	0458	.1131	.2771	57
4	.43942	.89828	.48917	.0443	.1132	.2757	56
5	.43968	.89815	.48953	2.0427	1.1134	2.2744	55
6	.43994	.89803	.48989	.0412	.1135	.2730	54
7	.44020	.89790	.49025	.0397	.1137	.2717	53
8	.44046	.89777	.49062	.0382	.1139	.2703	52
9	.44072	.89764	.49098	.0367	.1140	.2690	51
10	.44098	.89751	.49134	2.0352	1.1142	2.2676	50
11	.44124	.89739	.49170	.0338	.1143	.2663	49
12	.44150	.89726	.49206	.0323	.1145	.2650	48
13	.44177	.89713	.49242	.0308	.1147	.2636	47
14	.44203	.89700	.49278	.0293	.1148	.2623	46
15	.44229	.89687	.49314	2.0278	1.1150	2.2610	45
16	.44255	.89674	.49351	.0263	.1151	.2596	44
17	.44281	.89661	.49387	.0248	.1153	.2583	43
18	.44307	.89649	.49423	.0233	.1155	.2570	42
19	.44333	.89636	.49459	.0219	.1156	.2556	41
20	.44359	.89623	.49495	2.0204	1.1158	2.2543	40
21	.44385	.89610	.49532	.0189	.1159	.2530	39
22	.44411	.89597	.49568	.0174	.1161	.2517	38
23	.44437	.89584	.49604	.0159	.1163	.2503	37
24	.44463	.89571	.49640	.0145	.1164	.2490	36
25	.44489	.89558	.49677	2.0130	1.1166	2.2477	35
26	.44516	.89545	.49713	.0115	.1167	.2464	34
27	.44542	.89532	.49749	.0101	.1169	.2451	33
28	.44568	.89519	.49785	.0086	.1171	.2438	32
29	.44594	.89506	.49822	.0071	.1172	2425	31
30	.44620	.89493	.49858	2.0057	1.1174	2.2411	30
31	.44646	.89480	.49894	.0042	.1176	.2398	29
32	.44672	.89467	.49931	.0028	.1177	.2385	28
33	.44698	.89454	.49967	.0013	.1179	.2372	27
34	.44724	.89441	.50003	1.9998	.1180	.2359	26
35	.44750	.89428	.50040	1.9984	1.1182	2.2348	25
36	.44776	.89415	.50076	.9969	.1184	.2333	24
37	.44802	.89402	.50113	.9955	.1185	.2320	23
38	.44828	.89389	.50149	.9940	.1187	.2307	22
39	.44854	.89376	.50185	.9926	.1189	.2294	21
40	.44880	.89363	.50222	1.9912	1.1190	2.2282	20
41	.44906	.89350	.50258	.9897	.1192	.2269	19
42	.44932	.89337	.50295	.9883	.1193	.2256	18
43	.44958	.89324	.50331	.9868	.1195	.2243	17
44	.44984	.89311	.50368	.9854	.1197	.2230	16
45	.45010	.89298	.50404	1.9840	1.1198	2.2217	15
46	.45036	.89285	.50441	.9825	.1200	.2204	14
47	.45062	.89272	.50477	.9811	.1202	.2192	13
48	.45088	.89258	.50514	.9797	.1203	.2179	12
49	.45114	.89245	.50550	.9782	.1205	.2166	11
50	.45140	.89232	.50587	1.9768	1.1207	2.2153	10
51	.45166	.89219	.50623	.9754	.1208	.2141	9
52	.45191	.89206	.50660	.9739	.1210	.2128	8
53	.45217	.89193	.50696	.9725	.1212	.2115	7
54	.45243	.89180	.50733	.9711	.1213	.2103	6
55	.45269	.89166	.50769	1.9697	1.1215	2.2090	5
56	.45295	.89153	.50806	.9683	.1217	.2077	4
57	.45321	.89140	.50843	.9668	.1218	.2065	3
58	.45347	.89127	.50879	.9654	.1220	.2052	2
59	.45373	.89114	.50916	.9640	.1222	.2039	1
60	.45399	.89101	.50952	1.9626	1.1223	2.2027	0
M	Cosine	Sine	Cotan.	Tangent	Cosec.	Secant	M

63°

27°

M	Sine	Cosine	Tangent	Cotan.	Secant	Cosec.	M
0	.45399	.89101	.50952	1.9626	1.1223	2.2027	60
1	.45425	.89087	.50989	.9612	.1225	.2014	59
2	.45451	.89074	.51026	.9598	.1226	.2002	58
3	.45477	.89061	.51062	9584	.1228	.1989	57
4	.45503	.89048	.51099	.9570	.1230	.1977	56
5	.45528	.89034	.51136	1.9556	1.1231	2.1964	55
6	.45554	.89021	.51172	.9542	.1233	.1952	54
7	.45580	.89008	.51209	.9528	.1235	.1939	53
8	.45606	.88995	.51246	.9514	.1237	.1927	52
9	.45632	.88981	.51283	.9500	.1238	.1914	51
10	.45658	.88968	.51319	1.9486	1.1240	2.1902	50
11	.45684	.88955	.51356	.9472	.1242	.1889	49
12	.45710	.88942	.51393	.9458	.1243	.1877	48
13	.45736	.88928	.51430	.9444	.1245	.1865	47
14	.45761	.88915	.51466	.9430	.1247	.1852	46
15	.45787	.88902	.51503	1.9416	1.1248	2.1840	45
16	.45813	.88888	.51540	.9402	.1250	.1828	44
17	.45839	.88875	.51577	.9388	.1252	.1815	43
18	.45865	.88862	.51614	.9375	.1253	.1803	42
19	.45891	.88848	.51651	.9361	.1255	.1791	41
20	.45917	.88835	.51687	1.9347	1.1257	2.1778	40
21	.45942	.88822	.51724	.9333	.1258	.1766	39
22	.45968	.88808	.51761	.9319	.1260	.1754	38
23	.45994	.88795	.51798	.9306	.1262	.1742	37
24	.46020	.88781	.51835	.9292	.1264	.1730	36
25	.46046	.88768	.51872	1.9278	1.1265	2.1717	35
26	.46072	.88755	.51909	.9264	.1267	.1705	34
27	.46097	.88741	.51946	.9251	.1269	.1693	33
28	.46123	.88728	.51983	.9237	.1270	.1681	32
29	.46149	.88714	.52020	.9223	.1272	1669	31
30	.46175	.88701	.52057	1.9210	1.1274	2.1657	30
31	.46201	.88688	.52094	9196	.1275	.1645	29
32	.46226	.88674	.52131	.9182	.1277	.1633	28
33	.46252	.88661	.52168	.9169	.1279	.1620	27
34	.46278	.88647	.52205	.9155	.1281	.1608	26
35	.46304	.88634	.52242	1.9142	1.1282	2.1596	25
36	.46330	.88620	.52279	.9128	.1284	.1584	24
37	.46355	.88607	.52316	.9115	.1286	.1572	23
38	.46381	.88593	.52353	.9101	.1287	.1560	22
39	.46407	.88580	.52390	.9088	.1289	.1548	21
40	.46433	.88566	.52427	1.9074	1.1291	2.1536	20
41	.46458	.88553	.52464	.9061	.1293	.1525	19
42	.46484	.88539	.52501	.9047	.1294	.1513	18
43	.46510	.88526	.52538	.9034	.1296	.1501	17
44	.46536	.88512	.52575	.9020	.1298	.1489	16
45	.46561	.88499	.52612	1.9007	1.1299	2.1477	15
46	.46587	.88485	.52650	.8993	.1301	.1465	14
47	.46613	.88472	.52687	.8980	.1303	.1453	13
48	.46639	.88458	.52724	.8967	.1305	.1441	12
49	.46664	.88444	.52761	.8953	.1306	.1430	11
50	.46690	.88431	.52798	1.8940	1.1308	2.1418	10
51	.46716	.88417	.52836	8927	.1310	.1406	9
52	.46741	.88404	.52873	.8913	.1312	.1394	8
53	.46767	.88390	.52910	.8900	.1313	.1382	7
54	.46793	.88376	.52947	.8887	.1315	.1371	6
55	.46819	.88363	.52984	1.8873	1.1317	2.1359	5
56	.46844	.88349	.53022	.8860	.1319	.1347	4
57	.46870	.88336	.53059	.8847	.1320	.1335	3
58	.46896	.88322	.53096	.8834	.1322	.1324	2
59	.46921	.88308	.53134	.8820	.1324	.1312	1
60	.46947	.88295	.53171	1.8807	1.1326	2.1300	0
M	Cosine	Sine	Cotan.	Tangent	Cosec.	Secant	M

62°

28°

M	Sine	Cosine	Tangent	Cotan.	Secant	Cosec.	M
0	.46947	.88295	.53171	1.8807	1.1326	2.1300	60
1	.46973	.88281	.53208	.8794	.1327	.1289	59
2	.46998	.88267	.53245	.8781	.1329	.1277	58
3	.47024	.88254	.53283	.8768	.1331	.1266	57
4	.47050	.88240	.53320	.8754	.1333	.1254	56
5	.47075	.88226	.53358	1.8741	1.1334	2.1242	55
6	.47101	.88213	.53395	.8728	.1336	.1231	54
7	.47127	.88199	.53432	.8715	.1338	.1219	53
8	.47152	.88185	.53470	.8702	.1340	.1208	52
9	.47178	.88171	.53507	.8689	.1341	.1196	51
10	.47204	.88158	.53545	1.8676	1.1343	2.1185	50
11	.47229	.88144	.53582	.8663	.1345	.1173	49
12	.47255	.88130	.53619	.8650	.1347	.1162	48
13	.47281	.88117	.53657	.8637	.1349	.1150	47
14	.47306	.88103	.53694	.8624	.1350	.1139	46
15	.47332	.88089	.53732	1.8611	1.1352	2.1127	45
16	.47357	.88075	.53769	.8598	.1354	.1116	44
17	.47383	.88061	.53807	.8585	.1356	.1104	43
18	.47409	.88048	.53844	.8572	.1357	.1093	42
19	.47434	.88034	.53882	.8559	.1359	.1082	41
20	.47460	.88020	.53919	1.8546	1.1361	2.1070	40
21	.47486	.88006	.53957	.8533	.1363	.1059	39
22	.47511	.87992	.53995	.8520	.1365	.1048	38
23	.47537	.87979	.54032	.8507	.1366	.1036	37
24	.47562	.87965	.54070	.8495	.1368	.1025	36
25	.47588	.87951	.54107	1.8482	1.1370	2.1014	35
26	.47613	.87937	.54145	.8469	.1372	.1002	34
27	.47639	.87923	.54183	.8456	.1373	.0991	33
28	.47665	.87909	.54220	.8443	.1375	.0980	32
29	.47690	.87895	.54258	.8430	.1377	0969	31
30	.47716	.87882	.54295	1.8418	1.1379	2.0957	30
31	.47741	.87868	.54333	8405	.1381	.0946	29
32	.47767	.87854	.54371	.8392	.1382	.0935	28
33	.47792	.87840	.54409	.8379	.1384	.0924	27
34	.47818	.87826	.54446	.8367	.1386	.0912	26
35	.47844	.87812	.54484	1.8354	1.1388	2.0901	25
36	.47869	.87798	.54522	.8341	.1390	.0890	24
37	.47895	.87784	.54559	.8329	.1391	.0879	23
38	.47920	.87770	.54597	.8316	.1393	.0868	22
39	.47946	.87756	.54635	.8303	.1395	.0857	21
40	.47971	.87742	.54673	1.8291	1.1397	2.0846	20
41	.47997	.87728	.54711	.8278	.1399	.0835	19
42	.48022	.87715	.54748	.8265	.1401	.0824	18
43	.48048	.87701	.54786	.8253	.1402	.0812	17
44	.48073	.87687	.54824	.8240	.1404	.0801	16
45	.48099	.87673	.54862	1.8227	1.1406	2.0790	15
46	.48124	.87659	.54900	.8215	.1408	.0779	14
47	.48150	.87645	.54937	.8202	.1410	.0768	13
48	.48175	.87631	.54975	.8190	.1411	.0757	12
49	.48201	.87617	.55013	.8177	.1413	.0746	11
50	.48226	.87603	.55051	1.8165	1.1415	2.0735	10
51	.48252	.87588	.55089	8152	.1417	.0725	9
52	.48277	.87574	.55127	.8140	.1419	.0714	8
53	.48303	.87560	.55165	.8127	.1421	.0703	7
54	.48328	.87546	.55203	.8115	.1422	.0692	6
55	.48354	.87532	.55241	1.8102	1.1424	2.0681	5
56	.48379	.87518	.55279	.8090	.1426	.0670	4
57	.48405	.87504	.55317	.8078	.1428	.0659	3
58	.48430	.87490	.55355	.8065	.1430	.0648	2
59	.48455	.87476	.55393	.8053	.1432	.0637	1
60	.48481	.87462	.55431	1.8040	1.1433	2.0627	0
M	Cosine	Sine	Cotan.	Tangent	Cosec.	Secant	M

61°

29°

M	Sine	Cosine	Tangent	Cotan.	Secant	Cosec.	M
0	.48481	.87462	.55431	1.8040	1.1433	2.0627	60
1	.48506	.87448	.55469	.8028	.1435	.0616	59
2	.48532	.87434	.55507	.8016	.1437	.0605	58
3	.48557	.87420	.55545	8003	.1439	.0594	57
4	.48583	.87405	.55583	.7991	.1441	.0583	56
5	.48608	.87391	.55621	1.7979	1.1443	2.0573	55
6	.48633	.87377	.55659	.7966	.1445	.0562	54
7	.48659	.87363	.55697	.7954	.1446	.0551	53
8	.48684	.87349	.55735	.7942	.1448	.0540	52
9	.48710	.87335	.55774	.7930	.1450	.0530	51
10	.48735	.87320	.55812	1.7917	1.1452	2.0519	50
11	.48760	.87306	.55850	.7905	.1454	.0508	49
12	.48786	.87292	.55888	.7893	.1456	.0498	48
13	.48811	.87278	.55926	.7881	.1458	.0487	47
14	.48837	.87264	.55964	.7868	.1459	.0476	46
15	.48862	.87250	.56003	1.7856	1.1461	2.0466	45
16	.48887	.87235	.56041	.7844	.1463	.0455	44
17	.48913	.87221	.56079	.7832	.1465	.0444	43
18	.48938	.87207	.56117	.7820	.1467	.0434	42
19	.48964	.87193	.56156	.7808	.1469	.0423	41
20	.48989	.87178	.56194	1.7795	1.1471	2.0413	40
21	.49014	.87164	.56232	.7783	.1473	.0402	39
22	.49040	.87150	.56270	.7771	.1474	.0392	38
23	.49065	.87136	.56309	.7759	.1476	.0381	37
24	.49090	.87121	.56347	.7747	.1478	.0370	36
25	.49116	.87107	.56385	1.7735	1.1480	2.0360	35
26	.49141	.87093	.56424	.7723	.1482	.0349	34
27	.49166	.87078	.56462	.7711	.1484	.0339	33
28	.49192	.87064	.56500	.7699	.1486	.0329	32
29	.49217	.87050	.56539	.7687	.1488	0318	31
30	.49242	.87035	.56577	1.7675	1.1489	2.0308	30
31	.49268	.87021	.56616	7663	.1491	.0297	29
32	.49293	.87007	.56654	.7651	.1493	.0287	28
33	.49318	.86992	.56692	.7639	.1495	.0276	27
34	.49343	.86978	.56731	.7627	.1497	.0266	26
35	.49369	.86964	.56769	1.7615	1.1499	2.0256	25
36	.49394	.86949	.56808	.7603	.1501	.0245	24
37	.49419	.86935	.56846	.7591	.1503	.0235	23
38	.49445	.86921	.56885	.7579	.1505	.0224	22
39	.49470	.86906	.56923	.7567	.1507	.0214	21
40	.49495	.86892	.56962	1.7555	1.1508	2.0204	20
41	.49521	.86877	.57000	.7544	.1510	.0194	19
42	.49546	.86863	.57039	.7532	.1512	.0183	18
43	.49571	.86849	.57077	.7520	.1514	.0173	17
44	.49596	.86834	.57116	.7508	.1516	.0163	16
45	.49622	.86820	.57155	1.7496	1.1518	2.0152	15
46	.49647	.86805	.57193	.7484	.1520	.0142	14
47	.49672	.86791	.57232	.7473	.1522	.0132	13
48	.49697	.86776	.57270	.7461	.1524	.0122	12
49	.49723	.86762	.57309	.7449	.1526	.0111	11
50	.49748	.86748	.57348	1.7437	1.1528	2.0101	10
51	.49773	.86733	.57386	7426	.1530	.0091	9
52	.49798	.86719	.57425	.7414	.1531	.0081	8
53	.49823	.86704	.57464	.7402	.1533	.0071	7
54	.49849	.86690	.57502	.7390	.1535	.0061	6
55	.49874	.86675	.57541	1.7379	1.1537	2.0050	5
56	.49899	.86661	.57580	.7367	.1539	.0040	4
57	.49924	.86646	.57619	.7355	.1541	.0030	3
58	.49950	.86632	.57657	.7344	.1543	.0020	2
59	.49975	.86617	.57696	.7332	.1545	.0010	1
60	.50000	.86603	.57735	1.7320	1.1547	2.0000	0
M	Cosine	Sine	Cotan.	Tangent	Cosec.	Secant	M

60°

30°

M	Sine	Cosine	Tangent	Cotan.	Secant	Cosec.	M
0	.50000	.86603	.57735	1.7320	1.1547	2.0000	60
1	.50025	.86588	.57774	.7309	.1549	1.9990	59
2	.50050	.86573	.57813	.7297	.1551	.9980	58
3	.50075	.86559	.57851	7286	.1553	.9970	57
4	.50101	.86544	.57890	.7274	.1555	.9960	56
5	.50126	.86530	.57929	1.7262	1.1557	1.9950	55
6	.50151	.86515	.57968	.7251	.1559	.9940	54
7	.50176	.86500	.58007	.7239	.1561	.9930	53
8	.50201	.86486	.58046	.7228	.1562	.9920	52
9	.50226	.86471	.58085	.7216	.1564	.9910	51
10	.50252	.86457	.58123	1.7205	1.1566	1.9900	50
11	.50277	.86442	.58162	.7193	.1568	.9890	49
12	.50302	.86427	.58201	.7182	.1570	.9880	48
13	.50327	.86413	.58240	.7170	.1572	.9870	47
14	.50352	.86398	.58279	.7159	.1574	.9860	46
15	.50377	.86383	.58318	1.7147	1.1576	1.9850	45
16	.50402	.86369	.58357	.7136	.1578	.9840	44
17	.50428	.86354	.58396	.7124	.1580	.9830	43
18	.50453	.86339	.58435	.7113	.1582	.9820	42
19	.50478	.86325	.58474	.7101	.1584	.9811	41
20	.50503	.86310	.58513	1.7090	1.1586	1.9801	40
21	.50528	.86295	.58552	.7079	.1588	.9791	39
22	.50553	.86281	.58591	.7067	.1590	.9781	38
23	.50578	.86266	.58630	.7056	.1592	.9771	37
24	.50603	.86251	.58670	.7044	.1594	.9761	36
25	.50628	.86237	.58709	1.7033	1.1596	1.9752	35
26	.50653	.86222	.58748	.7022	.1598	.9742	34
27	.50679	.86207	.58787	.7010	.1600	.9732	33
28	.50704	.86192	.58826	.6999	.1602	.9722	32
29	.50729	.86178	.58865	.6988	.1604	9713	31
30	.50754	.86163	.58904	1.6977	1.1606	1.9703	30
31	.50779	.86148	.58944	.6965	.1608	.9693	29
32	.50804	.86133	.58983	.6954	.1610	.9683	28
33	.50829	.86118	.59022	.6943	.1612	.9674	27
34	.50854	.86104	.59061	.6931	.1614	.9664	26
35	.50879	.86089	.59100	1.6920	1.1616	1.9654	25
36	.50904	.86074	.59140	.6909	.1618	.9645	24
37	.50929	.86059	.59179	.6898	.1620	.9635	23
38	.50954	.86044	.59218	.6887	.1622	.9625	22
39	.50979	.86030	.59258	.6875	.1624	.9616	21
40	.51004	.86015	.59297	1.6864	1.1626	1.9606	20
41	.51029	.86000	.59336	.6853	.1628	.9596	19
42	.51054	.85985	.59376	.6842	.1630	.9587	18
43	.51079	.85970	.59415	.6831	.1632	.9577	17
44	.51104	.85955	.59454	.6820	.1634	.9568	16
45	.51129	.85941	.59494	1.6808	1.1636	1.9558	15
46	.51154	.85926	.59533	.6797	.1638	.9549	14
47	.51179	.85911	.59572	.6786	.1640	.9539	13
48	.51204	.85896	.59612	.6775	.1642	.9530	12
49	.51229	.85881	.59651	.6764	.1644	.9520	11
50	.51254	.85866	.59691	1.6753	1.1646	1.9510	10
51	.51279	.85851	.59730	6742	.1648	.9501	9
52	.51304	.85836	.59770	.6731	.1650	.9491	8
53	.51329	.85821	.59809	.6720	.1652	.9482	7
54	.51354	.85806	.59849	.6709	.1654	.9473	6
55	.51379	.85791	.59888	1.6698	1.1656	1.9463	5
56	.51404	.85777	.59928	.6687	.1658	.9454	4
57	.51429	.85762	.59967	.6676	.1660	.9444	3
58	.51454	.85747	.60007	.6665	.1662	.9435	2
59	.51479	.85732	.60046	.6654	.1664	.9425	1
60	.51504	.85717	.60086	1.6643	1.1666	1.9416	0
M	Cosine	Sine	Cotan.	Tangent	Cosec.	Secant	M

59°

31°

M	Sine	Cosine	Tangent	Cotan.	Secant	Cosec.	M
0	.51504	.85717	.60086	1.6643	1.1666	1.9416	60
1	.51529	.85702	.60126	.6632	.1668	.9407	59
2	.51554	.85687	.60165	.6621	.1670	.9397	58
3	.51578	.85672	.60205	6610	.1672	.9388	57
4	.51603	.85657	.60244	.6599	.1674	.9378	56
5	.51628	.85642	.60284	1.6588	1.1676	1.9369	55
6	.51653	.85627	.60324	.6577	.1678	.9360	54
7	.51678	.85612	.60363	.6566	.1681	.9350	53
8	.51703	.85597	.60403	.6555	.1683	.9341	52
9	.51728	.85582	.60443	.6544	.1685	.9332	51
10	.51753	.85566	.60483	1.6534	1.1687	1.9322	50
11	.51778	.85551	.60522	.6523	.1689	.9313	49
12	.51803	.85536	.60562	.6512	.1691	.9304	48
13	.51827	.85521	.60602	.6501	.1693	.9295	47
14	.51852	.85506	.60642	.6490	.1695	.9285	46
15	.51877	.85491	.60681	1.6479	1.1697	1.9276	45
16	.51902	.85476	.60721	.6469	.1699	.9267	44
17	.51927	.85461	.60761	.6458	.1701	.9258	43
18	.51952	.85446	.60801	.6447	.1703	.9248	42
19	.51977	.85431	.60841	.6436	.1705	.9239	41
20	.52002	.85416	.60881	1.6425	1.1707	1.9230	40
21	.52026	.85400	.60920	.6415	.1709	.9221	39
22	.52051	.85385	.60960	.6404	.1712	.9212	38
23	.52076	.85370	.61000	.6393	.1714	.9203	37
24	.52101	.85355	.61040	.6383	.1716	.9193	36
25	.52126	.85340	.61080	1.6372	1.1718	1.9184	35
26	.52151	.85325	.61120	.6361	.1720	.9175	34
27	.52175	.85309	.61160	.6350	.1722	.9166	33
28	.52200	.85294	.61200	.6340	.1724	.9157	32
29	.52225	.85279	.61240	.6329	.1726	9148	31
30	.52250	.85264	.61280	1.6318	1.1728	1.9139	30
31	.52275	.85249	.61320	6308	.1730	.9130	29
32	.52299	.85234	.61360	.6297	.1732	.9121	28
33	.52324	.85218	.61400	.6286	.1734	.9112	27
34	.52349	.85203	.61440	.6276	.1737	.9102	26
35	.52374	.85188	.61480	1.6265	1.1739	1.9093	25
36	.52398	.85173	.61520	.6255	.1741	.9084	24
37	.52423	.85157	.61560	.6244	.1743	.9075	23
38	.52448	.85142	.61601	.6233	.1745	.9066	22
39	.52473	.85127	.61641	.6223	.1747	.9057	21
40	.52498	.85112	.61681	1.6212	1.1749	1.9048	20
41	.52522	.85096	.61721	.6202	.1751	.9039	19
42	.52547	.85081	.61761	.6191	.1753	.9030	18
43	.52572	.85066	.61801	.6181	.1756	.9021	17
44	.52597	.85050	.61842	.6170	.1758	.9013	16
45	.52621	.85035	.61882	1.6160	1.1760	1.9004	15
46	.52646	.85020	.61922	.6149	.1762	.8995	14
47	.52671	.85004	.61962	.6139	.1764	.8986	13
48	.52695	.84989	.62003	.6128	.1766	.8977	12
49	.52720	.84974	.62043	.6118	.1768	.8968	11
50	.52745	.84959	.62083	1.6107	1.1770	1.8959	10
51	.52770	.84943	.62123	6097	.1772	.8950	9
52	.52794	.84928	.62164	.6086	.1775	.8941	8
53	.52819	.84912	.62204	.6076	.1777	.8932	7
54	.52844	.84897	.62244	.6066	.1779	.8924	6
55	.52868	.84882	.62285	1.6055	1.1781	1.8915	5
56	.52893	.84866	.62325	.6045	.1783	.8906	4
57	.52918	.84851	.62366	.6034	.1785	.8897	3
58	.52942	.84836	.62406	.6024	.1787	.8888	2
59	.52967	.84820	.62446	.6014	.1790	.8879	1
60	.52992	.84805	.62487	1.6003	1.1792	1.8871	0
M	Cosine	Sine	Cotan.	Tangent	Cosec.	Secant	M

58°

32°

M	Sine	Cosine	Tangent	Cotan.	Secant	Cosec.	M
0	.52992	.84805	.62487	1.6003	1.1792	1.8871	60
1	.53016	.84789	.62527	.5993	.1794	.8862	59
2	.53041	.84774	.62568	.5983	.1796	.8853	58
3	.53066	.84758	.62608	5972	.1798	.8844	57
4	.53090	.84743	.62649	.5962	.1800	.8836	56
5	.53115	.84728	.62689	1.5952	1.1802	1.8827	55
6	.53140	.84712	.62730	.5941	.1805	.8818	54
7	.53164	.84697	.62770	.5931	.1807	.8809	53
8	.53189	.84681	.62811	.5921	.1809	.8801	52
9	.53214	.84666	.62851	.5910	.1811	.8792	51
10	.53238	.84650	.62892	1.5900	1.1813	1.8783	50
11	.53263	.84635	.62933	.5890	.1815	.8775	49
12	.53288	.84619	.62973	.5880	.1818	.8766	48
13	.53312	.84604	.63014	.5869	.1820	.8757	47
14	.53337	.84588	.63055	.5859	.1822	.8749	46
15	.53361	.84573	.63095	1.5849	1.1824	1.8740	45
16	.53386	.84557	.63136	.5839	.1826	.8731	44
17	.53411	.84542	.63177	.5829	.1828	.8723	43
18	.53435	.84526	.63217	.5818	.1831	.8714	42
19	.53460	.84511	.63258	.5808	.1833	.8706	41
20	.53484	.84495	.63299	1.5798	1.1835	1.8697	40
21	.53509	.84479	.63339	.5788	.1837	.8688	39
22	.53533	.84464	.63380	.5778	.1839	.8680	38
23	.53558	.84448	.63421	.5768	.1841	.8671	37
24	.53583	.84433	.63462	.5757	.1844	.8663	36
25	.53607	.84417	.63503	1.5747	1.1846	1.8654	35
26	.53632	.84402	.63543	.5737	.1848	.8646	34
27	.53656	.84386	.63584	.5727	.1850	.8637	33
28	.53681	.84370	.63625	.5717	.1852	.8629	32
29	.53705	.84355	.63666	.5707	.1855	8620	31
30	.53730	.84339	.63707	1.5697	1.1857	1.8611	30
31	.53754	.84323	.63748	5687	.1859	.8603	29
32	.53779	.84308	.63789	.5677	.1861	.8595	28
33	.53803	.84292	.63830	.5667	.1863	.8586	27
34	.53828	.84276	.63871	.5657	.1866	.8578	26
35	.53852	.84261	.63912	1.5646	1.1868	1.8569	25
36	.53877	.84245	.63953	.5636	.1870	.8561	24
37	.53901	.84229	.63994	.5626	.1872	.8552	23
38	.53926	.84214	.64035	.5616	.1874	.8544	22
39	.53950	.84198	.64076	.5606	.1877	.8535	21
40	.53975	.84182	.64117	1.5596	1.1879	1.8527	20
41	.53999	.84167	.64158	.5586	.1881	.8519	19
42	.54024	.84151	.64199	.5577	.1883	.8510	18
43	.54048	.84135	.64240	.5567	.1886	.8502	17
44	.54073	.84120	.64281	.5557	.1888	.8493	16
45	.54097	.84104	.64322	1.5547	1.1890	1.8485	15
46	.54122	.84088	.64363	.5537	.1892	.8477	14
47	.54146	.84072	.64404	.5527	.1894	.8468	13
48	.54171	.84057	.64446	.5517	.1897	.8460	12
49	.54195	.84041	.64487	.5507	.1899	.8452	11
50	.54220	.84025	.64528	1.5497	1.1901	1.8443	10
51	.54244	.84009	.64569	5487	.1903	.8435	9
52	.54268	.83993	.64610	.5477	.1906	.8427	8
53	.54293	.83978	.64652	.5467	.1908	.8418	7
54	.54317	.83962	.64693	.5458	.1910	.8410	6
55	.54342	.83946	.64734	1.5448	1.1912	1.8402	5
56	.54366	.83930	.64775	.5438	.1915	.8394	4
57	.54391	.83914	.64817	.5428	.1917	.8385	3
58	.54415	.83899	.64858	.5418	.1919	.8377	2
59	.54439	.83883	.64899	.5408	.1921	.8369	1
60	.54464	.83867	.64941	1.5399	1.1924	1.8361	0
M	Cosine	Sine	Cotan.	Tangent	Cosec.	Secant	M

57°

33°

M	Sine	Cosine	Tangent	Cotan.	Secant	Cosec.	M
0	.54464	.83867	.64941	1.5399	1.1924	1.8361	60
1	.54488	.83851	.64982	.5389	.1926	.8352	59
2	.54513	.83835	.65023	.5379	.1928	.8344	58
3	.54537	.83819	.65065	5369	.1930	.8336	57
4	.54561	.83804	.65106	.5359	.1933	.8328	56
5	.54586	.83788	.65148	1.5350	1.1935	1.8320	55
6	.54610	.83772	.65189	.5340	.1937	.8311	54
7	.54634	.83756	.65231	.5330	.1939	.8303	53
8	.54659	.83740	.65272	.5320	.1942	.8295	52
9	.54683	.83724	.65314	.5311	.1944	.8287	51
10	.54708	.83708	.65355	1.5301	1.1946	1.8279	50
11	.54732	.83692	.65397	.5291	.1948	.8271	49
12	.54756	.83676	.65438	.5282	.1951	.8263	48
13	.54781	.83660	.65480	.5272	.1953	.8255	47
14	.54805	.83644	.65521	.5262	.1955	.8246	46
15	.54829	.83629	.65563	1.5252	1.1958	1.8238	45
16	.54854	.83613	.65604	.5243	.1960	.8230	44
17	.54878	.83597	.65646	.5233	.1962	.8222	43
18	.54902	.83581	.65688	.5223	.1964	.8214	42
19	.54926	.83565	.65729	.5214	.1967	.8206	41
20	.54951	.83549	.65771	1.5204	1.1969	1.8198	40
21	.54975	.83533	.65813	.5195	.1971	.8190	39
22	.54999	.83517	.65854	.5185	.1974	.8182	38
23	.55024	.83501	.65896	.5175	.1976	.8174	37
24	.55048	.83485	.65938	.5166	.1978	.8166	36
25	.55072	.83469	.65980	1.5156	1.1980	1.8158	35
26	.55097	.83453	.66021	.5147	.1983	.8150	34
27	.55121	.83437	.66063	.5137	.1985	.8142	33
28	.55145	.83421	.66105	.5127	.1987	.8134	32
29	.55169	.83405	.66147	.5118	.1990	8126	31
30	.55194	.83388	.66188	1.5108	1.1992	1.8118	30
31	.55218	.83372	.66230	5099	.1994	.8110	29
32	.55242	.83356	.66272	.5089	.1997	.8102	28
33	.55266	.83340	.66314	.5080	.1999	.8094	27
34	.55291	.83324	.66356	.5070	.2001	.8086	26
35	.55315	.83308	.66398	1.5061	1.2004	1.8078	25
36	.55339	.83292	.66440	.5051	.2006	.8070	24
37	.55363	.83276	.66482	.5042	.2008	.8062	23
38	.55388	.83260	.66524	.5032	.2010	.8054	22
39	.55412	.83244	.66566	.5023	.2013	.8047	21
40	.55436	.83228	.66608	1.5013	1.2015	1.8039	20
41	.55460	.83211	.66650	.5004	.2017	.8031	19
42	.55484	.83195	.66692	.4994	.2020	.8023	18
43	.55509	.83179	.66734	.4985	.2022	.8015	17
44	.55533	.83163	.66776	.4975	.2024	.8007	16
45	.55557	.83147	.66818	1.4966	1.2027	1.7999	15
46	.55581	.83131	.66860	.4957	.2029	.7992	14
47	.55605	.83115	.66902	.4947	.2031	.7984	13
48	.55629	.83098	.66944	.4938	.2034	.7976	12
49	.55654	.83082	.66986	.4928	.2036	.7968	11
50	.55678	.83066	.67028	1.4919	1.2039	1.7960	10
51	.55702	.83050	.67071	4910	.2041	.7953	9
52	.55726	.83034	.67113	.4900	.2043	.7945	8
53	.55750	.83017	.67155	.4891	.2046	.7937	7
54	.55774	.83001	.67197	.4881	.2048	.7929	6
55	.55799	.82985	.67239	1.4872	1.2050	1.7921	5
56	.55823	.82969	.67282	.4863	.2053	.7914	4
57	.55847	.82952	.67324	.4853	.2055	.7906	3
58	.55871	.82936	.67366	.4844	.2057	.7898	2
59	.55895	.82920	.67408	.4835	.2060	.7891	1
60	.55919	.82904	.67451	1.4826	1.2062	1.7883	0
M	Cosine	Sine	Cotan.	Tangent	Cosec.	Secant	M

56°

34°

M	Sine	Cosine	Tangent	Cotan.	Secant	Cosec.	M
0	.55919	.82904	.67451	1.4826	1.2062	1.7883	60
1	.55943	.82887	.67493	.4816	.2064	.7875	59
2	.55967	.82871	.67535	.4807	.2067	.7867	58
3	.55992	.82855	.67578	.4798	.2069	.7860	57
4	.56016	.82839	.67620	.4788	.2072	.7852	56
5	.55040	.82822	.67663	1.4779	1.2074	1.7844	55
6	.56064	.82806	.67705	.4770	.2076	.7837	54
7	.56088	.82790	.67747	.4761	.2079	.7829	53
8	.56112	.82773	.67790	.4751	.2081	.7821	52
9	.56136	.82757	.67832	.4742	.2083	.7814	51
10	.56160	.82741	.67875	1.4733	1.2086	1.7806	50
11	.56184	.82724	.67917	.4724	.2088	.7798	49
12	.56208	.82708	.67960	.4714	.2091	.7791	48
13	.56232	.82692	.68002	.4705	.2093	.7783	47
14	.56256	.82675	.68045	.4696	.2095	.7776	46
15	.56280	.82659	.68087	1.4687	1.2098	1.7768	45
16	.56304	.82643	.68130	.4678	.2100	.7760	44
17	.56328	.82626	.68173	.4669	.2103	.7753	43
18	.56353	.82610	.68215	.4659	.2105	.7745	42
19	.56377	.82593	.68258	.4650	.2107	.7738	41
20	.56401	.82577	.68301	1.4641	1.2110	1.7730	40
21	.56425	.82561	.68343	.4632	.2112	.7723	39
22	.56449	.82544	.68386	.4623	.2115	.7715	38
23	.56473	.82528	.68429	.4614	.2117	.7708	37
24	.56497	.82511	.68471	.4605	.2119	.7700	36
25	.56521	.82495	.68514	1.4595	1.2122	1.7693	35
26	.56545	.82478	.68557	.4586	.2124	.7685	34
27	.56569	.82462	.68600	.4577	.2127	.7678	33
28	.56593	.82445	.68642	.4568	.2129	.7670	32
29	.56617	.82429	.68685	.4559	.2132	7663	31
30	.56641	.82413	.68728	1.4550	1.2134	1.7655	30
31	.56664	.82396	.68771	4541	.2136	.7648	29
32	.56688	.82380	.68814	.4532	.2139	.7640	28
33	.56712	.82363	.68857	.4523	.2141	.7633	27
34	.56736	.82347	.68899	.4514	.2144	.7625	26
35	.56760	.82330	.68942	1.4505	1.2146	1.7618	25
36	.56784	.82314	.68985	.4496	.2149	.7610	24
37	.56808	.82297	.69028	.4487	.2151	.7603	23
38	.56832	.82280	.69071	.4478	.2153	.7596	22
39	.56856	.82264	.69114	.4469	.2156	.7588	21
40	.56880	.82247	.69157	1.4460	1.2158	1.7581	20
41	.56904	.82231	.69200	.4451	.2161	.7573	19
42	.56928	.82214	.69243	.4442	.2163	.7566	18
43	.56952	.82198	.69286	.4433	.2166	.7559	17
44	.56976	.82181	.69329	.4424	.2168	.7551	16
45	.57000	.82165	.69372	1.4415	1.2171	1.7544	15
46	.57023	.82148	.69415	.4406	.2173	.7537	14
47	.57047	.82131	.69459	.4397	.2175	.7529	13
48	.57071	.82115	.69502	.4388	.2178	.7522	12
49	.57095	.82098	.69545	.4379	.2180	.7514	11
50	.57119	.82082	.69588	1.4370	1.2183	1.7507	10
51	.57143	.82065	.69631	4361	.2185	.7500	9
52	.57167	.82048	.69674	.4352	.2188	.7493	8
53	.57191	.82032	.69718	.4343	.2190	.7485	7
54	.57214	.82015	.69761	.4335	.2193	.7478	6
55	.57238	.81998	.69804	1.4326	1.2195	1.7471	5
56	.57262	.81982	.69847	.4317	.2198	.7463	4
57	.57286	.81965	.69891	.4308	.2200	.7456	3
58	.57310	.81948	.69934	.4299	.2203	.7449	2
59	.57334	.81932	.69977	.4290	.2205	.7442	1
60	.57358	.81915	.70021	1.4281	1.2208	1.7434	0
M	Cosine	Sine	Cotan.	Tangent	Cosec.	Secant	M

55°

35°

M	Sine	Cosine	Tangent	Cotan.	Secant	Cosec.	M
0	.57358	.81915	.70021	1.4281	1.2208	1.7434	60
1	.57381	.81898	.70064	.4273	.2210	.7427	59
2	.57405	.81882	.70107	.4264	.2213	.7420	58
3	.57429	.81865	.70151	.4255	.2215	.7413	57
4	.57453	.81848	.70194	.4246	.2218	.7405	56
5	.57477	.81832	.70238	1.4237	1.2220	1.7398	55
6	.57500	.81815	.70281	.4228	.2223	.7391	54
7	.57524	.81798	.70325	.4220	.2225	.7384	53
8	.57548	.81781	.70368	.4211	.2228	.7377	52
9	.57572	.81765	.70412	.4202	.2230	.7369	51
10	.57596	.81748	.70455	1.4193	1.2233	1.7362	50
11	.57619	.81731	.70499	.4185	.2235	.7355	49
12	.57643	.81714	.70542	.4176	.2238	.7348	48
13	.57667	.81698	.70586	.4167	.2240	.7341	47
14	.57691	.81681	.70629	.4158	.2243	.7334	46
15	.57714	.81664	.70673	1.4150	1.2245	1.7327	45
16	.57738	.81647	.70717	.4141	.2248	.7319	44
17	.57762	.81630	.70760	.4132	.2250	.7312	43
18	.57786	.81614	.70804	.4123	.2253	.7305	42
19	.57809	.81597	.70848	.4115	.2255	.7298	41
20	.57833	.81580	.70891	1.4106	1.2258	1.7291	40
21	.57857	.81563	.70935	.4097	.2260	.7284	39
22	.57881	.81546	.70979	.4089	.2263	.7277	38
23	.57904	.81530	.71022	.4080	.2265	.7270	37
24	.57928	.81513	.71066	.4071	.2268	.7263	36
25	.57952	.81496	.71110	1.4063	1.2270	1.7256	35
26	.57975	.81479	.71154	.4054	.2273	.7249	34
27	.57999	.81462	.71198	.4045	.2276	.7242	33
28	.58023	.81445	.71241	.4037	.2278	.7234	32
29	.58047	.81428	.71285	.4028	.2281	7227	31
30	.58070	.81411	.71329	1.4019	1.2283	1.7220	30
31	.58094	.81395	.71373	4011	.2286	.7213	29
32	.58118	.81378	.71417	.4002	.2288	.7206	28
33	.58141	.81361	.71461	.3994	.2291	.7199	27
34	.58165	.81344	.71505	.3985	.2293	.7192	26
35	.58189	.81327	.71549	1.3976	1.2296	1.7185	25
36	.58212	.81310	.71593	.3968	.2298	.7178	24
37	.58236	.81293	.71637	.3959	.2301	.7171	23
38	.58259	.81276	.71681	.3951	.2304	.7164	22
39	.58283	.81259	.71725	.3942	.2306	.7157	21
40	.58307	.81242	.71769	1.3933	1.2309	1.7151	20
41	.58330	.81225	.71813	.3925	.2311	.7144	19
42	.58354	.81208	.71857	.3916	.2314	.7137	18
43	.58378	.81191	.71901	.3908	.2316	.7130	17
44	.58401	.81174	.71945	.3899	.2319	.7123	16
45	.58425	.81157	.71990	1.3891	1.2322	1.7116	15
46	.58448	.81140	.72034	.3882	.2324	.7109	14
47	.58472	.81123	.72078	.3874	.2327	.7102	13
48	.58496	.81106	.72122	.3865	.2329	.7095	12
49	.58519	.81089	.72166	.3857	.2332	.7088	11
50	.58543	.81072	.72211	1.3848	1.2335	1.7081	10
51	.58566	.81055	.72255	3840	.2337	.7075	9
52	.58590	.81038	.72299	.3831	.2340	.7068	8
53	.58614	.81021	.72344	.3823	.2342	.7061	7
54	.58637	.81004	.72388	.3814	.2345	.7054	6
55	.58661	.80987	.72432	1.3806	1.2348	1.7047	5
56	.58684	.80970	.72477	.3797	.2350	.7040	4
57	.58708	.80953	.72521	.3789	.2353	.7033	3
58	.58731	.80936	.72565	.3781	.2355	.7027	2
59	.58755	.80919	.72610	.3772	.2358	.7020	1
60	.58778	.80902	.72654	1.3764	1.2361	1.7013	0
M	Cosine	Sine	Cotan.	Tangent	Cosec.	Secant	M

54°

36°

M	Sine	Cosine	Tangent	Cotan.	Secant	Cosec.	M
0	.58778	.80902	.72654	1.3764	1.2361	1.7013	60
1	.58802	.80885	.72699	.3755	.2363	.7006	59
2	.58825	.80867	.72743	.3747	.2366	.6999	58
3	.58849	.80850	.72788	.3738	.2368	.6993	57
4	.58873	.80833	.72832	.3730	.2371	.6986	56
5	.58896	.80816	.72877	1.3722	1.2374	1.6979	55
6	.58920	.80799	.72921	.3713	.2376	.6972	54
7	.58943	.80782	.72966	.3705	.2379	.6965	53
8	.58967	.80765	.73010	.3697	.2382	.6959	52
9	.58990	.80747	.73055	.3688	.2384	.6952	51
10	.59014	.80730	.73100	1.3680	1.2387	1.6945	50
11	.59037	.80713	.73144	.3672	.2389	.6938	49
12	.59060	.80696	.73189	.3663	.2392	.6932	48
13	.59084	.80679	.73234	.3655	.2395	.6925	47
14	.59107	.80662	.73278	.3647	.2397	.6918	46
15	.59131	.80644	.73323	1.3638	1.2400	1.6912	45
16	.59154	.80627	.73368	.3630	.2403	.6905	44
17	.59178	.80610	.73412	.3622	.2405	.6898	43
18	.59201	.80593	.73457	.3613	.2408	.6891	42
19	.59225	.80576	.73502	.3605	.2411	.6885	41
20	.59248	.80558	.73547	1.3597	1.2413	1.6878	40
21	.59272	.80541	.73592	.3588	.2416	.6871	39
22	.59295	.80524	.73637	.3580	.2419	.6865	38
23	.59318	.80507	.73681	.3572	.2421	.6858	37
24	.59342	.80489	.73726	.3564	.2424	.6851	36
25	.59365	.80472	.73771	1.3555	1.2427	1.6845	35
26	.59389	.80455	.73816	.3547	.2429	.6838	34
27	.59412	.80437	.73861	.3539	.2432	.6831	33
28	.59435	.80420	.73906	.3531	.2435	.6825	32
29	.59459	.80403	.73951	.3522	.2437	6818	31
30	.59482	.80386	.73996	1.3514	1.2440	1.6812	30
31	.59506	.80368	.74041	3506	.2443	.6805	29
32	.59529	.80351	.74086	.3498	.2445	.6798	28
33	.59552	.80334	.74131	.3489	.2448	.6792	27
34	.59576	.80316	.74176	.3481	.2451	.6785	26
35	.59599	.80299	.74221	1.3473	1.2453	1.6779	25
36	.59622	.80282	.74266	.3465	.2456	.6772	24
37	.59646	.80264	.74312	.3457	.2459	.6766	23
38	.59669	.80247	.74357	.3449	.2461	.6759	22
39	.59692	.80230	.74402	.3440	.2464	.6752	21
40	.59716	.80212	.74447	1.3432	1.2467	1.6746	20
41	.59739	.80195	.74492	.3424	.2470	.6739	19
42	.59762	.80177	.74538	.3416	.2472	.6733	18
43	.59786	.80160	.74583	.3408	.2475	.6726	17
44	.59809	.80143	.74628	.3400	.2478	.6720	16
45	.59832	.80125	.74673	1.3392	1.2480	1.6713	15
46	.59856	.80108	.74719	.3383	.2483	.6707	14
47	.59879	.80090	.74764	.3375	.2486	.6700	13
48	.59902	.80073	.74809	.3367	.2488	.6694	12
49	.59926	.80056	.74855	.3359	.2491	.6687	11
50	.59949	.80038	.74900	1.3351	1.2494	1.6681	10
51	.59972	.80021	.74946	3343	.2497	.6674	9
52	.59995	.80003	.74991	.3335	.2499	.6668	8
53	.60019	.79986	.75037	.3327	.2502	.6661	7
54	.60042	.79968	.75082	.3319	.2505	.6655	6
55	.60065	.79951	.75128	1.3311	1.2508	1.6648	5
56	.60088	.79933	.75173	.3303	.2510	.6642	4
57	.60112	.79916	.75219	.3294	.2513	.6636	3
58	.60135	.79898	.75264	.3286	.2516	.6629	2
59	.60158	.79881	.75310	.3278	.2519	.6623	1
60	.60181	.79863	.75355	1.3270	1.2521	1.6616	0
M	Cosine	Sine	Cotan.	Tangent	Cosec.	Secant	M

53°

37°

M	Sine	Cosine	Tangent	Cotan.	Secant	Cosec.	M
0	.60181	.79863	.75355	1.3270	1.2521	1.6616	60
1	.60205	.79846	.75401	.3262	.2524	.6610	59
2	.60228	.79828	.75447	.3254	.2527	.6603	58
3	.60251	.79811	.75492	.3246	.2530	.6597	57
4	.60274	.79793	.75538	.3238	.2532	.6591	56
5	.60298	.79776	.75584	1.3230	1.2535	1.6584	55
6	.60320	.79758	.75629	.3222	.2538	.6578	54
7	.60344	.79741	.75675	.3214	.2541	.6572	53
8	.60367	.79723	.75721	.3206	.2543	.6565	52
9	.60390	.79706	.75767	.3198	.2546	.6559	51
10	.60413	.79688	.75812	1.3190	1.2549	1.6552	50
11	.60437	.79670	.75858	.3182	.2552	.6546	49
12	.60460	.79653	.75904	.3174	.2554	.6540	48
13	.60483	.79635	.75950	.3166	.2557	.6533	47
14	.60506	.79618	.75996	.3159	.2560	.6527	46
15	.60529	.79600	.76042	1.3151	1.2563	1.6521	45
16	.60552	.79582	.76088	.3143	.2565	.6514	44
17	.60576	.79565	.76134	.3135	.2568	.6508	43
18	.60599	.79547	.76179	.3127	.2571	.6502	42
19	.60622	.79530	.76225	.3119	.2574	.6496	41
20	.60645	.79512	.76271	1.3111	1.2577	1.6489	40
21	.60668	.79494	.76317	.3103	.2579	.6483	39
22	.60691	.79477	.76364	.3095	.2582	.6477	38
23	.60714	.79459	.76410	.3087	.2585	.6470	37
24	.60737	.79441	.76456	.3079	.2588	.6464	36
25	.60761	.79424	.76502	1.3071	1.2591	1.6458	35
26	.60784	.79406	.76548	.3064	.2593	.6452	34
27	.60807	.79388	.76594	.3056	.2596	.6445	33
28	.60830	.79371	.76640	.3048	.2599	.6439	32
29	.60853	.79353	.76686	.3040	.2602	6433	31
30	.60876	.79335	.76733	1.3032	1.2605	1.6427	30
31	.60899	.79318	.76779	3024	.2607	.6420	29
32	.60922	.79300	.76825	.3016	.2610	.6414	28
33	.60945	.79282	.76871	.3009	.2613	.6408	27
34	.60968	.79264	.76918	.3001	.2616	.6402	26
35	.60991	.79247	.76964	1.2993	1.2619	1.6396	25
36	.61014	.79229	.77010	.2985	.2622	.6389	24
37	.61037	.79211	.77057	.2977	.2624	.6383	23
38	.61061	.79193	.77103	.2970	.2627	.6377	22
39	.61084	.79176	.77149	.2962	.2630	.6371	21
40	.61107	.79158	.77196	1.2954	1.2633	1.6365	20
41	.61130	.79140	.77242	.2946	.2636	.6359	19
42	.61153	.79122	.77289	.2938	.2639	.6352	18
43	.61176	.79104	.77335	.2931	.2641	.6346	17
44	.61199	.79087	.77382	.2923	.2644	.6340	16
45	.61222	.79069	.77428	1.2915	1.2647	1.6334	15
46	.61245	.79051	.77475	.2907	.2650	.6328	14
47	.61268	.79033	.77521	.2900	.2653	.6322	13
48	.61290	.79015	.77568	.2892	.2656	.6316	12
49	.61314	.78998	.77614	.2884	.2659	.6309	11
50	.61337	.78980	.77661	1.2876	1.2661	1.6303	10
51	.61360	.78962	.77708	2869	.2664	.6297	9
52	.61383	.78944	.77754	.2861	.2667	.6291	8
53	.61405	.78926	.77801	.2853	.2670	.6285	7
54	.61428	.78908	.77848	.2845	.2673	.6279	6
55	.61451	.78890	.77895	1.2838	1.2676	1.6273	5
56	.61474	.78873	.77941	.2830	.2679	.6267	4
57	.61497	.78855	.77988	.2822	.2681	.6261	3
58	.61520	.78837	.78035	.2815	.2684	.6255	2
59	.61543	.78819	.78082	.2807	.2687	.6249	1
60	.61566	.78801	.78128	1.2799	1.2690	1.6243	0
M	Cosine	Sine	Cotan.	Tangent	Cosec.	Secant	M

52°

38°

M	Sine	Cosine	Tangent	Cotan.	Secant	Cosec.	M
0	.61566	.78801	.78128	1.2799	1.2690	1.6243	60
1	.61589	.78783	.78175	.2792	.2693	.6237	59
2	.61612	.78765	.78222	.2784	.2696	.6231	58
3	.61635	.78747	.78269	.2776	.2699	.6224	57
4	.61658	.78729	.78316	.2769	.2702	.6218	56
5	.61681	.78711	.78363	1.2761	1.2705	1.6212	55
6	.61703	.78693	.78410	.2753	.2707	.6206	54
7	.61726	.78675	.78457	.2746	.2710	.6200	53
8	.61749	.78657	.78504	.2738	.2713	.6194	52
9	.61772	.78640	.78551	.2730	.2716	.6188	51
10	.61795	.78622	.78598	1.2723	1.2719	1.6182	50
11	.61818	.78604	.78645	.2715	.2722	.6176	49
12	.61841	.78586	.78692	.2708	.2725	.6170	48
13	.61864	.78568	.78739	.2700	.2728	.6164	47
14	.61886	.78550	.78786	.2692	.2731	.6159	46
15	.61909	.78532	.78834	1.2685	1.2734	1.6153	45
16	.61932	.78514	.78881	.2677	.2737	.6147	44
17	.61955	.78496	.78928	.2670	.2739	.6141	43
18	.61978	.78478	.78975	.2662	.2742	.6135	42
19	.62001	.78460	.79022	.2655	.2745	.6129	41
20	.62023	.78441	.79070	1.2647	1.2748	1.6123	40
21	.62046	.78423	.79117	.2639	.2751	.6117	39
22	.62069	.78405	.79164	.2632	.2754	.6111	38
23	.62092	.78387	.79212	.2624	.2757	.6105	37
24	.62115	.78369	.79259	.2617	.2760	.6099	36
25	.62137	.78351	.79306	1.2609	1.2763	1.6093	35
26	.62160	.78333	.79354	.2602	.2766	.6087	34
27	.62183	.78315	.79401	.2594	.2769	.6081	33
28	.62206	.78297	.79449	.2587	.2772	.6077	32
29	.62229	.78279	.79496	.2579	.2775	6070	31
30	.62251	.78261	.9543	1.2572	1.2778	1.6064	30
31	.62274	.78243	.79591	2564	.2781	.6058	29
32	.62297	.78224	.79639	.2557	.2784	.6052	28
33	.62320	.78206	.79686	.2549	.2787	.6046	27
34	.62342	.78188	.79734	.2542	.2790	.6040	26
35	.62365	.78170	.79781	1.2534	1.2793	1.6034	25
36	.62388	.78152	.79829	.2527	.2795	.6029	24
37	.62411	.78134	.79876	.2519	.2798	.6023	23
38	.62433	.78116	.79924	.2512	.2801	.6017	22
39	.62456	.78097	.79972	.2504	.2804	.6011	21
40	.62479	.78079	.80020	1.2497	1.2807	1.6005	20
41	.62501	.78061	.80067	.2489	.2810	.6000	19
42	.62524	.78043	.80115	.2482	.2813	.5994	18
43	.62547	.78025	.80163	.2475	.2816	.5988	17
44	.62570	.78007	.80211	.2467	.2819	.5982	16
45	.62592	.77988	.80258	1.2460	1.2822	1.5976	15
46	.62615	.77970	.80306	.2452	.2825	.5971	14
47	.62638	.77952	.80354	.2445	.2828	.5965	13
48	.62660	.77934	.80402	.2437	.2831	.5959	12
49	.62683	.77915	.80450	.2430	.2834	.5953	11
50	.62706	.77897	.80498	1.2423	1.2837	1.5947	10
51	.62728	.77879	.80546	2415	.2840	.5942	9
52	.62751	.77861	.80594	.2408	.2843	.5936	8
53	.62774	.77842	.80642	.2400	.2846	.5930	7
54	.62796	.77824	.80690	.2393	.2849	.5924	6
55	.62819	.77806	.80738	1.2386	1.2852	1.5919	5
56	.62841	.77788	.80786	.2378	.2855	.5913	4
57	.62864	.77769	.80834	.2371	.2858	.5907	3
58	.62887	.77751	.80882	.2364	.2861	.5901	2
59	.62909	.77733	.80930	.2356	.2864	.5896	1
60	.62932	.77715	.80978	1.2349	1.2867	1.5890	0
M	Cosine	Sine	Cotan.	Tangent	Cosec.	Secant	M

51°

39°

M	Sine	Cosine	Tangent	Cotan.	Secant	Cosec.	M
0	.62932	.77715	.80978	1.2349	1.2867	1.5890	60
1	.62955	.77696	.81026	.2342	.2871	.5884	59
2	.62977	.77678	.81075	.2334	.2874	.5879	58
3	.63000	.77660	.81123	.2327	.2877	.5873	57
4	.63022	.77641	.81171	.2320	.2880	.5867	56
5	.63045	.77623	.81219	1.2312	1.2883	1.5862	55
6	.63067	.77605	.81268	.2305	.2886	.5856	54
7	.63090	.77586	.81316	.2297	.2889	.5850	53
8	.63113	.77568	.81364	.2290	.2892	.5845	52
9	.63135	.77549	.81413	.2283	.2895	.5839	51
10	.63158	.77531	.81461	1.2276	1.2898	1.5833	50
11	.63180	.77513	.81509	.2268	.2901	.5828	49
12	.63203	.77494	.81558	.2261	.2904	.5822	48
13	.63225	.77476	.81606	.2254	.2907	.5816	47
14	.63248	.77458	.81655	.2247	.2910	.5811	46
15	.63270	.77439	.81703	1.2239	1.2913	1.5805	45
16	.63293	.77421	.81752	.2232	.2916	.5799	44
17	.63315	.77402	.81800	.2225	.2919	.5794	43
18	.63338	.77384	.81849	.2218	.2922	.5788	42
19	.63360	.77365	.81898	.2210	.2926	.5783	41
20	.63383	.77347	.81946	1.2203	1.2929	1.5777	40
21	.63405	.77329	.81995	.2196	.2932	.5771	39
22	.63428	.77310	.82043	.2189	.2935	.5766	38
23	.63450	.77292	.82092	.2181	.2938	.5760	37
24	.63473	.77273	.82141	.2174	.2941	.5755	36
25	.63495	.77255	.82190	1.2167	1.2944	1.5749	35
26	.63518	.77236	.82238	.2160	.2947	.5743	34
27	.63540	.77218	.82287	.2152	.2950	.5738	33
28	.63563	.77199	.82336	.2145	.2953	.5732	32
29	.63585	.77181	.82385	.2138	.2956	5727	31
30	.63608	.77162	82434	1.2131	1.2960	1.5721	30
31	.63630	.77144	.82482	2124	.2963	.5716	29
32	.63653	.77125	.82531	.2117	.2966	.5710	28
33	.63675	.77107	.82580	.2109	.2969	.5705	27
34	.63697	.77088	.82629	.2102	.2972	.5699	26
35	.63720	.77070	.82678	1.2095	1.2975	1.5694	25
36	.63742	.77051	.82727	.2088	.2978	.5688	24
37	.63765	.77033	.82776	.2081	.2981	.5683	23
38	.63787	.77014	.82825	.2074	.2985	.5677	22
39	.63810	.76996	.82874	.2066	.2988	.5672	21
40	.63832	.76977	.82923	1.2059	1.2991	1.5666	20
41	.63854	.76958	.82972	.2052	.2994	.5661	19
42	.63877	.76940	.83022	.2045	.2997	.5655	18
43	.63899	.76921	.83071	.2038	.3000	.5650	17
44	.63921	.76903	.83120	.2031	.3003	.5644	16
45	.63944	.76884	.83169	1.2024	1.3006	1.5639	15
46	.63966	.76865	.83218	.2016	.3010	.5633	14
47	.63989	.76847	.83267	.2009	.3013	.5628	13
48	.64011	.76828	.83317	.2002	.3016	.5622	12
49	.64033	.76810	.83366	.1995	.3019	.5617	11
50	.64056	.76791	.83415	1.1988	1.3022	1.5611	10
51	.64078	.76772	.83465	1981	.3025	.5606	9
52	.64100	.76754	.83514	.1974	.3029	.5600	8
53	.64123	.76735	.83563	.1967	.3032	.5595	7
54	.64145	.76716	.83613	.1960	.3035	.5590	6
55	.64167	.76698	.83662	1.1953	1.3038	1.5584	5
56	.64189	.76679	.83712	.1946	.3041	.5579	4
57	.64212	.76660	.83761	.1939	.3044	.5573	3
58	.64234	.76642	.83811	.1932	.3048	.5568	2
59	.64256	.76623	.83860	.1924	.3051	.5563	1
60	.64279	.76604	.83910	1.1917	1.3054	1.5557	0
M	Cosine	Sine	Cotan.	Tangent	Cosec.	Secant	M

50°

40°

M	Sine	Cosine	Tangent	Cotan.	Secant	Cosec.	M
0	.64279	.76604	.83910	1.1917	1.3054	1.5557	60
1	.64301	.76586	.83959	.1910	.3057	.5552	59
2	.64323	.76567	.84009	.1903	.3060	.5546	58
3	.64345	.76548	.84059	.1896	.3064	.5541	57
4	.64368	.76530	.84108	.1889	.3067	.5536	56
5	.64390	.76511	.84158	1.1882	1.3070	1.5530	55
6	.64412	.76492	.84208	.1875	.3073	.5525	54
7	.64435	.76473	.84257	.1868	.3076	.5520	53
8	.64457	.76455	.84307	.1861	.3080	.5514	52
9	.64479	.76436	.84357	.1854	.3083	.5509	51
10	.64501	.76417	.84407	1.1847	1.3086	1.5503	50
11	.64523	.76398	.84457	.1840	.3089	.5498	49
12	.64546	.76380	.84506	.1833	.3092	.5493	48
13	.64568	.76361	.84556	.1826	.3096	.5487	47
14	.64590	.76342	.84606	.1819	.3099	.5482	46
15	.64612	.76323	.84656	1.1812	1.3102	1.5477	45
16	.64635	.76304	.84706	.1805	.3105	.5471	44
17	.64657	.76286	.84756	.1798	.3109	.5466	43
18	.64679	.76267	.84806	.1791	.3112	.5461	42
19	.64701	.76248	.84856	.1785	.3115	.5456	41
20	.64723	.76229	.84906	1.1778	1.3118	1.5450	40
21	.64745	.76210	.84956	.1771	.3121	.5445	39
22	.64768	.76191	.85006	.1764	.3125	.5440	38
23	.64790	.76173	.85056	.1757	.3128	.5434	37
24	.64812	.76154	.85107	.1750	.3131	.5429	36
25	.64834	.76135	.85157	1.1743	1.3134	1.5424	35
26	.64856	.76116	.85207	.1736	.3138	.5419	34
27	.64878	.76097	.85257	.1729	.3141	.5413	33
28	.64900	.76078	.85307	.1722	.3144	.5408	32
29	.64923	.76059	.85358	.1715	.3148	5403	31
30	.64945	.76041	85408	1.1708	1.3151	1.5398	30
31	.64967	.76022	.85458	1702	.3154	.5392	29
32	.64989	.76003	.85509	.1695	.3157	.5387	28
33	.65011	.75984	.85559	.1688	.3161	.5382	27
34	.65033	.75965	.85609	.1681	.3164	.5377	26
35	.65055	.75946	.85660	1.1674	1.3167	1.5371	25
36	.65077	.75927	.85710	.1667	.3170	.5366	24
37	.65100	.75908	.85761	.1660	.3174	.5361	23
38	.65121	.75889	.85811	.1653	.3177	.5356	22
39	.65144	.75870	.85862	.1647	.3180	.5351	21
40	.65166	.75851	.85912	1.1640	1.3184	1.5345	20
41	.65188	.75832	.85963	.1633	.3187	.5340	19
42	.65210	.75813	.86013	.1626	.3190	.5335	18
43	.65232	.75794	.86064	.1619	.3193	.5330	17
44	.65254	.75775	.86115	.1612	.3197	.5325	16
45	.65276	.75756	.86165	1.1605	1.3200	1.5319	15
46	.65298	.75737	.86216	.1599	.3203	.5314	14
47	.65320	.75718	.86267	.1592	.3207	.5309	13
48	.65342	.75700	.86318	.1585	.3210	.5304	12
49	.65364	.75680	.86368	.1578	.3213	.5290	11
50	.65386	.75661	.86419	1.1571	1.3217	1.5294	10
51	.65408	.75642	.86470	1565	.3220	.5289	9
52	.65430	.75623	.86521	.1558	.3223	.5283	8
53	.65452	.75604	.86572	.1551	.3227	.5278	7
54	.65474	.75585	.86623	.1544	.3230	.5273	6
55	.65496	.75566	.86674	1.1537	1.3233	1.5268	5
56	.65518	.75547	.86725	.1531	.3237	.5263	4
57	.65540	.75528	.86775	.1524	.3240	.5258	3
58	.65562	.75509	.86826	.1517	.3243	.5253	2
59	.65584	.75490	.86878	.1510	.3247	.5248	1
60	.65606	.75471	.86929	1.1504	1.3250	1.5242	0
M	Cosine	Sine	Cotan.	Tangent	Cosec.	Secant	M

49°

41°

M	Sine	Cosine	Tangent	Cotan.	Secant	Cosec.	M
0	.65606	.75471	.86929	1.1504	1.3250	1.5242	60
1	.65628	.75452	.86980	.1497	.3253	.5237	59
2	.65650	.75433	.87031	.1490	.3257	.5232	58
3	.65672	.75414	.87082	.1483	.3260	.5227	57
4	.65694	.75394	.87133	.1477	.3263	.5222	56
5	.65716	.75375	.87184	1.1470	1.3267	1.5217	55
6	.65737	.75356	.87235	.1463	.3270	.5212	54
7	.65759	.75337	.87287	.1456	.3274	.5207	53
8	.65781	.75318	.87338	.1450	.3277	.5202	52
9	.65803	.75299	.87389	.1443	.3280	.5197	51
10	.65825	.75280	.87441	1.1436	1.3284	1.5192	50
11	.65847	.75261	.87492	.1430	.3287	.5187	49
12	.65869	.75241	.87543	.1423	.3290	.5182	48
13	.65891	.75222	.87595	.1416	.3294	.5177	47
14	.65913	.75203	.87646	.1409	.3297	.5171	46
15	.65934	.75184	.87698	1.1403	1.3301	1.5166	45
16	.65956	.75165	.87749	.1396	.3304	.5161	44
17	.65978	.75146	.87801	.1389	.3307	.5156	43
18	.66000	.75126	.87852	.1383	.3311	.5151	42
19	.66022	.75107	.87904	.1376	.3314	.5146	41
20	.66044	.75088	.87955	1.1369	1.3318	1.5141	40
21	.66066	.75069	.88007	.1363	.3321	.5136	39
22	.66087	.75049	.88058	.1356	.3324	.5131	38
23	.66109	.75030	.88110	.1349	.3328	.5126	37
24	.66131	.75011	.88162	.1343	.3331	.5121	36
25	.66153	.74992	.88213	1.1336	1.3335	1.5116	35
26	.66175	.74973	.88265	.1329	.3338	.5111	34
27	.66197	.74953	.88317	.1323	.3342	.5106	33
28	.66218	.74934	.88369	.1316	.3345	.5101	32
29	.66240	.74915	.88421	.1309	.3348	5096	31
30	.66262	.74895	88472	1.1303	1.3352	1.5092	30
31	.66284	.74876	.88524	1296	.3355	.5087	29
32	.66305	.74857	.88576	.1290	.3359	.5082	28
33	.66327	.74838	.88628	.1283	.3362	.5077	27
34	.66349	.74818	.88680	.1276	.3366	.5072	26
35	.66371	.74799	.88732	1.1270	1.3369	1.5067	25
36	.66393	.74780	.88784	.1263	.3372	.5062	24
37	.66414	.74760	.88836	.1257	.3376	.5057	23
38	.66436	.74741	.88888	.1250	.3379	.5052	22
39	.66458	.74722	.88940	.1243	.3383	.5047	21
40	.66479	.74702	.88992	1.1237	1.3386	1.5042	20
41	.66501	.74683	.89044	.1230	.3390	.5037	19
42	.66523	.74664	.89097	.1224	.3393	.5032	18
43	.66545	.74644	.89149	.1217	.3397	.5027	17
44	.66566	.74625	.89201	.1211	.3400	.5022	16
45	.66588	.74606	.89253	1.1204	1.3404	1.5018	15
46	.66610	.74586	.89306	.1197	.3407	.5013	14
47	.66631	.74567	.89358	.1191	.3411	.5008	13
48	.66653	.74548	.89410	.1184	.3414	.5003	12
49	.66675	.74528	.89463	.1178	.3418	.4998	11
50	.66697	.74509	.89515	1.1171	1.3421	1.4993	10
51	.66718	.74489	.89567	1165	.3425	.4988	9
52	.66740	.74470	.89620	.1158	.3428	.4983	8
53	.66762	.74450	.89672	.1152	.3432	.4979	7
54	.66783	.74431	.89725	.1145	.3435	.4974	6
55	.66805	.74412	.89777	1.1139	1.3439	1.4969	5
56	.66826	.74392	.89830	.1132	.3442	.4964	4
57	.66848	.74373	.89882	.1126	.3446	.4959	3
58	.66870	.74353	.89935	.1119	.3449	.4954	2
59	.66891	.74334	.89988	.1113	.3453	.4949	1
60	.66913	.74314	.90040	1.1106	1.3456	1.4945	0
M	Cosine	Sine	Cotan.	Tangent	Cosec.	Secant	M

48°

42°

M	Sine	Cosine	Tangent	Cotan.	Secant	Cosec.	M
0	.66913	.74314	.90040	1.1106	1.3456	1.4945	60
1	.66935	.74295	.90093	.1100	.3460	.4940	59
2	.66956	.74275	.90146	.1093	.3463	.4935	58
3	.66978	.74256	.90198	.1086	.3467	.4930	57
4	.66999	.74236	.90251	.1080	.3470	.4925	56
5	.67021	.74217	.90304	1.1074	1.3474	1.4921	55
6	.67043	.74197	.90357	.1067	.3477	.4916	54
7	.67064	.74178	.90410	.1061	.3481	.4911	53
8	.67086	.74158	.90463	.1054	.3485	.4906	52
9	.67107	.74139	.90515	.1048	.3488	.4901	51
10	.67129	.74119	.90568	1.1041	1.3492	1.4897	50
11	.67150	.74100	.90621	.1035	.3495	.4892	49
12	.67172	.74080	.90674	.1028	.3499	.4887	48
13	.67194	.74061	.90727	.1022	.3502	.4882	47
14	.67215	.74041	.90780	.1015	.3506	.4877	46
15	.67237	.74022	.90834	1.1009	1.3509	1.4873	45
16	.67258	.74002	.90887	.1003	.3513	.4868	44
17	.67280	.73983	.90940	.0996	.3517	.4863	43
18	.67301	.73963	.90993	.0990	.3520	.4858	42
19	.67323	.73943	.91046	.0983	.3524	.4854	41
20	.67344	.73924	.91099	1.0977	1.3527	1.4849	40
21	.67366	.73904	.91153	.0971	.3531	.4844	39
22	.67387	.73885	.91206	.0964	.3534	.4839	38
23	.67409	.73865	.91259	.0958	.3538	.4835	37
24	.67430	.73845	.91312	.0951	.3542	.4830	36
25	.67452	.73826	.91366	1.0945	1.3545	1.4825	35
26	.67473	.73806	.91419	.0939	.3549	.4821	34
27	.67495	.73787	.91473	.0932	.3552	.4816	33
28	.67516	.73767	.91526	.0926	.3556	.4811	32
29	.67537	.73747	.91580	.0919	.3560	4806	31
30	.67559	.73728	91633	1.0913	1.3563	1.4802	30
31	.67580	.73708	.91687	0907	.3567	.4797	29
32	.67602	.73688	.91740	.0900	.3571	.4792	28
33	.67623	.73669	.91794	.0894	.3574	.4788	27
34	.67645	.73649	.91847	.0888	.3578	.4783	26
35	.67666	.73629	.91901	1.0881	1.3581	1.4778	25
36	.67688	.73610	.91955	.0875	.3585	.4774	24
37	.67709	.73590	.92008	.0868	.3589	.4769	23
38	.67730	.73570	.92062	.0862	.3592	.4764	22
39	.67752	.73551	.92116	.0856	.3596	.4760	21
40	.67773	.73531	.92170	1.0849	1.3600	1.4755	20
41	.67794	.73511	.92223	.0843	.3603	.4750	19
42	.67816	.73491	.92277	.0837	.3607	.4746	18
43	.67837	.73472	.92331	.0830	.3611	.4741	17
44	.67859	.73452	.92385	.0824	.3614	.4736	16
45	.67880	.73432	.92439	1.0818	1.3618	1.4732	15
46	.67901	.73412	.92493	.0812	.3622	.4727	14
47	.67923	.73393	.92547	.0805	.3625	.4723	13
48	.67944	.73373	.92601	.0799	.3629	.4718	12
49	.67965	.73353	.92655	.0793	.3633	.4713	11
50	.67987	.73333	.92709	1.0786	1.3636	1.4709	10
51	.68008	.73314	.92763	0780	.3640	.4704	9
52	.68029	.73294	.92817	.0774	.3644	.4699	8
53	.68051	.73274	.92871	.0767	.3647	.4695	7
54	.68072	.73254	.92926	.0761	.3651	.4690	6
55	.68093	.73234	.92980	1.0755	1.3655	1.4686	5
56	.68115	.73215	.93034	.0749	.3658	.4681	4
57	.68136	.73195	.93088	.0742	.3662	.4676	3
58	.68157	.73175	.93143	.0736	.3666	.4672	2
59	.68178	.73155	.93197	.0730	.3669	.4667	1
60	.68200	.73135	.93251	1.0724	1.3673	1.4663	0
M	Cosine	Sine	Cotan.	Tangent	Cosec.	Secant	M

47°

43°

M	Sine	Cosine	Tangent	Cotan.	Secant	Cosec.	M
0	.68200	.73135	.93251	1.0724	1.3673	1.4663	60
1	.68221	.73115	.93306	.0717	.3677	.4658	59
2	.68242	.73096	.93360	.0711	.3681	.4654	58
3	.68264	.73076	.93415	.0705	.3684	.4649	57
4	.68285	.73056	.93469	.0699	.3688	.4644	56
5	.68306	.73036	.93524	1.0692	1.3692	1.4640	55
6	.68327	.73016	.93578	.0686	.3695	.4635	54
7	.68349	.72996	.93633	.0680	.3699	.4631	53
8	.68370	.72976	.93687	.0674	.3703	.4626	52
9	.68391	.72956	.93742	.0667	.3707	.4622	51
10	.68412	.72937	.93797	1.0661	1.3710	1.4617	50
11	.68433	.72917	.93851	.0655	.3714	.4613	49
12	.68455	.72897	.93906	.0649	.3718	.4608	48
13	.68476	.72877	.93961	.0643	.3722	.4604	47
14	.68497	.72857	.94016	.0636	.3725	.4599	46
15	.68518	.72837	.94071	1.0630	1.3729	1.4595	45
16	.68539	.72817	.94125	.0624	.3733	.4590	44
17	.68561	.72797	.94180	.0618	.3737	.4586	43
18	.68582	.72777	.94235	.0612	.3740	.4581	42
19	.68603	.72757	.94290	.0605	.3744	.4577	41
20	.68624	.72737	.94345	1.0599	1.3748	1.4572	40
21	.68645	.72717	.94400	.0593	.3752	.4568	39
22	.68666	.72697	.94455	.0587	.3756	.4563	38
23	.68688	.72677	.94510	.0581	.3759	.4559	37
24	.68709	.72657	.94565	.0575	.3763	.4554	36
25	.68730	.72637	.94620	1.0568	1.3767	1.4550	35
26	.68751	.72617	.94675	.0562	.3771	.4545	34
27	.68772	.72597	.94731	.0556	.3774	.4541	33
28	.68793	.72577	.94786	.0550	.3778	.4536	32
29	.68814	.72557	.94841	.0544	.3782	4532	31
30	.68835	.72537	94896	1.0538	1.3786	1.4527	30
31	.68856	.72517	.94952	0532	.3790	.4523	29
32	.68878	.72497	.95007	.0525	.3794	.4518	28
33	.68899	.72477	.95062	.0519	.3797	.4514	27
34	.68920	.72457	.95118	.0513	.3801	.4510	26
35	.68941	.72437	.95173	1.0507	1.3805	1.4505	25
36	.68962	.72417	.95229	.0501	.3809	.4501	24
37	.68983	.72397	.95284	.0495	.3813	.4496	23
38	.69004	.72377	.95340	.0489	.3816	.4492	22
39	.69025	.72357	.95395	.0483	.3820	.4487	21
40	.69046	.72337	.95451	1.0476	1.3824	1.4483	20
41	.69067	.72317	.95506	.0470	.3828	.4479	19
42	.69088	.72297	.95562	.0464	.3832	.4474	18
43	.69109	.72277	.95618	.0458	.3836	.4470	17
44	.69130	.72256	.95673	.0452	.3839	.4465	16
45	.69151	.72236	.95729	1.0446	1.3843	1.4461	15
46	.69172	.72216	.95785	.0440	.3847	.4457	14
47	.69193	.72196	.95841	.0434	.3851	.4452	13
48	.69214	.72176	.95896	.0428	.3855	.4448	12
49	.69235	.72156	.95952	.0422	.3859	.4443	11
50	.69256	.72136	.96008	1.0416	1.3863	1.4439	10
51	.69277	.72115	.96064	0410	.3867	.4435	9
52	.69298	.72095	.96120	.0404	.3870	.4430	8
53	.69319	.72075	.96176	.0397	.3874	.4426	7
54	.69340	.72055	.96232	.0391	.3878	.4422	6
55	.69361	.72035	.96288	1.0385	1.3882	1.4417	5
56	.69382	.72015	.96344	.0379	.3886	.4413	4
57	.69403	.71994	.96400	.0373	.3890	.4408	3
58	.69424	.71974	.96456	.0367	.3894	.4404	2
59	.69445	.71954	.96513	.0361	.3898	.4400	1
60	.69466	.71934	.96569	1.0355	1.3902	1.4395	0
M	Cosine	Sine	Cotan.	Tangent	Cosec.	Secant	M

46°

44°

M	Sine	Cosine	Tangent	Cotan.	Secant	Cosec.	M
0	.69466	.71934	.96569	1.0355	1.3902	1.4395	60
1	.69487	.71914	.96625	.0349	.3905	.4391	59
2	.69508	.71893	.96681	.0343	.3909	.4387	58
3	.69528	.71873	.96738	.0337	.3913	.4382	57
4	.69549	.71853	.96794	.0331	.3917	.4378	56
5	.69570	.71833	.96850	1.0325	1.3921	1.4374	55
6	.69591	.71813	.96907	.0319	.3925	.4370	54
7	.69612	.71792	.96963	.0313	.3929	.4365	53
8	.69633	.71772	.97020	.0307	.3933	.4361	52
9	.69654	.71752	.97076	.0301	.3937	.4357	51
10	.69675	.71732	.97133	1.0295	1.3941	1.4352	50
11	.69696	.71711	.97189	.0289	.3945	.4348	49
12	.69716	.71691	.97246	.0283	.3949	.4344	48
13	.69737	.71671	.97302	.0277	.3953	.4339	47
14	.69758	.71650	.97359	.0271	.3957	.4335	46
15	.69779	.71630	.97416	1.0265	1.3960	1.4331	45
16	.69800	.71610	.97472	.0259	.3964	.4327	44
17	.69821	.71589	.97529	.0253	.3968	.4322	43
18	.69841	.71569	.97586	.0247	.3972	.4318	42
19	.69862	.71549	.97643	.0241	.3976	.4314	41
20	.69883	.71529	.97700	1.0235	1.3980	1.4310	40
21	.69904	.71508	.97756	.0229	.3984	.4305	39
22	.69925	.71488	.97813	.0223	.3988	.4301	38
23	.69945	.71468	.97870	.0218	.3992	.4297	37
24	.69966	.71447	.97927	.0212	.3996	.4292	36
25	.69987	.71427	.97984	1.0206	1.4000	1.4288	35
26	.70008	.71406	.98041	.0200	.4004	.4284	34
27	.70029	.71386	.98098	.0194	.4008	.4280	33
28	.70049	.71366	.98155	.0188	.4012	.4276	32
29	.70070	.71345	.98212	.0182	.4016	4271	31
30	.70091	.71325	98270	1.0176	1.4020	1.4267	30
31	.70112	.71305	.98327	0170	.4024	.4263	29
32	.70132	.71284	.98384	.0164	.4028	.4259	28
33	.70153	.71264	.98441	.0158	.4032	.4254	27
34	.70174	.71243	.98499	.0152	.4036	.4250	26
35	.70194	.71223	.98556	1.0146	1.4040	1.4246	25
36	.70215	.71203	.98613	.0141	.4044	.4242	24
37	.70236	.71182	.98671	.0135	.4048	.4238	23
38	.70257	.71162	.98728	.0129	.4052	.4233	22
39	.70277	.71141	.98786	.0123	.4056	.4229	21
40	.70298	.71121	.98843	1.0117	1.4060	1.4225	20
41	.70319	.71100	.98901	.0111	.4065	.4221	19
42	.70339	.71080	.98958	.0105	.4069	.4217	18
43	.70360	.71059	.99016	.0099	.4073	.4212	17
44	.70381	.71039	.99073	.0093	.4077	.4208	16
45	.70401	.71018	.99131	1.0088	1.4081	1.4204	15
46	.70422	.70998	.99189	.0082	.4085	.4200	14
47	.70443	.70977	.99246	.0076	.4089	.4196	13
48	.70463	.70957	.99304	.0070	.4093	.4192	12
49	.70484	.70936	.99362	.0064	.4097	.4188	11
50	.70505	.70916	.99420	1.0058	1.4101	1.4183	10
51	.70525	.70895	.99478	0052	.4105	.4179	9
52	.70546	.70875	.99536	.0047	.4109	.4175	8
53	.70566	.70854	.99593	.0041	.4113	.4171	7
54	.70587	.70834	.99651	.0035	.4117	.4167	6
55	.70608	.70813	.99709	1.0029	1.4122	1.4163	5
56	.70628	.70793	.99767	.0023	.4126	.4159	4
57	.70649	.70772	.99826	.0017	.4130	.4154	3
58	.70669	.70752	.99884	.0012	.4134	.4150	2
59	.70690	.70731	.99942	.0006	.4138	.4146	1
60	.70711	.70711	1.0000	1.0000	1.4142	1.4142	0
M	Cosine	Sine	Cotan.	Tangent	Cosec.	Secant	M

45°

[Note how the "Trig Tables" coincide with the "Scientific Calculator". A good way to learn how to use the trig tables in the book, is to use them in conjunction with a scientific calculator. Remember, if the batteries in the scientific calculator go dead - these tables will be real important, and there will be times when a scientific calculator will not be readily available.]

Note: Make sure the Calculator is in the DEG mode.
The display should show: DEG 0.

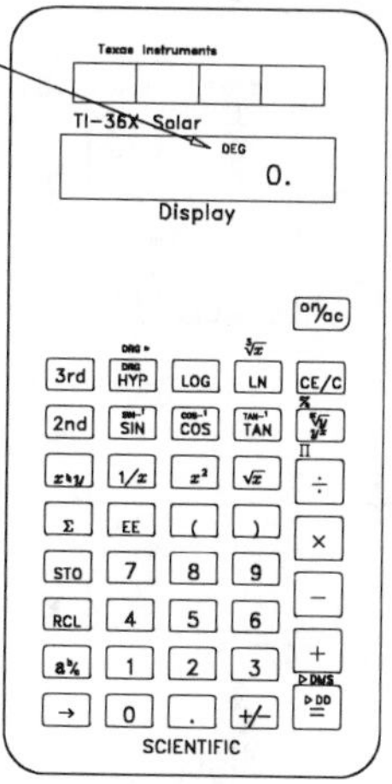

<u>How to Calculate the Functions of Angles:</u>

1. Calculate the Sine, Cosine, and Tangent of angles.

Example 1. Find the SIN of 45° angle.

Select: [4] [5] [SIN]

The display should show: DEG 0.707106781 (Display)

Example 2. Find the COS of 52° angle.

Select: [5] [2] [COS]

The display should show: DEG 0.615661475 (Display)

Example 3. Find the TAN of 15° angle.

Select: [1] [5] [TAN]

The display should show: DEG 0.267949192 (Display)

<u>Calculate the Angles from the Functions of Angles:</u>

2. Calculate the Angle from the Sine, Cosine, and Tangent angle functions.

Example 1. Find the Angle from the SIN. 0.707106781 (Carry Decimal 3–5 Places).

Select: [.] [7] [0] [7] [1] [2nd] [SIN]

The display should show: 45° — DEG 44.99945053 (Display)

Example 2. Find the Angle from the COS. 0.615661475 (Carry Decimal 3–5 Places).

Select: [.] [6] [1] [5] [6] [2nd] [COS]

The display should show: 52° — DEG 52.0044697 (Display)

Example 3. Find the Angle from the TAN. 0.267949192 (Carry Decimal 3–5 Places).

Select: [.] [2] [6] [7] [9] [2nd] [TAN]

The display should show: 15° — DEG 14.99737025 (Display)

[Note how the Functions of Angles and the Angles themselves relate to each other. This can also be verified in the Trig Tables, pages 220 – 264.]

[Note: Most scientific calculators operate similar to the one above.]

N1	◂ Task No.	TAPE MEASURE

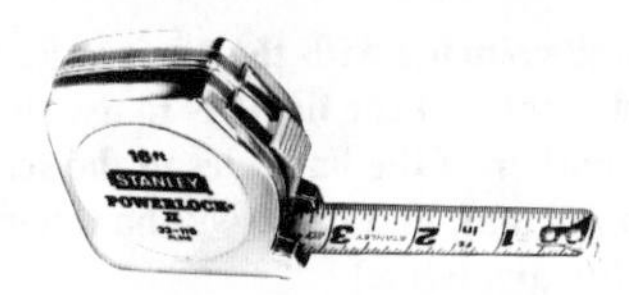

The American (English) Tape Measure

How the Tape Measure is Divided

[Note: 12 inches = 1 foot]

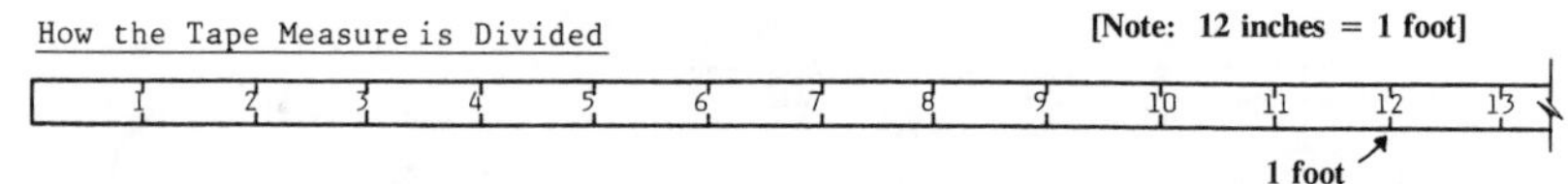

The Inch and Its Fractions

[3 feet = 1 yard]

[5280 ft. = 1 mile]

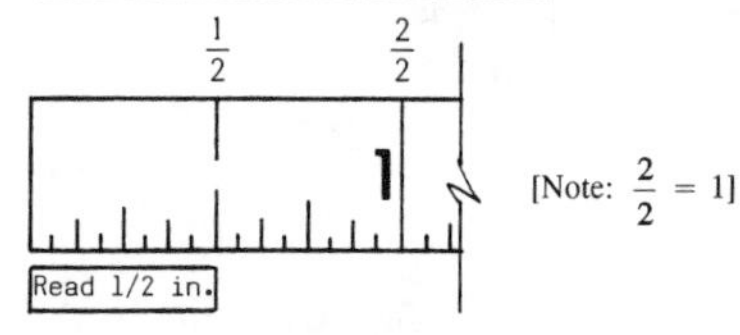

[Note: $\frac{2}{2} = 1$]

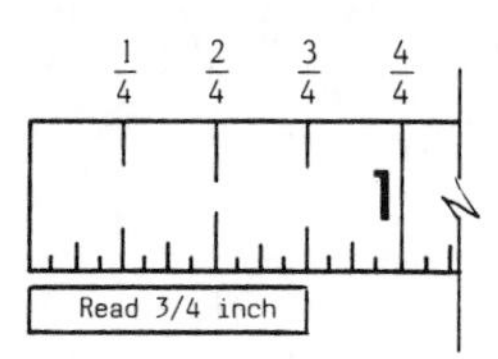

[Note: $\frac{2}{4} = \frac{1}{2}$ and $\frac{4}{4} = 1$]

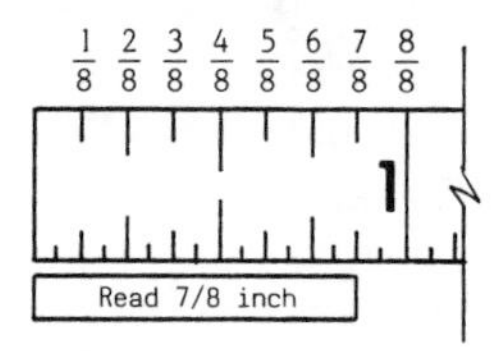

[Note: $\frac{2}{8} = \frac{1}{4}$, $\frac{4}{8} = \frac{1}{2}$, $\frac{6}{8} = \frac{3}{4}$, and $\frac{8}{8} = 1$]

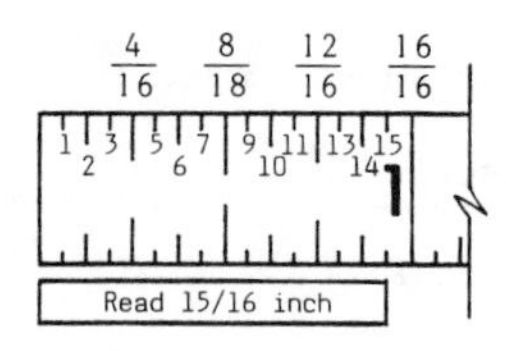

[Note: $\frac{2}{16} = \frac{1}{8}$, $\frac{4}{16} = \frac{1}{4}$, $\frac{6}{16} = \frac{3}{8}$, $\frac{8}{16} = \frac{1}{2}$, $\frac{10}{16} = \frac{5}{8}$, $\frac{12}{16} = \frac{3}{4}$, $\frac{14}{16} = \frac{7}{8}$, and $\frac{16}{16} = 1$]

[Notice that the upper part of the tape measure is usually divided into 16 equal parts.

Usually, the 1/2" lines are longer than the 1/4" lines, which are longer than the 1/8" lines, and so on.

[Notice that the lower part of the tape measure is usually divided into 32 equal parts per inch. On most tape measures, this is true for the first 12 inches only. **See note page 269.]**

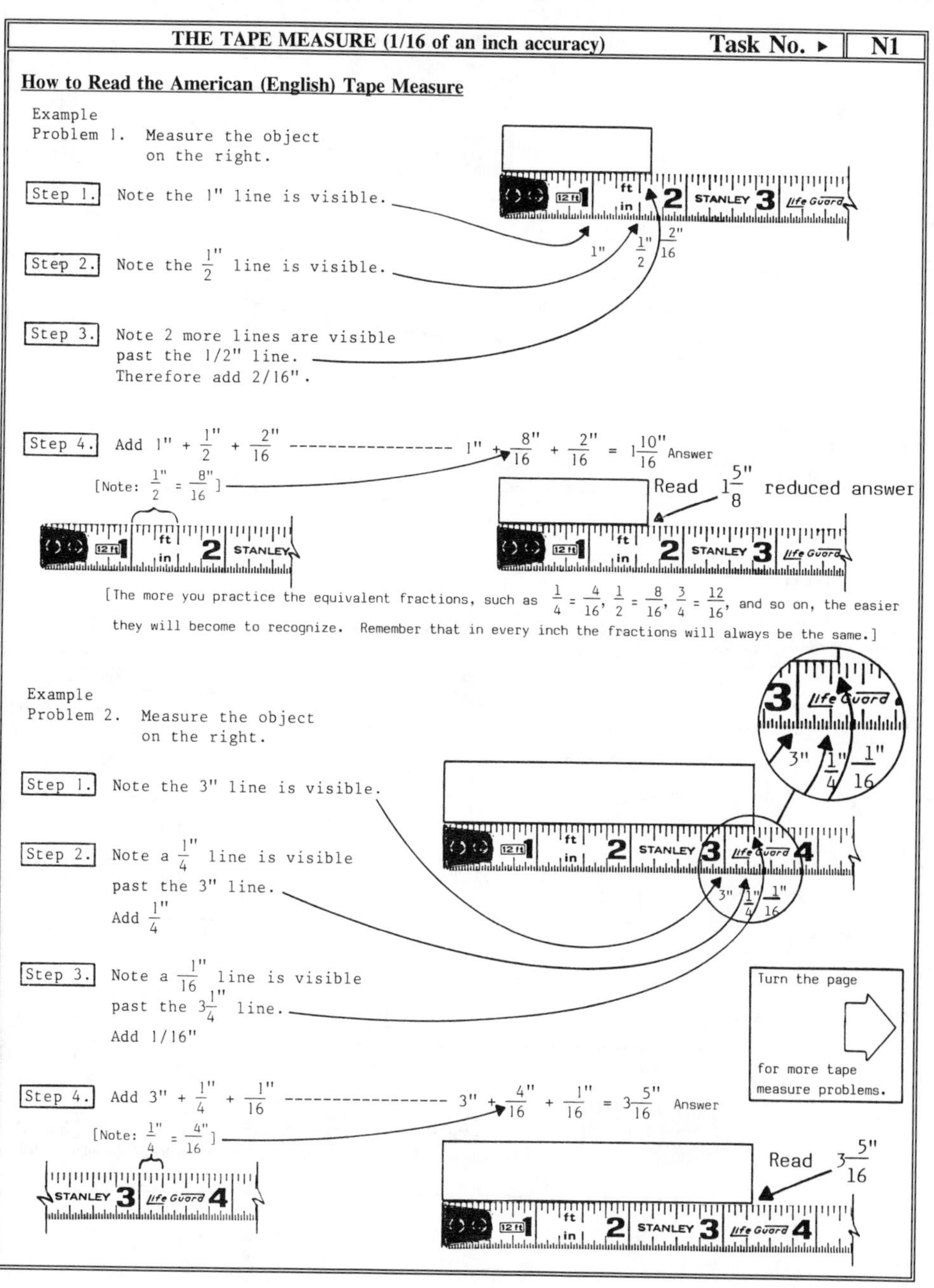

THE TAPE MEASURE (1/16 of an inch accuracy)	Task No. ▸	N1

How to Read the American (English) Tape Measure

Example Problem 1. Measure the object on the right.

Step 1. Note the 1" line is visible.

Step 2. Note the $\frac{1}{2}$" line is visible.

Step 3. Note 2 more lines are visible past the 1/2" line. Therefore add 2/16".

Step 4. Add 1" + $\frac{1}{2}$" + $\frac{2}{16}$" ---------------- 1" + $\frac{8}{16}$" + $\frac{2}{16}$" = $1\frac{10}{16}$" Answer

[Note: $\frac{1}{2}$" = $\frac{8}{16}$"]

Read $1\frac{5}{8}$" reduced answer

[The more you practice the equivalent fractions, such as $\frac{1}{4} = \frac{4}{16}$, $\frac{1}{2} = \frac{8}{16}$, $\frac{3}{4} = \frac{12}{16}$, and so on, the easier they will become to recognize. Remember that in every inch the fractions will always be the same.]

Example Problem 2. Measure the object on the right.

Step 1. Note the 3" line is visible.

Step 2. Note a $\frac{1}{4}$" line is visible past the 3" line. Add $\frac{1}{4}$"

Step 3. Note a $\frac{1}{16}$" line is visible past the $3\frac{1}{4}$" line. Add 1/16"

Step 4. Add 3" + $\frac{1}{4}$" + $\frac{1}{16}$" ---------------- 3" + $\frac{4}{16}$" + $\frac{1}{16}$" = $3\frac{5}{16}$" Answer

[Note: $\frac{1}{4}$" = $\frac{4}{16}$"]

Read $3\frac{5}{16}$"

Turn the page for more tape measure problems.

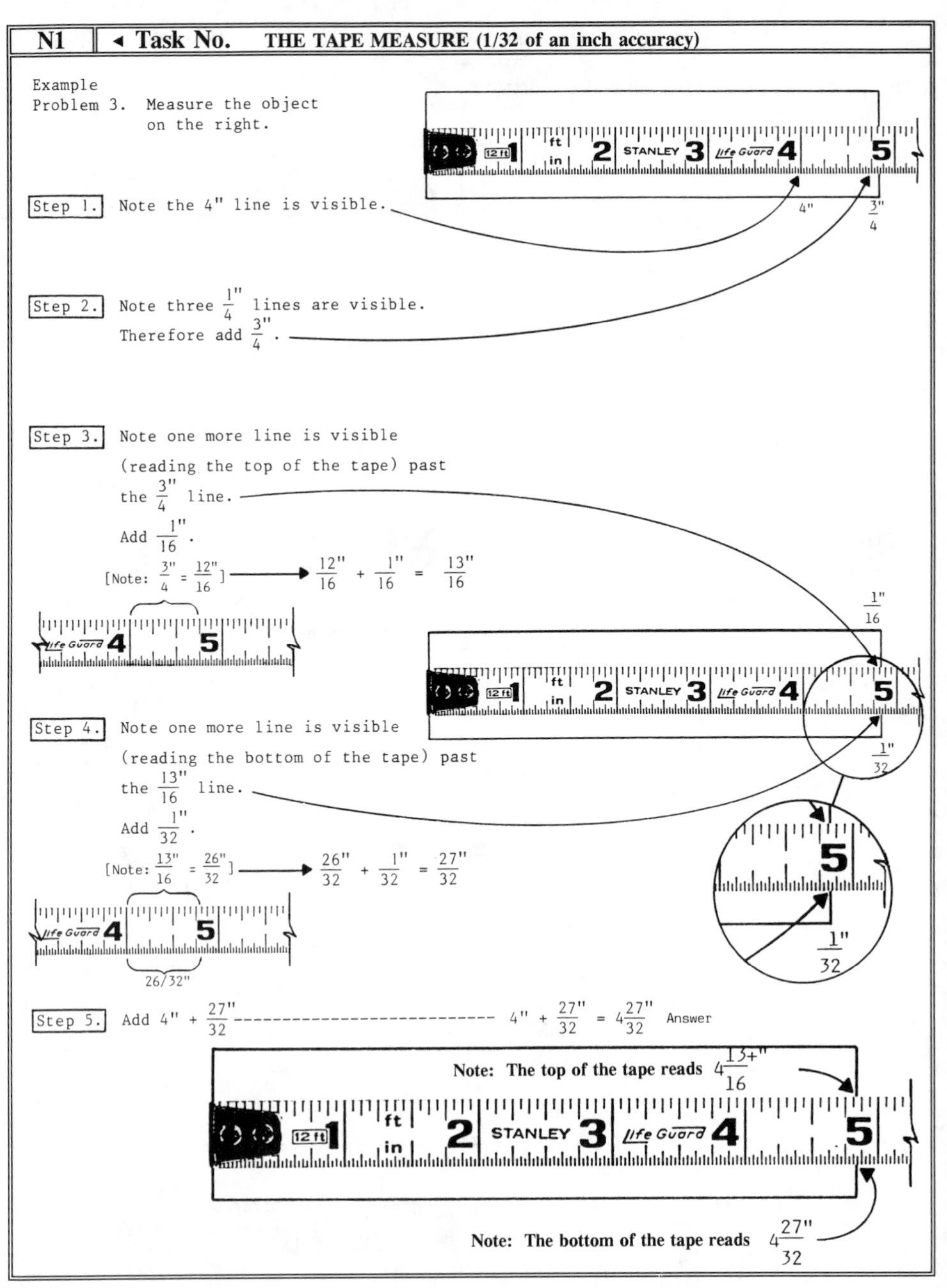

N1 ◄ **Task No.** **THE TAPE MEASURE (1/32 of an inch accuracy)**

Example
Problem 3. Measure the object on the right.

Step 1. Note the 4" line is visible.

Step 2. Note three $\frac{1''}{4}$ lines are visible.
Therefore add $\frac{3''}{4}$.

Step 3. Note one more line is visible (reading the top of the tape) past the $\frac{3''}{4}$ line.
Add $\frac{1''}{16}$.

[Note: $\frac{3''}{4} = \frac{12''}{16}$] ⟶ $\frac{12''}{16} + \frac{1''}{16} = \frac{13''}{16}$

Step 4. Note one more line is visible (reading the bottom of the tape) past the $\frac{13''}{16}$ line.
Add $\frac{1''}{32}$.

[Note: $\frac{13''}{16} = \frac{26''}{32}$] ⟶ $\frac{26''}{32} + \frac{1''}{32} = \frac{27''}{32}$

Step 5. Add $4'' + \frac{27''}{32}$ ------------------------ $4'' + \frac{27''}{32} = 4\frac{27''}{32}$ Answer

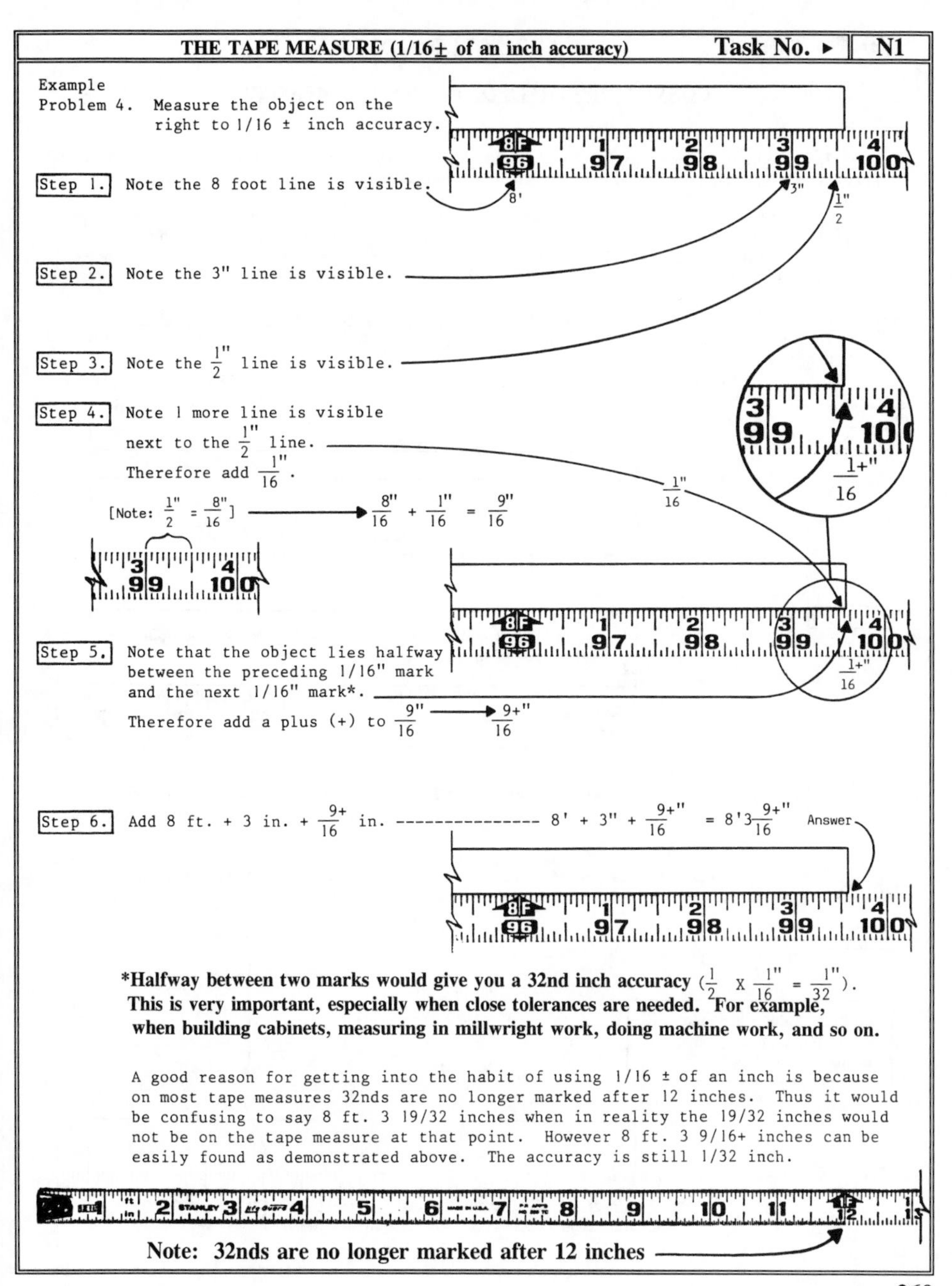
THE TAPE MEASURE (1/16± of an inch accuracy)
Task No. ▸ N1
Example
Problem 4. Measure the object on the right to 1/16 ± inch accuracy.
Step 1. Note the 8 foot line is visible.
8'
3"
1"/2
Step 2. Note the 3" line is visible.
Step 3. Note the 1"/2 line is visible.
Step 4. Note 1 more line is visible next to the 1"/2 line.
Therefore add 1"/16.
[Note: 1"/2 = 8"/16] 8"/16 + 1"/16 = 9"/16
1"/16
1+"/16
Step 5. Note that the object lies halfway between the preceding 1/16" mark and the next 1/16" mark*.
Therefore add a plus (+) to 9"/16 → 9+"/16
Step 6. Add 8 ft. + 3 in. + 9+/16 in. --------------- 8' + 3" + 9+"/16 = 8'3 9+"/16
Answer
*Halfway between two marks would give you a 32nd inch accuracy (1/2 X 1"/16 = 1"/32).
This is very important, especially when close tolerances are needed. For example, when building cabinets, measuring in millwright work, doing machine work, and so on.
A good reason for getting into the habit of using 1/16 ± of an inch is because on most tape measures 32nds are no longer marked after 12 inches. Thus it would be confusing to say 8 ft. 3 19/32 inches when in reality the 19/32 inches would not be on the tape measure at that point. However 8 ft. 3 9/16+ inches can be easily found as demonstrated above. The accuracy is still 1/32 inch.
Note: 32nds are no longer marked after 12 inches

N1A | ◂ Task No. **APPLY-IT MODULE**

CONVERT DECIMALS OF INCHES AND MEASURE

This module demonstrates the why and the how of conversion from decimals to fractions -- a math skill essential to everyone in trade and technical occupations.

Currently in America the most accepted system of linear measure is the English system--inches, feet, yards, etc., therefore it is a must to know how to convert decimals of those units to fractions. The problem listed below will show you why and how!

WHY? Because realistically in life measurements will not always come out in even increments, but they will have to be laid out correctly for the parts to fit. Example:

Divide the object below into 3 equal spaces so that you can drill, weld, or nail the parts to within 1/16 ± of an inch accuracy.

Step 1. Divide 7" by 3.

$$3\overline{)7.00} = 2.33"$$

6
10
9
10

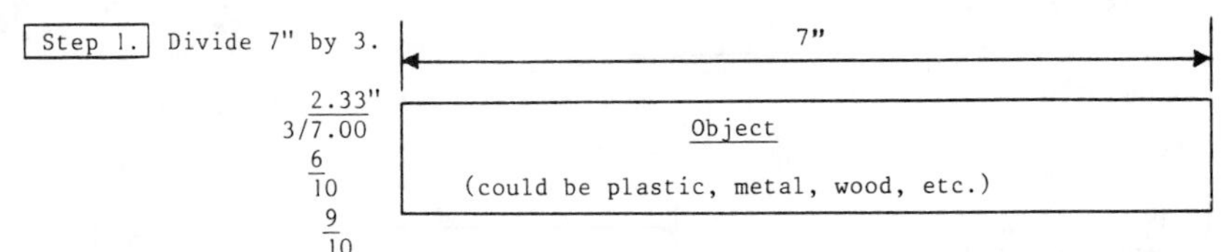

Step 2. Measure with a tape measure.

There are 3 spaces of 2.33" each. Mark the spaces below to 1/16 ± of an inch accuracy.

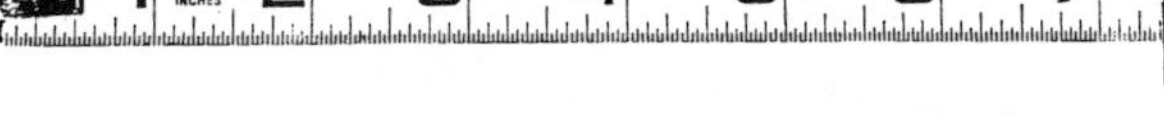

Note: The dimension 2.33" cannot be read on an American (English) tape measure, therefore it must be converted. See the next step.

HOW? To convert decimals to fractions, multiply the decimal by the desired denominator. [Or use a decimal equivalent chart.]

Step 1. Multiply the decimal only times 16 for 16ths. ---------

See Task Number E7, for more information on how to convert decimals to fractions

2 .33" × 16: 198, 33, 5.28 = $\frac{5+"}{16}$ → $2\frac{5+"}{16}$ First dimension

2.33" + 2.33" = 4.66"

4 .66" × 16: 396, 66, 10.56 = $\frac{10+"}{16}$ or $\frac{5+"}{8}$ → $4\frac{5+"}{8}$ Second dimension

Step 2. Measure with a tape measure and mark.

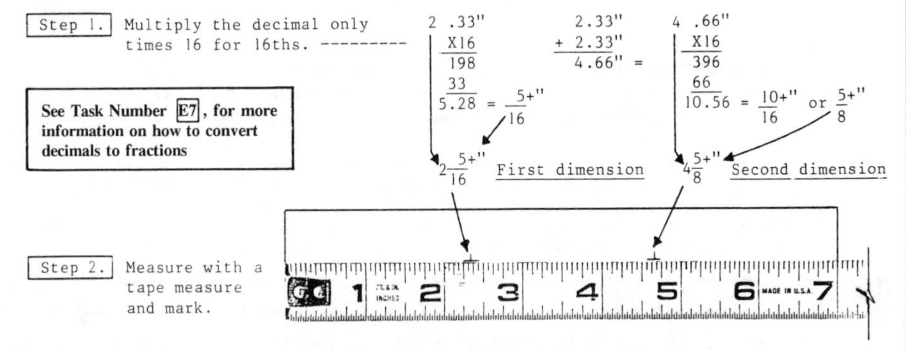

[Note below that you could multiply by different desired denominators, but 1/16 ± of an inch is most common because most tape measures after 12 inches do not show 32nds. Therefore it would be confusing to mention 32nds. However, millwrights, machinists, engineers, etc. would use 32nds or 64ths for greater accuracy.]

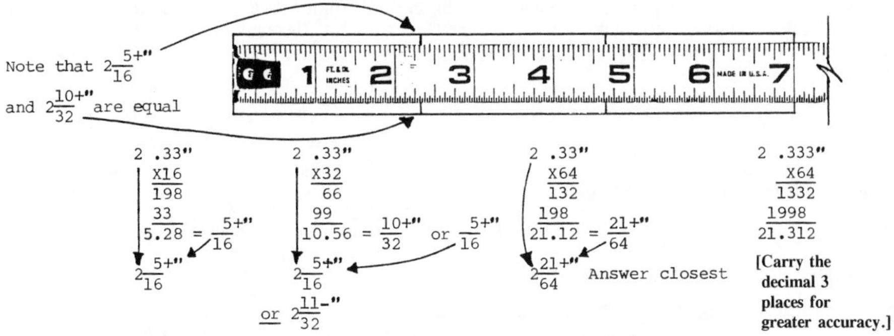

OR Use a decimal equivalent chart [decimals of an inch]

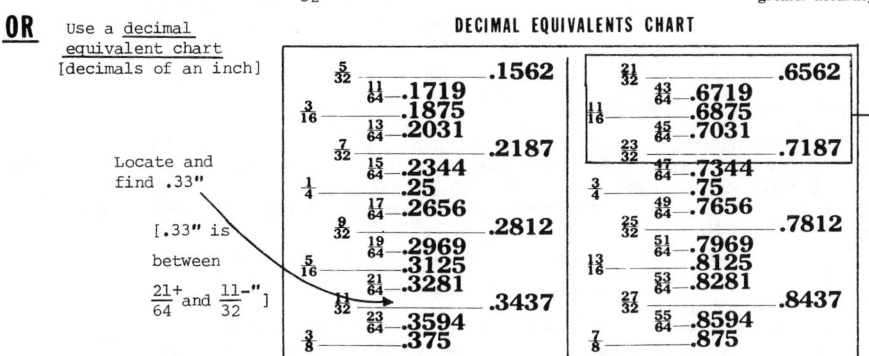

DECIMAL EQUIVALENTS CHART

Fraction	Decimal	Fraction	Decimal
5/32	.1562	21/32	.6562
11/64	.1719	43/64	.6719
3/16	.1875	11/16	.6875
13/64	.2031	45/64	.7031
7/32	.2187	23/32	.7187
15/64	.2344	47/64	.7344
1/4	.25	3/4	.75
17/64	.2656	49/64	.7656
9/32	.2812	25/32	.7812
19/64	.2969	51/64	.7969
5/16	.3125	13/16	.8125
21/64	.3281	53/64	.8281
11/32	.3437	27/32	.8437
23/64	.3594	55/64	.8594
3/8	.375	7/8	.875

Exercise on how to convert using both the decimal equivalent chart and by multiplying by the desired denominator.

Example: Change .67 to a fraction so you can read the dimension on an American (English) tape measure.

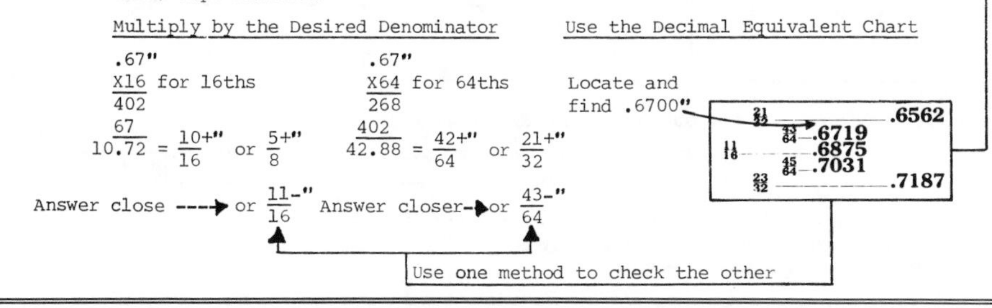

Checking Measurements

Calculating and marking a list of measurements, as on the preceding two pages, is also called making a story pole. Because once the measurements are actually marked on an object, they can be physically checked for accuracy by actually measuring between each mark (the marks tell you the story if they are correct or incorrect). The example shown below demonstrates this concept. In reality the measurement can be checked mathematically. However, errors could still be made when measuring (laying out) and marking them out. It is a good idea to get into the habit of using one method to check the other method. This significantly decreases your chances for errors.

Notice:

5/16+" is between 5/16" and 3/8".
This makes the accuracy 1/32".

Mathematically Check Each Measurement:

Measurement 1.	2.333"	= 2 5/16+"
	+2.333"	
Measurement 2.	4.666"	= 4 5/8+"
	+2.333"	
Measurement 3.	6.999"	= 7" ----

Mathematically these measurements should be correct because they compute to be a total of 6.999" or 7", the length of the object. However, you could have measured them wrong when actually marking them. To avoid this use the method shown below.

Physically Check Each Measurement:

Check the above dimensions by making sure the measurement between each space is 2 5/16+" by moving the tape measure for each specific space. **This method is OK to use for checking. However, when actually laying out each space, the chances for errors would increase each time the tape measure was moved. That is the very reason why the tape measure should be locked or held in place securely while marking all the measurements.**

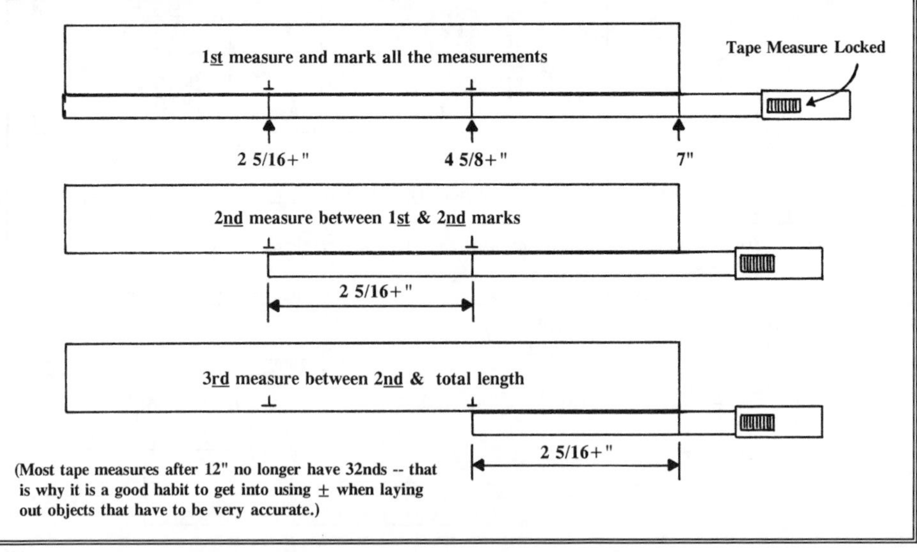

Task No. ▸ N1A

Checking Measurements Using Math & A Decimal Equivalent Chart

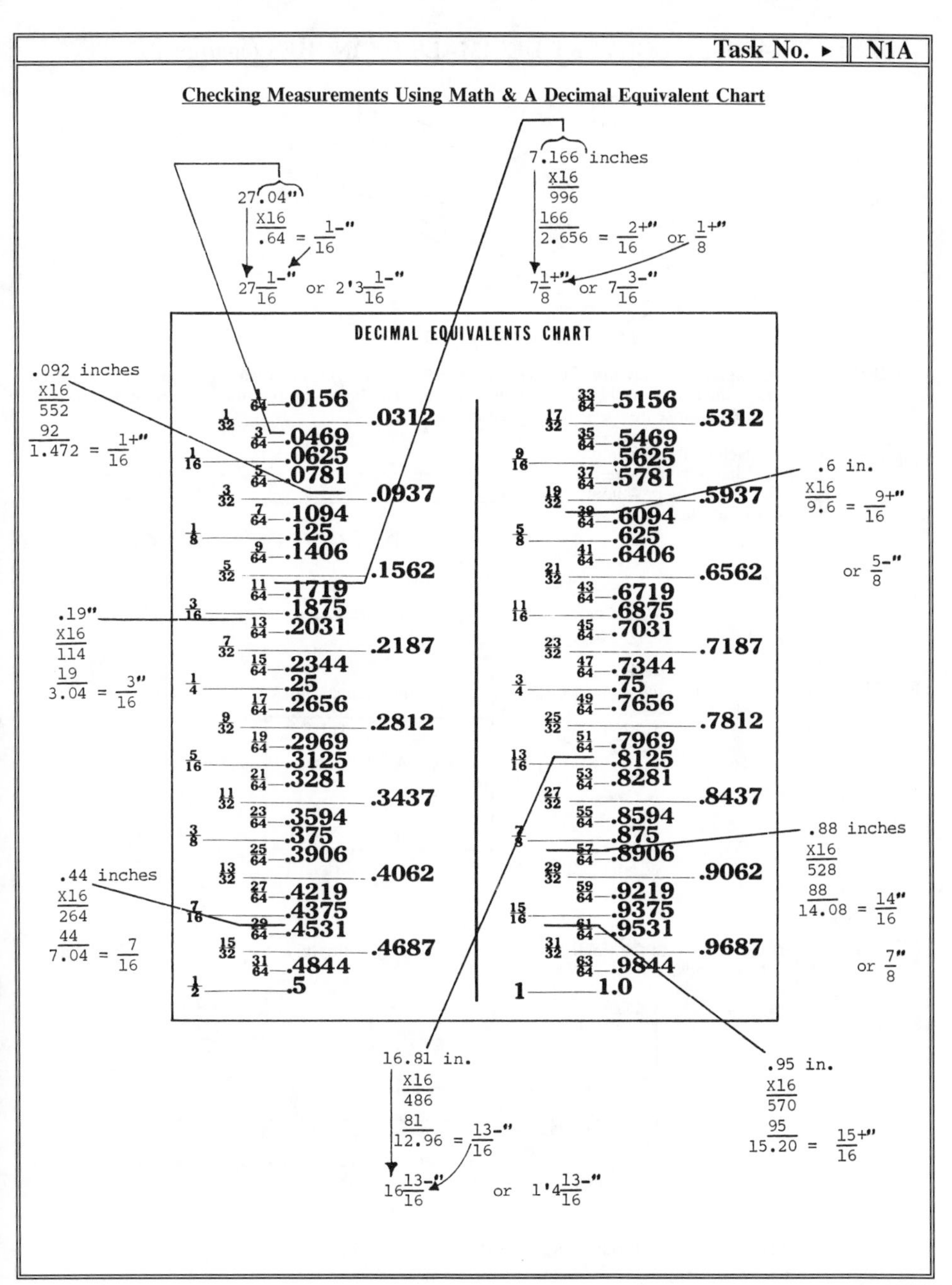

DECIMAL EQUIVALENTS CHART

Fraction	Decimal	Fraction	Decimal
1/64	.0156	33/64	.5156
1/32	.0312	17/32	.5312
3/64	.0469	35/64	.5469
1/16	.0625	9/16	.5625
5/64	.0781	37/64	.5781
3/32	.0937	19/32	.5937
7/64	.1094	39/64	.6094
1/8	.125	5/8	.625
9/64	.1406	41/64	.6406
5/32	.1562	21/32	.6562
11/64	.1719	43/64	.6719
3/16	.1875	11/16	.6875
13/64	.2031	45/64	.7031
7/32	.2187	23/32	.7187
15/64	.2344	47/64	.7344
1/4	.25	3/4	.75
17/64	.2656	49/64	.7656
9/32	.2812	25/32	.7812
19/64	.2969	51/64	.7969
5/16	.3125	13/16	.8125
21/64	.3281	53/64	.8281
11/32	.3437	27/32	.8437
23/64	.3594	55/64	.8594
3/8	.375	7/8	.875
25/64	.3906	57/64	.8906
13/32	.4062	29/32	.9062
27/64	.4219	59/64	.9219
7/16	.4375	15/16	.9375
29/64	.4531	61/64	.9531
15/32	.4687	31/32	.9687
31/64	.4844	63/64	.9844
1/2	.5	1	1.0

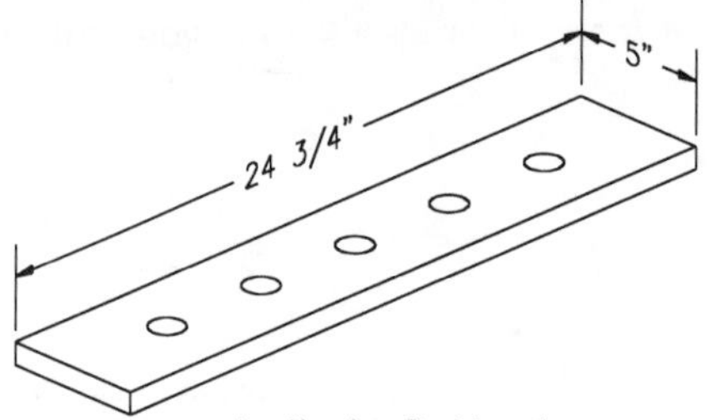

Application Problem 1.

DIRECTIONS: Calculate and lay out 5 center lines for 5 (1/2" dia.) holes with an American (English) tape measure for the object shown above. The holes, center to center, must be the same distance apart ± 1/16 of an inch.

Step 1. Draw a sketch of the problem with dimension lines. Note there are 5 holes in this problem -- how many spaces are there?

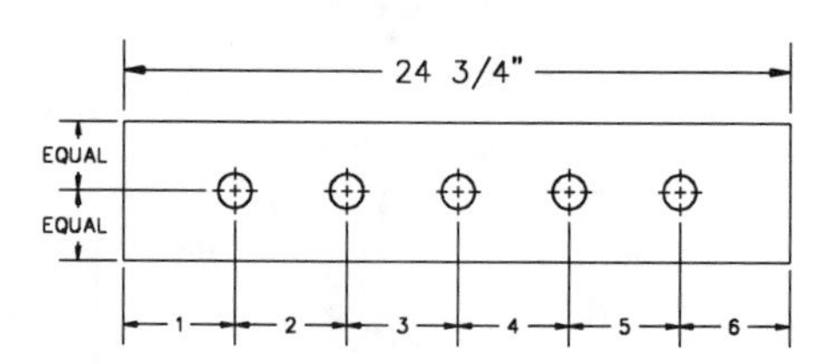

Step 2. Divide the length by the number of spaces. (Carry the decimal 3 places if necessary.)

```
   4.125"
6/24.750
  24
    7
    6
    15
    12
     30
     30
```

Check by Multiplying

```
 4.125"
X    6
24.750
Answer OK
```

Step 3. Convert the decimal to the English system of measure by multiplying the decimal* only by 16 for 16ths, 32 for 32nds, and 64 for 64ths.

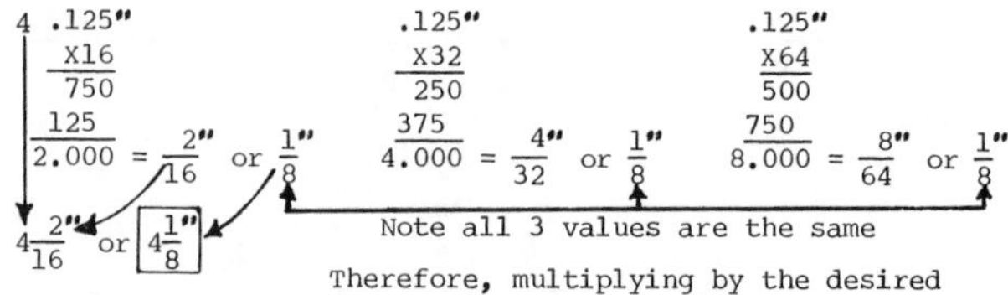

*The whole number is already in inches and can be easily found on the tape measure.

Step 4. Make a list (story pole) of each separate dimension to be read on the tape measure.

Note: <u>Use a decimal equivalent chart to convert from decimals to inches faster.</u>

1.	4.125"	$= 4\frac{1}{8}''$	1st reading
	+4.125		
2.	8.250	$= 8\frac{1}{4}''$	2nd reading
	+4.125		
3.	12.375	$= 12\frac{3}{8}''$	3rd reading
	+ 4.125"		
4.	16.500	$= 16\frac{1}{2}''$	4th reading
	+ 4.125		
5.	20.625	$= 20\frac{5}{8}''$	5th reading
	+ 4.125		
6.	+24.750	$= 24\frac{3}{4}''$	6th reading

Step 5. Meaure each dimension with the tape measure. Mark each dimension with a "T" mark.

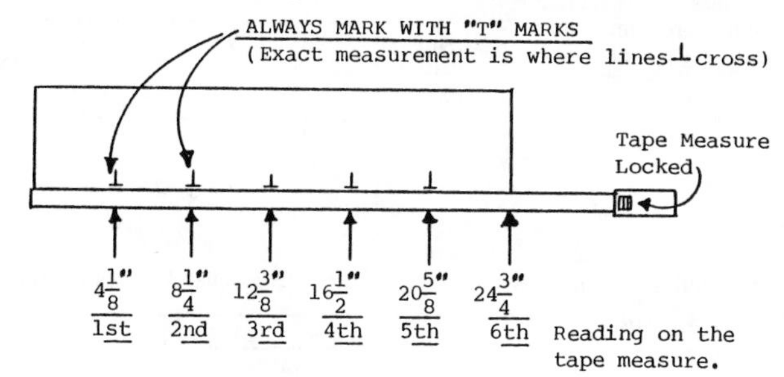

Step 6. Use a small square to extend for the width and mark.

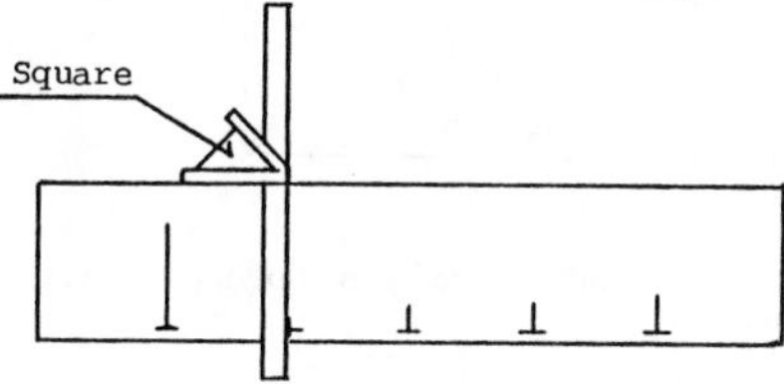

Step 7. Calculate the center spacing for the width and mark.

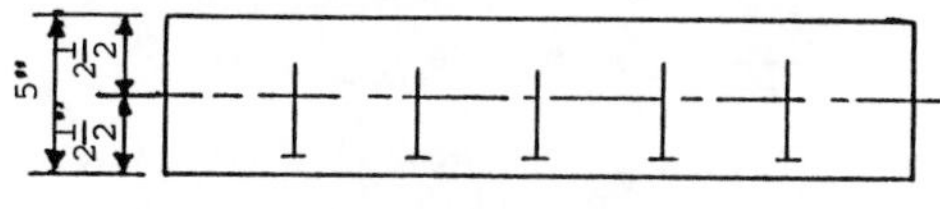

$$\frac{1}{2} \times 5'' = \frac{1}{2} \times \frac{5}{1} = \frac{5}{2} = 2\frac{1}{2}$$

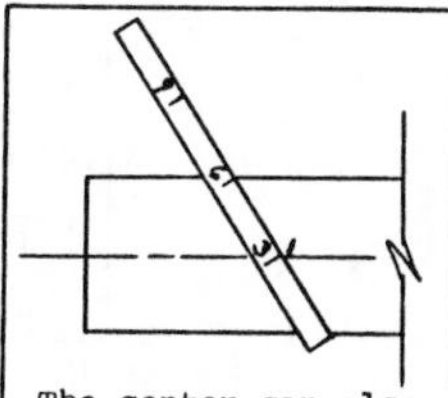

The center can also be found by slanting the rule to the closest even number divisible by 2

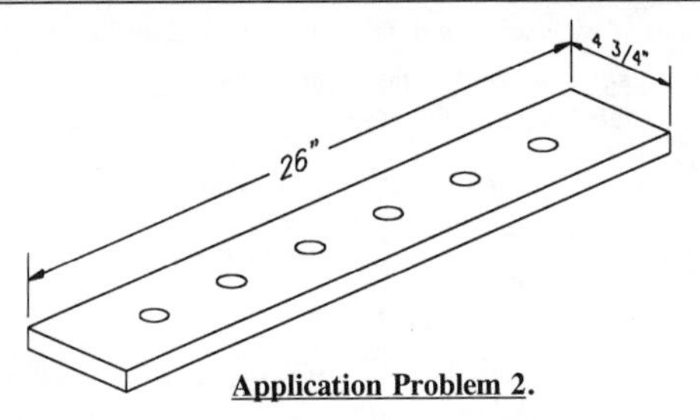

Application Problem 2.

DIRECTIONS: Calculate and lay out 6 center lines for 6 (3/8" dia.) holes with an American (English) tape measure for the object shown above. The holes, center to center, must be the same distance apart ± 1/16 of an inch.

Step 1. Draw a sketch of the problem with dimension lines. Note there are 6 holes in this problem -- how many spaces are there?

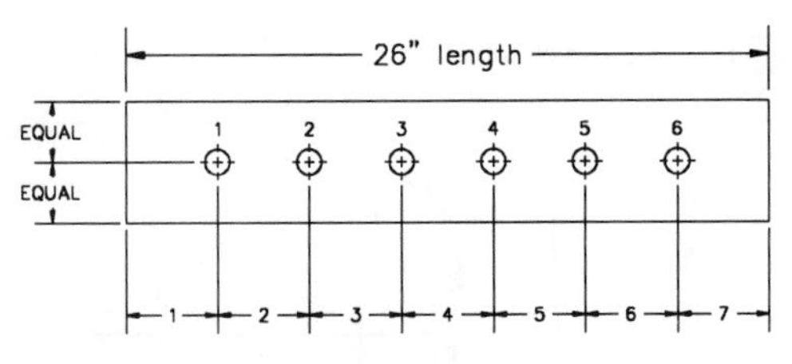

Step 2. Divide the length by the number of spaces. (Carry the decimal 3 places if necessary.)

```
   3.714"
7/26.000
  21
   50
   49
    10
     7
     30
     28
      2
```

```
Check by Multiplying
     3.714"
   X     7
    25.998
   +     2
    26.000
 Answer OK
```

Step 3. Convert the decimal to the English system of measure by multiplying the decimal* only by 16 for 16ths, 32 for 32nds, and 64 for 64ths.

```
3  .714"            .714"              .714"
    X16              X32                X64
   4284             1428               2856
   714              2142               4284
  11.424 = 11+"/16  22.848 = 23-"/32   45.696 = 45+"/64

3 11+"/16 Answer Close
3 23-"/32 Answer Closer
3 45+"/64 Answer Closest
```

*The whole number is already in inches and can be easily found on the tape measure.

Step 4. Make a list (story pole) of each separate dimension to be read on the tape measure.

Note: Use a decimal equivalent chart to convert from decimals to inches faster.

1. 3.714" = $3\frac{11}{16}$+" 1st reading
   +3.714
2. 7.428" = $7\frac{7}{16}$" 2nd reading
   +3.714
3. 11.142 = $11\frac{1}{8}$" 3rd reading
   + 3.714
4. 14.856 = $14\frac{7}{8}$-" 4th reading
   + 3.714
5. 18.570 = $18\frac{9}{16}$" 5th reading
   + 3.714
6. 22.284 = $22\frac{1}{4}$+" 6th reading
   + 3.714
7. 25.998" = 26" 7th reading

Step 5. Measure each dimension with the tape measure. Mark each dimension with a "T" mark.

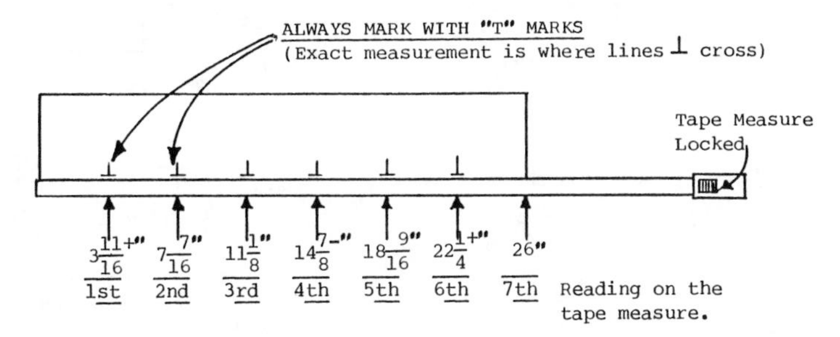

Step 6. Use a small square to extend for the width mark.

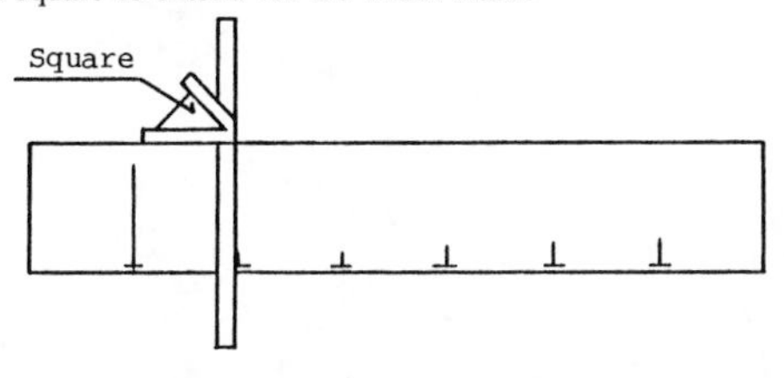

Step 7. Calculate the center spacing for the width and mark.

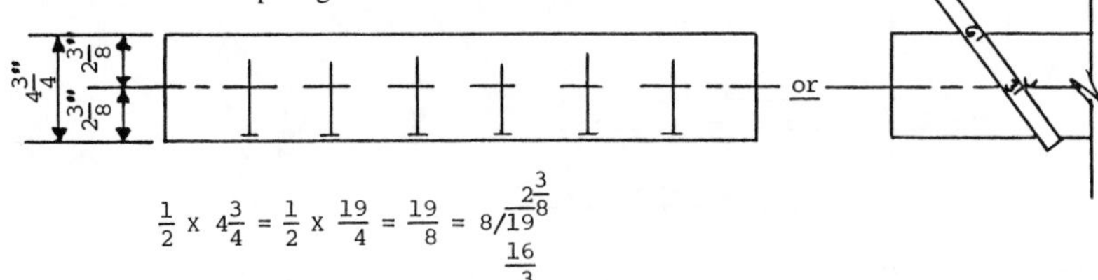

$$\frac{1}{2} \times 4\frac{3}{4} = \frac{1}{2} \times \frac{19}{4} = \frac{19}{8} = 8\overline{)19}\; 2\frac{3}{8}$$

$$\frac{16}{3}$$

N1B	◂ Task No.	APPLY-IT MODULE

CONVERT DECIMALS OF FEET AND MEASURE

APPLY-IT MODULE [N1A] demonstrated how to calculate and measure an object when it was in all inches. This apply-it module demonstrates how to calculate and measure an object when it is in both feet and inches by using three different methods. Use the method(s) that work best for you.

DIRECTIONS: Calculate and lay out (measure) 4 center lines for the 4 up-right objects shown on the right. The objects, center to center, must be the same distance apart ± 1/16 of an inch.

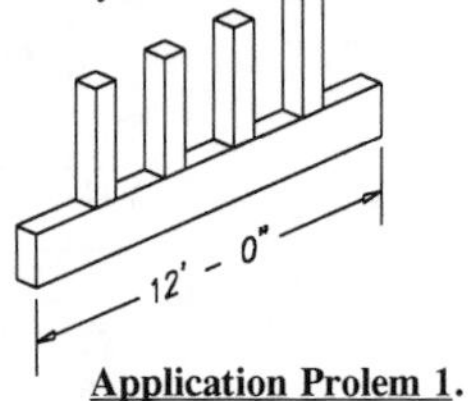

Application Prolem 1.

Method 1 -- Solve by using all feet.

Step 1. Draw a sketch of the object and dimension.

Note: There are 4 uprights, but there are 5 spaces.

12'-0"
1 2 3 4
1 2 3 4 5

Step 2. Divide by the number of spaces. (Carry the decimal 3 places if necessary.)

$12.0 \div 5 = 2.4'$

CHECK
2.4' × 5 = 12.0' answer OK

Step 3. Convert the decimal so you can read the dimension on an American (English) tape measure. Multiply by the desired denominator.

[Multiply by 12 when the sign is feet.]

[Multiply by 16 when the sign is inches.]

2.4' × 12 = 4.8"
.8 × 16 = 12.8 = 13/16-"
2'4 13/16-"

Step 4. Make a list (story pole) of each separate dimension and layout.

1. 2'4 13/16-"
2. 4'9 5/8-"
3. 7'2 7/16-"
4. 9'7 1/4-"
5. 12'0"

2ND: 2.4' + 2.4' = 4.8'; .8 × 12 = 9.6"; 4'9 5/8-"

3RD: 4.8' + 2.4' = 7.2'; .2 × 12 = 2.4"; 7'2 3/8+"

4TH: 7.2' + 2.4' = 9.6'; .6 × 12 = 7.2"; 9'7 3/16+"

5TH: 9.6' + 2.4' = 12.0' ANSWER OK

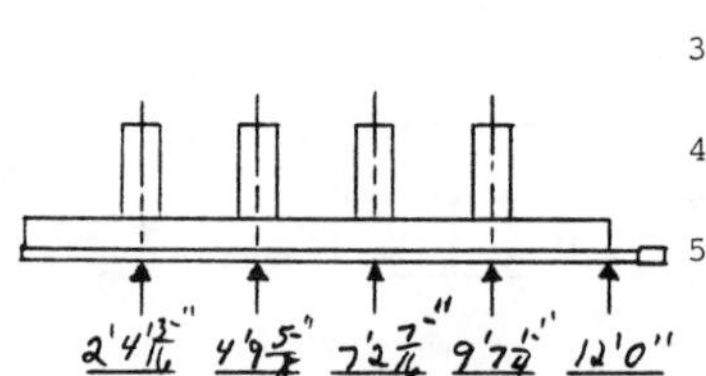

Reading on the tape measure.

[On the job you would check the above dimensions to make sure that they were 2'4 13/16-" apart.]

Method 2 -- Solve by using feet and inches.

Step 1. The same as on Method 1.

Step 2. Divide by the number of spaces. (Carry the decimal 3 places if necessary.)

$5 \overline{)12'\,0''} = 2'\,4.8''$
10
2' = +24''
20
40
40

CHECK
2' 4.8''
x 5
10' 24.0'' = 12'
OK

Step 3. Convert the decimal so you can read the dimension on an American (English) tape measure. Multiply by the desired denominator.

2' 4.8''
x16
12.8 = 13-/16''
↓
$2'4\frac{13}{16}-''$

Step 4. Make a list (story pole) of each separate dimension and lay out.

[NOTE: **Use the decimal equivalent chart inside the front cover to convert to fractions of an inch faster.]**

1. $2'4\frac{13}{16}-''$
2. $4'9\frac{5}{8}-''$
3. $7'2\frac{7}{16}-''$
4. $9'7\frac{1}{4}-''$
5. 12'0''

2' 4.8''
+2' 4.8''
4' 9.6'' = $4'9\frac{5}{8}-''$
+2' 4.8
6' 14.4 = $7'2\frac{7}{16}-''$
+2' 4.8
8' 19.2 = $9'7\frac{1}{4}-''$
+2' 4.8
10' 24.0'' = 12'-0''
answer OK

Method 3 -- Solve by using all inches.

Step 1. The same as on Method 1.

Step 2. Divide by the number of spaces. (Carry the decimal 3 places if necessary.)

CHANGE 12'-0'' = 12 x12 = 144''

$5 \overline{)144} = 28.8''$
10
44
40
40
40

CHECK
28.8
x 5
144.0 = 12'0''
answer OK

Step 3. Convert the decimal so you can read the dimension on an American (English) tape measure. Multiply by the desired denominator.

28.8''
x16
12.8 = 13-/16''
↓
$28\frac{13}{16}-'' = 2'4\frac{13}{16}-''$

Step 4. Make a list (story pole) of each separate dimension and lay out.

1. $2'4\frac{13}{16}-''$
2. $4'9\frac{5}{8}-''$
3. $7'2\frac{7}{16}-''$
4. $9'7\frac{1}{4}-''$
5. 12'0''

28.8''
+28.8''
57.6 = $4'9\frac{5}{8}-''$
+28.8
86.4 = $7'2\frac{7}{16}-''$
+28.8
115.2 = $9'7\frac{1}{4}-''$
+28.8
144.0 = 12'0'' answer OK

[NOTE: **Most of the longer tape measures are in both feet and inches, <u>not just total inches</u>. Therefore, it is a good idea to practice all the three methods demonstrated so you will be able to solve them easily later, no matter which measuring instrument you are given to work with.]**

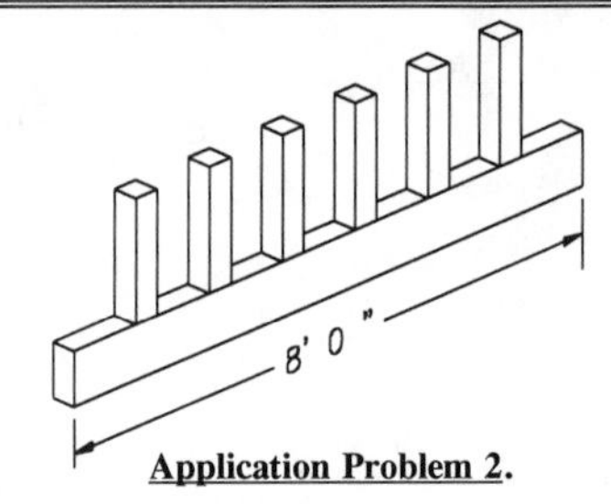

Application Problem 2.

DIRECTIONS: Calculate and measure 6 center lines for the 6 up-right objects shown above. The objects, center to center, must be the same distance apart ± 1/16 of an inch.

Step 1. Draw a sketch of the problem with dimension lines. Note there are 6 up-right objects and 6 center lines -- how many spaces between the objects are there?

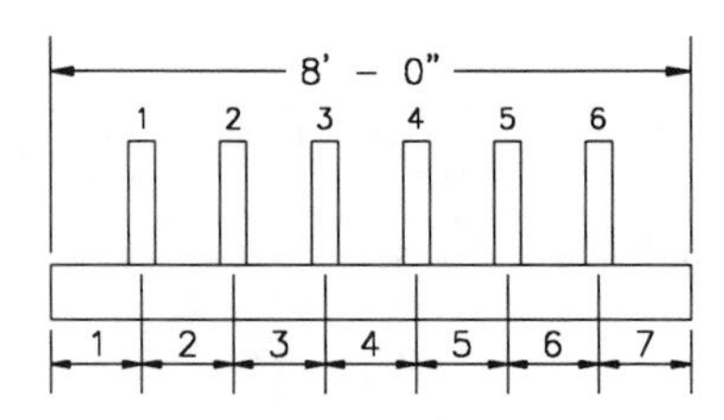

Step 2. Divide the length by the number of spaces. (Carry the decimal 3 places if necessary.)

```
   1.142'         Check by Multiplying
 7/8.000
   7                 1.142
   10              X     7
    7                7.994
    30             +     6
    28               8.000
     20
     14            Answer OK
      6
```

OR

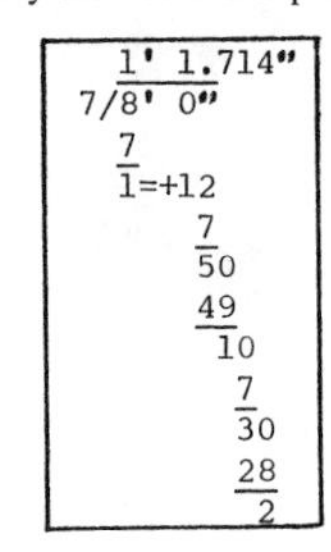
```
   1' 1.714"
 7/8' 0"
   7
   1=+12
      7
      50
      49
       10
        7
       30
       28
        2
```

Step 3. Convert the decimals so you can read the dimension on an American (English) tape measure. Multiply by the desired denominators.

[Multiply by 12 when the decimal is feet]

[Multiply by 16 when the decimal inches]

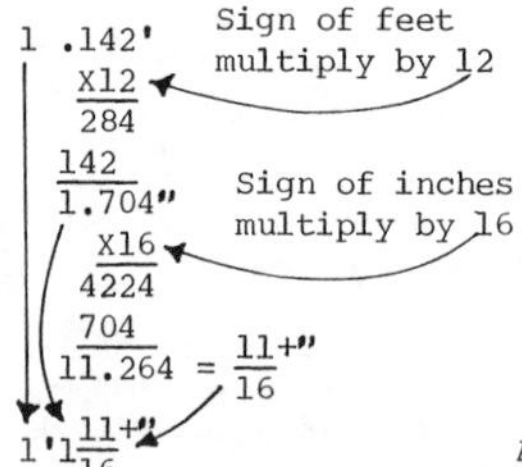
```
 1 .142'     Sign of feet
   X12       multiply by 12
   284
  142
  1.704"     Sign of inches
   X16       multiply by 16
  4224
  704
  11.264 = 11+"/16

Answer  1'1 11+"/16
```

```
 .704"
  X64
 2814
4224
45.064 = 45"/64

Answer closer 1'1 45"/64
```

Step 4. Make a list (story pole) of each separate dimension to be read onthe tape measure. Assume that the tape measure is <u>not</u> to be moved for each specific dimension.

NOTE: **<u>Use a decimal equivalent chart to convert from decimals of inches faster</u>.**

1. 1.142' × 12 = 284 + 142 = 1.704" → $1'1\frac{11}{16}+"$
2. 1.142' + 1.142 = 2 .284' × 12 = 568 + 284 = 3.408" → $2'3\frac{3}{8}+"$
3. 2.284' + 1.142' = 3 .426' × 12 = 852 + 426 = 5.112" → $3'5\frac{1}{8}"$
4. 3.426' + 1.142' = 4 .568' × 12 = 1136 + 568 = 6.816" → $4'6\frac{13}{16}"$
5. 4.568' + 1.142' = 5 .710' × 12 = 1420 + 710 = 8.52" → $5'8\frac{9}{16}-"$
6. 5.710' + 1.142' = 6 .852' × 12 = 1704 + 852 = 10.224" → $6'10\frac{1}{4}-"$
7. 6.852' + 1.142' = 7.994' = 8'0" Answer OK

<u>Conversion Using All Inches</u>

1st	*13.714"	=	1'1 11/16+"
	+13.714"		
2nd	27.428"	=	2'3 7/16-"
	+13.714"		
3rd	41.142"	=	3'5 1/8"
	+13.714"		
4th	54.856"	=	4'6 13/16+"
	+13.714"		
5th	68.570"	=	5'8 9/16"
	+13.714"		
6th	82.284"	=	6'10 1/4+"
	+13.714"		
7th	95.998"	=	96"

Note that this method might seen easier <u>but</u> in reality more steps are involve because you have to divide by 12 each time, especially if the tape measure is in both feet and inches.

<u>USE THE METHOD THAT WORKS BEST FOR YOU.</u>

8'0" = 96" 7/96.00 = *13.714"

Step 5. Measure each dimension with the tape measure. Mark each dimension with a "T" mark.

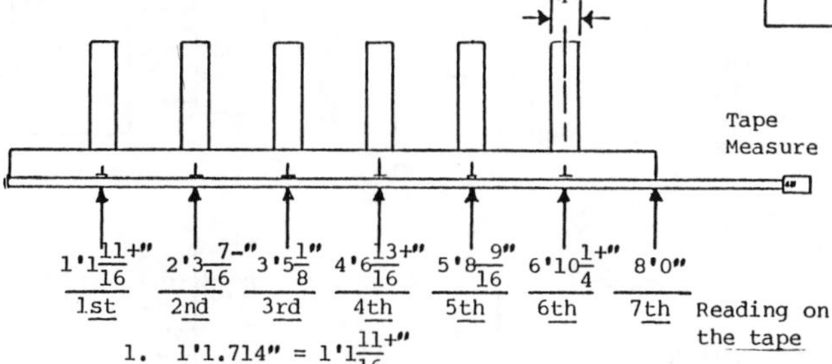

1. 1'1.714" = $1'1\frac{11}{16}+"$
   +1'1.714"
2. 2'3.428" = $2'3\frac{7}{16}-"$
   +1'1.714"
3. 3'5.142" = $3'5\frac{1}{8}"$
   +1'1.714"
4. 4'6.856" = $4'6\frac{13}{16}+"$
   +1'1.714"
5. 5'8.570" = $5'8\frac{9}{16}"$
   +1'1.714"
6. 6'10.284 = $6'10\frac{1}{4}+"$
   +1'1.714"
7. 7'11.998"= 8'0"

***If the decimal were carried out 4 places the answer would be closer yet, however, ± 1/16 of an inch (1/32) is close enough for most practical purposes. Carrying the decimal 3 places in all inches is closer because the increments are smaller.**

N2	◂ Task No.	THE MICROMETER

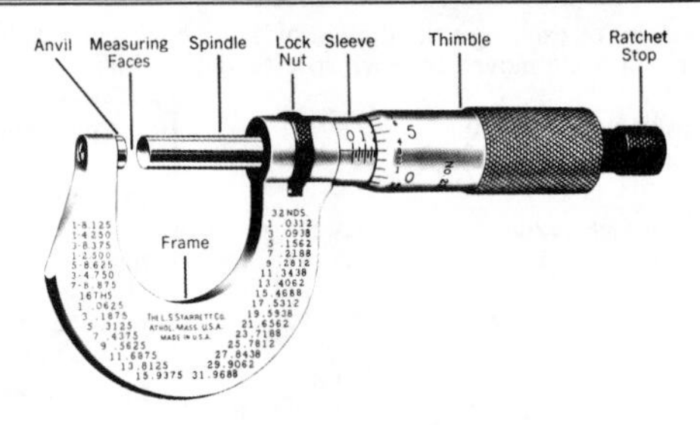

THE MICROMETER
[Marked in Thousandths (.001") of an Inch]

How to Read the Micrometer

The sleeve of the micrometer is divided into 40 equal spaces per inch. Each line is spaced 1/40 or .025 inches.

The 40 lines on the sleeve correspond with the 40 threads per inch on the thimble.

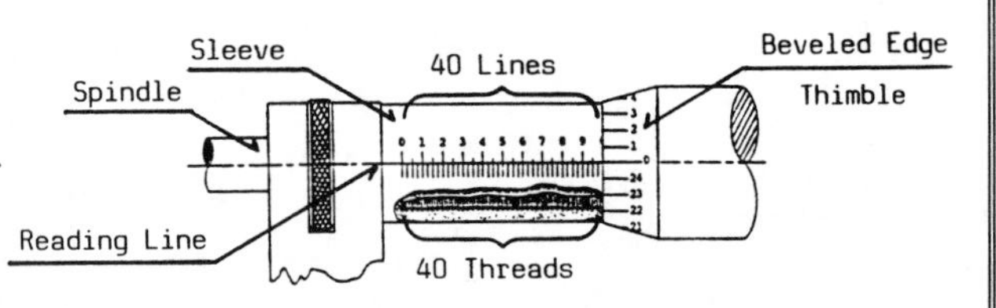

The beveled edge of the thimble is divided into 25 equal parts. Therefore, each mark is .001 inch because one complete revolution of the thimble advances the spindle face away from the anvil face one thread: 1/40 or .025 inch. This would make one of the 25 marks .001 inch (.025/25).

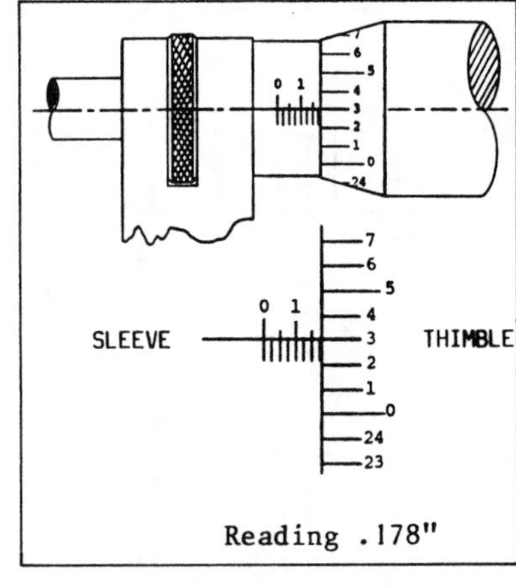

Example
Problem 1. Read the micrometer setting on the right.

Step 1 — Note the "1" line on the sleeve is visible, representing ------------------------ .100"

Step 2 — Note there are 3 additional lines visible, each representing .025" ------------- .075"

[3 X .025" = .075"]

Step 3 — Note line "3" on the thimble coincides with the reading line on the sleeve, each line representing .001" ----------------- .003"

[3 X .001" = .003"]

Step 4 — Add Step 1, Step 2, and Step 3.

.100"
.075"
+ .003"

The micrometer reading is ------------------ .178" **[Read one hundred seventy-eight thousandths of an inch.]**

[An easy way to remember: Think of the various units as if you were making change from a ten dollar bill. Count the figures on the sleeve as dollars, the vertical lines on the sleeve as quarters, and the divisions on the thimble as cents. Add up your change and put the decimal point in front of the figures.]

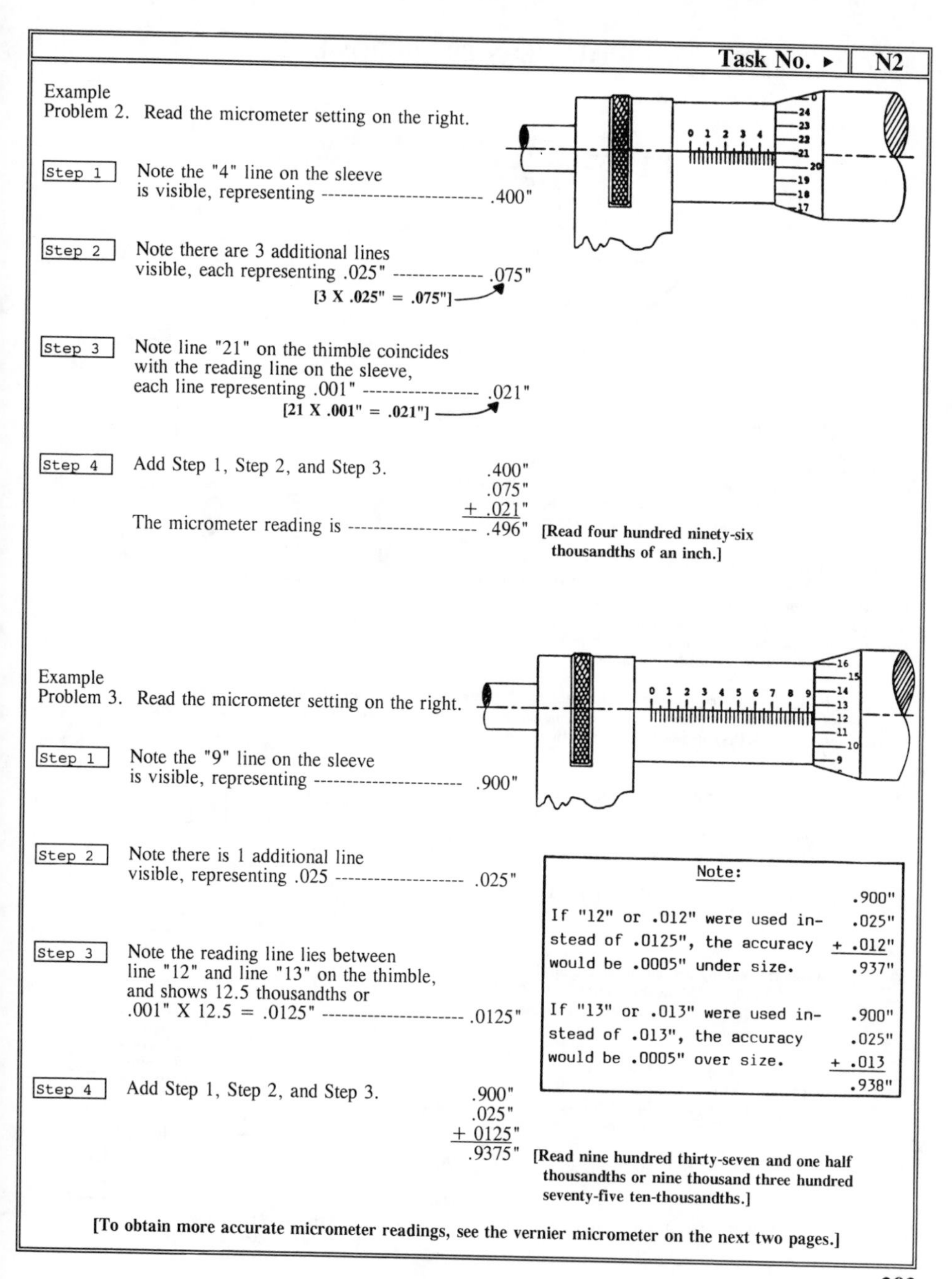

Task No. ▸ N2

Example
Problem 2. Read the micrometer setting on the right.

Step 1 — Note the "4" line on the sleeve is visible, representing ------------------------ .400"

Step 2 — Note there are 3 additional lines visible, each representing .025" -------------- .075"
[3 X .025" = .075"]

Step 3 — Note line "21" on the thimble coincides with the reading line on the sleeve, each line representing .001" ------------------ .021"
[21 X .001" = .021"]

Step 4 — Add Step 1, Step 2, and Step 3.

.400"
.075"
+ .021"

The micrometer reading is ------------------ .496" **[Read four hundred ninety-six thousandths of an inch.]**

Example
Problem 3. Read the micrometer setting on the right.

Step 1 — Note the "9" line on the sleeve is visible, representing ---------------------- .900"

Step 2 — Note there is 1 additional line visible, representing .025 ------------------- .025"

Step 3 — Note the reading line lies between line "12" and line "13" on the thimble, and shows 12.5 thousandths or .001" X 12.5 = .0125" -------------------- .0125"

Step 4 — Add Step 1, Step 2, and Step 3.

.900"
.025"
+ 0125"
.9375" **[Read nine hundred thirty-seven and one half thousandths or nine thousand three hundred seventy-five ten-thousandths.]**

Note:

If "12" or .012" were used instead of .0125", the accuracy would be .0005" under size.

.900"
.025"
+ .012"
.937"

If "13" or .013" were used instead of .013", the accuracy would be .0005" over size.

.900"
.025"
+ .013
.938"

[To obtain more accurate micrometer readings, see the vernier micrometer on the next two pages.]

THE VERNIER MICROMETER

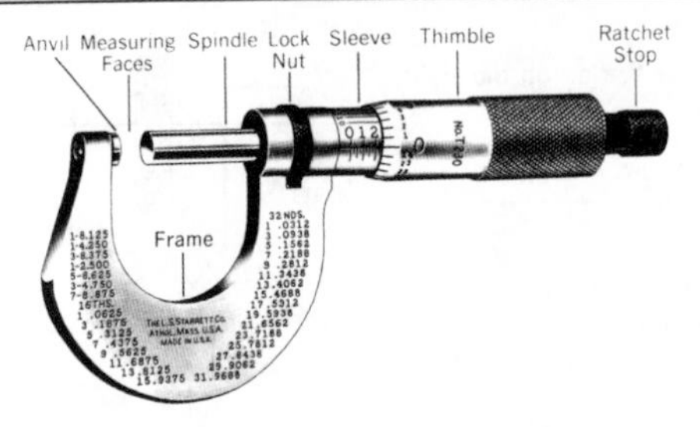

THE VERNIER MICROMETER

[Marked in Ten-Thousandths (.0001") of an Inch]

THIMBLE

How to Read the Vernier Micrometer

Micrometers marked in ten-thousandths of an inch are used like micrometers marked in thousandths as described on the preceding 2 pages, except that an additional reading in ten-thousandths, which is obtained from a vernier, is added to the thousandths reading.

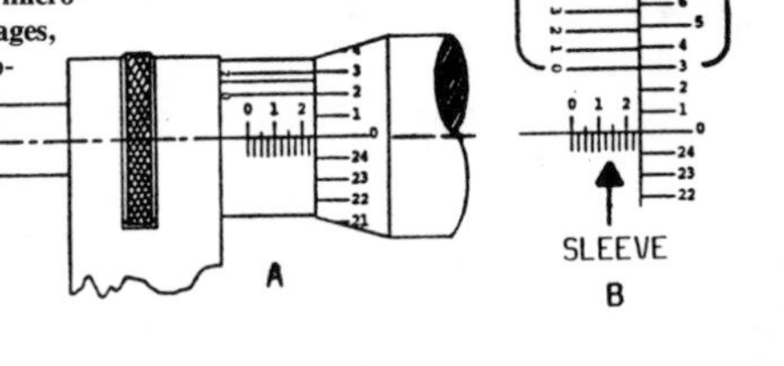

The vernier consists of ten divisions on the sleeve, shown in figure B, which occupy the same space as nine divisions on the thimble. Therefore, the difference between the width of one of the ten spaces on the vernier and one of the nine spaces on the thimble is one-tenth of a division on the thimble, or one-tenth of one-thousandth, which is one ten-thousandth.

To read a ten-thousandths micrometer, first obtain the thousandths reading, then see which of the lines on the vernier coincides with a line on the thimble. If it is the line marked "1" add one ten-thousandth, if it is the line marked "2" add two ten-thousandths, if it is the line marked "3" add three ten-thousandths, etc.

Example
Problem 1. Read the vernier micrometer setting on the right.

Step 1 Note the "2" line on the sleeve is visible, representing ---------------------------- .200"

Step 2 Note there are two additional lines visible, each representing .025" -------------- .050"
[2 X .025" = .050"]

Step 3 **[Note the reading line on the sleeve lies between the "0" and the "1" on the thimble indicating ten-thousandths of an inch are also to be added as read from the vernier.]**

Note the "7" line on the vernier coincides with a line on the thimble, representing .0007" --------------------------- .0007"
[7 X .0001" = .0007"]

THIMBLE

SLEEVE

Step 4 Add Step 1, Step 2, and Step 3.

.200"
.050"
+ .0007"

The vernier micrometer reading is ---------- .2507" **[Read two thousand five hundred seven ten-thousandths of an inch.]**

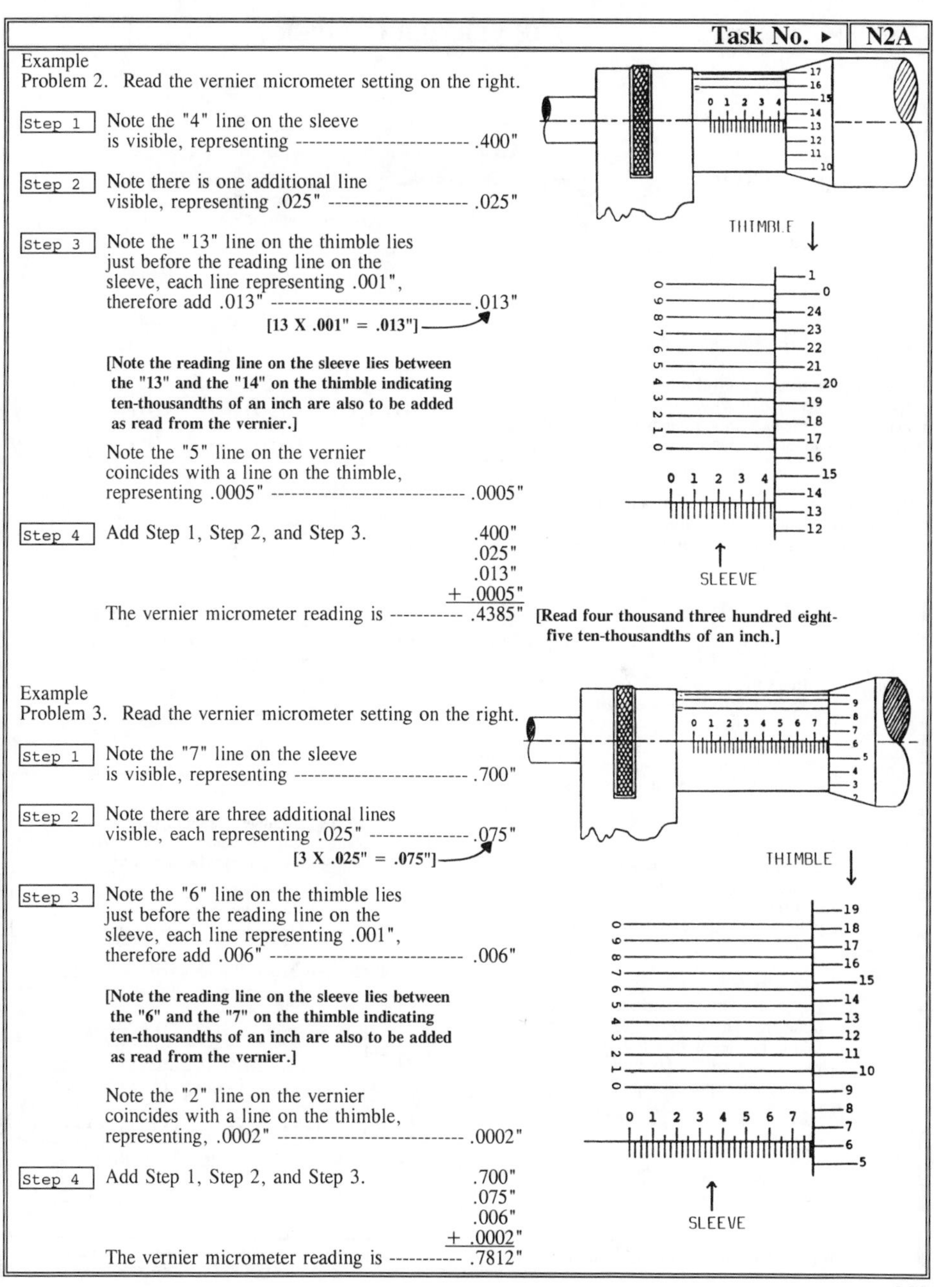

Task No. ▸ N2A

Example
Problem 2. Read the vernier micrometer setting on the right.

Step 1 Note the "4" line on the sleeve is visible, representing ------------------------- .400"

Step 2 Note there is one additional line visible, representing .025" --------------------- .025"

Step 3 Note the "13" line on the thimble lies just before the reading line on the sleeve, each line representing .001", therefore add .013" ----------------------------- .013"

[13 X .001" = .013"]

[Note the reading line on the sleeve lies between the "13" and the "14" on the thimble indicating ten-thousandths of an inch are also to be added as read from the vernier.]

Note the "5" line on the vernier coincides with a line on the thimble, representing .0005" ---------------------------- .0005"

Step 4 Add Step 1, Step 2, and Step 3.

.400"
.025"
.013"
+ .0005"

The vernier micrometer reading is ---------- .4385"

[Read four thousand three hundred eight-five ten-thousandths of an inch.]

Example
Problem 3. Read the vernier micrometer setting on the right.

Step 1 Note the "7" line on the sleeve is visible, representing ------------------------- .700"

Step 2 Note there are three additional lines visible, each representing .025" -------------- .075"

[3 X .025" = .075"]

Step 3 Note the "6" line on the thimble lies just before the reading line on the sleeve, each line representing .001", therefore add .006" --------------------------- .006"

[Note the reading line on the sleeve lies between the "6" and the "7" on the thimble indicating ten-thousandths of an inch are also to be added as read from the vernier.]

Note the "2" line on the vernier coincides with a line on the thimble, representing, .0002" ------------------------- .0002"

Step 4 Add Step 1, Step 2, and Step 3.

.700"
.075"
.006"
+ .0002"

The vernier micrometer reading is ---------- .7812"

N3	◂ Task No.	THE VERNIER CALIPER

THE VERNIER CALIPER

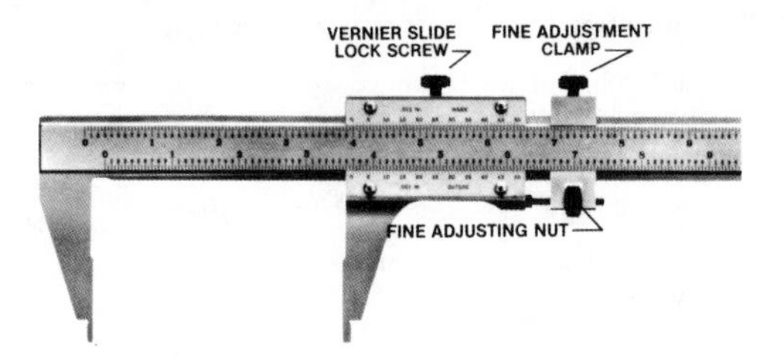

How to Read the Vernier Caliper

Every inch on the bar is divided into 20 equal spaces. Each line is spaced 1/20 or .050 inches. Every second mark between the inch lines is numbered and equals one-tenth of an inch or .100".

The vernier plate is divided into 50 equal spaces, each line representing .001". Every fifth line is numbered-5,10, 15, 20, ... 45, 50-for easy counting.

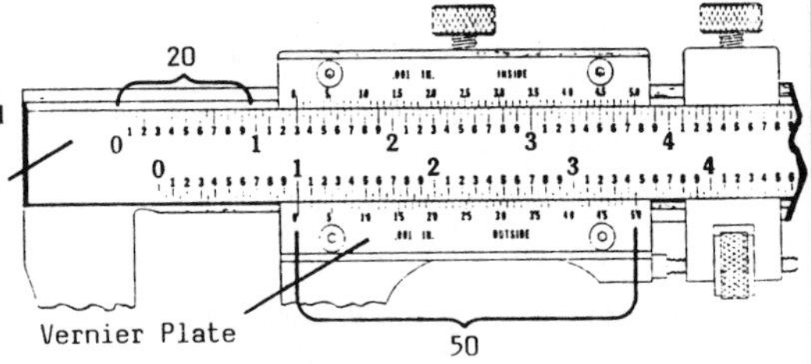

Example
Problem 1. Read the outside vernier caliper setting on the right.

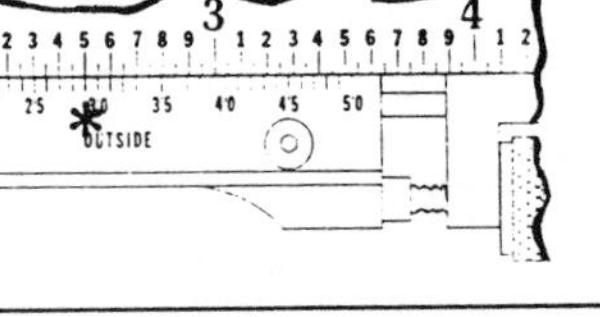

Step 1 Count how many inches are on the bar. Note -------------------- 1.000"

Step 2 Count how many tenths (.100") lie between the inch line on the bar and the zero line on the vernier plate. Note there are none --------------- .000"

Step 3 Count how many marks on the vernier plate from its zero line to the line that coincides with a line on the bar. Note there are 29 (as indicated by the star), each representing .001" ----------- .029"

[29 X .001" = .029"]

Step 5 Add Step 1, Step 2, Step 3, and Step 4.

1.000"
.000"
.050"
\+ .029"
1.079"

The outside vernier caliper reading is ------ 1.079" Read one and seventy-nine thousandths of an inch.

NOTE:

The inside caliper setting (upper vernier plate) is read the same, however, there will be .300" more at the same setting.

Example: 1.079" **Outside Reading**
\+ .300"
1.379" **Inside Reading**

.300" more than the outside reading.

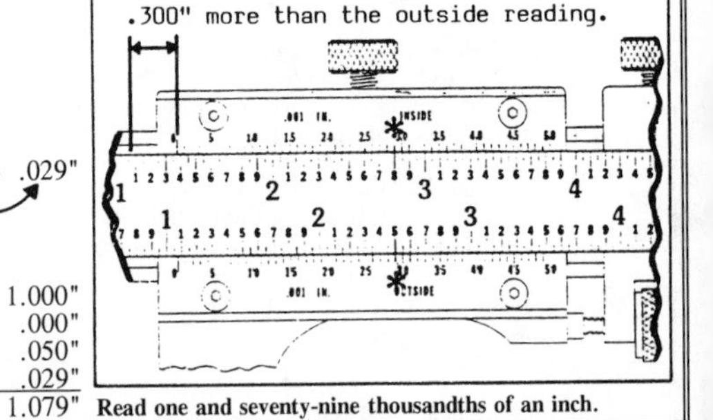

Task No. ▸ N3

Example
Problem 2. Read the vernier caliper setting on the right.

Step 1 Count how many inches are on the bar. Note ---------------- 1.000"

Step 2 Count how many tenths (.100") lie between the inch line on the bar and the zero line on the bar plate. Note there are 2 -------------- .200"

Step 3 Count how many twentieths (.050") lie between the inch line on the bar and the zero line on the vernier plate. Note ------------------- .050"

Step 4 Count how many marks on the vernier plate from its zero line to the line that coincides with a line on the bar (as indicated by the star). Note there are 25 --------------------- .025"

Step 5 Add Step 1, Step 2, Step 3, and Step 4.

1.000"
.200"
.050"
\+ .025"

The vernier caliper reading is ------ 1.275" **Read one and two-hundred seventy-five thousandths of an inch.**

Example
Problem 3. Read the vernier caliper setting on the right.

Step 1 Count how many inches are on the bar. Note ---------------- 2.000"

Step 2 Count how many tenths (.100") lie between the 2-inch line on the bar and the zero line on the vernier plate. Note there are 4 -------------- .040"

Step 3 Count how many twentieths (.050") lie between the 2-inch line on the bar and the zero line on the vernier plate. Note ----------------- .000"

Step 4 Count how many marks on the vernier plate from its zero line to the line that coincides with a line on the bar (as indicated by the star). Note there are 12 --------------------- .012"

Step 5 Add Step 1, Step 2, Step 3, and Step 4.

2.000"
.040"
.000"
\+ .012"

The vernier caliper reading is ------ 2.412" **Read two and four-hundred twelve thousandths of an inch.**

LAY OUT (CONSTRUCT) ANGLES

Lay Out an Angle Using a Protractor:

Example
Problem 1. Lay out a 42° angle using a straight edge, pencil, and a protractor.

Step 1. Draw a straight line with a straight edge and select the **vertex** (point at which the sides of the angle meet) as shown on the right.

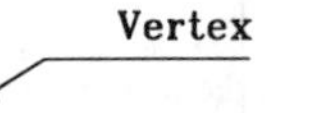

Step 2. Place a protractor over the line as shown on the right. Make sure the **zero points** on the protractor are in line with the line. The **vertex** point on the line and the vertex point on the protractor should also line up.

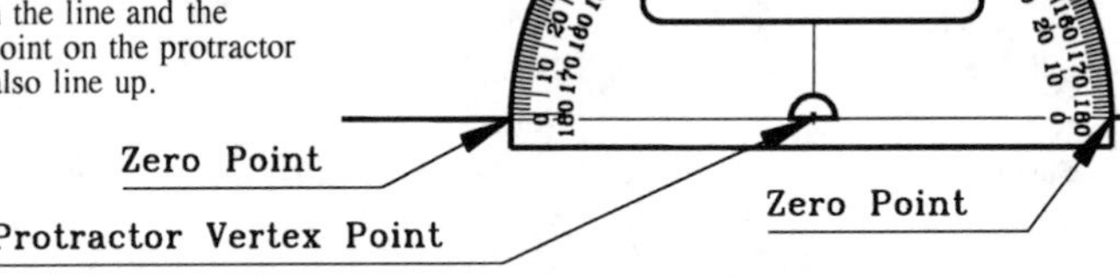

Step 3. Select the desired angle (in this case 42°) and mark it's location with a "T" mark as shown.

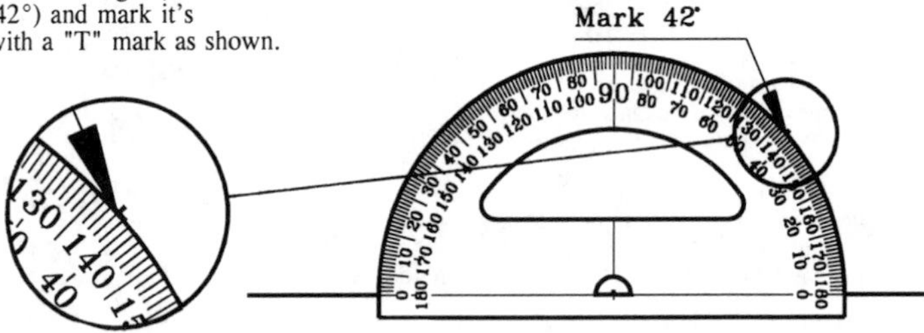

Step 4. Draw a straight line from the vertex of the angle through the "T" mark as shown on the right.

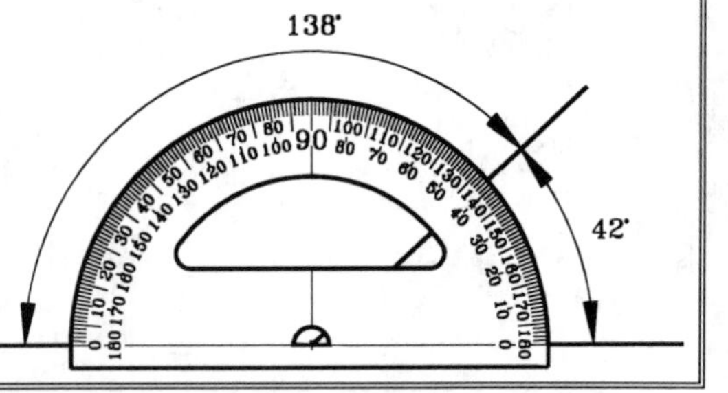

Lay Out an Angle Using a Compass:

Example
Problem 1. Lay out a right angle (90°) using a compass.

Step 1. Draw a straight line with a straight edge and select the **vertex** (point at which the sides of the angle meet) as shown on the right.

Vertex

Step 2. Using a compass, draw a 3" arc left and a 3" arc right of the vertex as shown. The points will be point **a** and point **b**.

Point a
Point b
3 in.
3 in.

Step 3. From points **a** and **b** draw 4" arcs to locate point **c** as shown.

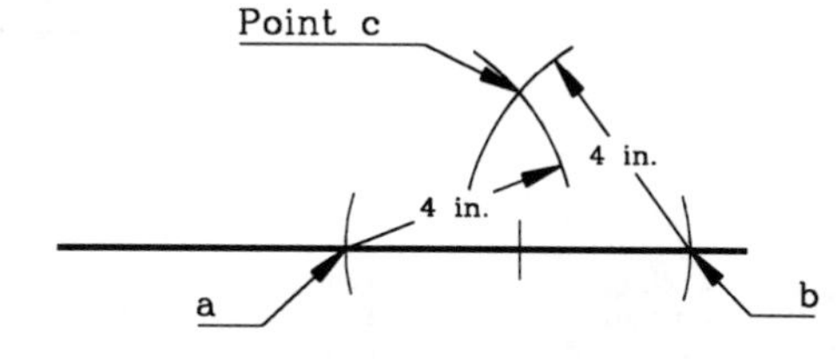

Step 4. Using a straight edge and a pencil, draw a straight line throught point **c** and the **vertex** point as shown.

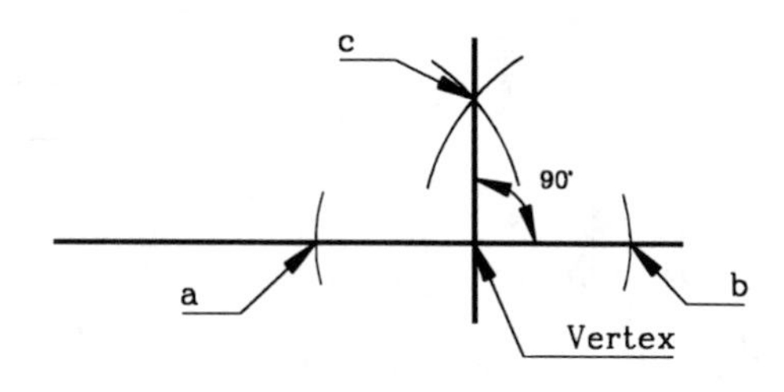

Step 5. Check the 90° angle by using a protractor.

Example
Problem 2. Lay out a 45° angle using a compass.

Step 1. Starting from the **vertex**, draw a 3" arc to form a new point, point **d**. Point **a** and point **d** <u>must be</u> the exact distance from the **vertex**.

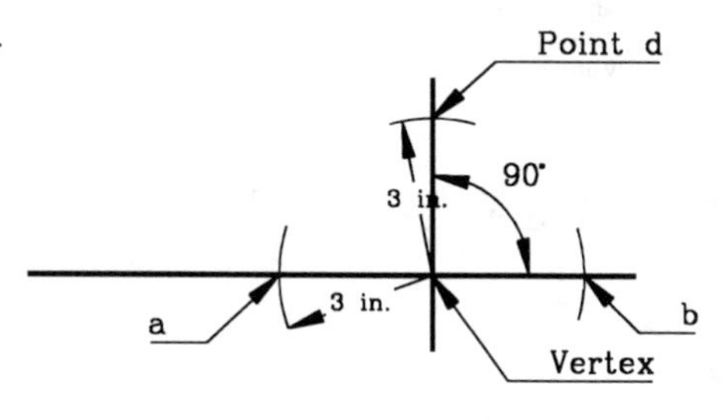

O1	◄ Task No. LAY OUT ANGLES (continued)

Step 2. Draw a 3.5" arc from both point **a** and point **d** to form point **e** as shown.

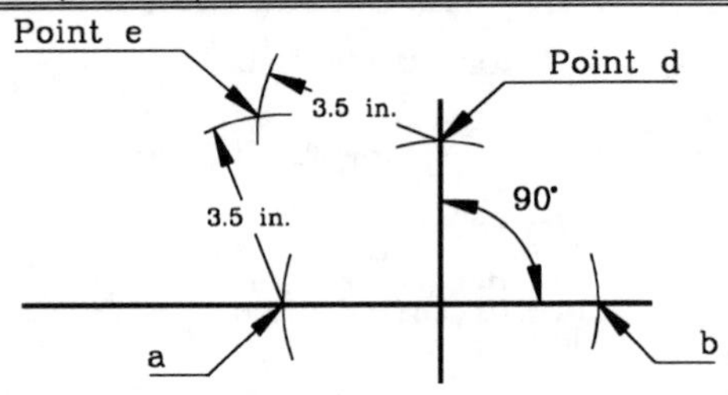

Step 3. Using a straight edge and a pencil, draw a straight line through point **e** and the **vertex** as shown.

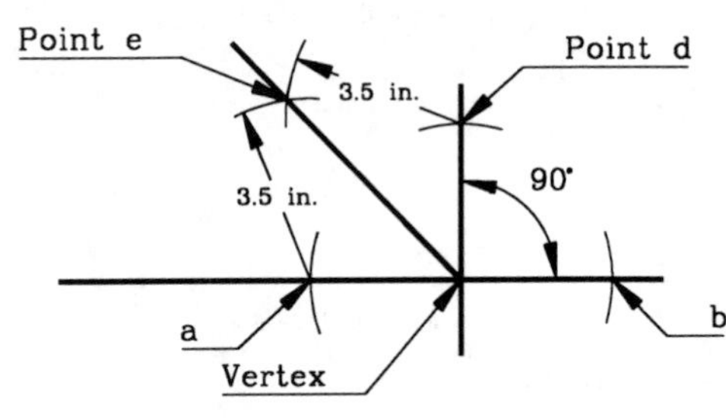

Step 4. Check the 45° angle by using a protractor. To draw a 22 1/2° angle, bisect (split in half) the 45° angle repeating the prior procedure.

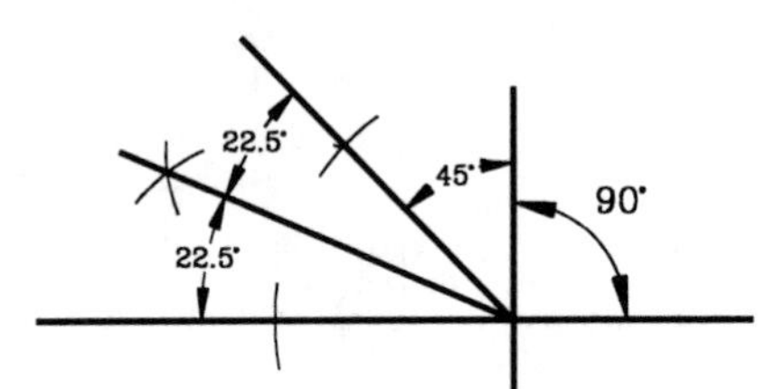

Bisect a Line or an Arc:

Example
Problem 1. Bisect (divide into two equal parts) a straight line.

Step 1. Draw a straight line with a straight edge as shown on the right. Label the ends of the line, point **A** and point **B**.

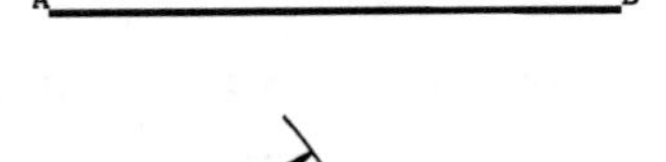

Step 2. Starting at point **A**, draw an arc (greater than half the distance of the line) above and below the line as shown on the right.

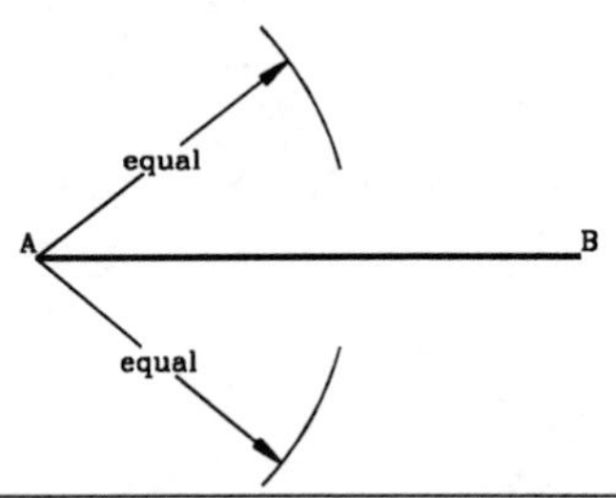

Step 3. Starting at point **B**, draw another arc the same size as the arc in step 2 above and below the line as shown. Where the two arcs cross above and below the line will be point **C** and point **D**.

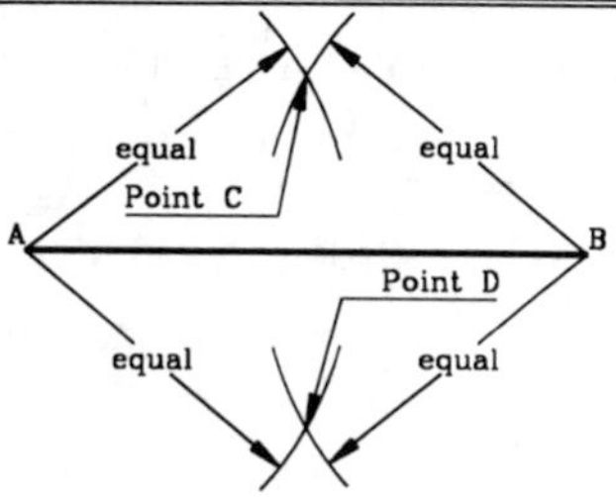

Step 4. Draw a straight line through point **C** and point **D** as shown on the right. The point where the two lines cross will be point **E**. The distance from point **A** to point **E** should be the same as the distance from point **E** to point **B**.

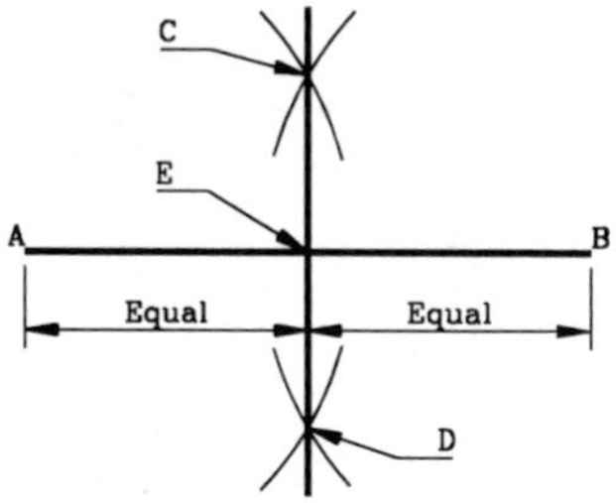

Example
Problem 2. Bisect an arc.

Step 1. Draw an arc of a circle. Label the ends of the arc, point **a** and point **b** as shown on the right.

Step 2. Starting at point **a**, draw an arc (greater than half the distance of the arc) above and below the arc as shown on the right.

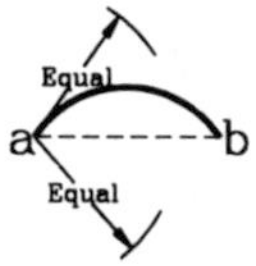

Step 3. Starting at point **b**, draw another arc the same size as the arc in step 2 above and and below the arc as shown on the right. Where the two arcs cross above and below the line will be point **c** and point **d**.

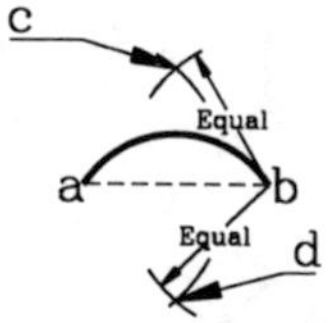

Step 4. Draw a straight line through point **c** and point **d** as shown on the right. The point where the two lines cross will be point **e**. The distance from point **a** to point **e** should be the same as the distance from point **e** to point **b**.

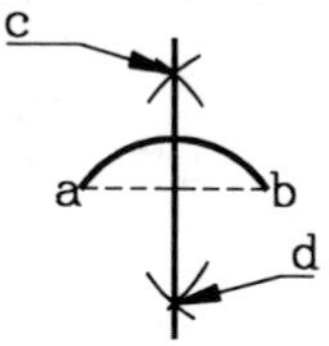

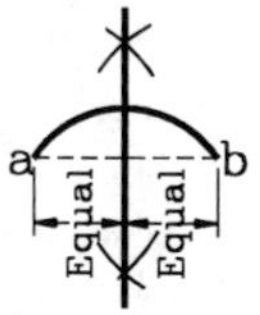

Copy an Angle Using a Layout Square:

Example
Problem 1. Transfer a given angle by using a layout square.

Step 1. Position the square with the vertex of the angle in line with the 12 inch mark on the blade of the square as shown below.

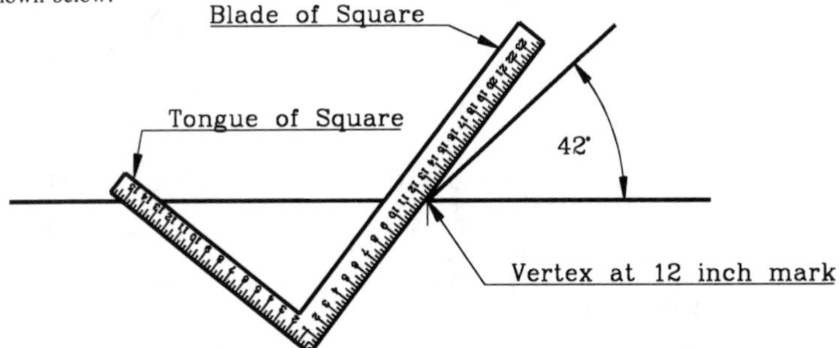

Step 2. Rotate the square pivoting on the vertex of the angle and the 12 inch mark on the blade, until the blade of the square and the side of the angle are exactly in line as shown below. Place a mark on the tongue of the square (with a pencil or marker) at the exact point where the line and square meet. In this case, mark 10 5/8-" (the best you can see with the naked eye).

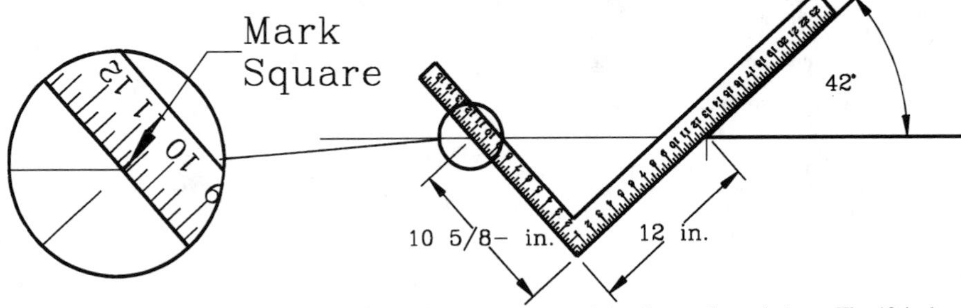

Step 3. Position the lay out square on a straight line or on a straight edge as shown below. The 12 inch mark on the blade and the 10 5/8-" mark on the tongue must line up exactly with the line or edge. Mark the angle on the blade of the square.

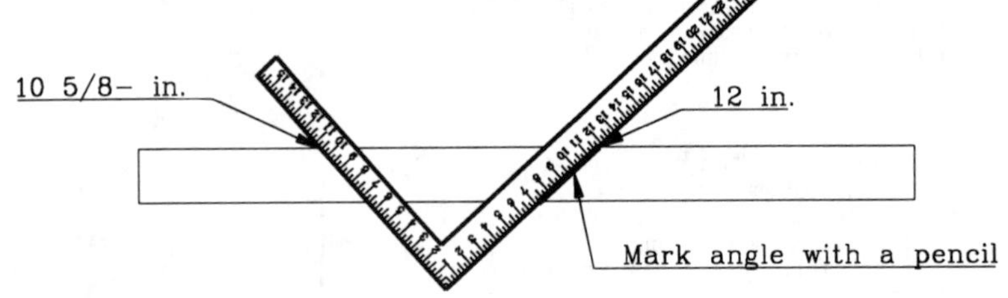

[Using trig, the tongue of the square above would measure 10.8" (Tan 42° = .90040 X 12 = 10.8"). For perfect angles, see the bottom of the next page.]

Task No. ▸ | O1

Lay Out Angles Using a Layout Square:

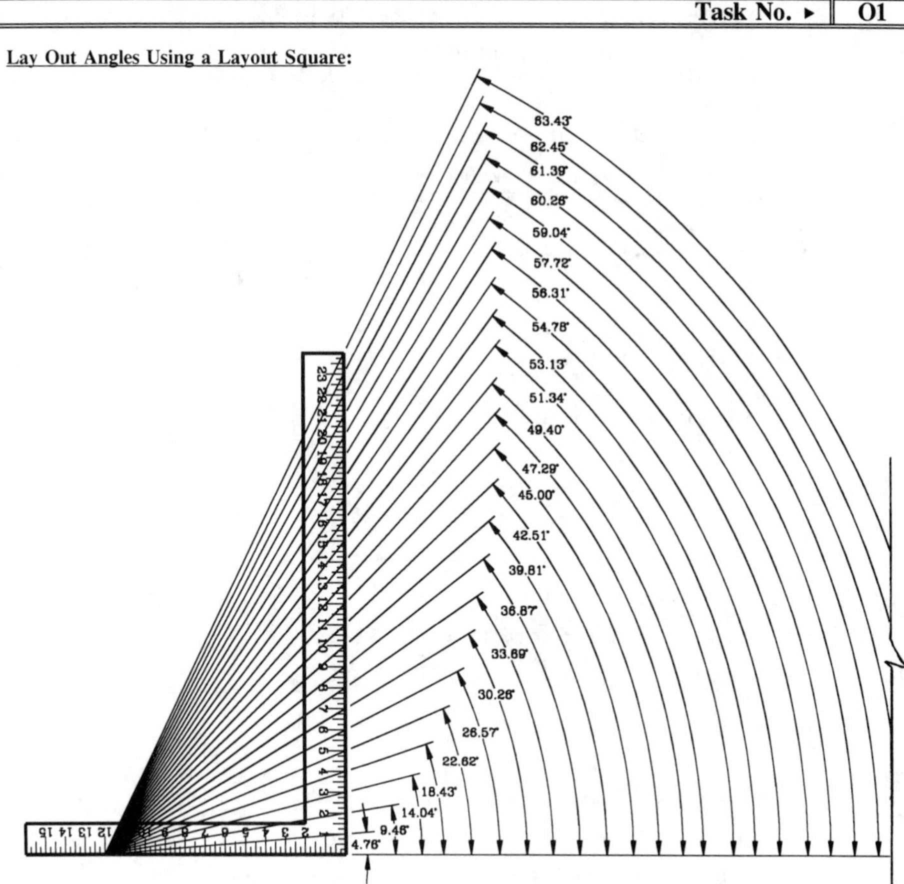

Directions: Using 12 on the tongue of the square as the vertex of an angle, pivot the square to even increments of 1 inch. The angle marked on the blade of the square will yield the degrees listed next to the even inch mark.

To get degrees on the 1/2 inch increments, take 1/2 of the difference between any two increments and add that difference to the preceding even increment. For example, 7 and 12 would be 30.26° and 8 and 12 would be 33.69°. The difference between the two would be 3.43°. 1/2 of 3.43° = 1.715°. 1.715° + 30.26° = 31.975°. Therefore, the angle for 7 1/2 and 12 would be approximately 32°.

[For perfect angles use the formula TOA. Formula: Tan = $\frac{\text{Side Opposite}}{\text{Side Adjacent}}$ Tan. = $\frac{7.5}{12}$ Tan. = .625

The top of page 252 shows Tangent .62487 to be 32°. Page 265 illustrates to select "32" than the "TAN" button = .624869352, or "2nd" than "TAN^{-1}" = 32°. Select the method that works best for you.]

LAY OUT EQUILATERAL TRIANGLES:
(THREE EQUAL ANGLES AND THREE EQUAL SIDES)

Example
Problem 1. Lay out an equilateral triangle by using a straight edge and a compass. The sides are 12 inches.

Step 1. Draw a 12 inch line with a straight edge (for longer lines use a chalkline or a transit) as shown on the right.

(See Task No. 09 how to use a homemade compass, if needed.)

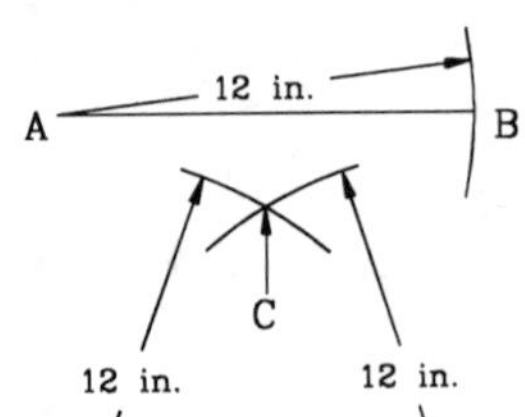

Step 2. Using a compass, draw a 12 inch arc from point **A** to approximately where point **C** will be located. Draw another 12 inch arc from point **B** to cross the first arc as shown on the right to form point **C**.

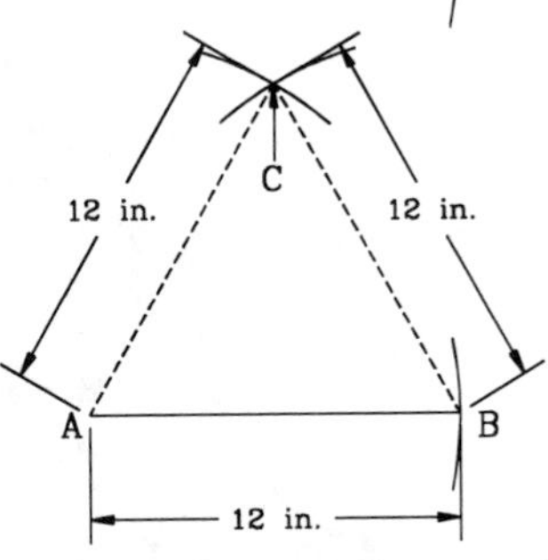

Step 3. Using a straight edge (chalkline or transit for larger triangles), draw a straight line from point **A** to point **C**, and from point **C** to point **B** to form the equilateral triangle as shown on the right.

Example
Problem 2. Lay out an equilateral triangle using the 3-4-5 squaring method or a transit. The sides are 12 ft. long.

Step 1. Draw a 12 foot line (line **AB**) using a chalkline and locate the midpoint, point **C**, as shown on the right below.

Step 2. Calculate the height of the triangle by multiplying the length of the side, 12 ft. times .866 as shown below. Draw a 10 ft. 4 3/4- in. line (perpendicular to line **AB**) from point **C** to point **D**, as shown on the right, by using the largest combination of 3-4-5 or a transit.

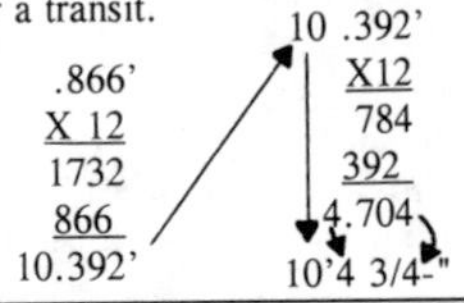

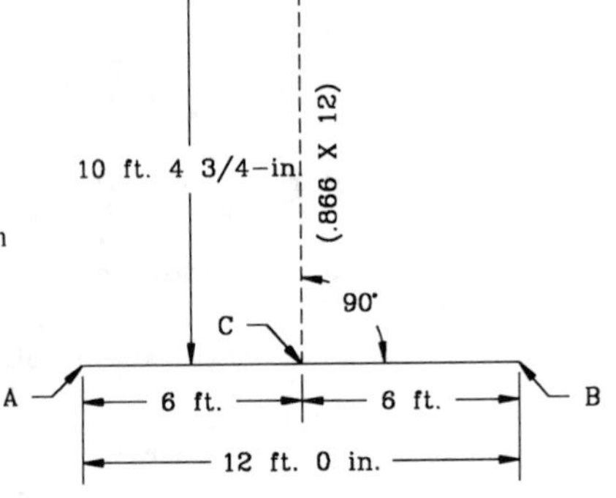

Step 3. Using a chalkline, draw a straight line from point **A** to point **D**, and from point **D** to point **B** to form the equilateral triangle as shown on the right.

Note how one unit with the height of .866' times 12 equals the total height of 10'4 3/4-".

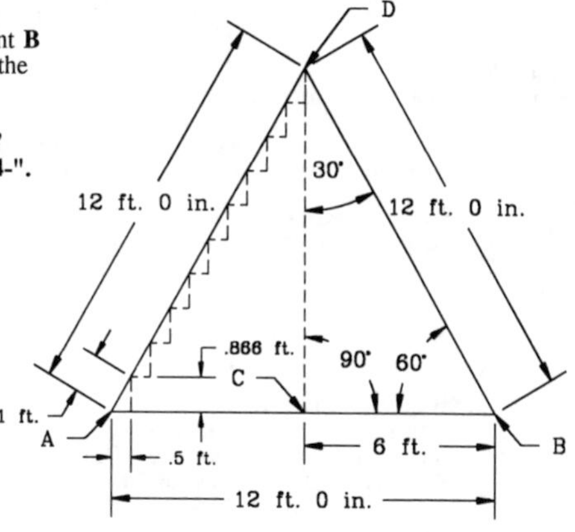

UNIT OF AN EQUILATERAL TRIANGLE
CIRCUMSCRIBED AND INSCRIBED WITHIN A CIRCLE:
(NOTE THE UNIT COULD BE INCHES, FEET, YARDS, METERS, OR SO FORTH)

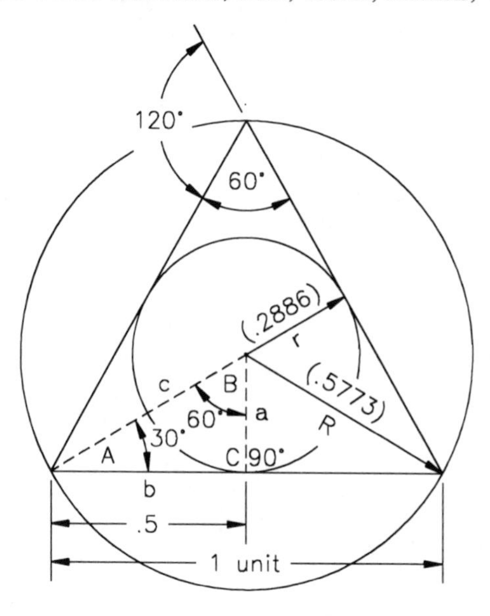

[Also, see Task No. K6B]

O2	◂ Task No.	LAY OUT TRIANGLES

LAY OUT ISOSCELES TRIANGLES:
(TWO EQUAL ANGLES AND TWO EQUAL SIDES)

Example
Problem 1. Lay out an isosceles triangle. The base is 12 inches and the two equal sides are 16 inches.

Step 1. Draw a 12 inch line with a straight edge (for longer lines use a chalkline) as shown the right.

(See Task No. 09 how to use a homemade compass, if needed.)

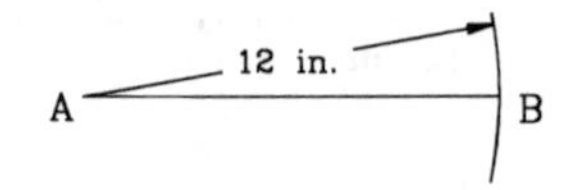

Step 2. Using a compass, draw a 16 inch arc from point **A** to approximately where point **C** will be located. Draw another 16 inch arc from point **B** to cross the first arc as shown on the right to form point **C**.

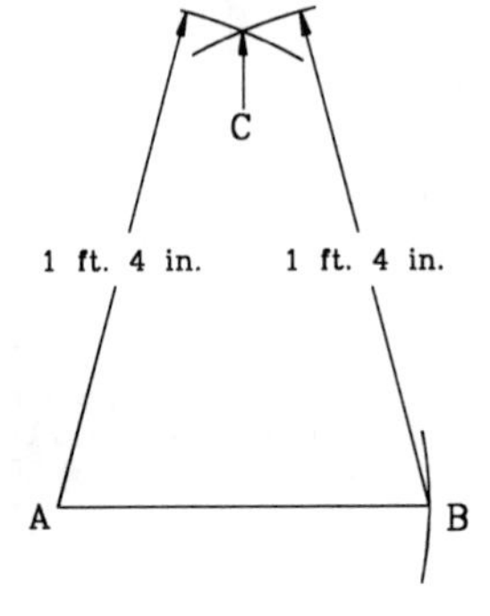

Step 3. Using a straight edge (chalkline or transit for larger triangles), draw a straight line from point **A** to point **C**, and from point **C** to point **B** to form the isosceles triangle as shown on the right.

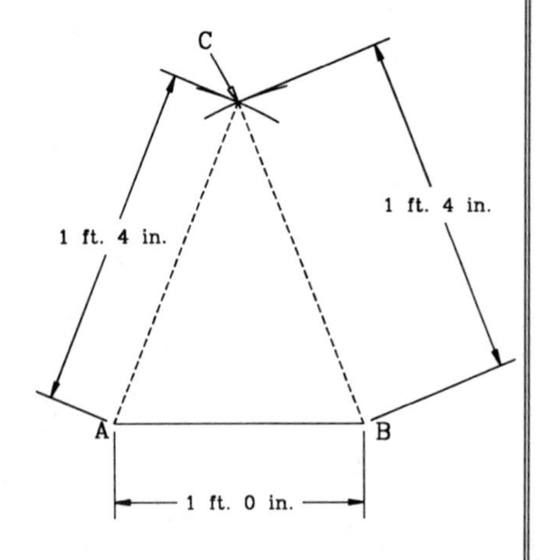

Task No. ▸ O2

Example
Problem 2. Lay out an isosceles triangle using the 3-4-5 squaring method or a transit. The base is 12 ft. and the two equal sides are 16 ft.

Step 1. Draw a 12 foot line with a chalkline and locate the midpoint, point **D**, as shown on the right.

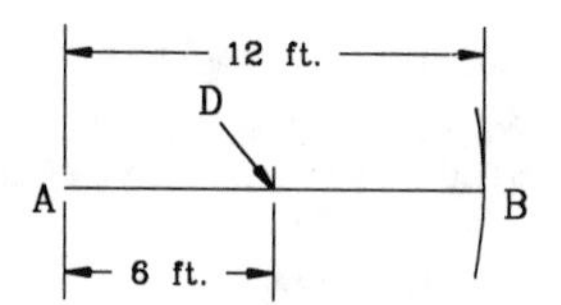

Step 2. Calculate the height of the triangle by using $a^2 + b^2 = c^2$ as shown below. Draw a right angle from point **D** to point **C** as shown on the right. For larger triangles use the 3-4-5 squaring method or a transit to form the right angle.

$$a^2 + b^2 = c^2$$
$$a^2 + 6^2 = 16^2$$
$$a^2 + 36 = 256$$
$$a^2 = 256 - 36$$
$$a^2 = 220$$
$$a = \sqrt{220}$$
$$a = 14.83 \text{ ft.}$$

14 .83'
X12
166
83
9 .96"
X16
576
96
15.36

14'9 15/16+"

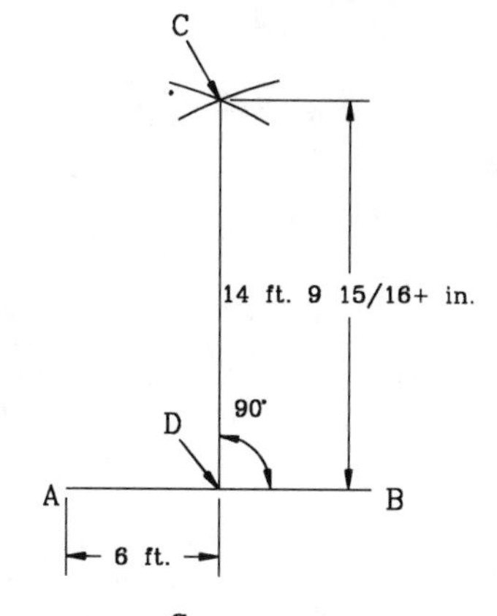

Step 3. Using a straight edge (chalkline or transit for larger triangles), draw a straight line from point **A** to point **C**, and from point **C** to point **B** to form the isosceles triangle as shown on the right.

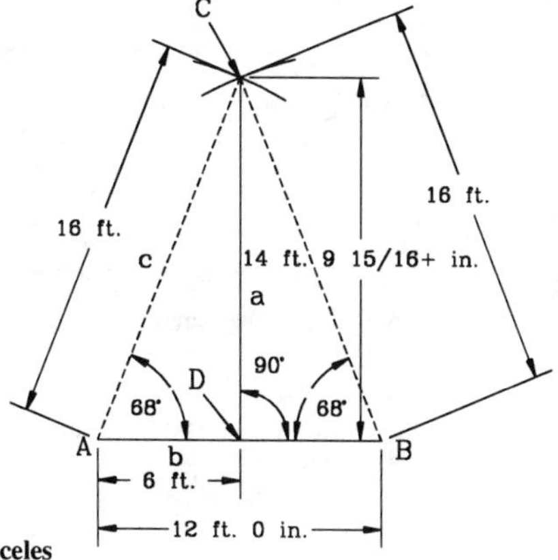

Note: Use the same steps as shown above on the isosceles triangle, but make the dimensions, inches (base 12" and two equal sides 16") instead of feet. Use a straight edge and a pencil instead of a chalkline.

[Also, see Task No. K6A]

LAY OUT SQUARES:
(FOUR EQUAL ANGLES AND FOUR EQUAL SIDES)

Example
Problem 1. Lay out a 12 inch square encompassed within a circle.

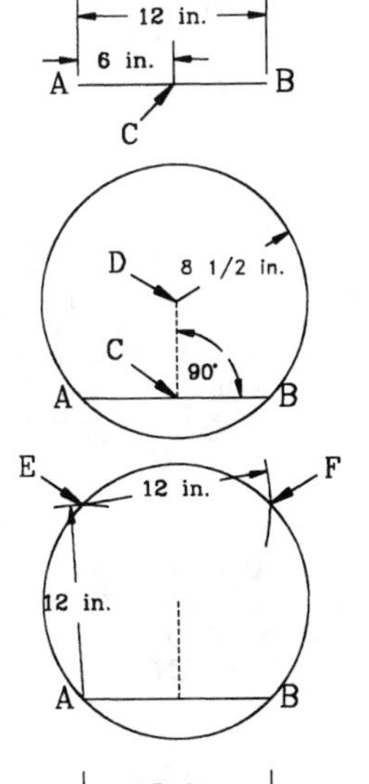

Step 1. Lay out a 12 inch side with a straight edge and locate the midpoint, point **C**, as shown on the right.

Step 2. Layout a 6 inch line (line **CD**) perpendicular to line **AB** by using a compass or a square as shown on the right.

Step 3. Multiply 12 times .707 to find the radius of the circumscribed circle as shown below. Starting at point **D**, draw a circle with a radius of 8 ½ inches as shown on the right.

Step 4. Using a compass, draw a 12 inch arc from point **A** to cross the circle to form point **E**. Draw another 12 inch arc from point **E** to cross the circle and form point **F** as shown on the right.

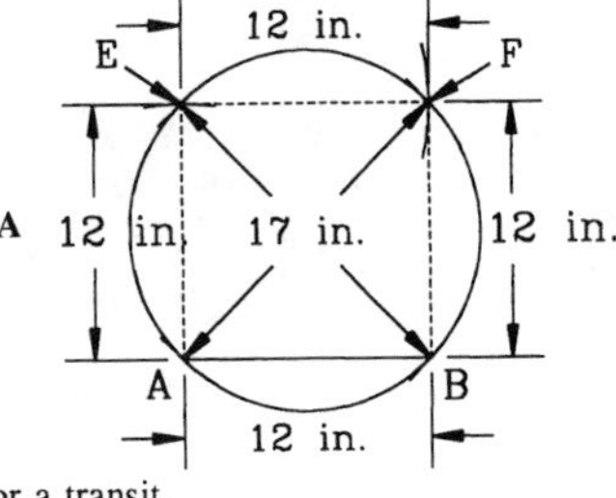

Step 5. Using a straight edge, draw a straight line from point **A** to point **E**, point **E** to point **F**, and from point **F** to point **B**.

Step 6. Check the square for accuracy by measuring the diagonals. The diagonals should measure exactly 17 inches (from point **A** to point **F**, and from point **B** to point **E**).

Example
Problem 2. Lay out a 12 foot square using the 3-4-5 squaring method or a transit.

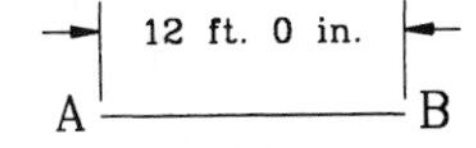

Step 1. Draw a 12 foot line (line **AB**) using a chalkline.

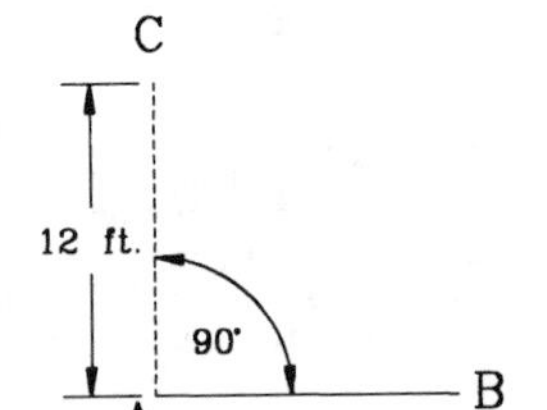

Step 2. Using a chalkline, lay out a 12 foot line (line **AC**) perpendicular to line **AB**.

[Also, see Task No. P1, Task No. P2, or Task No. R1]

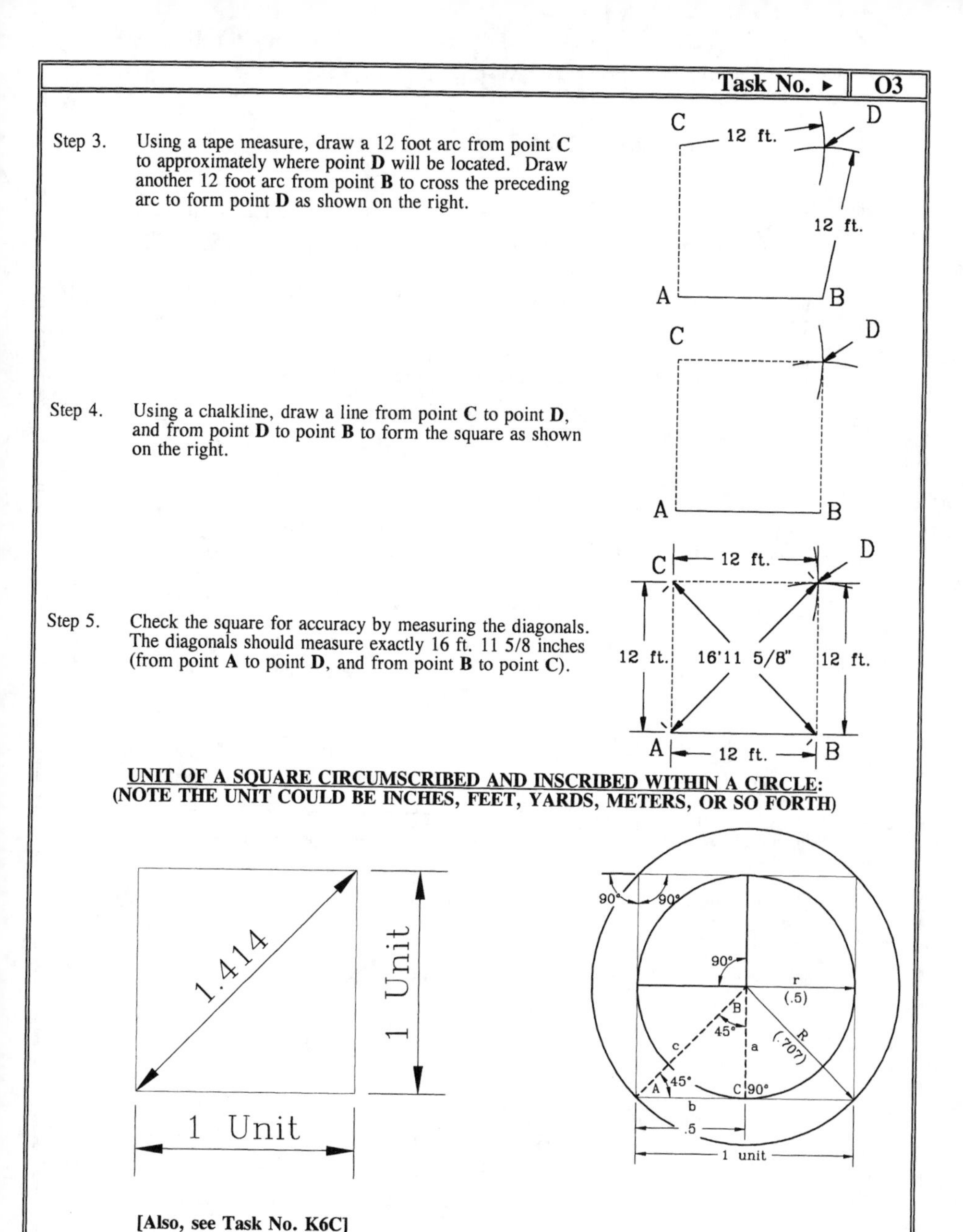

Step 3. Using a tape measure, draw a 12 foot arc from point **C** to approximately where point **D** will be located. Draw another 12 foot arc from point **B** to cross the preceding arc to form point **D** as shown on the right.

Step 4. Using a chalkline, draw a line from point **C** to point **D**, and from point **D** to point **B** to form the square as shown on the right.

Step 5. Check the square for accuracy by measuring the diagonals. The diagonals should measure exactly 16 ft. 11 5/8 inches (from point **A** to point **D**, and from point **B** to point **C**).

UNIT OF A SQUARE CIRCUMSCRIBED AND INSCRIBED WITHIN A CIRCLE:
(NOTE THE UNIT COULD BE INCHES, FEET, YARDS, METERS, OR SO FORTH)

[Also, see Task No. K6C]

LAY OUT RECTANGLES:
(FOUR EQUAL ANGLES AND FOUR PARALLEL SIDES)

Example
Problem 1. Lay out a 6 inch by 12 inch rectangle.

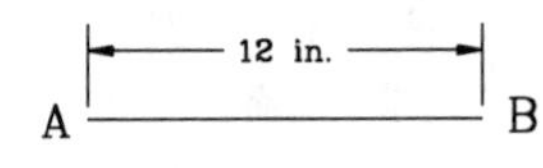

Step 1. Draw a 12 inch line (line **AB**) with a straight edge.

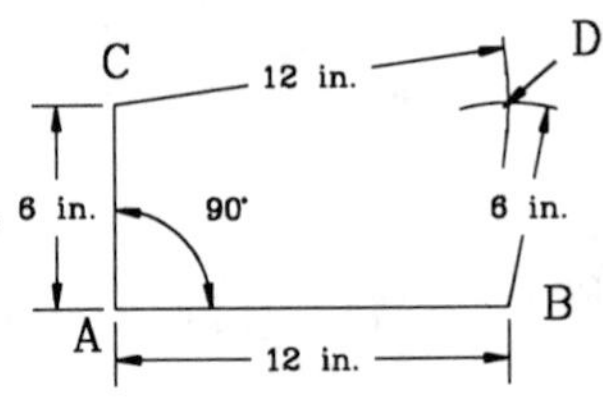

Step 2. Using a compass or a square, draw a 6 inch line (line **AC**) perpendicular to the 12 inch line as shown on right.

Step 3. Using a compass or a tape measure, draw a 12 inch arc from point **C** to approximately where point **D** will be located. located. Draw another 6 inch arc from point **B** to cross the preceding arc to form point **D** as shown on the right.

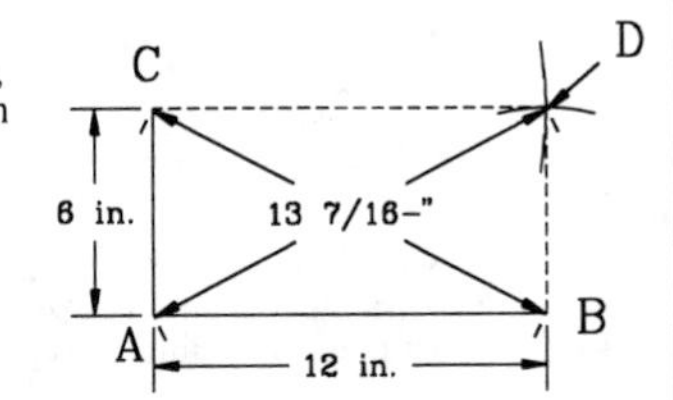

Step 4. Using a straight edge, draw a line from point **C** to point **D**, and from point **D** to point **B** to form the rectangle as shown on the right.

Step 5. Check the rectangle for accuracy by measuring the diagonals. The diagonals should measure exactly 13 7/16-" (from point **A** to point **D**, and from point **B** to point **C**).

Example
Problem 2. Lay out a 12 foot by 24 foot rectangle using the 3-4-5 squaring method or a transit.

Step 1. Draw a 24 foot line (line **AB**) using a chalkline.

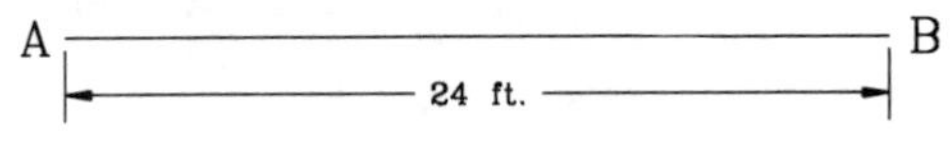

Step 2. Using a chalkline, lay out a 12 foot line (line **AC**) perpendicular to line **AB**. Use a tape measure to lay out the largest combination of 3-4-5 (12-16-20) to lay out the right angle. Or use a transit.

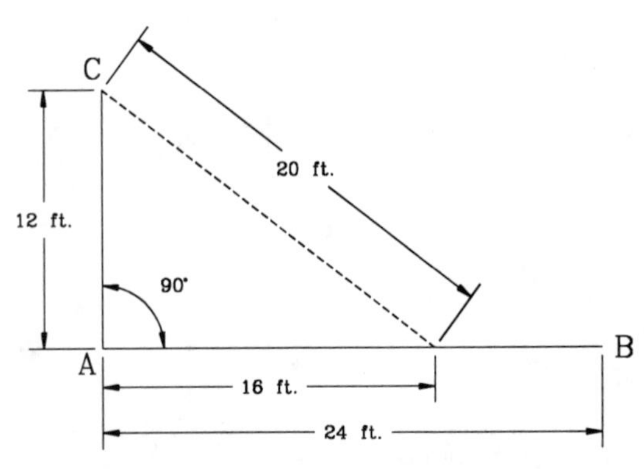

[Also, see Task No. P1, The 3-4-5 and 5-12-13 Squaring Method]

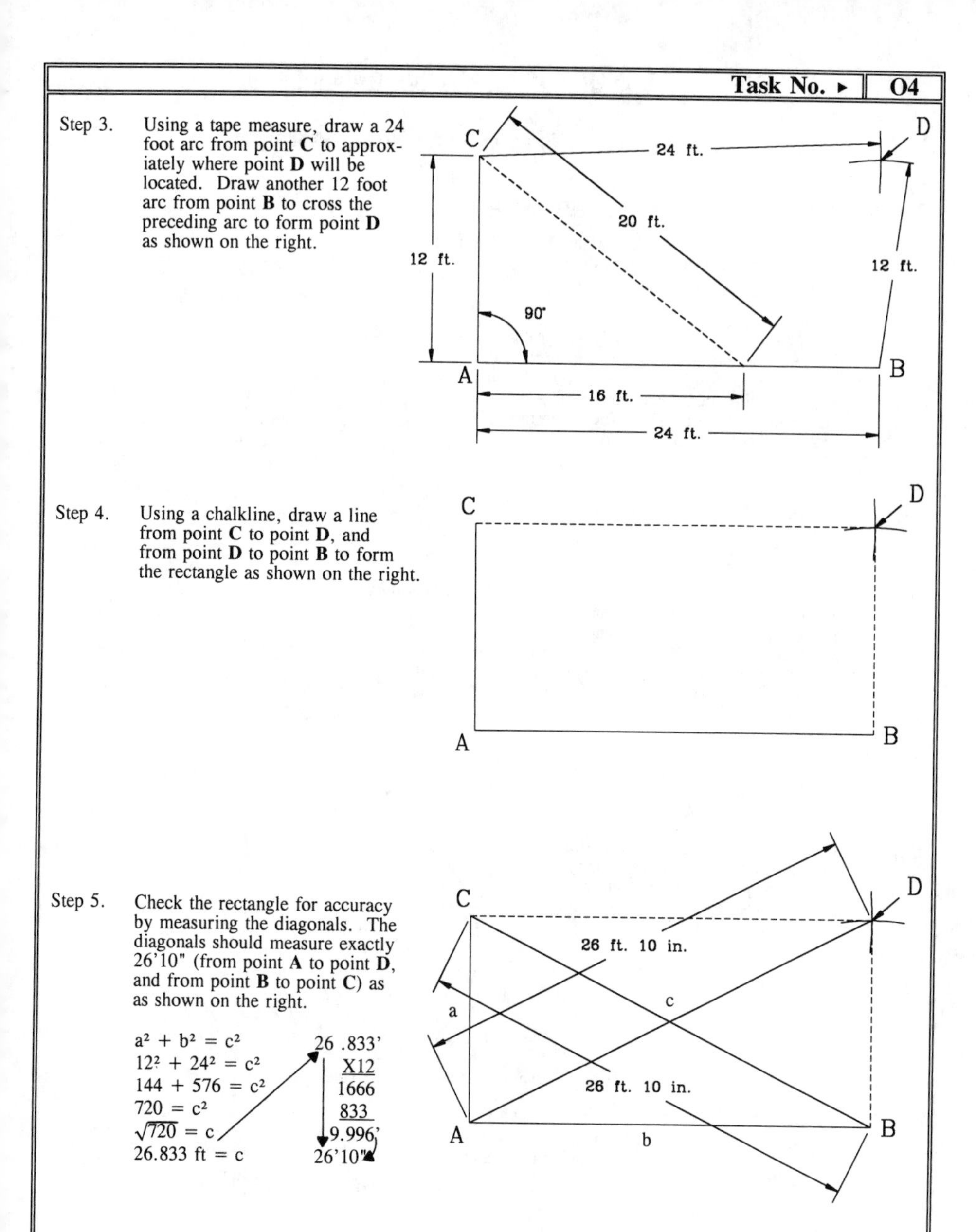

Task No. ▸ O4

Step 3. Using a tape measure, draw a 24 foot arc from point **C** to approxiately where point **D** will be located. Draw another 12 foot arc from point **B** to cross the preceding arc to form point **D** as shown on the right.

Step 4. Using a chalkline, draw a line from point **C** to point **D**, and from point **D** to point **B** to form the rectangle as shown on the right.

Step 5. Check the rectangle for accuracy by measuring the diagonals. The diagonals should measure exactly 26'10" (from point **A** to point **D**, and from point **B** to point **C**) as as shown on the right.

$a^2 + b^2 = c^2$
$12^2 + 24^2 = c^2$
$144 + 576 = c^2$
$720 = c^2$
$\sqrt{720} = c$
$26.833 \text{ ft} = c$

26 .833'
X12
1666
833
9.996'
26'10"

[Also, see Task No. K5A, The Rule of Pythagoras]

LAY OUT A PENTAGON:
(FIVE EQUAL ANGLES AND FIVE EQUAL SIDES)

Example
Problem 1. Lay out a pentagon by using a straight edge and a compass. The sides are 12 inches.

Step 1. Draw a 12 inch line with a straight edge (for longer lines use a chalkline or a transit) and find the center point as shown on the right.

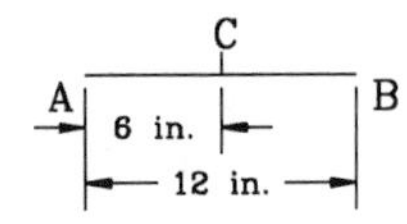

Step 2. Calculate the inscribed circle radius of the pentagon by multiplying the length of the side, 12 inches times .688 as shown below. Draw a 8 1/4 inch line (perpendicular to line **AB**) from point **C** to point **D**, as shown on the right, by using a compass or a square.

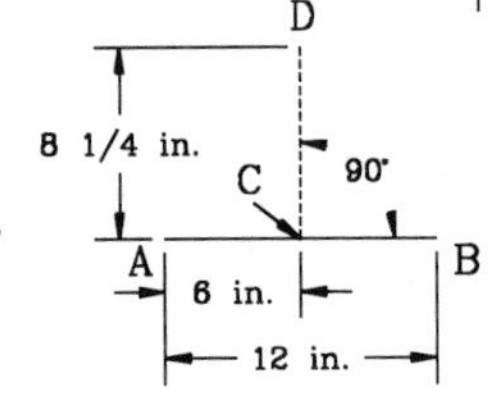


```
  .688"
  X 12
  1376
  688
 8.256    = 8 1/4+"
```

Step 3. Calculate the circumscribed circle radius of the pentagon by multiplying the length of the side, 12 inches times .851 as as shown below. Using a compass, draw a 10 1/4-" inch circle as shown on the right. Then draw 12 inch arcs from point **B** to Point **E**, from point **E** to point **F**, from point **F** to point **G**, and from point **A** to point **G**.

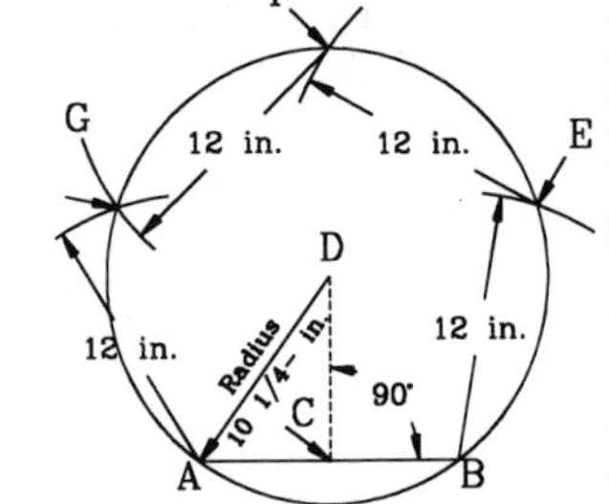


```
  .851"
  X 12
  1702
  851
 10.212    = 10 1/4-"
```

Step 4. Using a straight edge, draw a straight line from point **B** to point **E**, from point **E** to point **F**, from point **F** to point **G**, and from point **G** to point **A**. As shown on the right.

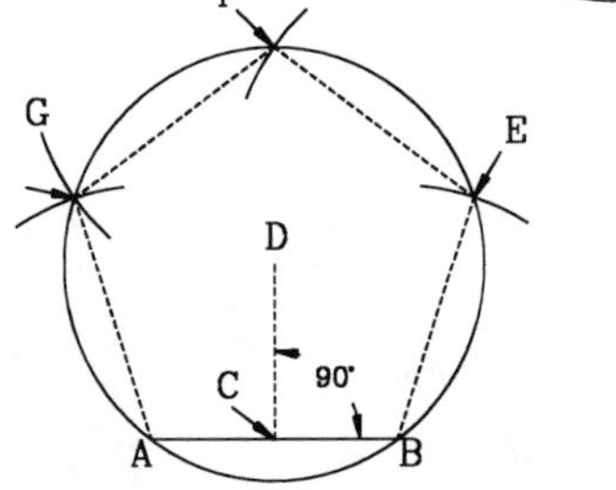

Example
Problem 2. Lay out a pentagon by using a protractor or a transit. The sides could be 12 inches or 12 feet.

Step 1. Draw a 12 foot (or 12 inch) line (line **AB**) by using a straight edge, chalkline, or a transit. Then draw a semi-circle (arc **BC**) and divide it into 5 equal 36° angles as shown on the right.

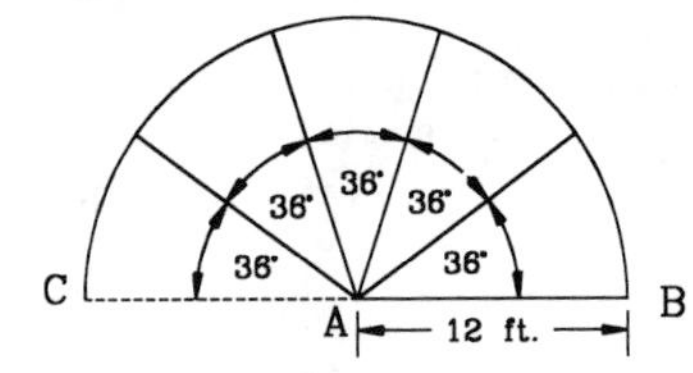

Step 2. Draw a 12 ft. arc from point **B** to approximately where point **D** will be located. Draw a 36° line from point **A** to cross the preceding arc to form point **D**. Draw another 12 ft. arc from point **D** to approximately where point **E** will be located. Draw a 72° line from point **A** to cross the preceding arc to form point **E**. Draw another 12 ft. arc from point **E** to point **F** to complete the pentagon points as shown on the right.

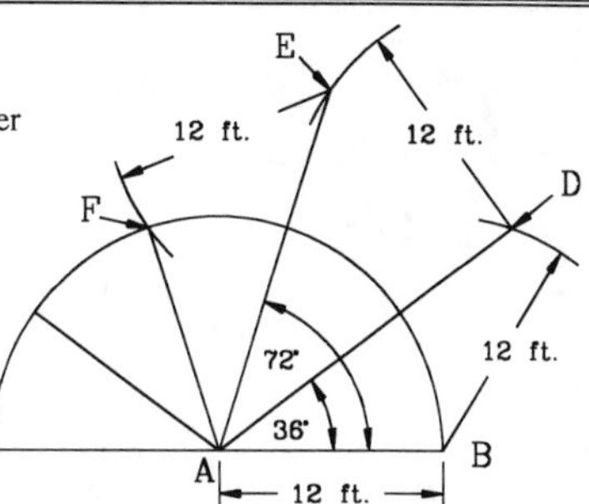

Step 3. Using a chalkline or straight edge, draw a straight line from point **B** to point **D**, from point **D**, to point **E**, and from point **E** to point **F**.

E
12 ft.
12 ft.
D
F
(19 ft. 4 15/16 in.)
19 ft. 4 15/16 in.
72°
12 ft.
36°
A
B
12 ft.
12 X 1.6176

UNIT OF A PENTAGON CIRCUMSCRIBED AND INSCRIBED WITHIN A CIRCLE:
(NOTE UNIT COULD BE INCHES, FEET, YARDS, METERS, OR SO FORTH)

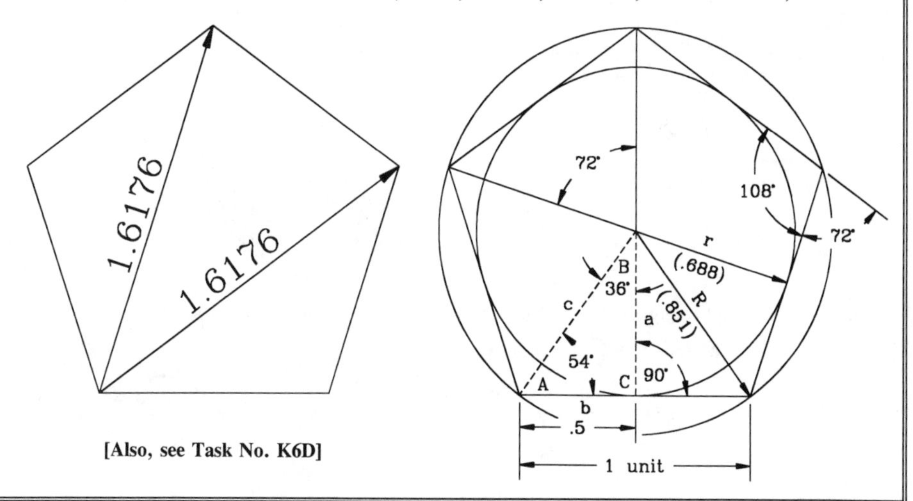

[Also, see Task No. K6D]

LAY OUT A HEXAGON:
(SIX EQUAL ANGLES AND SIX EQUAL SIDES)

Example
Problem 1. Lay out a hexagon by using a straight edge and a compass. The sides are 12 inches.

Step 1. Draw a 12 inch line with a straight edge (for longer lines use a a chalkline or a transit) and find the center point as shown on the right.

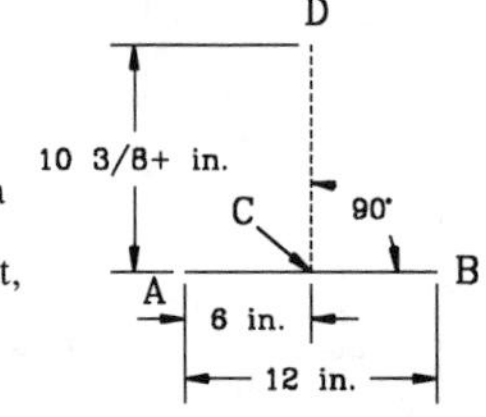

Step 2. Calculate the inscribed circle radius of the hexagon by multiplying the length of the side, 12 inches times .866 as shown below. Draw a 10 3/8+" inch line (perpendicular to line **AB**) from point **C** to point **D**, as shown on the right, by using a compass or a square.

.866"
X 12
1732
866
10.392 = 10 3/8"

Step 3. Note that the radius of the circumscribed circle is the same as the length of the side. Using a a compass, draw a 12" inch circle as shown on the right. Then draw 12 inch arcs from point **B** to Point **E**, from point **E** to point **F**, from point **F** to point **G**, from point **G** to point **H**, and from point **H** to point **A.**

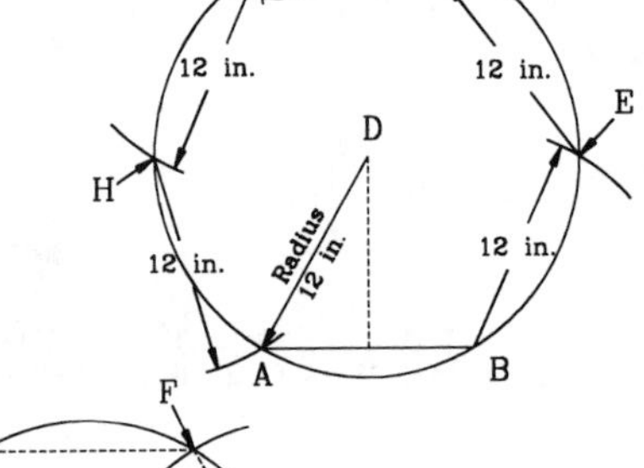

Step 4. Using a straight edge, draw a straight line from point **B** to point **E**, from point **E** to point **F**, from point **F** to point **G**, from point **G** to point **H**, from point **H** to point **A**. As shown on the right.

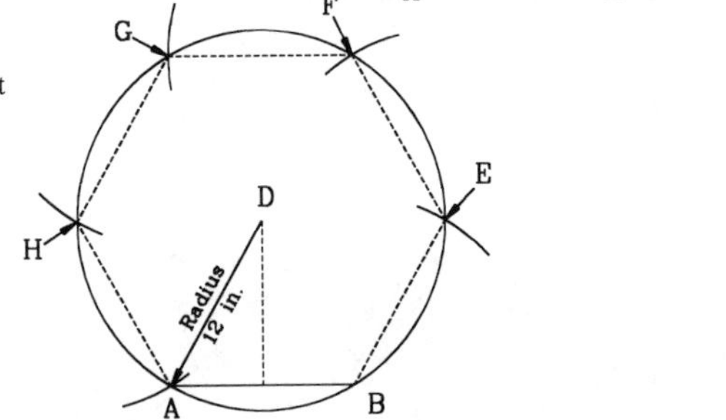

Example
Problem 2. Lay out a hexagon by using a protractor or a transit. The sides could be 12 inches or 12 feet.

Step 1. Draw a 12 foot (or 12 inch) line (line **AB**) by using a straight edge, chalkline, or a transit. Then draw a semi-circle (arc **BC**) and divide it into 6 equal 30° angles as shown on the right.

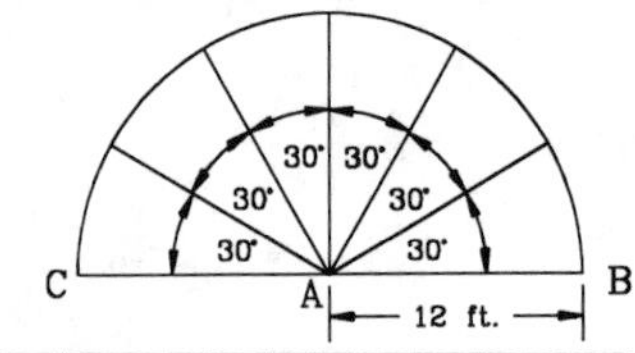

Step 2. Draw a 12 ft. arc from point **B** to approximately where point **D** will be located. Draw a 30° line from point **A** to cross the preceding arc to form point **D**. Draw another 12 ft. arc from point **D** to approximately where point **E** will be located. Draw a 60° line from point **A** to cross the preceding arc to form point **E**. Draw another 12 ft. arc from point **E** to approximately where point **F** will be located. Draw a 90° line from point **A** to cross the preceding arc to form point **F**. Draw another 12 ft. arc from point **G** to point **F** to complete the hexagon as shown on the right.

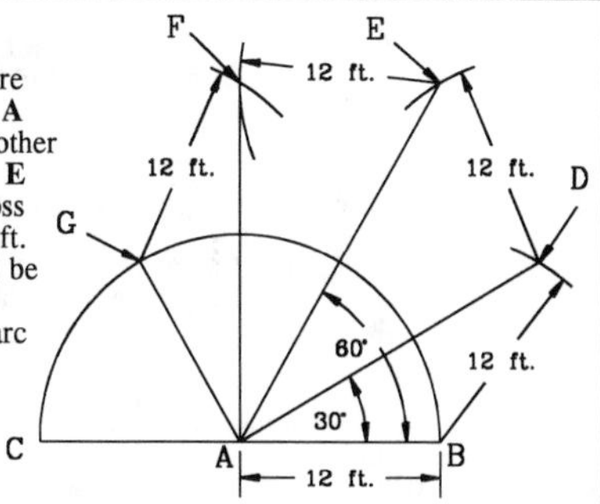

Step 3. Using a chalkline or straight edge, draw a straight line from point **B** to point **D**, from point **D** to point **E**, from point **E** to point **F**, from point **F** to point **G**, and from point **G** to point **A**.

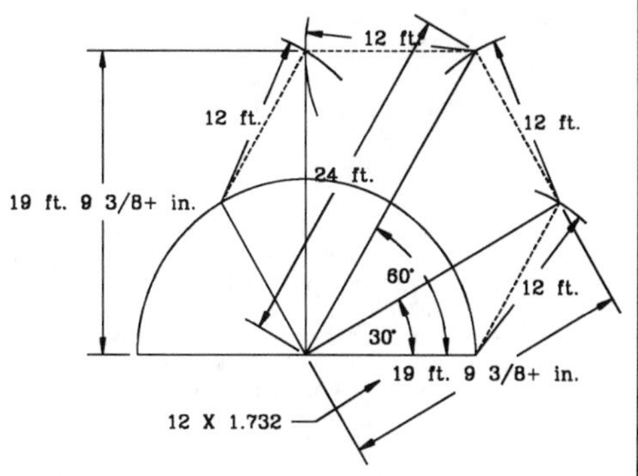

UNIT OF A HEXAGON CIRCUMSCRIBED AND INSCRIBED WITHIN A CIRCLE:
(NOTE UNIT COULD BE INCHES, FEET, YARDS, METERS, OR SO FORTH)

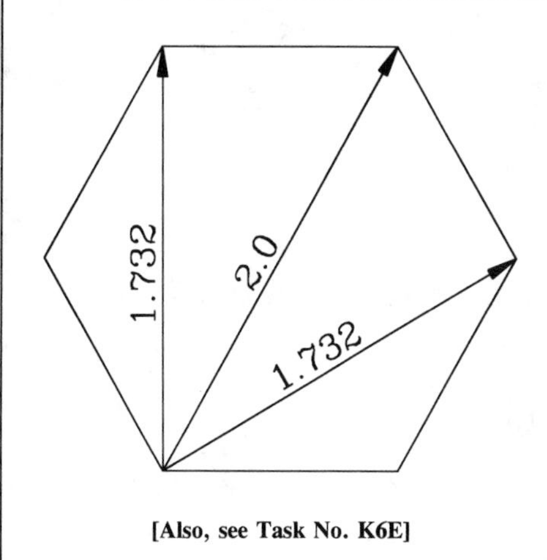

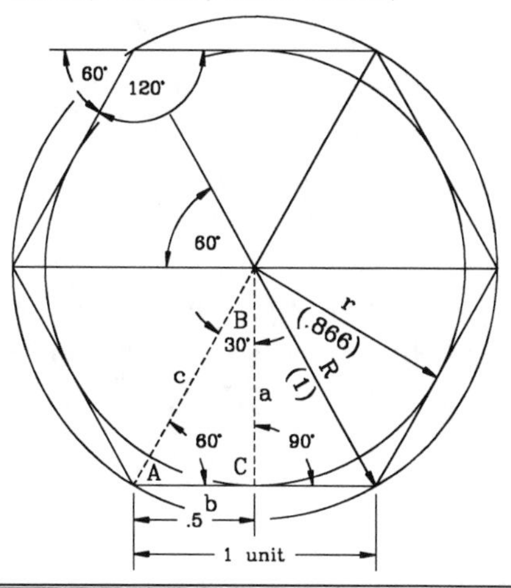

[Also, see Task No. K6E]

LAY OUT A HEPTAGON:
(SEVEN EQUAL ANGLES AND SEVEN EQUAL SIDES)

Example
Problem 1. Lay out a heptagon by using a straight edge and a compass. The sides are 12 inches.

Step 1. Draw a 12 inch line with a straight edge (for longer lines use a a chalkline or a transit) and find the center point as shown on the right.

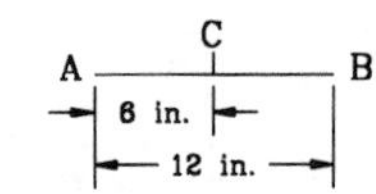

Step 2. Calculate the inscribed circle radius of the heptagon by multiplying the length of the side, 12 inches times 1.038 as shown below. Draw a 12 7/16+ inch line (perpendicular to line **AB**) from point **C** to point **D**, as shown on the right, by using a compass or a square.

```
 1.038"
  X 12
  2076
 1038
12.456    = 12 7/16+"
```

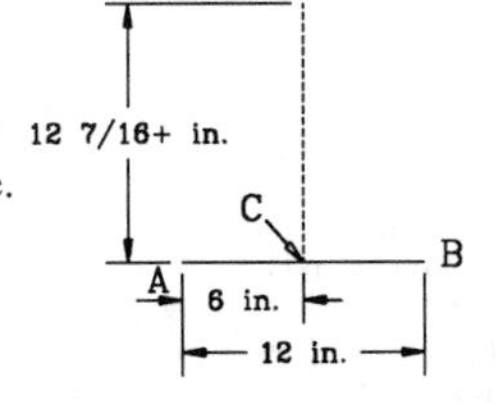

Step 3. Calculate the circumscribed circle radius of the heptagon by multiplying the length of the side, 12 inches times 1.152 as shown below. Using a compass, draw a 13 13/16 inch circle as shown on the right. Then draw 12 inch arcs from point **B** to point **E**, from point **E** to point **F**, from point **F** to point **G**, and from point **G** to point **H**, from point **H** to point **I**, and from point **A** to point **I**.

```
 1.152"
  X 12
  2304
 1152
13.824    = 13 13/16"
```

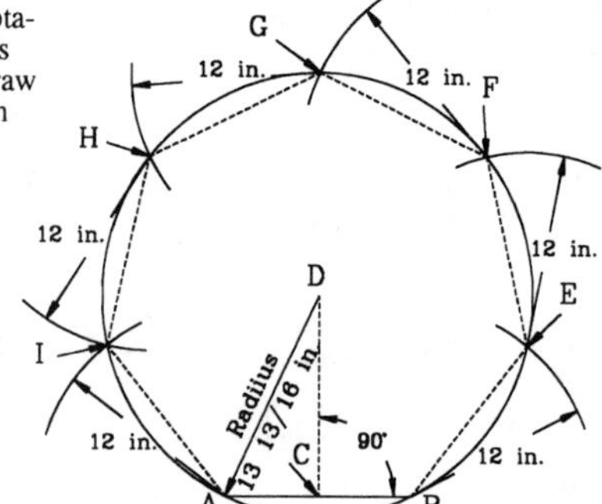

Step 4. Using a straight edge, draw a straight line from point **B** to point **E**, from point **E** to point **F**, from point **F** to point **G**, from point **G** to point **H**, from point **H** to point **I**, and from point **I** to point **A** as shown on the right.

Example
Problem 2. Lay out a heptagon by using a protractor or a transit. The sides could be 12 inches or 12 feet.

Step 1. Draw a 12 foot (or 12 inch) line (line **AB**) by using a straight edge, chalkline, or a transit. Then draw a semi-circle (arc **BC**) and divide it into 7 equal 25°42'51" (25°43' rounded off) angles as shown on the right.

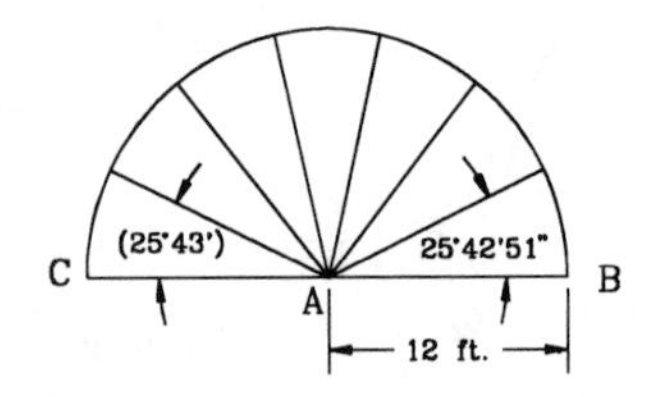

Step 2. Draw a 12 ft. arc from point **B** to approximately where point **D** will be located. Draw a 25°43' line from point **A** to cross the preceding arc to form point **D**. Draw another 12 ft. arc from point **D** to approximately where point **E** will be located. Draw a 51°26' line from point **A** to cross the preceding arc to form point **E**. Draw another 12 ft. arc from point **E** to approximately where point **F** will be located. Draw a 77°9' line from point **A** to cross the preceding arc to form point **F**. Draw another 12 ft. arc from point **H** to approximately where point **G** will be located. Draw a 102°52' line from point **A** to cross the preceding arc to form point **G**. Draw another 12 ft. arc from point **G** to point **F** to complete the heptagon as shown on the right.

Step 3. Using a chalkline or straight edge, draw a straight line from point **B** to point **D**, from point **D** to point **E**, from point **E** to point **F**, from point **F** to point **G**, from point **G** to point **H**, and from point **H** to point **A**.

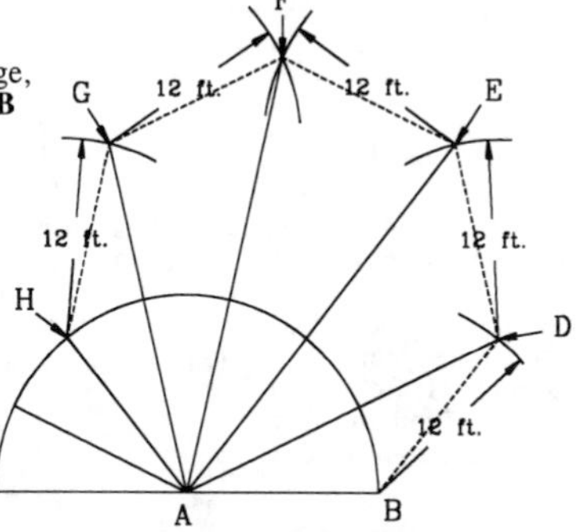

UNIT OF A HEPTAGON CIRCUMSCRIBED AND INSCRIBED WITHIN A CIRCLE:
(NOTE UNIT COULD BE INCHES, FEET, YARDS, METERS, OR SO FORTH)

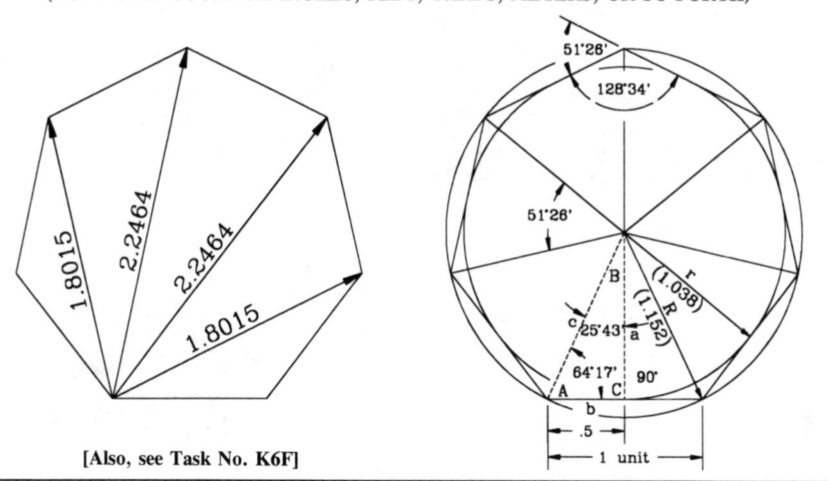

[Also, see Task No. K6F]

LAY OUT AN OCTAGON:
(EIGHT EQUAL ANGLES AND EIGHT EQUAL SIDES)

Example
Problem 1. Lay out an octagon by using a straight edge and a compass. The sides are 12 inches.

Step 1. Draw a 12 inch line with a straight edge (for longer lines use a a chalkline or a transit) and find the center point as shown on the right.

A C B
6 in.
12 in.

Step 2. Calculate the inscribed circle radius of the octagon by multiplying the length of the side, 12 inches times 1.207 as shown below. Draw a 14 1/2- inch line (perpendicular to line **AB**) from point **C** to point **D**, as shown on the right, by using a compass or a square.

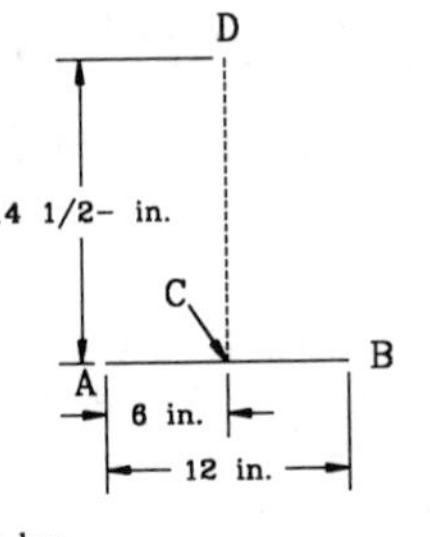


```
  1.207"
  X 12
  2414
 1207
14.484   = 14 1/2-"
```

Step 3. Calculate the circumscribed circle radius of the heptagon by multiplying the length of the side, 12 inches times 1.307 as shown below. Using a compass, draw a 15 5/8+ inch circle as shown on the right. Then draw 12 inch arcs from point **B** to point **E**, from point **E** to point **F**, from point **F** to point **G**, from point **G** to point **H**, from point **H** to point **I**, from point **I** to point **J**, and from point **A** to point **J**.

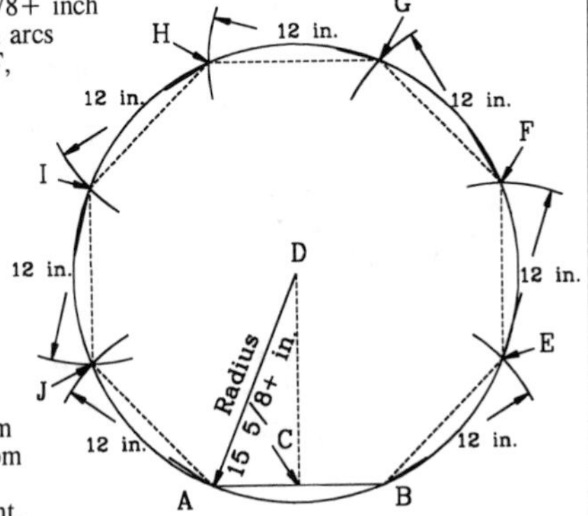


```
  1.307"
  X 12
  2614
 1307
15.684   = 15 5/8+"
```

Step 4. Using a straight edge, draw a straight line from point **B** to point **E**, from point **E** to point **F**, from point **F** to point **G**, from point **G** to point **H**, from point **H** to point **I**, and from point **I** to point **J**, and from point **J** to point **A** as shown on the right.

Example
Problem 2. Lay out an octagon by using a protractor or a transit. The sides could be 12 inches or 12 feet.

Step 1. Draw a 12 foot (or 12 inch) line (line **AB**) by using a straight edge, chalkline, or a transit. Then draw a semi-circle (arc **BC**) and divide it into 8 equal 22°30' angles as shown on the right.

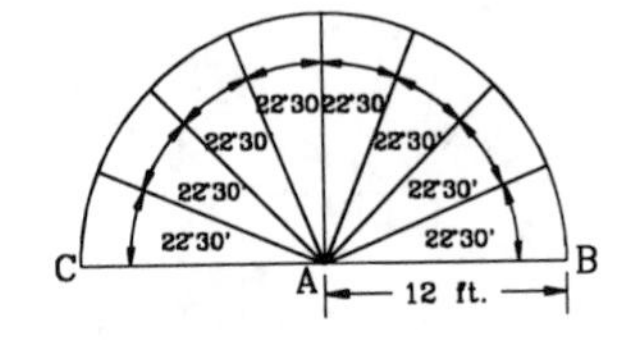

Step 2. Draw a 12 ft. arc from point **B** to approximately where point **D** will be located. Draw a 22°30' line from point **A** to cross the preceding arc to form point **D**. Draw another 12 ft. arc from point **D** to approximately where point **E** will be located. Draw a 45° line from point **A** to cross the preceding arc to form point **E**. Draw another 12 ft. arc from point **E** to approximately where point **F** will be located. Draw a 67°30' line from point **A** to cross the preceding arc to form point **F**. Draw another 12 ft. arc from point **F** to approximately where point **G** will be located. Draw another 12 ft. arc from point **I** to approximately where point **H** will be located. Draw a 112°30' line from point **A** to cross the preceding arc to form point **H**. Draw another 12 ft. arc from point **H** to cross the arc at point **G** to complete the octagon as shown on the right.

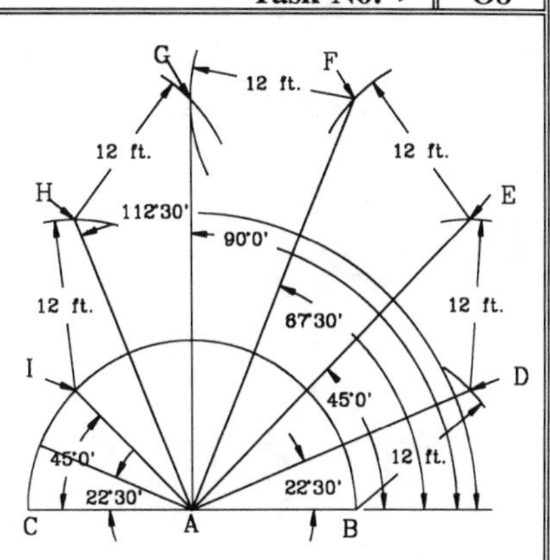

Step 3. Using a chalkline or straight edge, draw a straight line from point **B** to point **D**, from point **D** to point **E**, from point **E** to point **F**, from point **F** to point **G**, from point **G** to point **H**, from point **H** to point **I**, and from point **I** to point point **A** to complete the octagon.

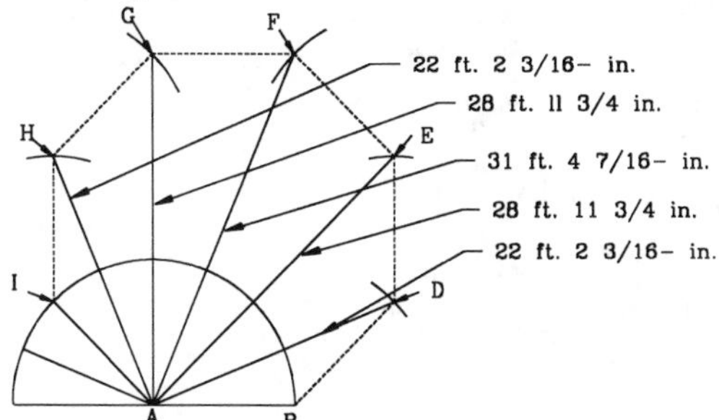

UNIT OF AN OCTAGON CIRCUMSCRIBED AND INSCRIBED WITHIN A CIRCLE:
(NOTE UNIT COULD BE INCHES, FEET, YARDS, METERS, OR SO FORTH)

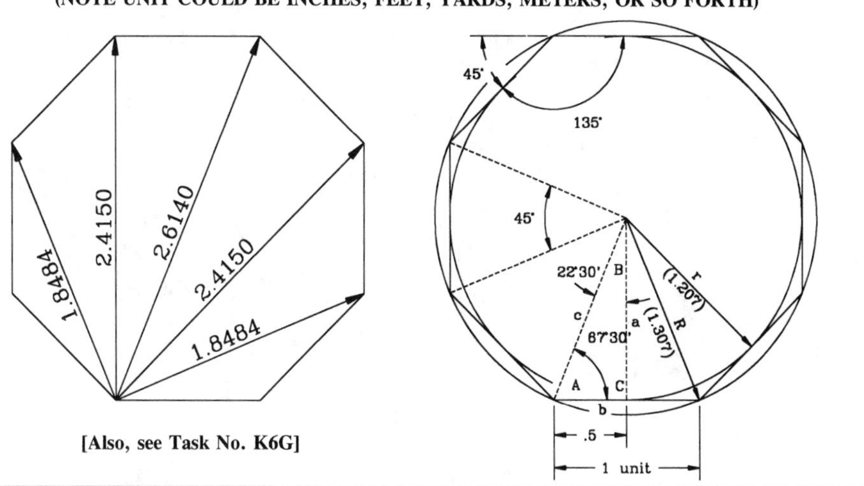

[Also, see Task No. K6G]

LAY OUT A CIRCLE:

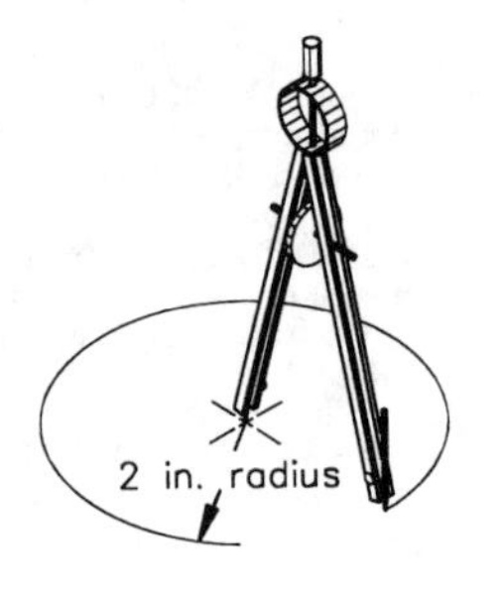

Lay Out a Circle With a Compass:

Example
Problem 1. Lay out a 4 inch diameter circle with a compass.

Step 1. This is a small circle, therefore a a standard compass can be used.

Example
Problem 2. Lay out a 16 inch circle with a homemade compass.

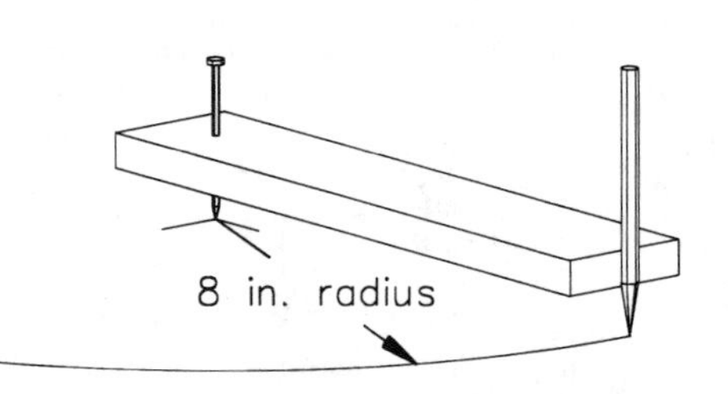

Step 1. This is a large circle, therefore a larger compass should be used. Make a homemade compass with a small board. Drill a hole on one end with a drill bit, nail (cut head off), or screw bit.

Step 2. Measure the radius (8 inch) of the circle from the drilled hole to the edge of the board and cut the board off at that point.

Step 3. Nail or screw the board in the circle center point and lay out the circle as shown. Use a sharp pencil. Check with a tape measure and make adjustments if necessary.

Lay Out a Circle With a Layout Square:

Example
Problem 1. Lay out an 8 inch circle with a layout square. Note, it is possible to lay out circles up to 13 7/8 inches with a layout square.

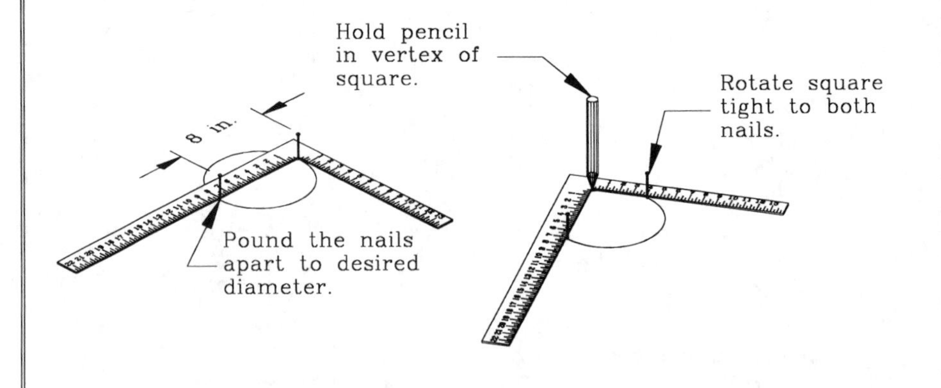

LAY OUT AN ELLIPSE — Task No. ▸ O10

LAY OUT AN ELLIPSE:

[An easy way to draw an ellipse is to use a template. Most templates range is size from 15° to 60°.

Examples:

Template angle:	**15°**	**20°**	**25°**	**30°**	**35°**	**40°**	**45°**	**50°**	**55°**	**60°**
Lay out angle:	**(55°)**	**(50°)**	**(45°)**	**(40°)**	**(35°)**	**(30°)**	**(25°)**	**(20°)**	**(15°)**	**(10°)**

To lay out a larger ellipse, or to lay out an ellipse without the use of a template, the method listed below can be used.

Lay Out an Ellipse With a Compass:

Example
Problem 1. Lay out a 3 inch ellipse with a compass.

Step 1. Lay out a rhombus with all sides equal to the diameter of the 3 inch circle.

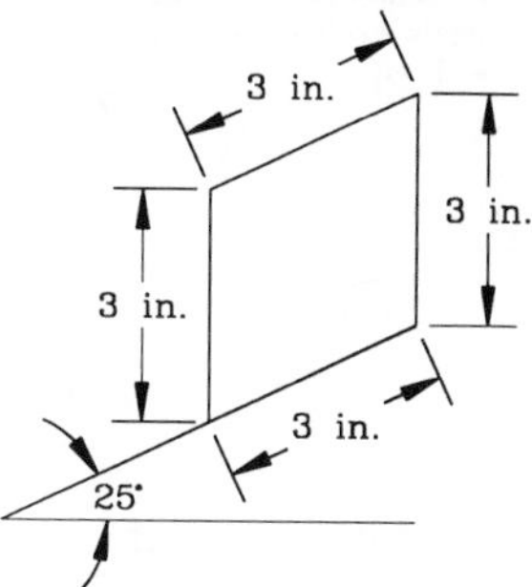

NOTE: An ellipse drawn on a 25° angle as shown on the right, will yield an ellipse of 45° on a template. The degrees in parenthesis (25°) above is the lay out equivalent angle for the specific ellipse shown.

Step 2. Find the center point of each side as shown on the right.

Step 3. Draw perpendicular lines to each side of the rhombus from the center points as shown on the right.

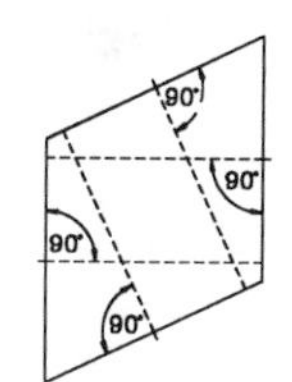

Step 4. The arc center points are the points (a, b, c, d) where the perpendicular lines meet.

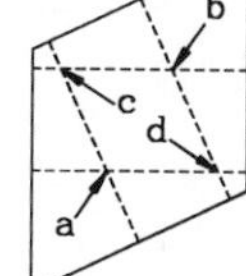
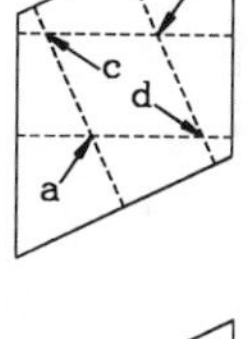

Step 5. Using a compass, draw the arcs and complete the ellipse as shown.

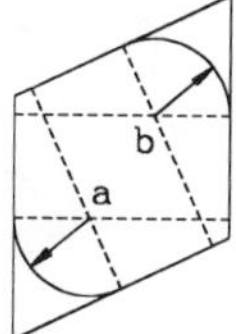

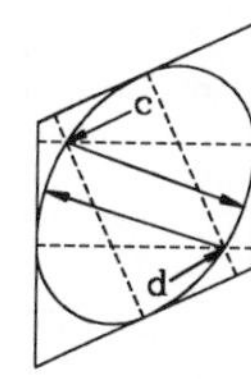

[Note: Any angle size ellipse can be laid out by using the above method. However there are other methods that can be used but are beyond the basic scope of this book.]

THE 3-4-5 and 5-12-13 SQUARING METHODS

[Also see Task H2A , how the 3-4-5 and the 5-12-13 squaring methods are determined. Task No. P1A following these two pages, actually demonstrates how to actually lay out and use the different combinations of 3-4-5 and 5-12-13.]

The 3-4-5 and 5-12-13 squaring methods are spin-offs of the **Rule of Pythagoras***, Task No. K5A, and are short cuts. Those triangles are good examples to use for squaring (laying out perfect right angles or rectangles) because there are a rare few right triangles with sides that measure out to whole numbers of inches, feet, meters, and so on. The short cuts make it quick and easy to measure them out and to increase all their sides by even multiples to fit many different situations. See the examples below.

NOTE: **The examples on the right can be squared fairly accurately by using the multiples of 3-4-5 in all feet.**

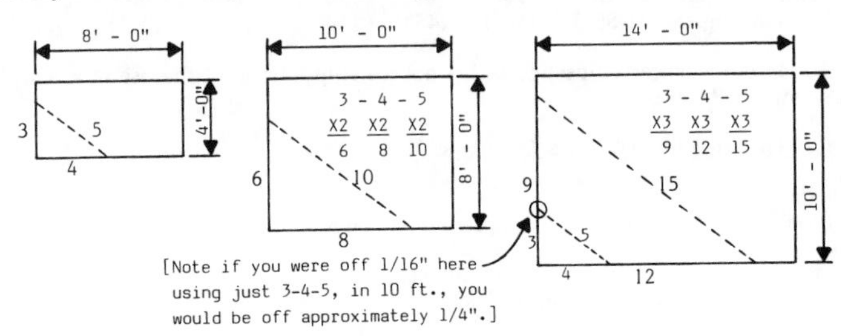

NOTE: **The same examples shown again on the right can be squared more accurately by using multiples of 3-4-5 in all inches.**

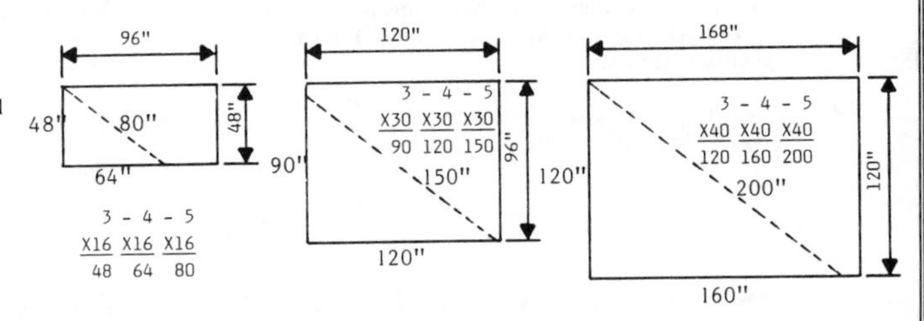

NOTE: **The exact diagonals of the same triangles can be found by using the Rule of Pythagoras. This might be the most accurate way, but look at the extra work involved, whereas, if care is taken when laying out, the 3-4-5 method can be almost as accurate. However, there will be times it will have to be done this way.**

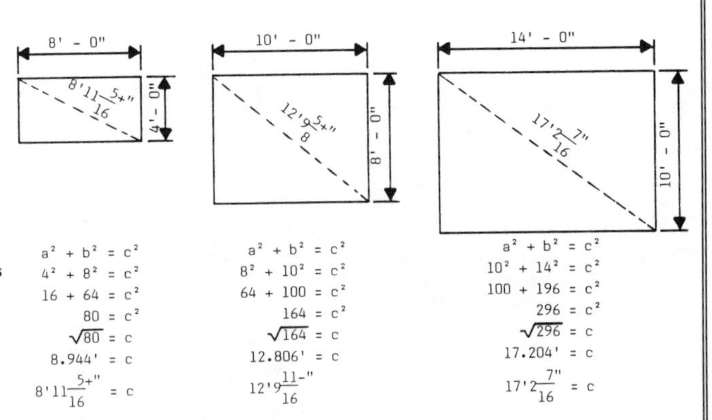

$$a^2 + b^2 = c^2$$
$$4^2 + 8^2 = c^2$$
$$16 + 64 = c^2$$
$$80 = c^2$$
$$\sqrt{80} = c$$
$$8.944' = c$$
$$8'11\frac{5+}{16}" = c$$

$$a^2 + b^2 = c^2$$
$$8^2 + 10^2 = c^2$$
$$64 + 100 = c^2$$
$$164 = c^2$$
$$\sqrt{164} = c$$
$$12.806' = c$$
$$12'9\frac{11-}{16}" = c$$

$$a^2 + b^2 = c^2$$
$$10^2 + 14^2 = c^2$$
$$100 + 196 = c^2$$
$$296 = c^2$$
$$\sqrt{296} = c$$
$$17.204' = c$$
$$17'2\frac{7}{16}" = c$$

[NOTE: All the above answers have to be converted from decimals of feet to inches by multiplying by 12, and from decimals to inches to 16ths by multiplying by 16 -- a lot of extra work.]

Many times when squaring foundations or other objects, obstacles such as cement blocks, motors, pumps, etc. can be in the way of a given 3-4-5 combinations. This is where a 5-12-13 combination might be substituted to square the objects. See the example below.

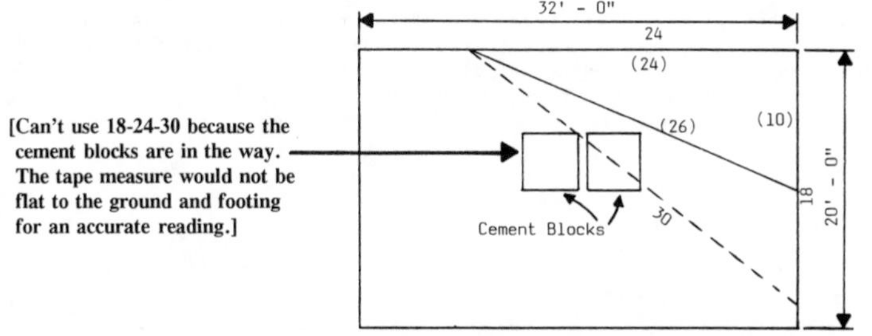

[NOTE: You could use a smaller 3-4-5 combination, but it would not reach out as far on the length side, thus giving you less accuracy.]

The 5-12-13 combination also works better than the 3-4-5 combination on longer rectangles or angles. See the examples below.

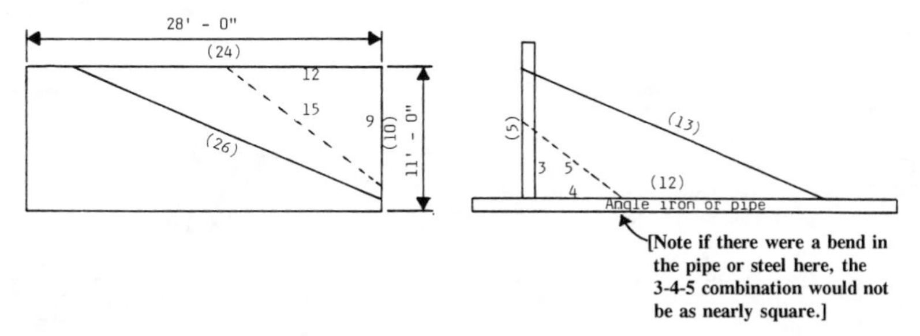

[Note how the accuracy is improved with the 5-12-13 combinations compared to the 3-4-5 combinations. In both of these cases, it would be impractical to find the exact diagonals by using the Rule of Pythagoras, unless very close tolerances were needed. You can see the time saved by using the above methods.]

* **Rule of Pythagoras: In any right triangle, the square of the hypotenuse is equal to the sum of the squares of the other two sides. By substituting the value of the sides of the right triangles into the Rule of Pythagoras, it can be easily seen that they indeed calculate out to whole numbers. Therefore, if you are in doubt about any of the above combinations or any others working out or being true, plug them into the rule (formula) and test them out.**

Examples:

	Check out 9-12-15.	Check out 10-24-26.
Formula:	$a^2 + b^2 = c^2$	$a^2 + b^2 = c^2$
Substitute:	$9^2 + 12^2 = 15^2$	$10^2 + 24^2 = 26^2$
Calculate:	$81 + 144 = 225$	$100 + 576 = 676$
Answer OK:	$225 = 225$	$676 = 676$

[Try any of the combinations and you will see they work out.]

How to Square an Object by using the 3-4-5 Squaring Method

The 3-4-5 squaring method is a spin-off technique of the rule of Pythagoras (The Law of the Right Triangle) that is commonly used to square large objects without the use of a metal square. Finding the exact diagonal of a right triangle by using the formula $a^2 + b^2 = c^2$ would be much harder and more time consuming. Finding the largest combination of 3-4-5 is much easier and precision is close to the same if every step is followed carefully.

Problem 1. (Sample) Fasten and square the objects on the right to form a perfect right angle (90°) by using the 3-4-5 squaring method. DO NOT USE A METAL SQUARE.

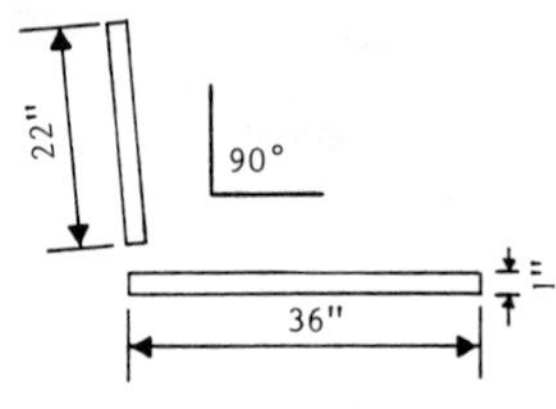

Step 1. Divide the short object by 3 to the nearest whole number (do not use the remainder).

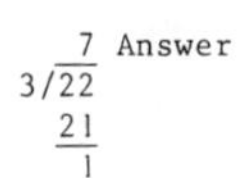

Step 2. Multiply the answer times each number in 3-4-5. Thus ------------------ 3 4 5

	3	4	5
	X7	X7	X7
The largest combination of 3-4-5 is ---▸	21	28	35

Step 3. 1st Place the two objects together as shown and measure and mark the short object at 21

2nd Measure and mark the long object at 28 as shown

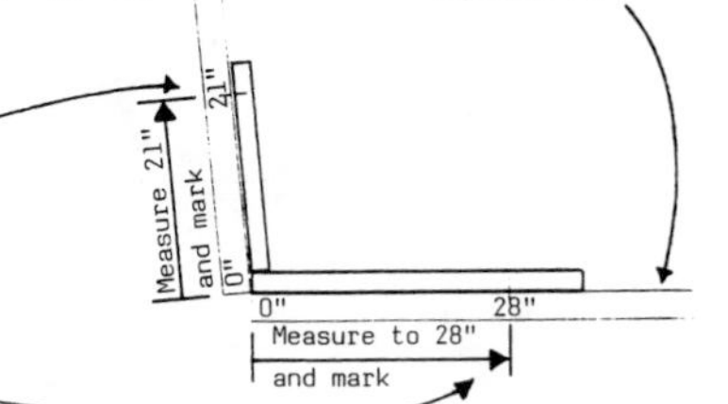

Step 4. Move the short and long objects until you measure exactly 35 between the 21 and the 28 marks. Fasten when you read exactly 35"

NOTE: If every step was done correctly you should have a perfect right angle (90°). Check with a metal square.

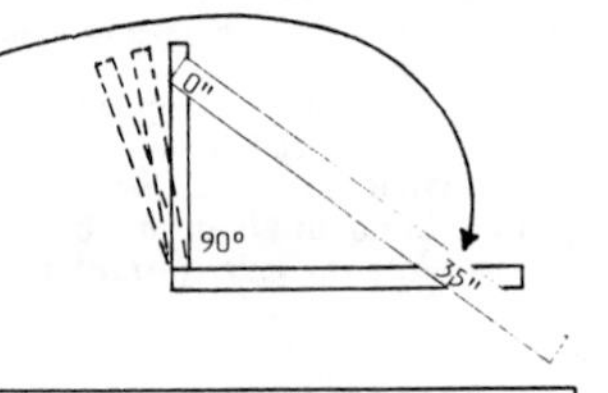

The above problem could be solved by using the Rule of Pythagoras ---------▸ $a^2 + b^2 = c^2$

NOTE: This method is used when the exact diagonal has to be found. However the 3-4-5 method is just as accurate in most cases.

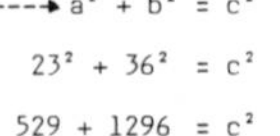

$$23^2 + 36^2 = c^2$$
$$529 + 1296 = c^2$$
$$1825 = c^2$$
$$\sqrt{1825} = c$$
$$42.72 = c$$

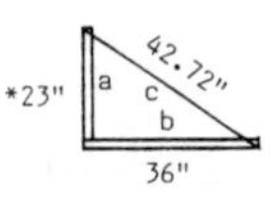

*1" added for the bottom thickness

Problem 2. (Sample) Given a 12" ruler, pencil and paper, lay out a rectangle 7 inches by 11 inches. Use the 3-4-5 squaring method.

Step 1. Divide the short side by 3 ------ (do not use the remainder)

```
   2  Answer
 3/7
   6
   1
```

Step 2. Multiply the answer 2 times each number in 3-4-5. Thus --

	3	4	5
	X2	X2	X2
Largest combination is ------▸	6	8	10

Step 3. 1st Draw a straight line representing the length on your sheet of paper as shown 1" off the bottom.

2nd Mark point "0", then measure 8" and mark.

Step 4. 3rd From point "0" measure 6" and carefully swing an arc as shown.

4th From the 8" point, measure 10" and carefully swing and arc through the 6" arc as shown.

Step 5. 5th Draw a straight line through point "0" and the point where the 6" arc and the 10" arc meet. (you now have a right angle)

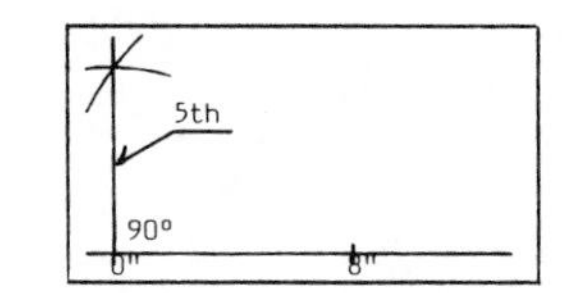

Step 6. 6th Measure and mark 7" on the short side as shown.

7th Measure and mark 11" on the long side as shown.

8th From the 11" mark, measure and swing a 7" arc as shown.

9th From the 7" mark, measure and swing an 11" arc as shown.

Step 7. 10th Finish the rectangle by drawing straight lines through the intersecting arcs and the 7" and 11" points as shown.

11th Check the accuracy of the rectangle by measuring the diagonals as shown. The diagonals should be exactly the same.

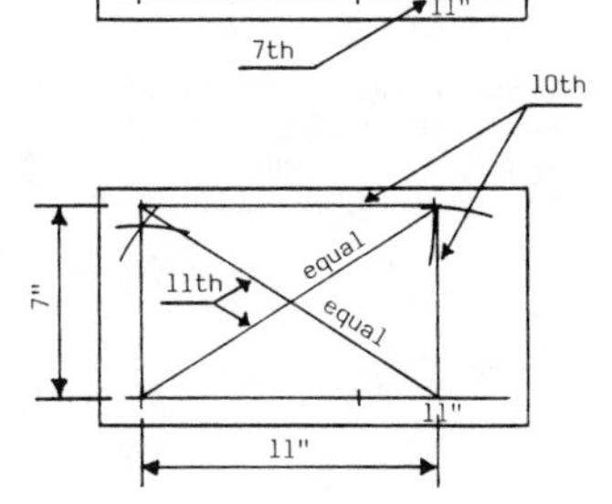

SUGGESTION: Using the floor and a chalkline lay out problem 2 above but use feet instead of inches.

P1A	◂ Task No. THE 3-4-5 AND 5-12-13 SQUARING METHOD (Continued)

Practical Applications of the 3-4-5 and 5-12-13 Squaring Method.

Example
Problem 1. Lay out and check a right angle.

Step 1. Draw or chalk a straight line (line **ab**). Select an approximate midpoint, point **c**. Starting from point **c**, swing and draw a 3 ft. arc as shown on the right.

Step 2. From point **c** swing and draw a 4 ft. arc to locate point **d** as shown on the right.

Step 3. From point **d** swing and draw a 5 ft. arc to locate point **e** as shown on the right.

Step 4. Draw or chalk a straight line through point point **c** and **e** as shown on the right.

Step 5. Check for square (90°) by measuring 4 ft. from point **c**, and mark point **f**. The measurement from point **f** to point **e** should be exactly 5 ft. If the measurement is off, double check and try again.

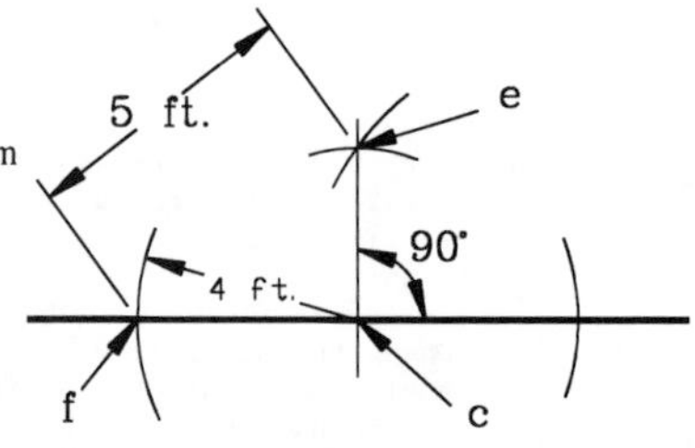

Example
Problem 2. Square an object (metal, wood, plastic, etc.) without the use of a square.

Note 47" doesn't divide evenly by 3, therefore drop to the nearest number divisible by 3 which is 45. Calculate:

$45 \div 3 = 15$	15	15	15
	X 3	X 4	X 5
	45	60	75

This will give you the greatest accuracy for this specific problem.

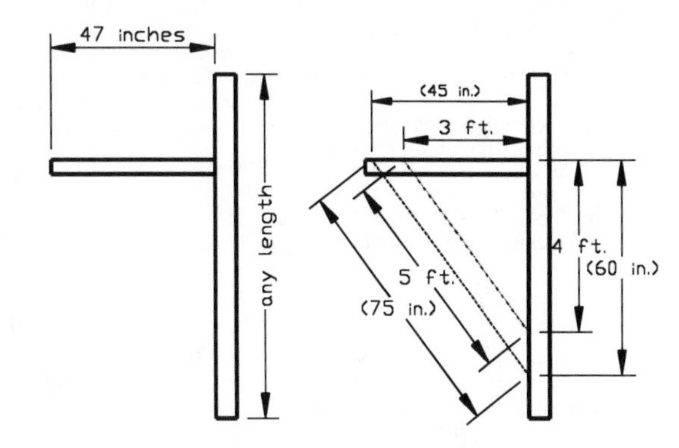

The largest combination in whole feet would be 3 ft.- 4 ft. - 5 ft.

Note if a greater reach was possible, you could use the largest combination of 5-12-13:

	5	12	13
	X 9	X 9	X 9
	45	108	117

However, in most cases the largest combination of 3-4-5 is sufficient for accuracy.

Example
Problem 3. Square a trailer frame.

Note the largest combination of 3-4-5 in all feet that could be used to square the frame would be 9-12-15. The largest combination of 3-4-5 in all inches that could be used to square the frame would be 120-160-200 or in feet and inches 10 ft. 0 in. - 13 ft. 4 in. - 16 ft. 8 in.

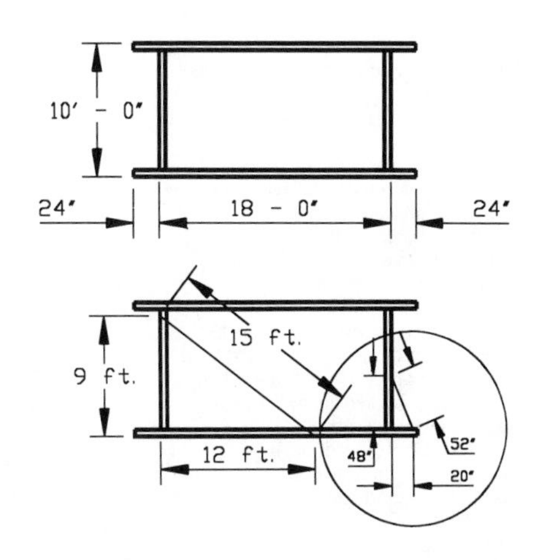

Note there could be obstacles inside the frame making it impossible to use either a 3-4-5 or 5-12-13 combination. In this case the 5-12-13 combination of 20-48-52 would be the largest combination to use outside the frame for accuracy.

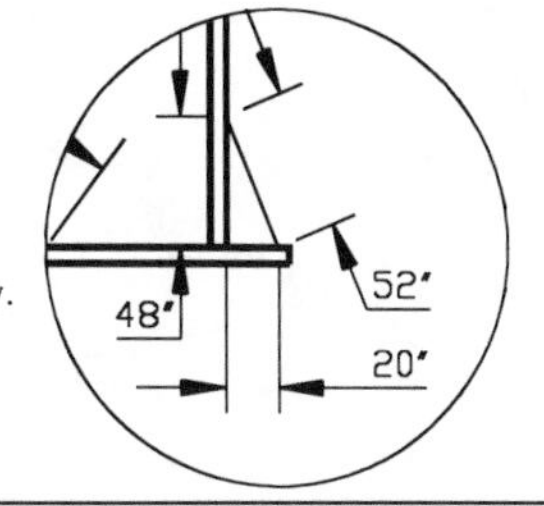

P2 | ◄ Task No. SQUARING WITH A TRANSIT

Task No. Q2 compares the level-transit with the builders' level. The most significant difference between the two is that the latter can only be used only to level. This module demonstrates how to actually lay out a simple right triangle and a rectangle using the level-transit. However, even after using a level-transit to lay out an angle or a rectangle (any geometric figure), they should always be checked out mathematically for accuracy. The procedure below explains the process. However if more information is needed to actually level up the instrument, refer back to Task No. Q2.

Step 1. Set up and level the level-transit with the plum bob over the vertex of the angle, **point A** as shown.

[Refer to Task No. Q2 if you need help leveling the instrument.]

Step 2. Transfer the instrument into a transit by unlocking the telescope, and line up the cross hairs on **point B** at the desired distance, 10 feet. Place a stake at this point with a nail center of the cross hairs and the correct distance from **point A**.

Step 3. Set the horizontal circle to zero. Then turn the telescope 90 degrees to the right.

Step 4. Line up the cross hairs on **point C** at the desired distance, 12 feet. Place a stake at this point with a nail center of the cross hairs and the correct distance from **point A**.

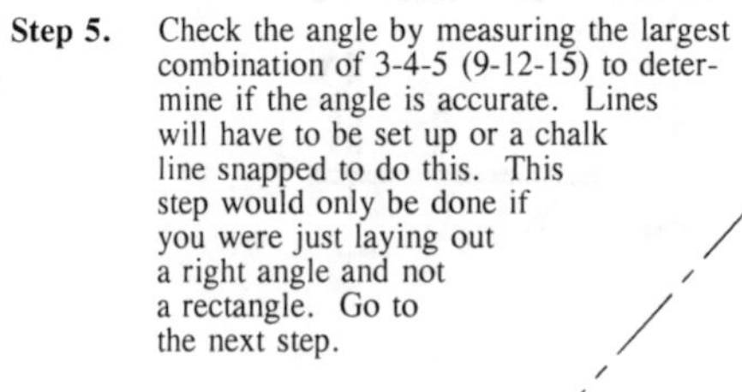

Step 5. Check the angle by measuring the largest combination of 3-4-5 (9-12-15) to determine if the angle is accurate. Lines will have to be set up or a chalk line snapped to do this. This step would only be done if you were just laying out a right angle and not a rectangle. Go to the next step.

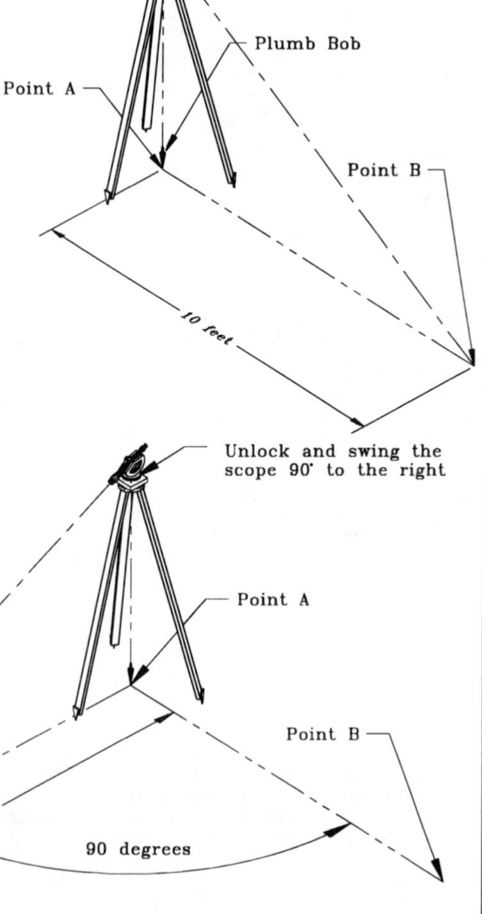

Step 6. Using two tape measures, measure from **point C** to **point D** and from **point B** to **point D**. Place a stake with a nail exactly where the two measurements cross.

Step 7. Check the rectangle for square by measuring diagonally (from corner to corner) from **point A** to **point D**, and from **point B** to **point C**.

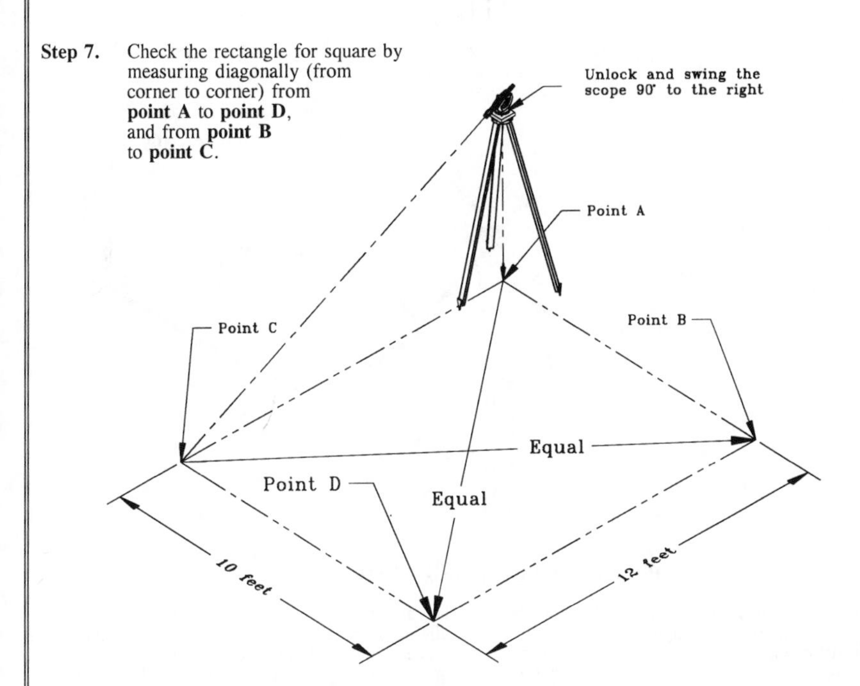

[Note: When measuring the diagonals, the measurements should be exactly the same. If not make the necessary adjustments until they are.]

Note the example on these two pages illustrates how to lay out an angle or a simple rectangle. However, even the more complex geometric figures are laid out using the same principles. The more complex figures could involve using the vernier to measure angles in degrees, minutes, and seconds. These methods are explained thoroughly in task number Q2, also. Generally, the accuracy of the work is contingent upon the quality of the instrument. However, even a less expensive instrument can yield quality workmanship by using mathematical layout techniques (triangulation) along with the instrument.

Q1	◂ Task No.	THE WATER LEVEL

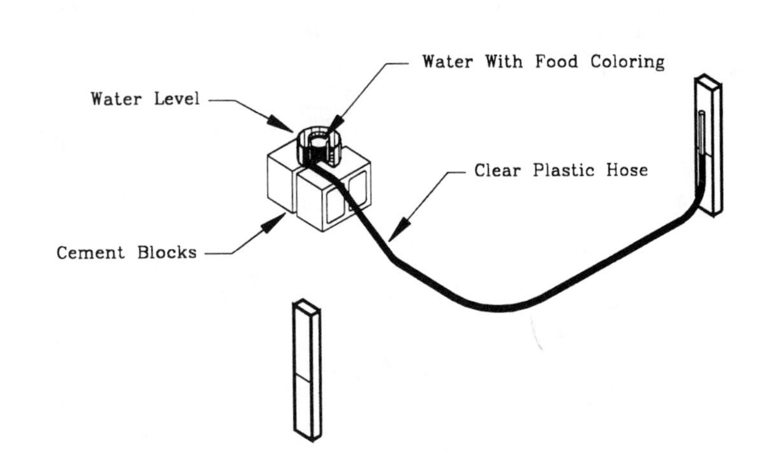

PREFACE TO LEVELING METHODS

Leveling methods are included in this book because they are equally important to squaring methods. As mentioned in squaring techniques before assembling or building practically any object, it must be laid out true in dimension and square (at 90° angles) before actually starting the construction of it. Otherwise, the parts that make up the object would not fit properly. The same is true for being plumb (level). Therefore, being square and level is mathematics, pure and simple. The more accurately an object is built, the better the fit and the better the craftsmanship and control of manufacturing it will be.

Of the four different leveling methods mentioned in this book, probably the most unique method is the water level. Some of the advantages of using the water level versus other methods also mentioned in this math book are: The water level cost about 1/20th of the price of the least expensive transit or builders level. It can be operated alone (a transit or builders level takes two people). It can level around blind corners, thus eliminating different set-ups. It is not a mechanical device, thus it eliminates the possibility of mechanical failures. However, it is important to note that there are times that the transit or level-transit would be more useful, for example, shooting grades, leveling longer distances, shooting angles, and so forth. Use the leveling instrument that works best for you.

The principle of the water level is plain and simple. Water seeks its own level due to gravity, period. The water level can be used to level foundations, ceilings, cabinets, doors (large and small), machines, grade stakes, plus many other uses too numerous to mention. The water level shown is called a HydroLevel and is factory made.

We extend our sincere thanks and appreciation to Pat and Patricia Mitchell, The Hydrolevel Co., P.O. Box 1378, Ocean Springs, Mississippi 39564. Hydrolevels can be purchased by writing; The Hydrolevel Co., and/or Chenier Educational Enterprises, Inc., P.O. Box 265, Wells, Michigan 49894.

How To Use The Water Level

Part I. How to Fill the Water Level Free of Air Bubbles.

Part II. How to Set Up and Secure The Water Level.

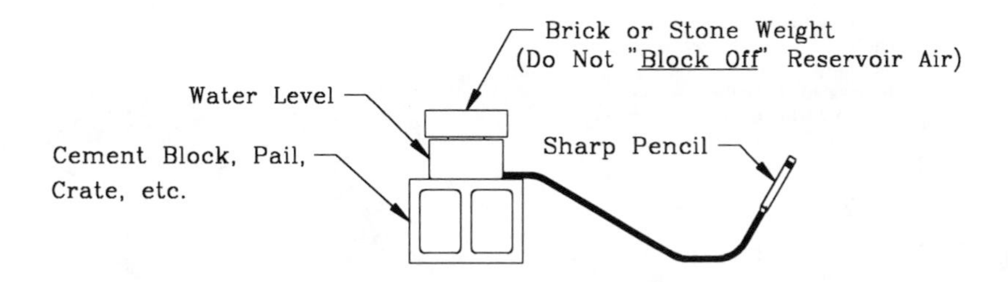

Part III. How to use the Water Level to Set Level Marks (elevations) and to Level Different Objects.

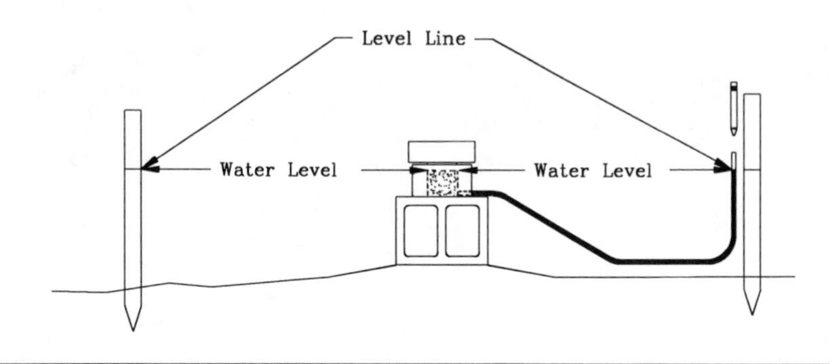

PART I **How to Fill the Water Level Free of air Bubbles.**

Step 1 Fill a gallon plastic milk jug or any other clear container with water. Then add food coloring (a dark color is the easiest to see) and let the water settle a few minutes to remove any air bubbles.

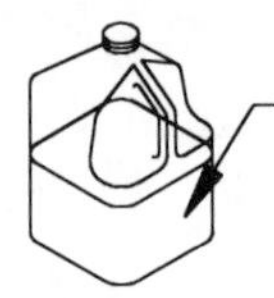

Water/food coloring in the summer, anti-freeze in the winter.

Step 2 Unravel the hose (as much as you will need) from cylinder outside the reservoir.

Step 3 Insert the hose to the bottom of the water jug as shown. **(Position the water jug on a chair, stool, saw horse, etc. as shown. When the water jug is elevated above the reservoir, the fluid will go into the reservoir faster.)**

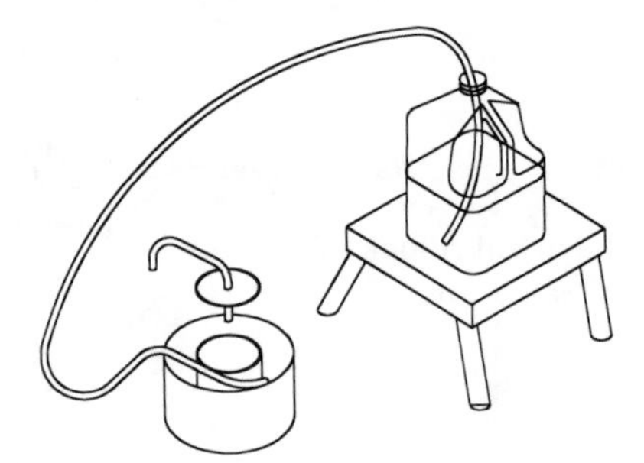

Step 4 Insert the red cover, with the small piece of clear hose attached to it, inside the reservoir as shown.

Step 5 Suck on the small hose until the fluid flows freely into the reservoir. Then remove the small hose and red cap.

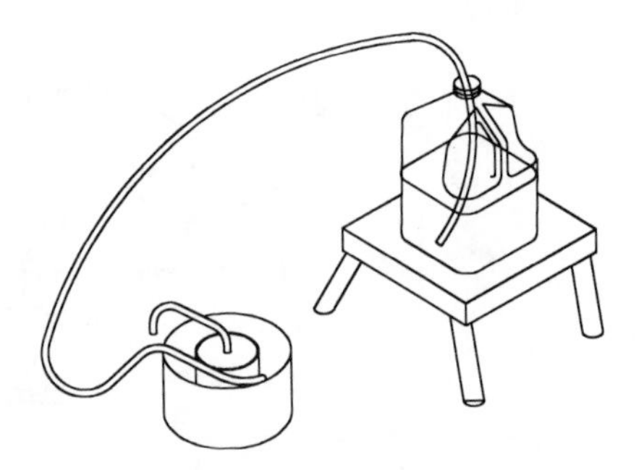

Step 6

Fill the reservoir as much as possible (1/4 to 3/8 inches from the top). The fuller the reservoir, the more water pressure. This, in turn, causes the water to settle in the hose quicker.

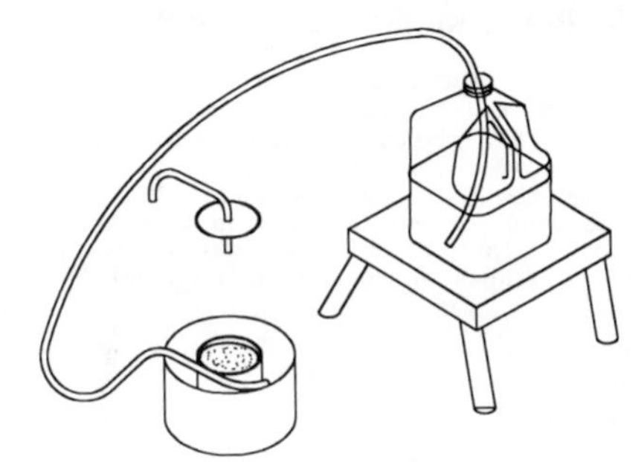

Step 7

Raise the water level to the same height of the water jug (this will equalize the pressure so you won't lose too much fluid). Then carefully pull the hose out of the water jug and lower the hose until the water stops to the top of the hose and cap with a pencil. If you happen to lose water up to this point in time, it will not make any difference. **However from this point on be careful not to lose too much water as it will change the level of the water in the reservoir thus causing an error in your elevations.**

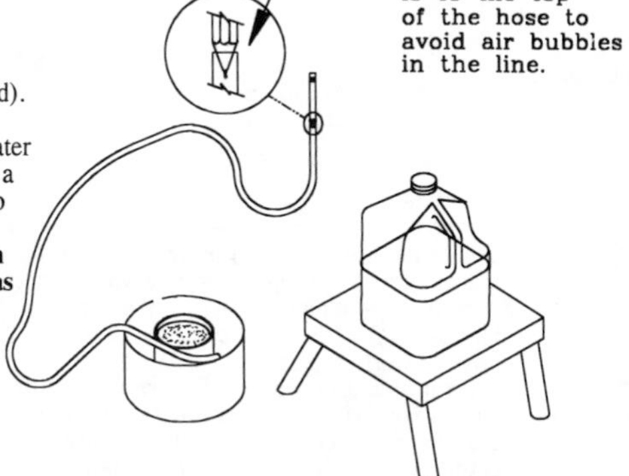

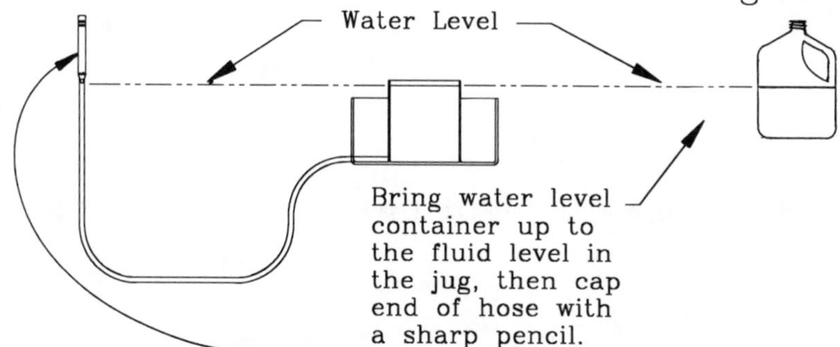

NOTE: IT MIGHT SEEM REAL EASY AND EVEN AWKWARD AT FIRST TO DO THE ABOVE SEQUENCE. HOWEVER, WITH A LITTLE BIT OF PRACTICE YOU WILL MASTER THE TECHNIQUES AND THE PAYOFF WILL BE INCREASED PROFICIENCY ON THE JOB.

PART II. How to Set Up and Secure the Water Level.

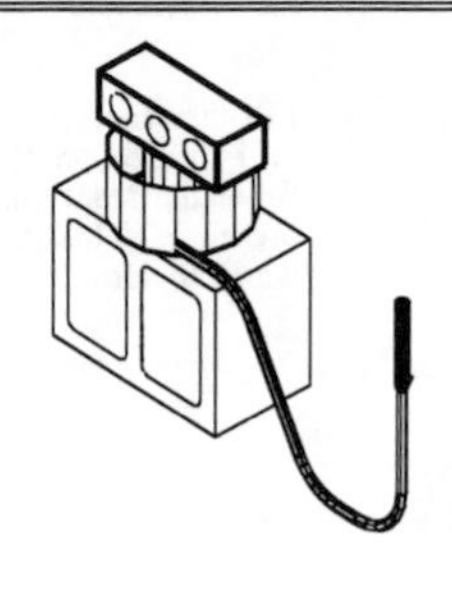

Step 1 Set the water level on cement blocks or any level surface in the center of your work area. The hose is 50 feet long, therefore, you can reach quite a distance.

(Note, set the water level high enough so that you don't have to bend too low to the ground to transfer the water-level marks to the the stakes, but not too high as to be above the stakes where you cannot mark.)

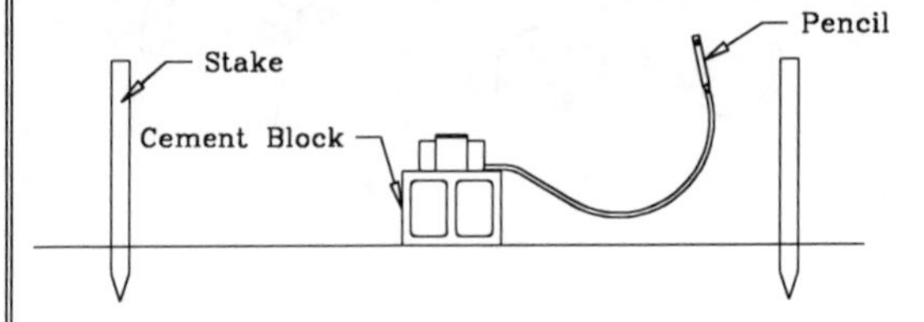

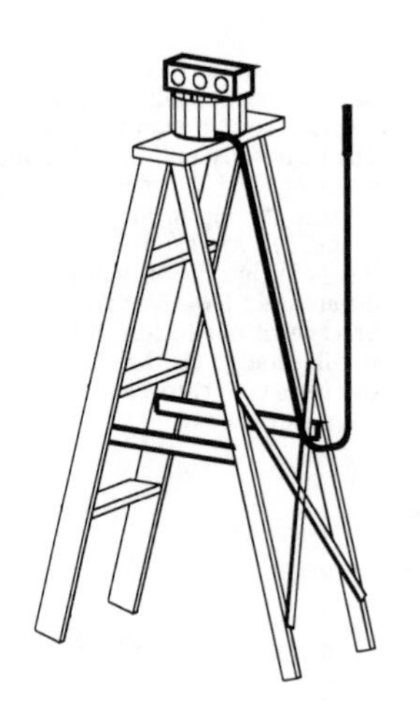

Step 2 Set the water level on a step ladder or elevated object to level for suspended ceilings, cabinets, or other objects.

Step 3 Secure the water level in place by setting a heavy object on top of the reservoir. Be careful not to choke off or seal the top of the reservoir as this will cause the water level to work improperly. Use a brick with holes in it and/or a rock. **Secure the object with a rope if needed for safety. It is very easy to trip over the clear plastic hose. This could be very dangerous if it is above your head.**

PART III. How to Use the Water Level to Set Level Marks (elevations) and to Level Different Objects.

Step 1 Uncap the pencil at the end of the hose at approximately the same level as the water level in the reservoir -- preferably above so the water will not run.

If a few drops of water leak out, it won't change the level enough to be concerned about.

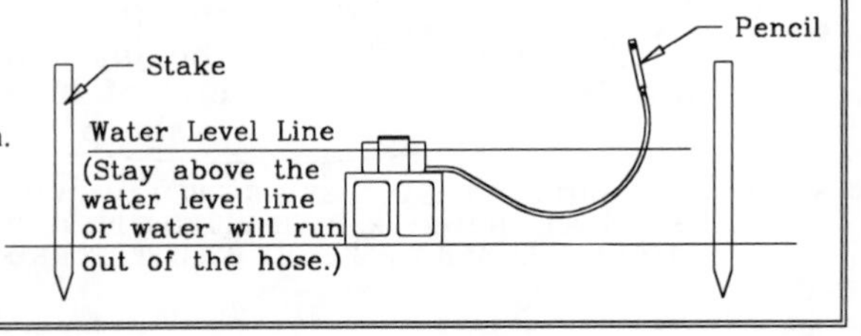

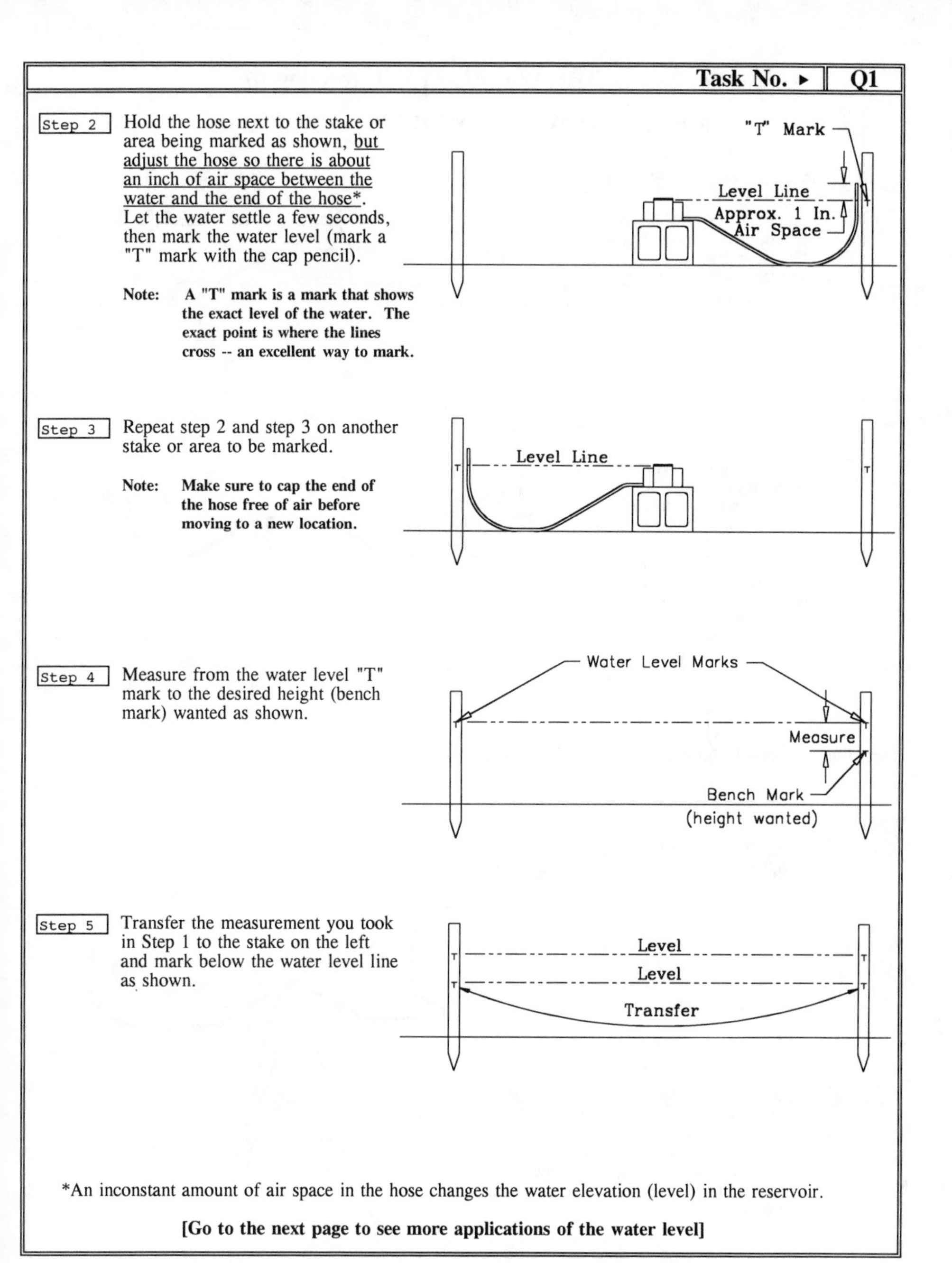

Task No. ▸ Q1

Step 2 — Hold the hose next to the stake or area being marked as shown, <u>but adjust the hose so there is about an inch of air space between the water and the end of the hose*</u>. Let the water settle a few seconds, then mark the water level (mark a "T" mark with the cap pencil).

Note: A "T" mark is a mark that shows the exact level of the water. The exact point is where the lines cross -- an excellent way to mark.

Step 3 — Repeat step 2 and step 3 on another stake or area to be marked.

Note: Make sure to cap the end of the hose free of air before moving to a new location.

Step 4 — Measure from the water level "T" mark to the desired height (bench mark) wanted as shown.

Step 5 — Transfer the measurement you took in Step 1 to the stake on the left and mark below the water level line as shown.

*An inconstant amount of air space in the hose changes the water elevation (level) in the reservoir.

[Go to the next page to see more applications of the water level]

[OTHER WAYS TO LEVEL WITH THE WATER LEVEL]

Leveling One Object with Other Object.

Step 1 Set the water level between two objects -- measure from the water level in the hose and from the top of the object on the right.

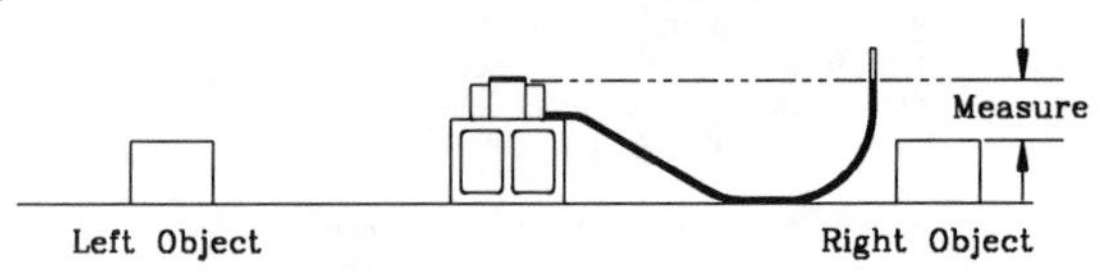

Step 2 Move the water hose above the object on the left -- if the dimension on the left is the same as the dimension on the right -- the objects are level with each other. If they are not level with each other, you now know what size shims to use to level them.

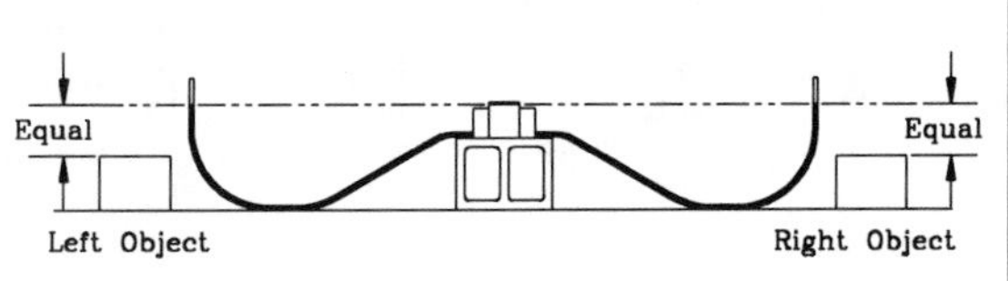

Leveling Ceilings, Cabinets, Garage Doors, and so on.

Step 1 Set up the water level on a step ladder or other elevated object. Mark the corners of the room as shown.

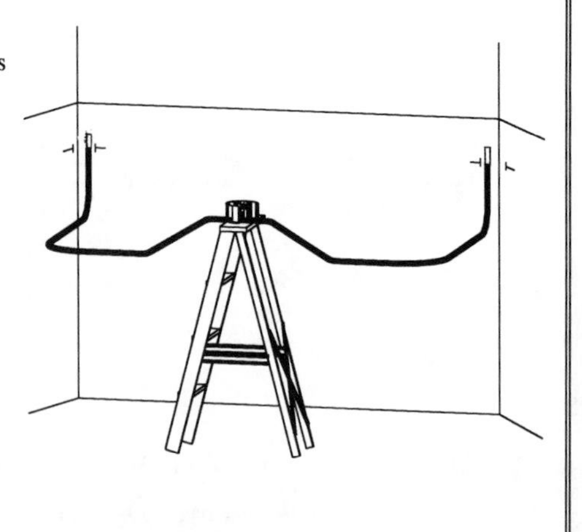

Step 2 Measure from the water level "T" mark to the desired height needed and mark with a "T" mark. Do this on all four corners. Chalk lines to connect all the desired height marks. You now have a perfectly level mark for your suspended ceiling angles.

Step 1 Set the water level up on a bench or stool. Mark water level marks on the walls as shown. Measure from the water level "T" marks to the floor in differ areas as shown. Record these measurements. You now know how to shim the base cabinets to get them perfectly level with each other. Level from the high point so you won't have to cut any of the cabinets. Use the same procedure on the wall cabinets.

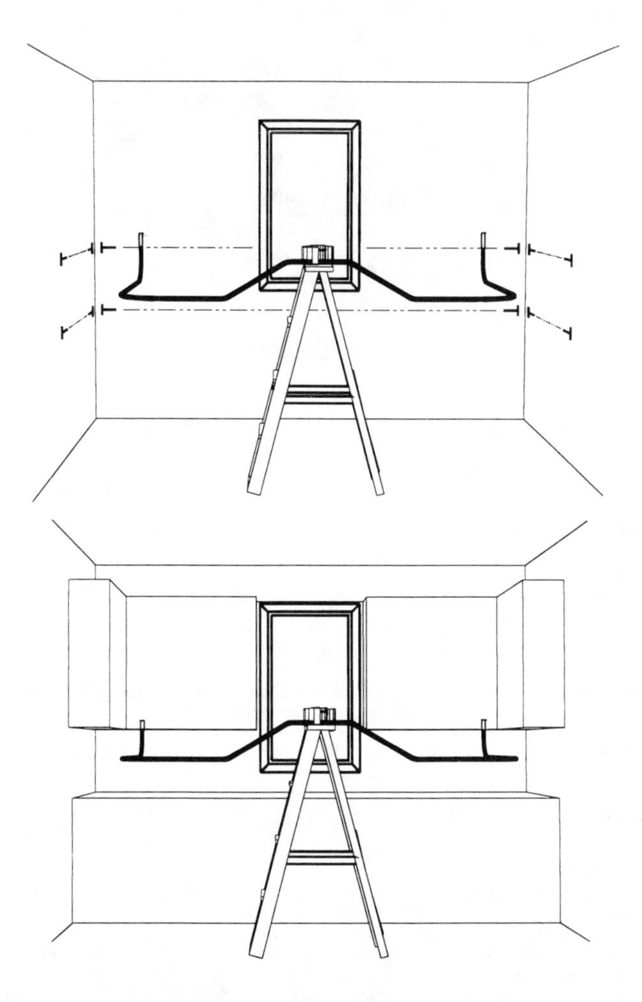

THE EXAMPLES LISTED ABOVE ILLUSTRATE A FEW DIFFERENT WAYS THE WATER CAN BE USED TO LEVEL OBJECTS. YOUR IMAGINATION CAN NOW RUN WILD TO SEE OTHER POSSIBILITIES TO LEVEL WITH THE WATER LEVEL.

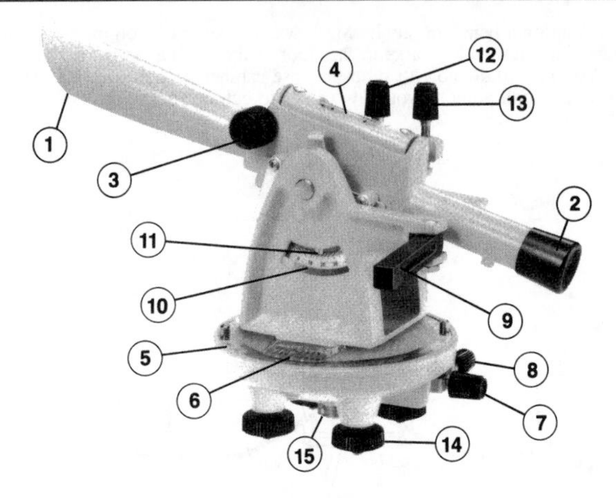

THE LEVEL-TRANSIT

1. Telescope
2. Eyepiece
3. Focusing knobs
4. Instrument level vial
5. Horizontal graduated circle
6. Horizontal vernier
7. Horizontal locking clamp
8. Horizontal tangent screw
9. Telescope lock lever
10. Vertical arc
11. Vertical arc pointer
12. Vertical locking clamp
13. Vertical tangent screw
14. Four leveling screws
15. Tripod mounting stud

The **level-transit** is a combination instrument. When used as **a level**, **the telescope lock lever (9) must be in a closed position**. When used as **a transit** the lock lever is in the open position. The picture above shows the telescope in the open position. The telescope moves up and down 45 degrees, and rotates 360 degrees, to measure vertical and horizontal angles.

The **vertical arc (10)** is divided in degrees and numbered every 10 degrees up to 45 degrees, for both upward and downward angles, and has an adjustable **index pointer (11)**.

The vertical clamp (12) holds the telescope at a vertical angle. Fine vertical settings can be made with the **vertical tangent (13)**. The vertical clamp must be hand tightened before the tangent will function.

Both the level-transit and the builders level have four **leveling screws (14)** for leveling the instrument. The instrument is mounted to the tripod by a tripod cup assembly which screws onto the **instrument mounting stud (15)**.

The level-transit is a precision instrument that will accurately handle both horizontal and vertical lay-out and leveling tasks. Accuracy is contingent on the type of level or transit you purchase. The instruments shown on these two pages are modestly priced with an accuracy range of ± 1/4 inch at 75 feet. The same company makes other instruments that are accurate to ± 1/16 of an inch at 200 feet. There are also instruments on the market with accuracy within thousandths of an inch. The instruments are used for setting in sophisticated machines and equipment that require very close tolerances. However, for most practical jobs the instruments shown in this book will handle most jobs.

The level-transit can be used to set construction grades, masonry work, create straight lines, lay out angles, determine grades, plumb walls and uprights, and so on. The builders level is limited to mostly leveling tasks, however both are similar in many ways. Therefore both are addressed in this module.

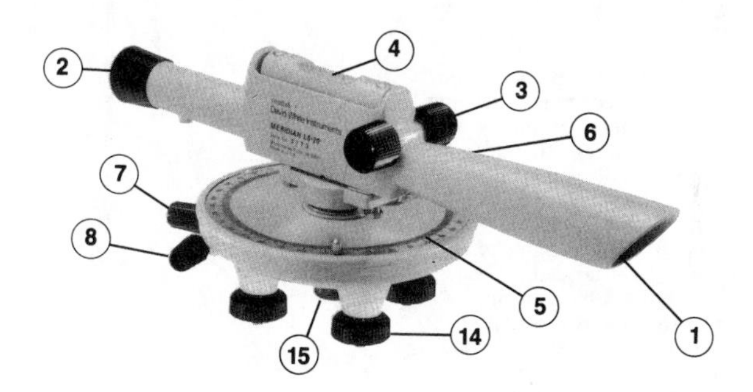

THE BUILDERS' LEVEL

1. Telescope
2. Eyepiece
3. Focusing knobs
4. Instrumental level vial
5. Horizontal graduated circle
6. Horizontal vernier
7. Horizontal locking clamp
8. Horizontal tangent screw
14. Four leveling screws
15. Tripod mounting stud

The **builders' level** is used mainly for leveling, determining grades, and establishing contour lines. It can also be used to lay out and measure angles, and make straight lines. However, it is much easier and faster to use the level-transit for the latter applications.

Compare the parts list on this page with the parts list on the preceding page. This will give you an idea of the basic difference between the level-transit and the builders level.

Use the instrument that works best for you, and what you can afford. Some applications can be duplicated by using the water level along with triangulation. It is the intent of this book to present a variety of different methods so you can choose the method or methods that work best for you.

Continue on to the following pages to learn how to set-up, level, read, and use builder levels and transit-levels.

We extent our sincere thanks and appreciation to David White, Inc., 11711 River Lane, P.O. Box 1007, Germantown, Wisconsin 53022-8207 for all their help in providing us with all the materials on their product line of level-transits, levels, leveling rods, plumb bobs and their other excellent products mentioned in this training module.

We also appreciate their help in developing, and the use of, their training materials encompassed in this training module.

HOW TO SET UP A TYPICAL LEVEL-TRANSIT

Each of the following steps is important in preparing to use your instrument.

Step 1. It is important that the tripod is set up firmly. Make sure that the tripod points are well into the ground and all leg locking mechanisms are firmly set.

When setting up on a smooth floor or paved surface, secure the points of the legs by chipping the concrete, attaching chains between the legs, or putting a brick in front of each leg. If setting up in dirt, apply your full weight to each leg to prevent settlement.

Check the tripod legs and tighten hex nuts with a wrench if needed. The legs should have about a 3 foot spread, **positioned so the top of the tripod head appears level**. If using a tripod with adjustable legs, be sure the leg clamp wing nuts are securely **hand tightened**.

"Eyeball level" tripod head to speed up the process of leveling the instrument.

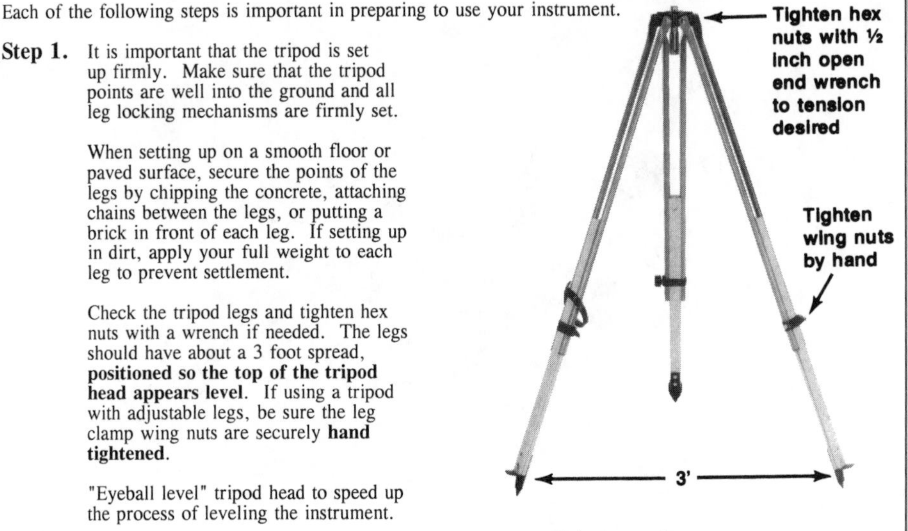

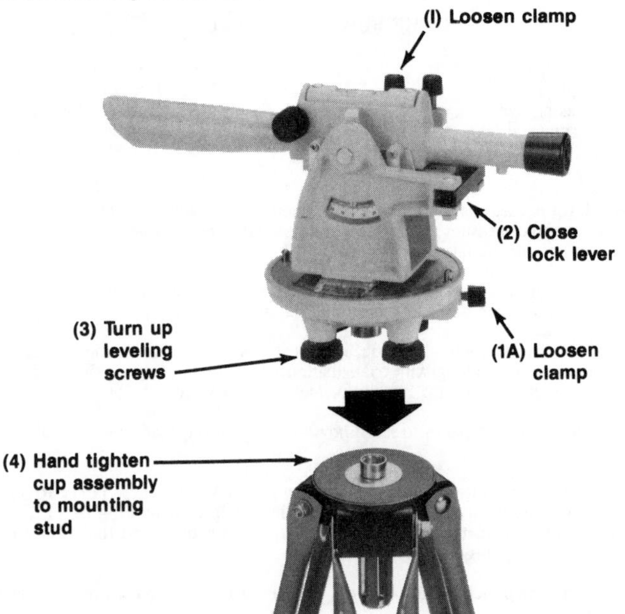

Step 2. Before setting up your instrument, loosen the clamps. Turn up the leveling screws far enough so that the tripod cup assembly can be securely **hand tightened** to the instrument mounting stud. If using the level-transit, be sure the telescope lock lever is in the closed position. Attach the instrument to the tripod securely, **hand tightening** the cup assembly to the mounting stud. If setting up over an exact point, read Step 3; otherwise, continue to Step 4.

Step 3. If setting up over a point, use a plumb bob to center on the exact point. To hang the plumb bob, attach cord to the plumb bob hook on the screwdriver-style handle of the tripod. Knot the cord as illustrated.

Move the tripod and instrument over the approximate point. (Be sure the tripod is set up firmly again, as described in Step 1.) Shift the instrument on the tripod head until the plumb bob is directly over the point. Then set the instrument leveling screws as described in Step 4.

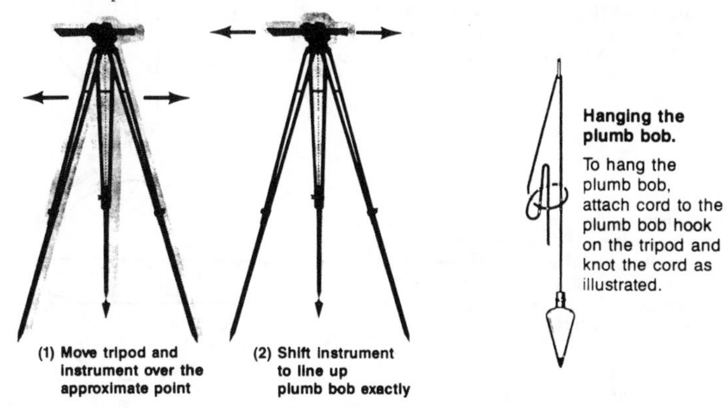

Step 4. Turn down the instrument leveling screws until firm contact is made with the tripod head. A word of caution: It is very possible to over-tighten the leveling screws. You want only a firm contact between the screws and the tripod head. If the instrument shifts on the tripod, turn down the screws more firmly **by hand**. If no shifting occurs, the instrument is ready for **Step 5**.

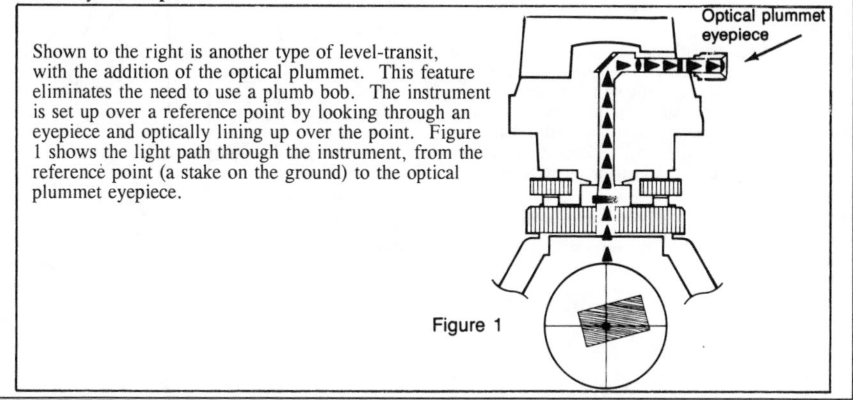

Shown to the right is another type of level-transit, with the addition of the optical plummet. This feature eliminates the need to use a plumb bob. The instrument is set up over a reference point by looking through an eyepiece and optically lining up over the point. Figure 1 shows the light path through the instrument, from the reference point (a stake on the ground) to the optical plummet eyepiece.

Figure 1

Step 5. Leveling the instrument so the vial bubble remains centered through a 360° rotation of the telescope is the most important operation in preparing to use your instrument. When leveling your instrument, be sure not to touch the tripod. Follow these instructions carefully.

a. Line up the telescope so that it is directly over one pair of leveling screws. Grasp these two leveling screws with the thumb and forefinger of each hand. **Turn both screws at the same time** by moving your thumbs toward each other or away from each other, until the bubble is centered.

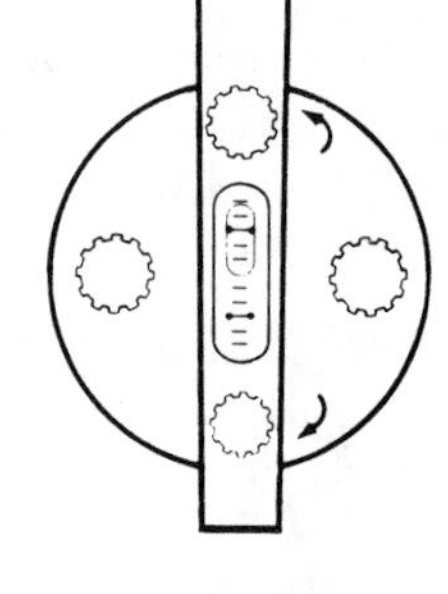

b. When the bubble is centered, **rotate the telescope 90 degrees over the second pair of leveling screws and repeat the thumbs in, thumbs out leveling procedure until the bubble is again centered.**

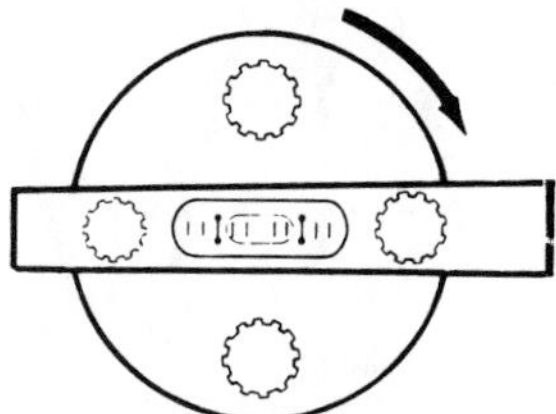

c. Shift back to the original position and check the level. Make minor adjustments with leveling screws if necessary, taking half the error out in position and the other half after rotating 180°. Repeat as necessary until the instrument is level.

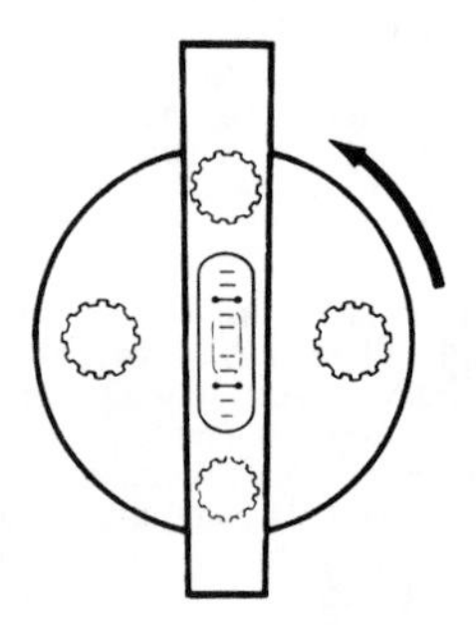

FOR A FINAL LEVEL CHECK, rotate the telescope over each of four leveling points to be sure the bubble remains centered.

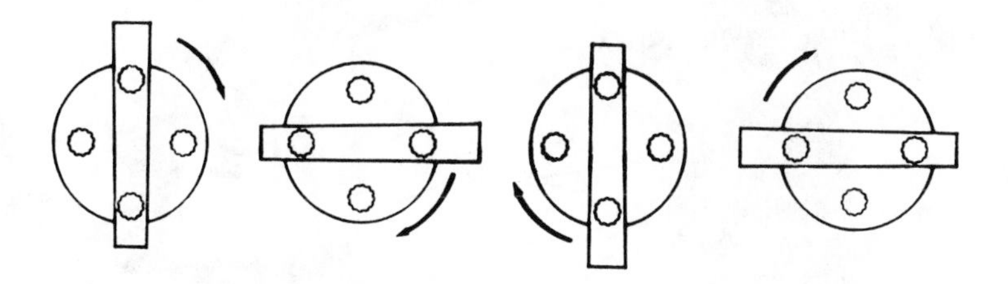

The Golden rule for quick and simple leveling is THUMBS IN, THUMBS OUT. Turn BOTH screws equally and simultaneously. Practice will help you get the feel of the screws and the movement of the bubble. It will also help to remember that the direction your left thumb moves is the direction the bubble will move.

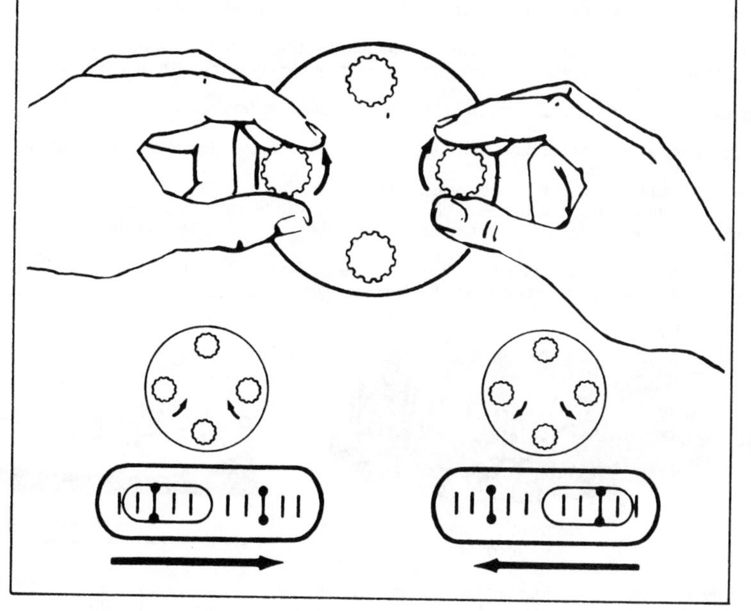

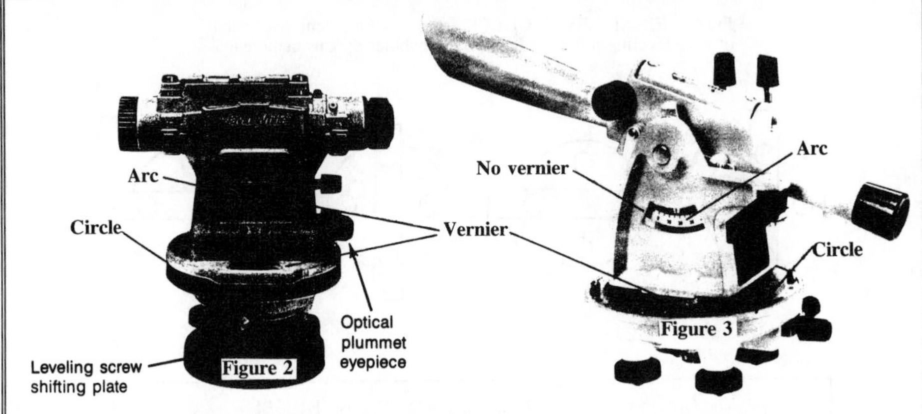

HOW TO READ THE CIRCLE, ARC, AND VERNIER

The 360° horizontal circle is divided into quadrants (0-90°). The circle is marked by degrees and numbered every 10 degrees. The horizontal circle is referred to as the circle or degree scale. On level-transits, the vertical arc also is a degree scale, and it is numbered every 10 degrees, up and down.

For very precise readings, Universal instruments are equipped with a vernier which divides each degree on the circle or arc into 12 equal parts of 5 minutes each. There are 60 minutes (60') in a degree.

The vernier scale is read in the same direction (right or left) as you're reading the degree scale.

The following examples will explain how to read the degree and vernier scales on the circle and arc.

Step 1. Note the point at which 0 on the vernier scale touches the circle. If the 0 coincides exactly with a degree line on the circle, your reading will be in exact degrees. There are no fractions of degrees, or minutes, to be added to the reading **(Figure 2a** and **Figure 3a)**.

The vertical arc is read in the same way - if the 0 on the vernier scale coincides exactly with a degree line, your reading will be in exact degrees.

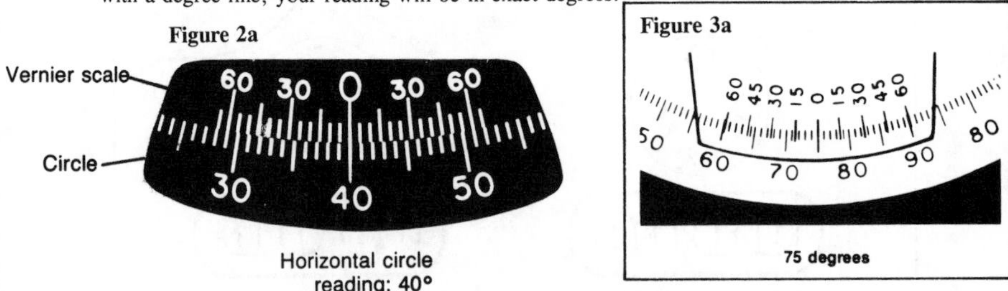

Step 2. If the 0 on the vernier does not coincide exactly with a degree line, your **degree reading** is the line which the 0 has just passed, reading up the degree scale, **plus** a fraction of the next degree. To determine the fraction, or minutes:

a. Start at 0 on the vernier and read up the vernier scale (in the same direction as you're reading the degree scale) until you find a minute line that coincides **exactly** with a degree line. In

figure 2b, 41° was the last degree line passed on the circle. Reading to the **right** on the vernier scale, the minute line which coincides exactly with a degree line is 25'.

Remember, each line on the vernier scale represents 5', so you will be reading 5', 10', etc.

Figure 2b

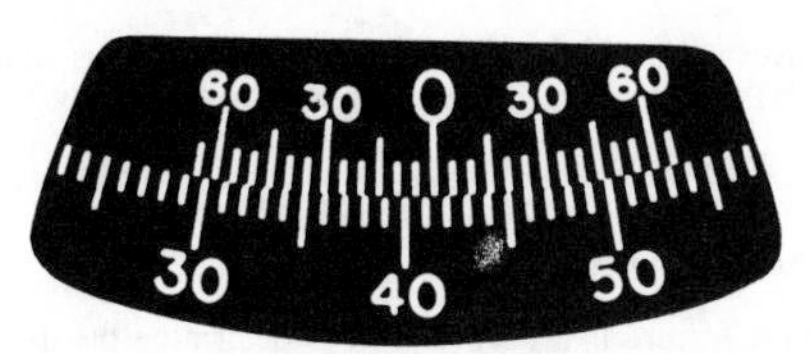

Horizontal circle reading: 41° 25'

Figure 3b

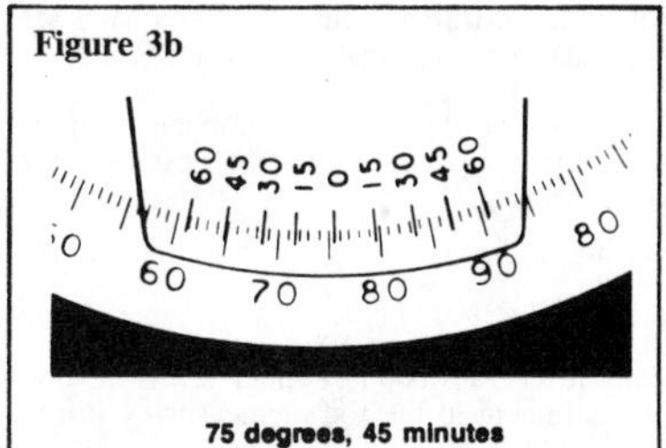

75 degrees, 45 minutes

b. In figure **2c**, the circle degree scale is being read to the **left**. Reading to the **left** on the vernier scale, the minute line which coincides exactly with a degree line is 45'.

Figure 2c

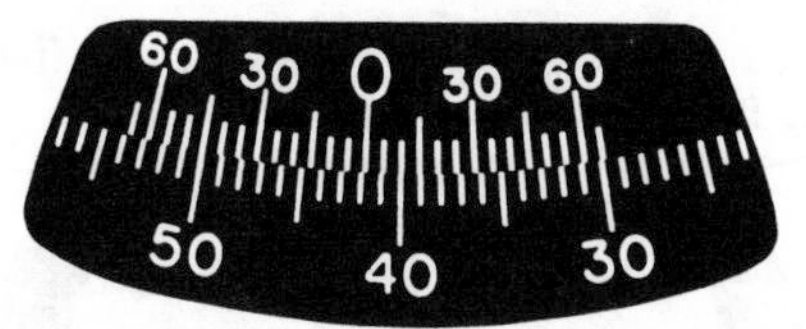

Horizontal circle reading: 41° 45'

c. The vertical arc is read in the same way. In **figure 2d**, 21° was the last degree line passed (when reading the arc scale to the **right**). Reading to the **right** on the vernier scale, the minute line which coincides exactly with a degree is 30'.

Figure 2d

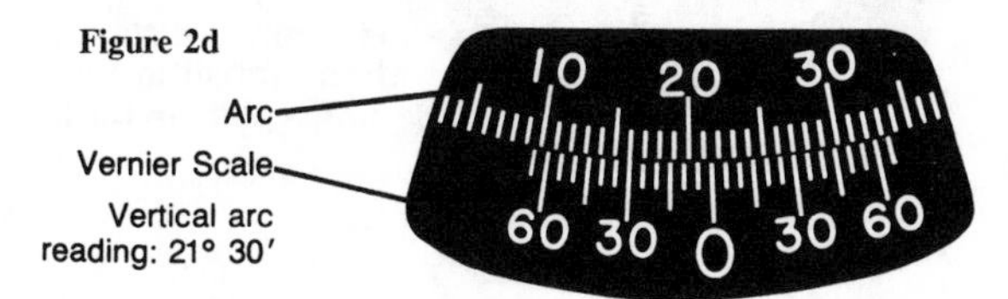

Vertical arc reading: 21° 30'

SIGHTING AND FOCUSING THE TELESCOPE

Aim the telescope at the object and sight first through the notches on the top of the level vial. Look through the telescope and focus the cross hairs. Then bring the object into focus.

Rotate the instrument until the cross hairs are positioned on or near the target. Tighten the horizontal clamp and make final settings with the tangent to bring the cross hairs exactly on point.

When sighting through the telescope, keep both eyes open. You will find that this eliminates squinting, will not tire your eyes and gives the best view through the telescope. Remember to avoid touching the tripod while sighting.

STADIA RETICLE

Many level-transits and builder levels have a unique feature that enables you to determine the distance from the instrument to the rod or measuring instrument.

All David White Universal instruments have a glass stadia reticle with two additional horizontal lines for use in determining distance. Stadia ratio is 1:100, which indicates that the rod measurement between the upper and lower stadia lines multiplied by 100 is the distance from the center of the instrument to the rod.

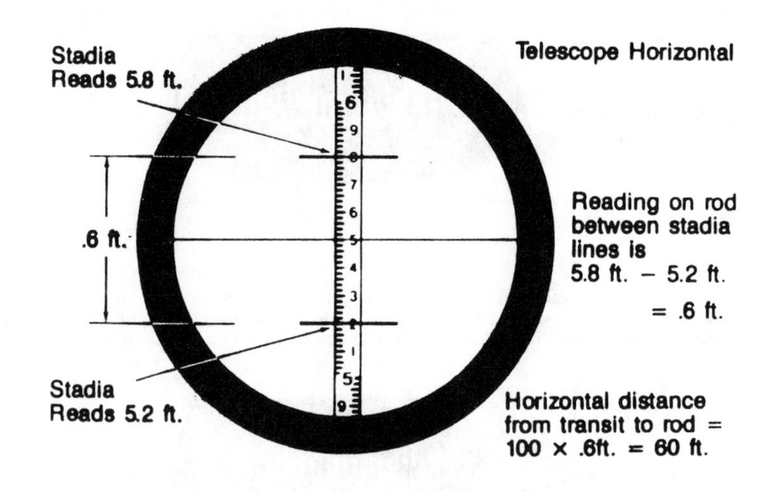

LEVELING RODS

A **Leveling Rod** is a measuring instrument made out of wood, fiberglass, or metal used in conjunction with a transit, transit-level, or a builders level. It can be used with or without movable targets (used to sight longer distances) and can vary in size from 8 feet to 25 feet fully extended. Unextended they are usually sectioned in length for easier handling and storage.

The different types of rods available are the builders' rod with graduations in feet, inches, and eighths; the engineers' rod with graduations in feet, tenths, and hundredths; and metric rods with graduations in meters, decimeters, and centimeters. Some of the different types of leveling rods are shown below:

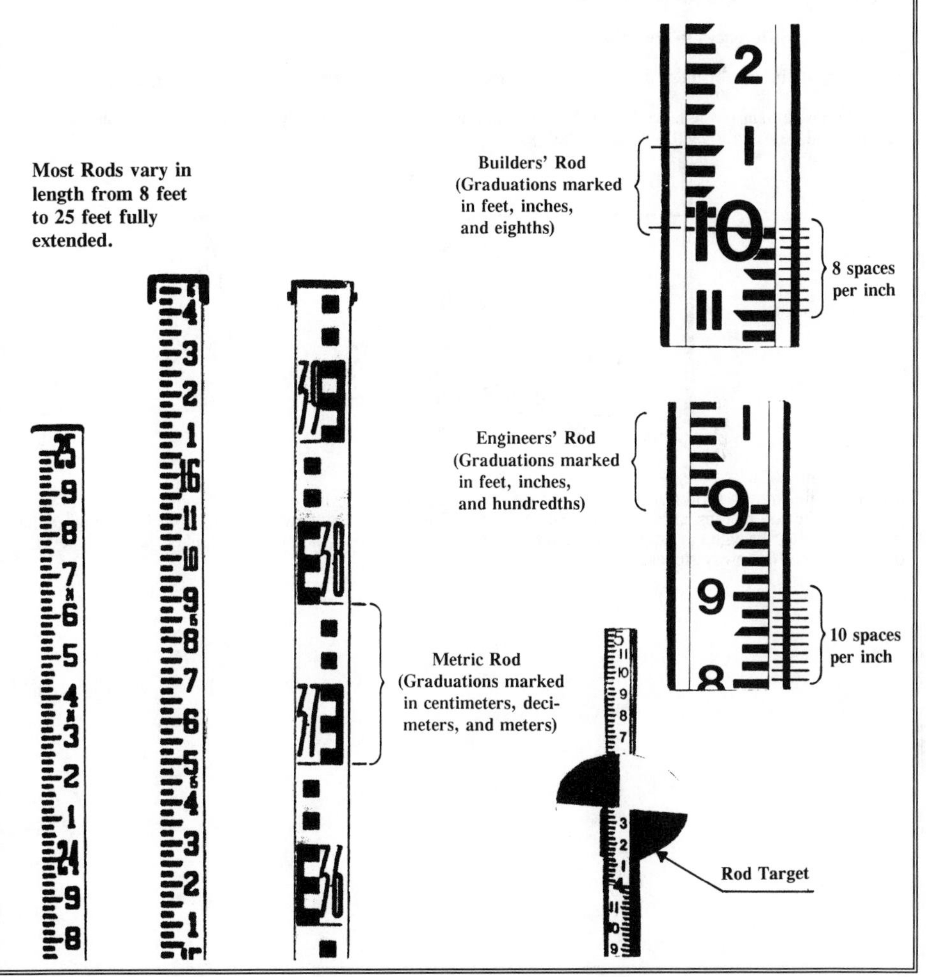

PRACTICAL APPLICATIONS OF THE LEVEL-TRANSIT

Levels and level-transits, as do all sighting instruments, operate on the principle that any point along a level line of sight is exactly level with any point along that line.

Levels and level-transits

The following jobs can be accurately performed with a level or level-transit used in the level position (with closed lock lever): grading for swimming pools, driveways, sidewalks, lawns, gardens; plotting contour plowing lines; laying out drainage ditches; setting fence lines; estimating cut and fill requirements; setting forms and footings; leveling walls and foundations; establishing drainage for landscaping; aligning trees and shrubs and building terraces and stone walls.

Determining differences in elevations

One of the main uses of these instruments is for measuring the differences in elevation for grading.

With the instrument leveled, we know that since the line of sight is perfectly straight, any point on that line of sight will be exactly level with any other point.

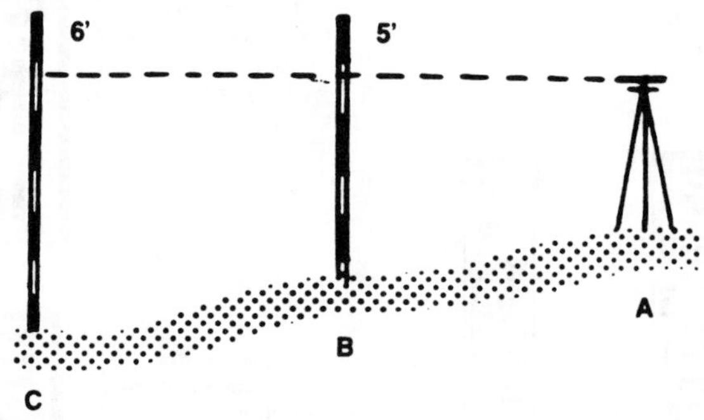

The above illustration shows how exactly we can check the difference in height (or elevation) between two points. If the rod reading at B is 5 feet and reading at C is 6 feet, we know that point B is 1 foot higher than point C. Using the same principle, you can easily check if a row of windows is straight, or a wall is level, or how much a driveway slopes.

Running straight lines with a level

Set up the instrument over Point A. A plumb bob should be held over Point B. Sight approximately on the plumb bob cord and turn the telescope so that the vertical cross hair coincides with it. To align the intermediate points, direct the person with the leveling rod to the right or left until the rod coincides with the vertical cross hair. It is important not to move the instrument during this operation. After all points have been set, check back on Point B to be sure that the instrument did not move.

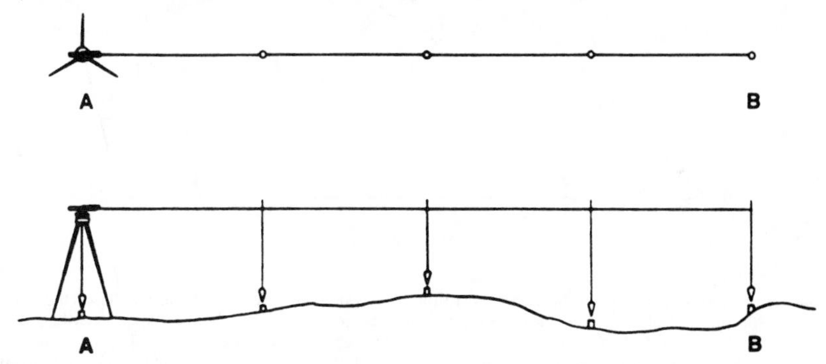

Determining contour lines

Contour lines, such as used for contour plowing, are lines connecting points of equal level. To determine contour lines, first level the instrument carefully. A sighting rod should be held at the beginning contour line about 100 feet from the instrument. Sight the rod and set a target on the rod at the point where the horizontal cross hair intersects the rod. Then move the rod to approximately the next place where a contour line stake is to be set and move the rod up or down the slope until the line of sight through the telescope again intersects the target. This determines a second point on the contour line. This step is repeated as many times as necessary.

If the person holding the rod is moving too far from the instrument, simply hold the rod in one of the positions determined from the original instrument position and move the instrument to another convenient location along the contour. Sight on the rod in this position and move the target up or down until it lines up with the cross hair. The line may then be continued in the same manner as before.

Measuring and laying out angles

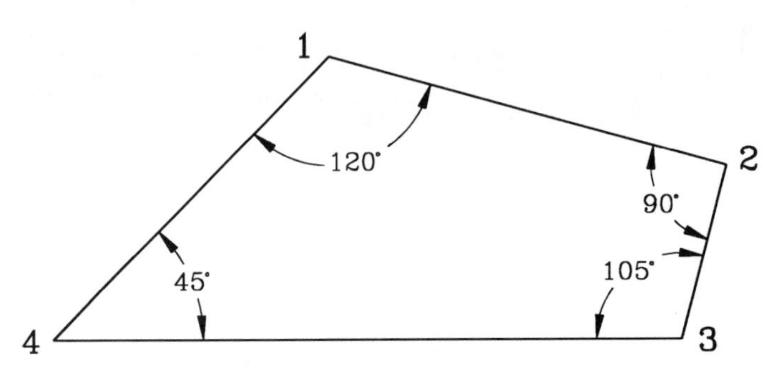

For measuring angles, attach a plumb bob cord to the hook on the screwdriver-style handle of the tripod, The point of the plumb bob will then indicate a point on the ground directly below the center of the instrument and, therefore, will also indicate the center of all angles to be measured. This point should be marked by a stake about two inches square with a tack indicating its center. Horizontal angles are always read at the vernier zero mark. (See "How to Read the Circle, Arc, and Vernier", page 334) The following example simply explains how to measure angles.

Set the instrument up at station 1. Place it so the plumb bob is directly over station 1. Now level the instrument as explained previously. Turn the telescope so that the vertical cross hair is directly in the center of the rod at station 2. Set the horizontal circle at zero to coincide with the vernier zero. Then turn the telescope to sight on station 4 and read the angle. (In this case it would be 120 degrees.)

Move the instrument and tripod to station 2 and level exactly as before. When the instrument has been leveled, sight back to read on station 1. Set the horizontal circle to zero, then sight the telescope to locate station 3 and read the angle (90 degrees). Move instrument and tripod to station 3 and level as before. Again sight back to the previous station (2) and set the circle at zero. Turn the telescope to sight on station 4. Your angle should be 105 degrees. The same procedure is followed to measure the angle at station 4.

You can prove the accuracy of your reading by adding the four inside angles together because the total of the inside angles of a quadrangle is always 360 degrees.

To lay out an angle, proceed in the same way as in measuring an angle. Set the instrument at station 1, level it, and set the circle at zero. Swing the telescope to the desired angle and move the rod to intersect the vertical cross hairs. This establishes your angle.

NOTE: Some levels have horizontal verniers which read to 15 minutes (¼ degree). For projects which need more accurate angle measurement for layout, a more precise instrument with a 5 minute vernier is recommended.

Task No. ▸	Q2

Laying out a swimming pool

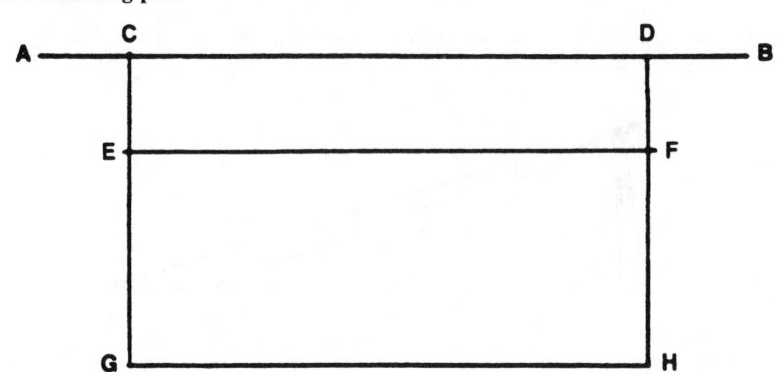

In the above figure, A-B is the lot line. The corner of the proposed swimming pool is E. Point C is the point where the line forming the side of the pool intersects this lot line. If E has not already been determined, set up and level the instrument directly over point C and line up the cross hairs on B. Set horizontal circle to zero. Turn the telescope 90 degrees to the right.

The vertical cross hair of the instrument will now cut across point E and point G. Measure the distance from the lot line to the corner of the pool, which is C-E. Also, the distance E-G is measured along this line. Place a stake at points C, E and G.

Next set up and level the instrument directly over point E and line up the cross hairs on G. Set reading to zero. Turn the telescope 90 degrees to the left to establish the line E-F. Measure out the distance and place a stake at point F. The distance, D-F, (from F to the lot line) will exactly equal E-C if the work is correct.

Next set up and level the instrument over point F and set the vertical cross hair at point E. Set reading to zero. Turn the telescope 90 degrees to the left to establish the line F-H. Measure out the distance and place a stake at point H.

Level-transits

The following example illustrates how to use a level-transit for laying out roads, streets, building lines, ditches, orchards, fences, hedges, fields, etc.

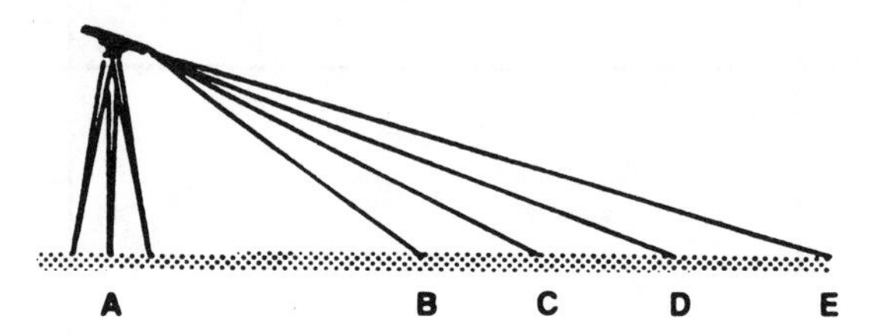

Running straight lines with a level-transit

(Although it is possible to run straight lines with a level, it is faster and more accurate to use a level-transit.)

To run a straight line between stakes A and E, position the instrument directly over A. After you level the instrument, release the lock that holds the telescope in the level position and swing the instrument until point E is aligned with the vertical cross hair. Tighten the horizontal clamp so the telescope can move only in a vertical plane. By pointing the telescope up or down, points B, C and D can be located.

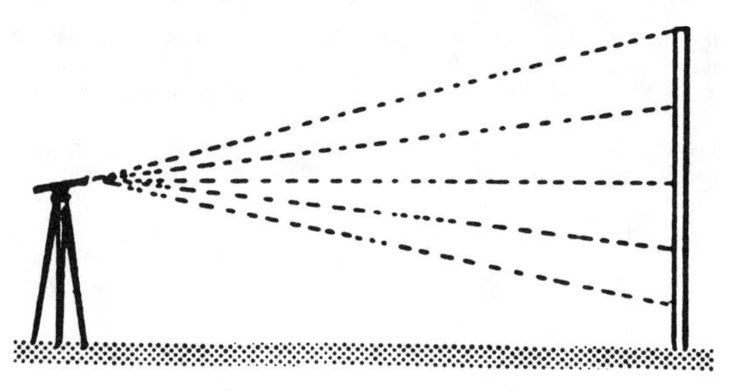

Establishing vertical lines and planes

It is necessary to use a level-transit instrument for taking vertical sights, such as lining up a building wall, aligning piers or fencing, flagpoles, T.V. antennas, plumbing windows or doorways, etc.

To establish vertical lines and planes, first level the instrument, then release the locking levers which hold the telescope in the level position. Swing the telescope vertically and horizontally until the line to be established is directly on the vertical cross hair. If the telescope is rotated up or down, each point cut by the vertical cross hair should be in a vertical plane with the starting point.

"This page intentionally blank."

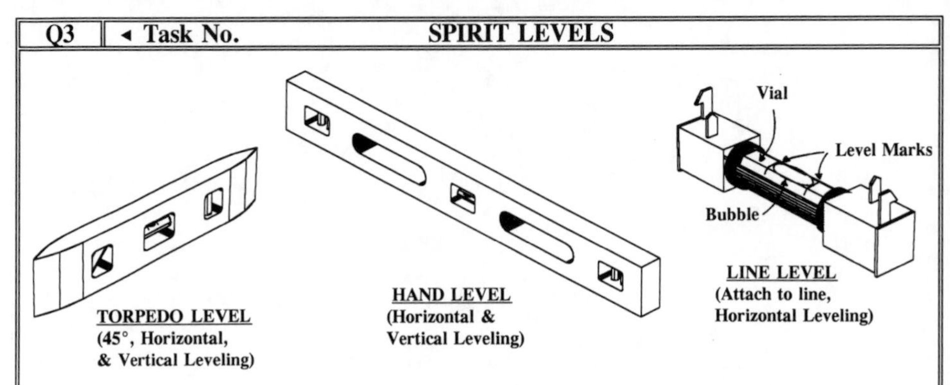

TORPEDO LEVEL
(45°, Horizontal, & Vertical Leveling)

HAND LEVEL
(Horizontal & Vertical Leveling)

LINE LEVEL
(Attach to line, Horizontal Leveling)

Different Kinds of Spirit Levels

The variety of spirit levels shown above are all similar in that they consist of glass or plastic vials filled with alcohol or ether to prevent freezing (this is where the name spirit level comes from). The vial is filled with liquid just enough so there is a bubble left. When the bubble is exactly between the two lines marked on the vial, supposedly the object will be level. However in many cases spirit levels do not measure true level because of quality of craftsmanship, damage due to handling, warping, and so on. To avoid these problems a spirit level can be checked out (this is demonstrated on page 4 of this module).

Usually, the more expensive the spirit level the more accurate it is. However even the more expensive levels can be untrue in level. Therefore it is advised that a level be checked out before purchasing.

Different Ways to Level With a Spirit Level:

Leveling Horizontally:

Method 1. Use a level directly to level objects horizontally (objects that are within reach of the level).

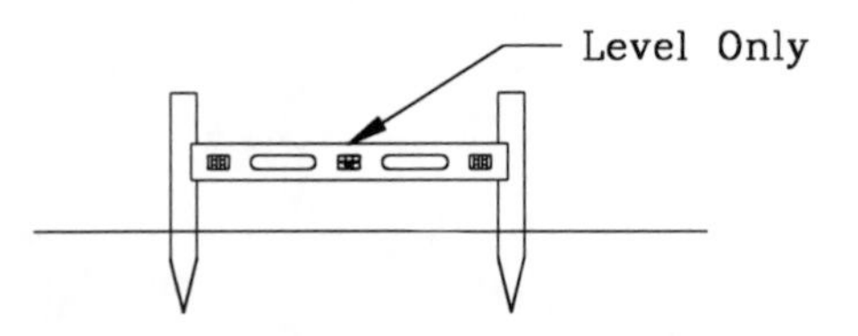

Method 2. Use a level with a straight edge to level objects longer than the level.

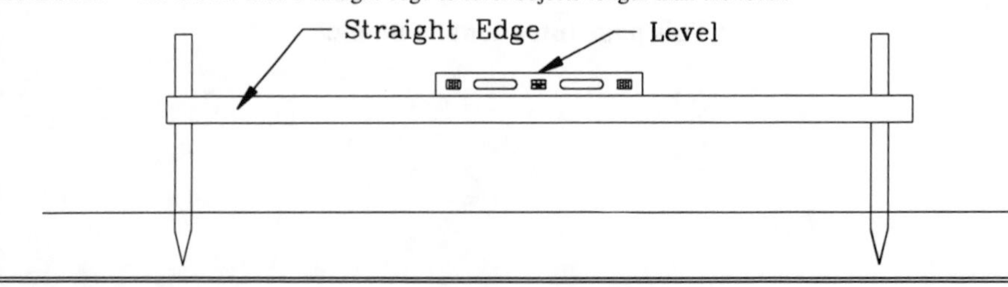

Method 3. Use a level with a straight edge and a pivot point to set grade stakes for small buildings or projects that don't have to be too accurate* and/or to set objects level with each other.

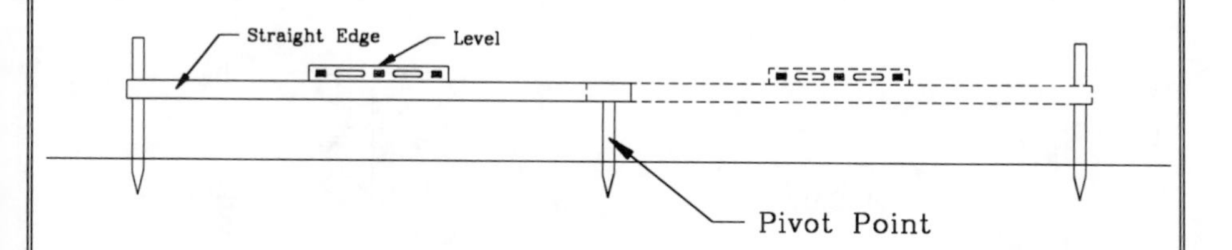

Method 4. Use a level with space blocks.

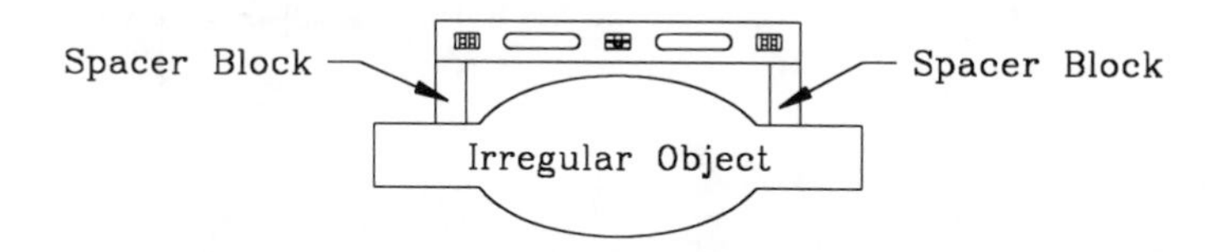

Method 5. Use a level with offset space blocks to level objects on a slant or taper.

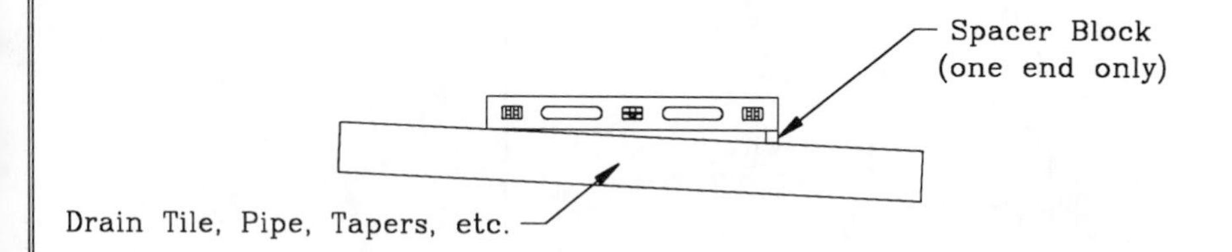

*To attain greater accuracy for leveling horizontally, see The Water Level Task No. Q1 or the Level-Transit Task No. Q2.

Leveling Vertically:

Method 1. Use a level directly to level objects vertically.

Method 2. Use a level with the help of a prop or a support.

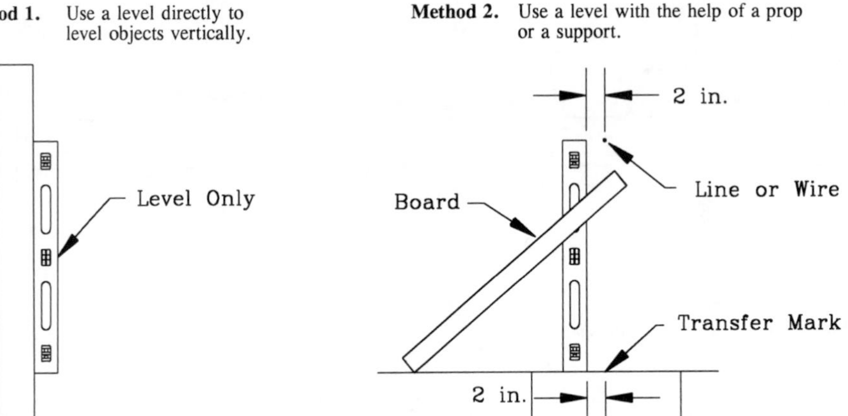

Method 3. Use a level with spacer blocks and/or with a level and straight edge with spacer blocks.

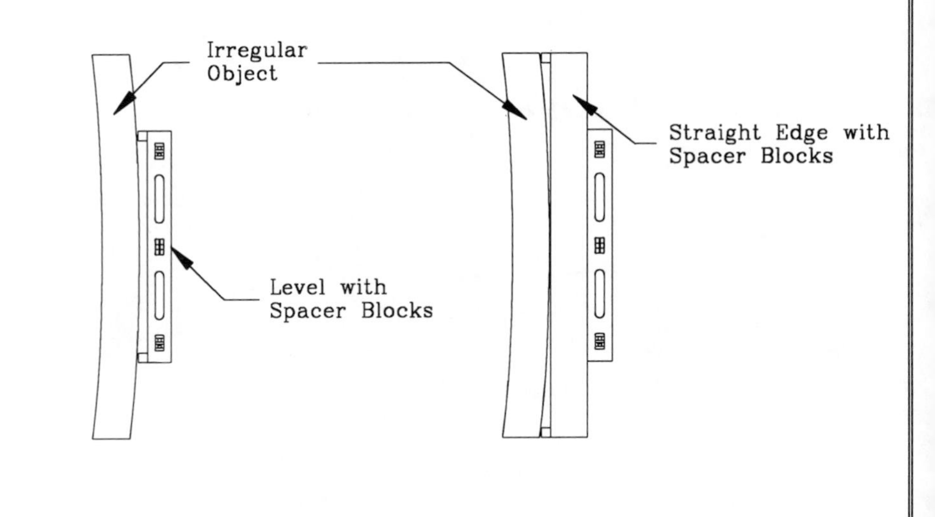

How To Check a Spirit Level for Accuracy:

Check a Level Vertically:

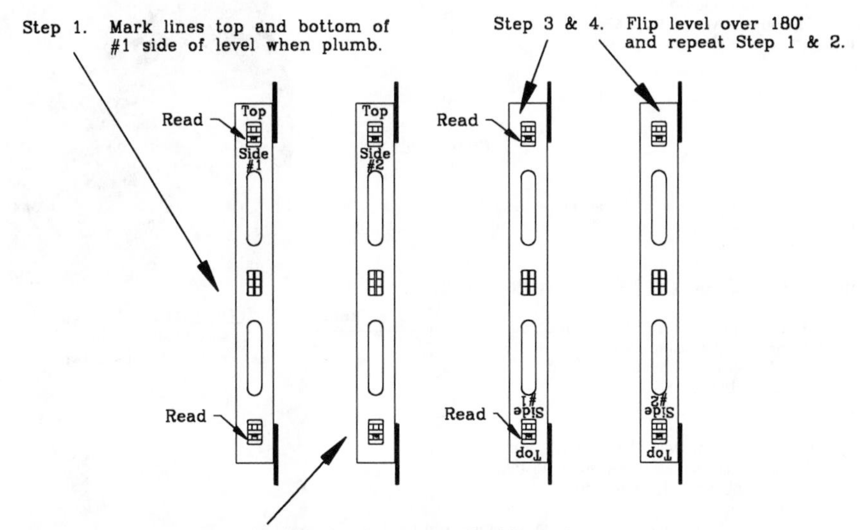

Check a Level Horizontally:

Step 1. Mark lines on the right and left of the #1 side of level when plumb.

Step 2. Flip level over to #2 side and plumb level, marks should match. If they do not match, the level is off*.

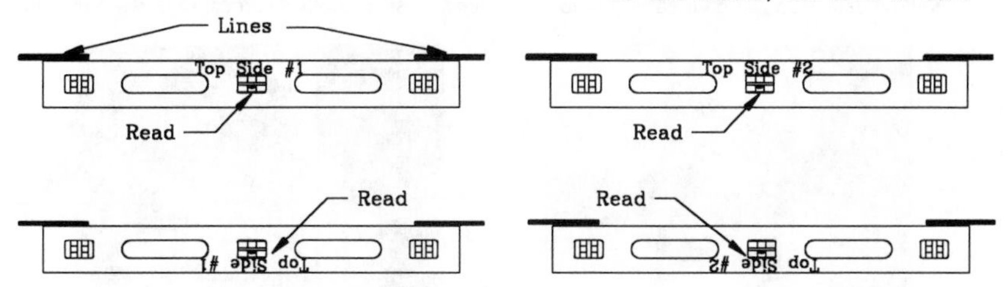

Step 3 & 4. Flip level over 180° and repeat Step 1. and Step 2.

*On levels that are not too far off plumb, mark identification marks to compensate for plumb (level) corrections on the specific side of the level. Then when leveling objects, you can tell if the object being leveled is plumb. It can be very difficult to find or maintain a level with all the vials reading 100% plumb.

Q4 | ◂ Task No. | THE PLUMB BOB

The **plumb bob** is included in this book because it is by far the most economical and accurate method to use to level (plumb) objects vertically. Most plumb bobs range is size from 6 to 24 ounces.

Plumb Bob

First Observation - Position Ⓐ
Make sure to sight in line with the tree stand legs as shown.

The following pictures, illustrations, and techniques demonstrate how the plumb bob can be used to plumb and position different vertical objects.

Plumb Christmas Trees, Tall Poles, Fence Posts, etc.:

Step 1. Hold the plumb bob 12 to 16 inches in front of you as shown in "Position A" on the right (leave 12-16 inches of line out).

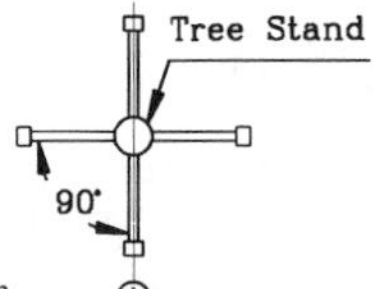

Step 2. Eyeball the line with the object. When the object coincides (lines up) with the line, the object will be plumb.

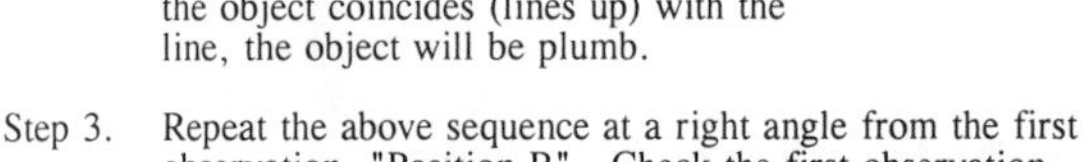

Step 3. Repeat the above sequence at a right angle from the first observation, "Position B". Check the first observation, "Position A", for plumb again. Repeat until you are satisfied the object is plumb.

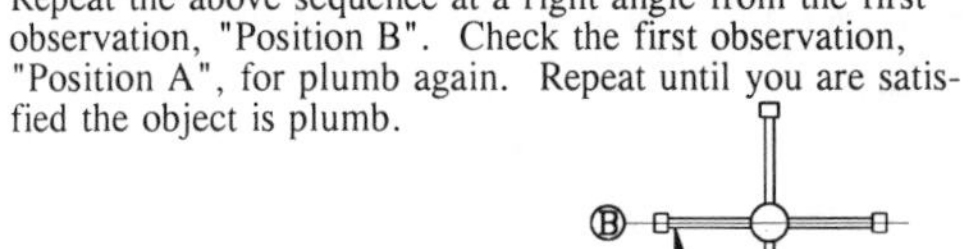

Second Observation - Position Ⓑ

Note: The above sequence can be used to plumb poles, fence post, and other vertical objects. See below.

Plumb a Telephone Pole.

Plumb Support Post and Fence Post.

	Task No. ▸	Q4

Plumb Walls and Vertical Objects:

Step 1. Secure the line an equal distance from the object being leveled as shown. Use a spacer at least 1/4 to 3/8 inches greater than 1/2 the thickness of the plumb bob.

Step 2. Lower plumb bob to within 1/2 to 3/4 inch from the floor or base of the object which is being plumbed (leveled).

Step 3. Secure the plumb bob using half hitches and/or slip knots (adjust height of plumb bob, if needed).

Step 4. Set plumb bob into a can of heavy oil* (20-30 weight) Fill can 1/2 to 3/4 full.

Step 5. Take measurements -- adjust object(s) until the measurements match measurements at the spacer positions.

*The oil will help stop the plumb bob from swinging, especially if there is a little breeze. Do not use a plumb bob if it is too windy.

Position Objects:

Step 1. Locate a definite point (such as a center point, edge of a shaft, pulley, etc.) and secure the line to the object as shown on right.

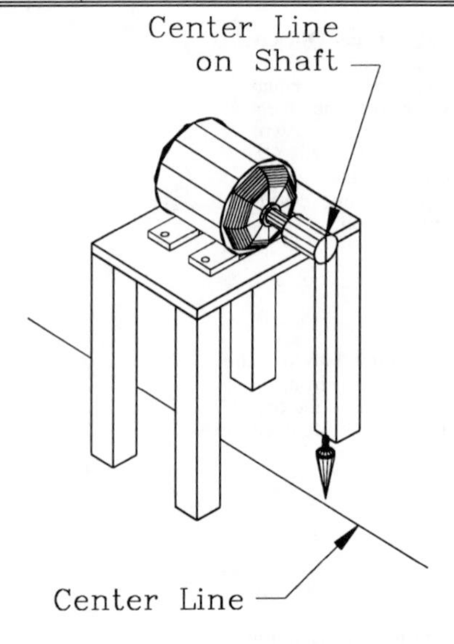

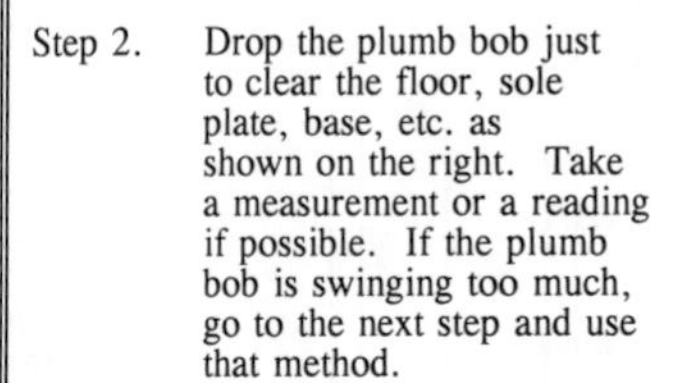

Step 2. Drop the plumb bob just to clear the floor, sole plate, base, etc. as shown on the right. Take a measurement or a reading if possible. If the plumb bob is swinging too much, go to the next step and use that method.

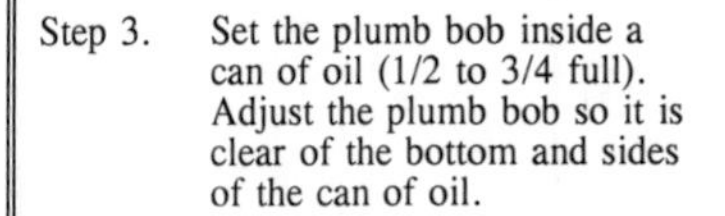

Step 3. Set the plumb bob inside a can of oil (1/2 to 3/4 full). Adjust the plumb bob so it is clear of the bottom and sides of the can of oil.

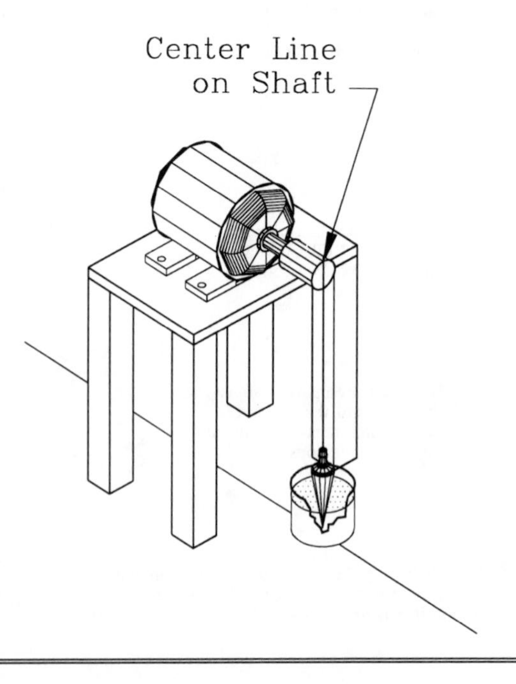

Step 4. Position a metal square and/or a combination square as shown on the right.

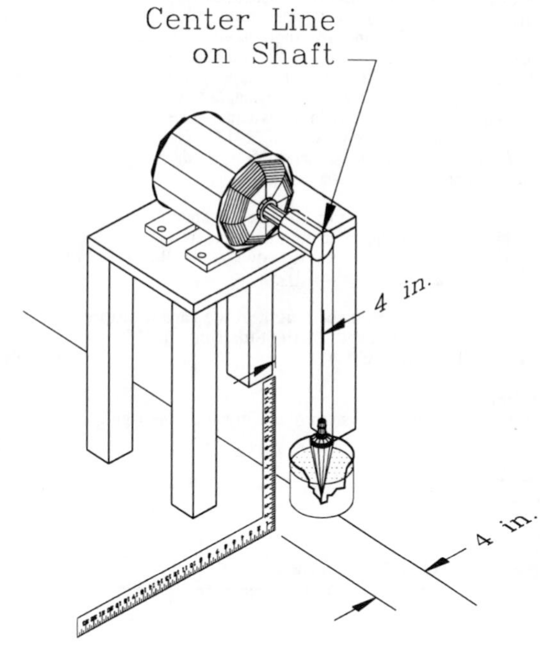

Step 5. Measure a distance between the square and designated point as shown. Then measure between the plump line and the square. When the measurements between are equal the object will be in line or plumb vertically.

[Note there are many different applications and ways to use a plumb bob not listed in this math book, however the methods on these two pages will give you some different ideas how a plumb bob can be used to obtain greater accuracy. Use the method(s) that work best for you.]

R1	◄ Task No.	BRACE MEASURE

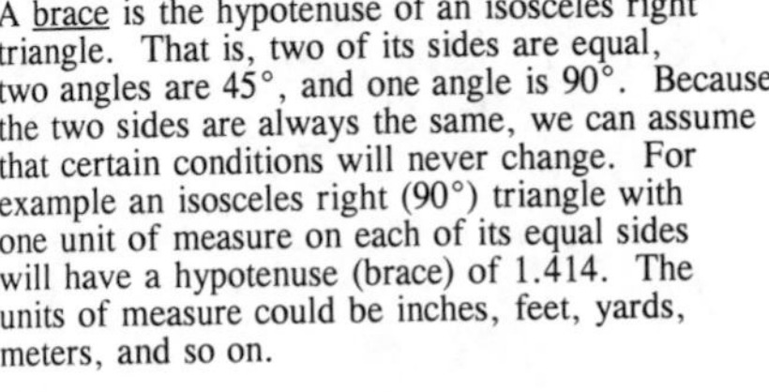

A brace is the hypotenuse of an isosceles right triangle. That is, two of its sides are equal, two angles are 45°, and one angle is 90°. Because the two sides are always the same, we can assume that certain conditions will never change. For example an isosceles right (90°) triangle with one unit of measure on each of its equal sides will have a hypotenuse (brace) of 1.414. The units of measure could be inches, feet, yards, meters, and so on.

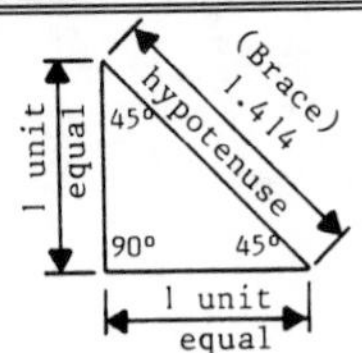

The hypotenuse or brace of an isosceles right triangle can be found by using the formula $a^2 + b^2 = c^2$. The formula is known as the Rule of Pythagoras, see Task No. K5A.

The Rule Of Pythagoras

FORMULA:	$a^2 + b^2 = c^2$
SUBSTITUTE:	$1^2 + 1^2 = c^2$
Calculate:	$1 + 1 = c^2$
	$2 = c^2$
	$\sqrt{2} = c$
	$1.414 = c$

RULE: To find the length of a Brace multiply the length of the Side times 1.414.

FORMULA: B = S(1.414)

Example

Problem 1. Find the length of the brace on the right.

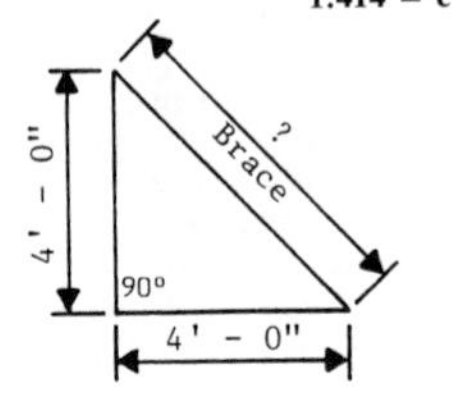

Step 1. Use the formula ------------- B = S(1.414)

Step 2. Substitute the known values into the formula ------------ B = 4' X 1.414

Step 3. Calculate the formula (do the operations indicated) -------- B =

```
  1.414
X     4
 5.656'
```

Answer in decimals of a foot

Solve by Changing to Inches.

4' = 48"

```
   1.414
 X   48"
  11312
  5656
 67.872"
```

equals 5'7 7/8-"

Note the answer is the same.

Step 4. Convert 5.656' so you can read the dimension on an American tape measure.

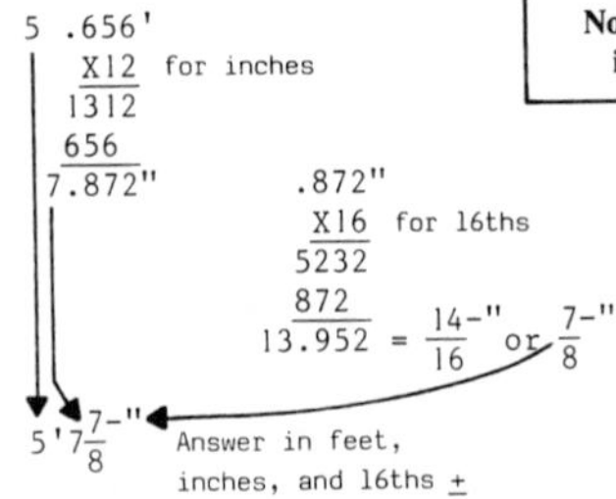

```
5 .656'
   X12   for inches
  1312
  656
 7.872"          .872"
                  X16   for 16ths
                 5232
                 872
               13.952 = 14-"/16 or 7-"/8

5'7 7/8-"   Answer in feet, inches, and 16ths ±
```

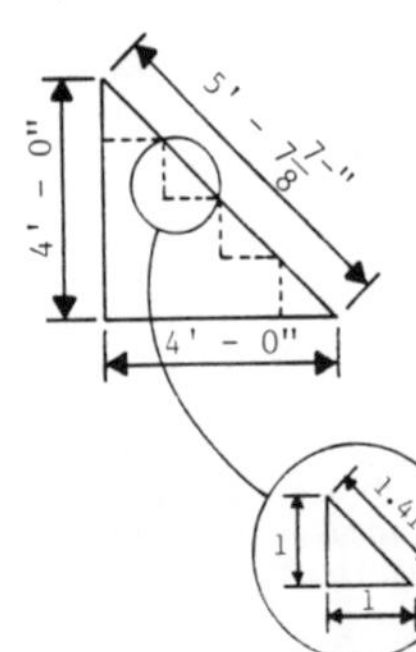

Note there are 4 units of brace encompassed within the isosceles right triangle, each 1.414 feet long.

Another Way to Use Brace Measure

Brace measure can also be used to square between parallel lines or to construct perpendiculars between fixed parallel objects. To square or to construct perpendiculars between parallel lines or fixed parallel objects follow the procedure shown below.

Example

Problem 2. Lay out a perfect 90° (perpendicular) line through line A and line B below.

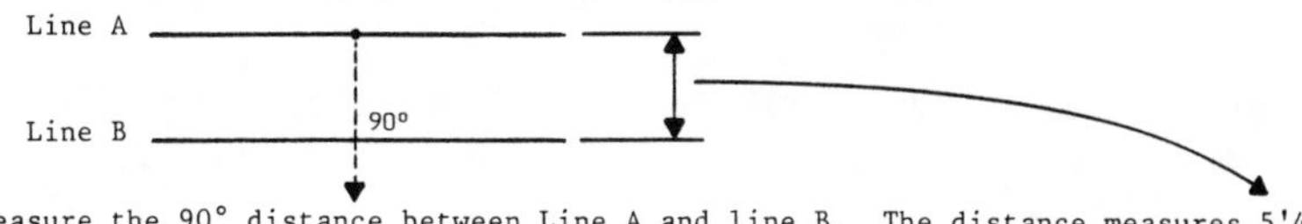

Step 1. Measure the 90° distance between Line A and line B. The distance measures $5'4\frac{1}{2}''$.

Step 2. Convert $5'4\frac{1}{2}''$ to decimals of a foot. $5'(4\frac{1}{2}'') = \frac{4\frac{1}{2}}{12}$ or $\frac{4.5}{12}$ or $12\overline{)4.5}$ = .375' → 5.375'

Step 3. Multiply 5.375' times 1.414.

```
   5.375'
 X 1.414
   21500
   5375
 21500
 5375
7.600250'
```

Step 4. Convert 7.6' so you can read the dimension on an American tape measure.

[Multiply .6' by 12 to change to inches]

[Multiply .2" by 16 to change to 16ths]

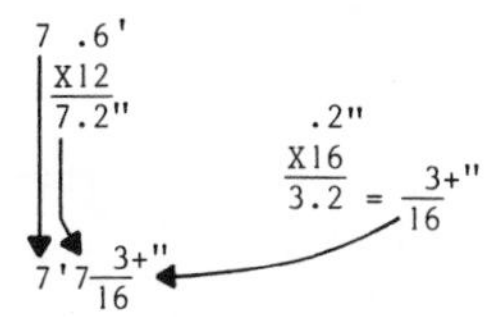

Step 5. Measure 7'7 3/16+" from point c and mark on line B. The mark will become point d.

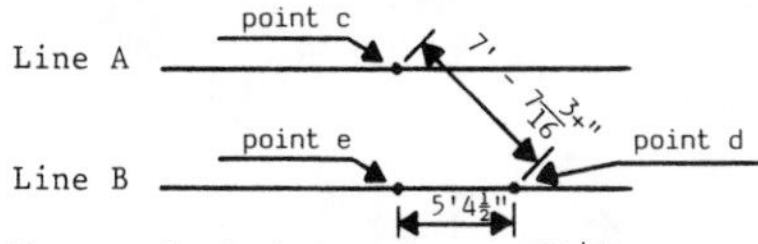

Step 6. Measure back from point d 5'4½" and mark. The mark will become point e.

Step 7. Draw a line through point c and point e. The line should be a perfect right angle (90°).

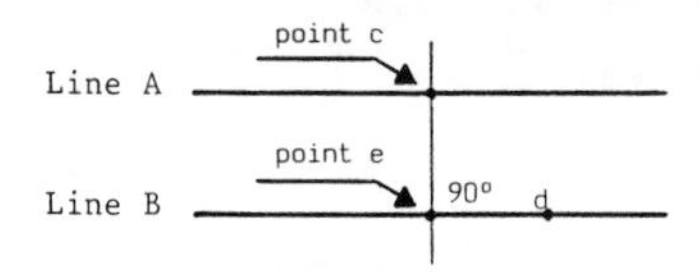

Step 8. Check by measuring back from point e 5'4½" and mark. This mark will be point f. If the distance between point f and point c is 7'7 3/16+", everything was done correctly.

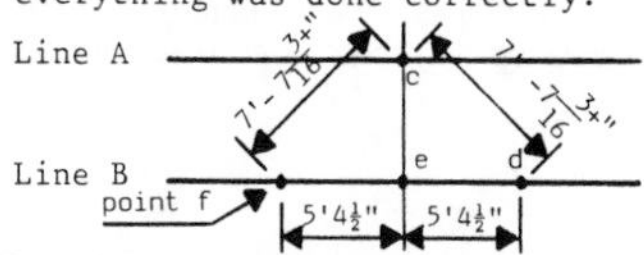

BRACE MEASURE

This page is included to give you a better idea how brace measure is determined and why it works. It demonstrates how a true brace is calculated followed by how to apply brace measure concepts to real on-the-job applications.

Brace Measure

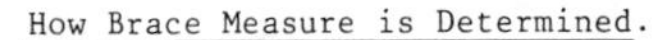
How Brace Measure is Determined.

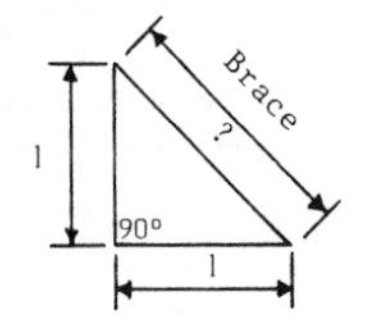

Why Brace Measure Always Works.

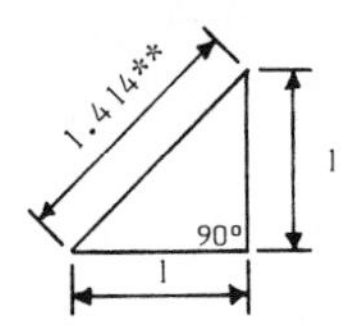

One unit of brace is determined by using the Rule of Pythagoras*.

Formula: $a^2 + b^2 = c^2$

Substitute: $1^2 + 1^2 = c^2$

Calculate: $1 + 1 = c^2$

$1^2 = 1 \times 1 = 1$

$1 + 1 = 2 \longrightarrow 2 = c^2$

$\sqrt{2} = c$

$1.414 = c$

Brace measure will always work because a true brace is the hypotenuse of an isoceles right triangle (has two equal sides and one angle is 90°).

****The hypotenuse of one unit of brace will always be 1.414, but 1.414 could be inches, feet, meters, etc.**

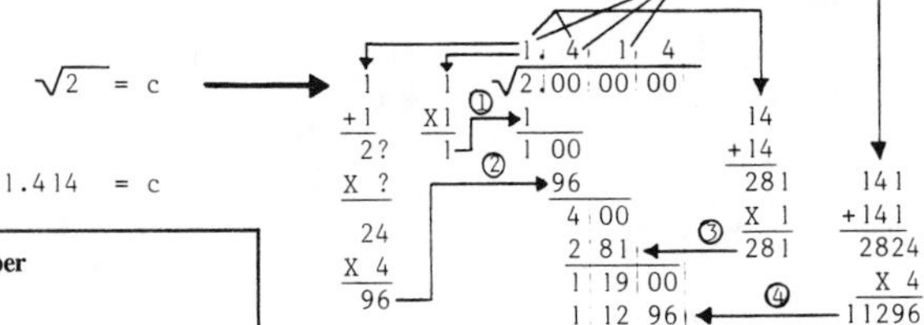

K5A ◄ Task Number

***The Rule of Pythagoras:** **In any right triangle, the square of the hypotenuse is equal to the sum of the squares of the other two sides.**

[For more information on finding square roots, see Task Number H2 .]

The next two pages demonstrate how to find the brace when the sides are either in inches, whole feet, or in feet and inches. Please note once again the importance of conversion. You could not measure out the brace if you did not know how to convert decimals to fractions and, vice-versa, fractions to decimals.

In most trade oriented occupations braces are a standard means of support. Therefore, the skilled tradesperson should be able to calculate them readily. A true brace is the hypotenuse of an isosceles right triangle. An example of a true brace is shown below.

Example:

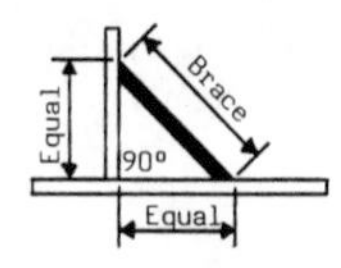

True Brace

Note

Brace triangles have two equal sides and are made up of one 90° angle.

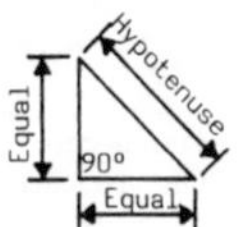

The following method is a shortcut used when calculating brace measure ($a^2 + b^2 = c^2$ could be used but this would be very time consuming). Notice that anytime a true brace is to be found certain factors are always the same. For example there are always two equal sides and one angle is always 90°. Therefore the hypotenuse (brace) of one of these units will always be 1.414.

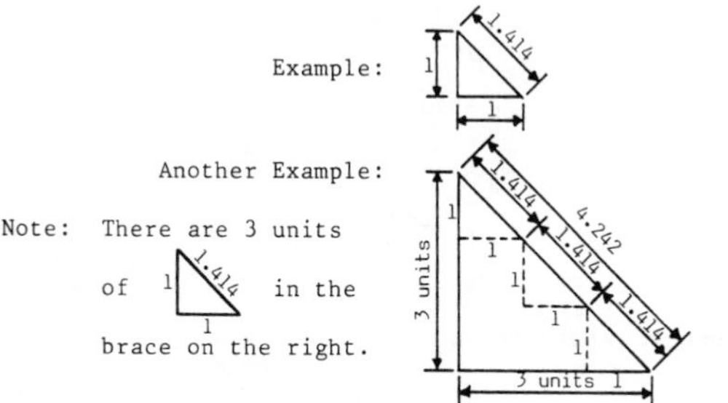

Note

To find the total length of the brace on the left multiply the length of a side (3) times the hypotenuse (brace) of one unit (1.414).

Thus: 3 X 1.414 = 4.242

Therefore,
the **RULE:** **To find the length of any Brace (hypotenuse of an isosceles right triangle) multiply the length of one Side times the hypotenuse of one unit (1.414).**

FORMULA: **B = S1.414**

Example
Problem 1. Find the length of a brace in inches.

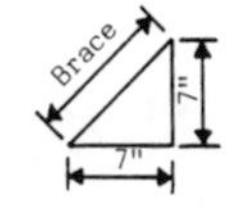

Step 1. Use the formula --------------- B = S1.414

Step 2. Substitute the known values into the formula -------------- B = 7 X 1.414

Step 3. Calculate the formula (do the operations indicated) --------- B = 1.414 X 7 = 9.898"

Step 4. Convert the decimal so you can read the dimension on an American tape measure.

[Note how conversion problems are always popping up.]

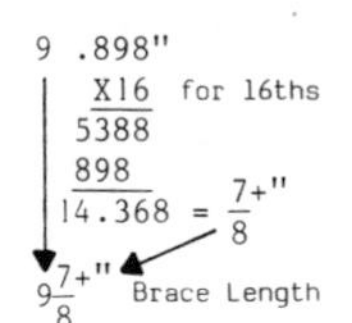

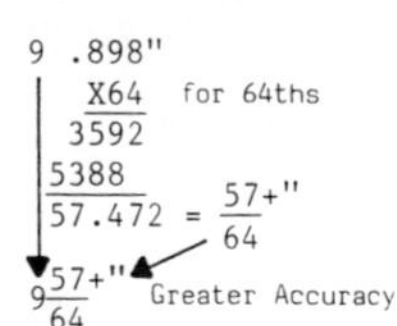

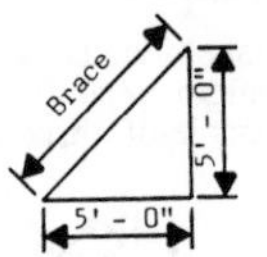

Example
Problem 2. Find the length of a brace in feet.

Step 1. Use the formula ---------------- B = S1.414

Step 2. Substitute the known values into the formula --------------- B = 5 X 1.414

Step 3. Calculate the formula (do the operations indicated) ---------- B = 1.414 X 5' = 7.070'

Step 4. Convert the decimal so you can read the dimension on an American tape measure.

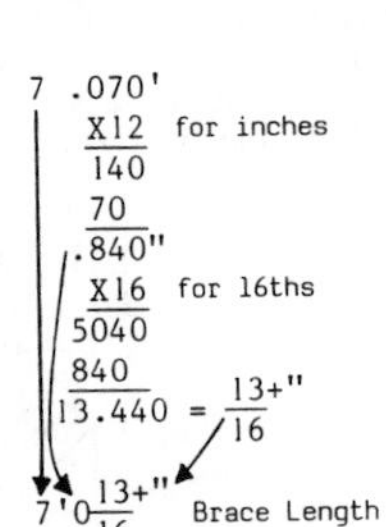

```
7 .070'
   X12   for inches
   140
   70
  .840"
   X16   for 16ths
  5040
  840
 13.440 = 13+"/16

7'0 13+"/16   Brace Length
```

Example
Problem 3. Find the length of a brace in both feet and inches.

Step 1. Use the formula --------------- B = S1.414

Step 2. Substitute the known values into the formula -------------- B = 7.416' X 1.414

[7'5" = $7\frac{5}{12}$ = 7.416']

Step 3. Calculate the formula (do the operations indicated) --------- B =

```
     7.416'
   X 1.414
     29664
     7416
   29664
   7416
 10.486224'
```

Step 4. Convert the decimal so you can read the dimension on an American tape measure.

```
10 .4862'
    X12   for inches
   9724
  4862
  5.8344"
     X16  for 16ths
   50064
   8344
  13.3504 = 13+"/16
```

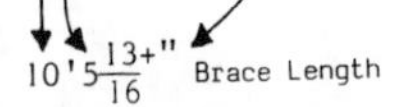

10'5 13+"/16   Brace Length

More Practical Applications of Brace Measure:

Example
Problem 4. Locate points when objects (center to center) are equal distance and parallel to each other.

Step 1. Change 3'7 1/2" to all inches and decimals of an inch (3'7 1/2" = 36" + 7.5" or 43.5"). Multiply that answer (43.5) times 1.414.

```
   1.414
 X 43.5
   7070
  4242
 5656
 61.5090" or 5' 1 1/2"
```

Step 2. Starting at point b measure (swing an arc) 5'1 1/2" and mark point c as shown.

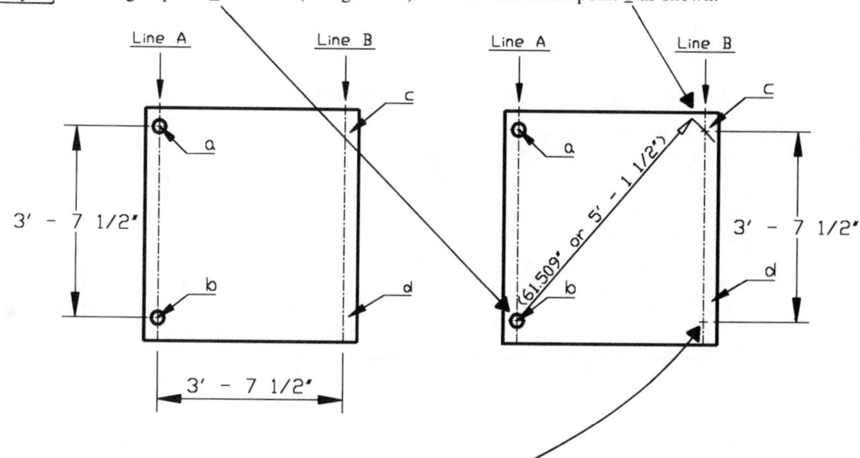

Step 3. Starting at point c measure 3'7 1/2" and mark point d as shown.

Step 4. Check for accuracy by measuring from point d to point a. The measurement should be exactly 5'1 1/2". If this measurement is off, retrace the steps and take care in measuring and marking until it it exact!

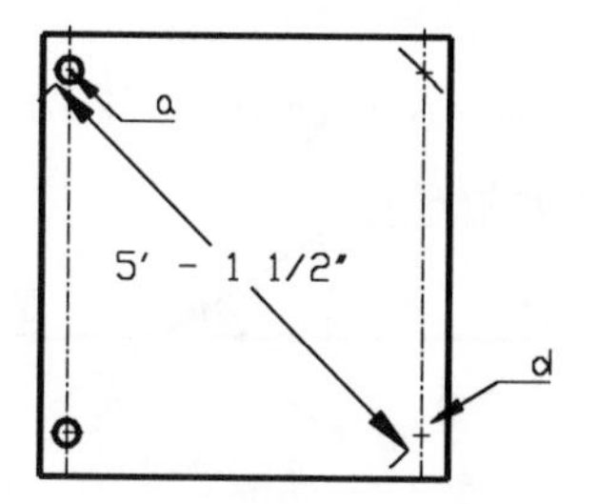

Example
Problem 5. Find the diameter of a circle when the side of an inscribed square is known.

Step 1. Change 4 1/4" to a decimal (4 1/4 = 4.25").

Step 2. Multiply 4.25 times 1.414.

```
   1.414
X   4.25
    7070
   2828
  5656
  6.00950 = 6" diameter
```

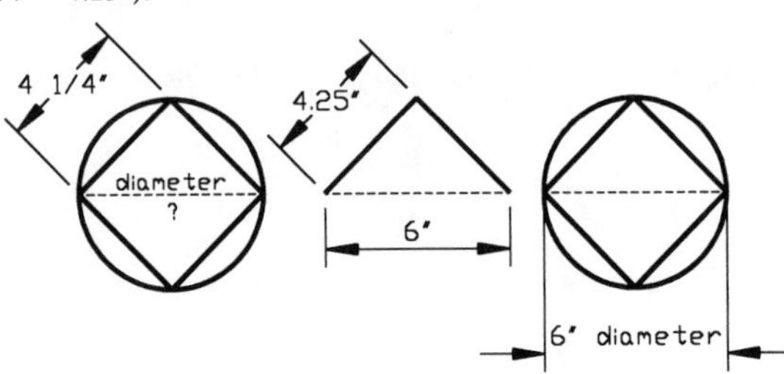

Example
Problem 6. Draw (lay out) a perpendicular line between parallel lines and check it.

Step 1. Change 4'6" to all inches and/or to decimals of a foot (4'6" = 54", or 4 6/12' = 4.5'). Note by changing 4'6" to both inches and feet, you are checking the answer). Multiply these answers times 1.414.

```
  1.414
X    54
   5656
  7070
  76.356" - 6'4 3/8-"
```

```
  1.414
X   4.5
   7070
  5656
  6.3630' = 6 .363'
               X12
               726
              363
              4.356"
          6'4 3/8-"
```

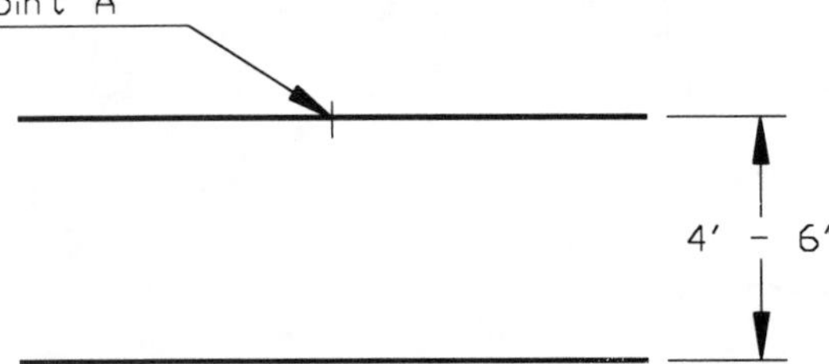

Step 2. Starting at point A measure (swing an arc) 6'4 3/8-" and mark point B as shown.

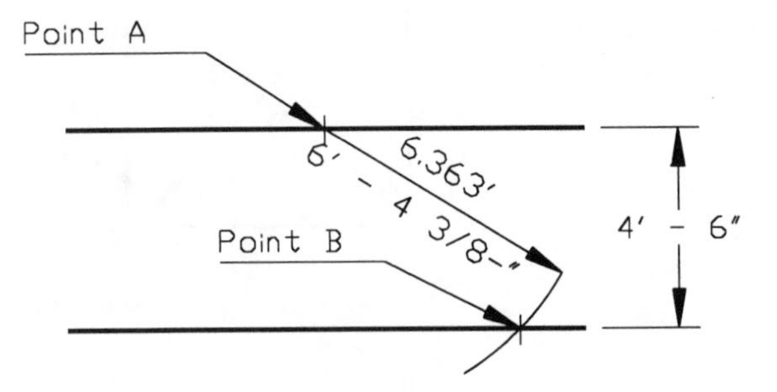

Step 3. Starting at point B measure 4'6" and mark point C as shown.

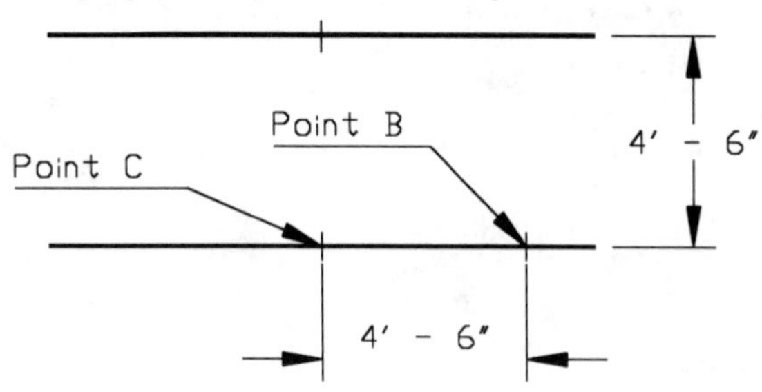

Step 4. Draw a line through point C and point A as shown. This line should be 90° (perpendicular) to the two parallel lines.

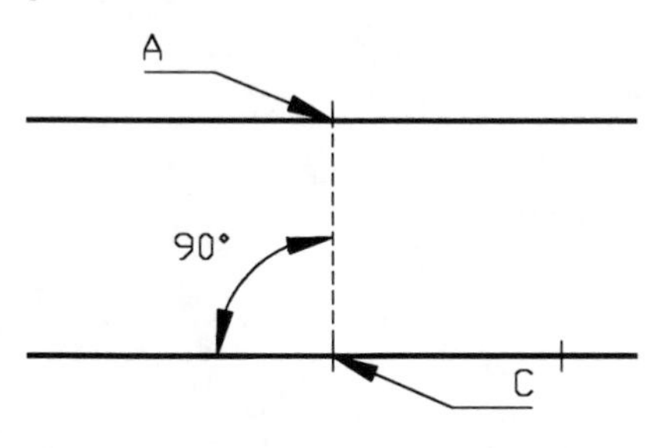

Step 5. Check for accuracy by measuring from point C to point D 4'6" as shown. The distance from point D to point A should be exactly 6'4 3/8-".

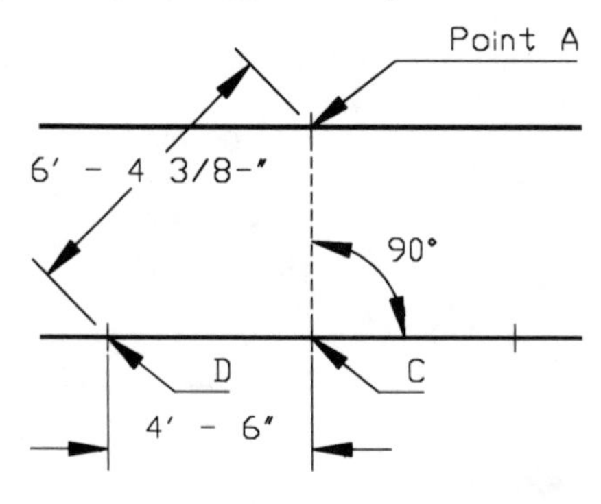

Find the Side When the Brace is Known

Task No. R1 on the preceding six pages demonstrated finding the brace when the side of an isosceles right triangle was known. Task No. R2 is the opposite, find the side when the brace is known. The conditions are the same (two equal sides, two 45° angles, and one 90° angle). However to calculate the length of the side, the sine of a 45° angle has to be known. The sine of a 45° angle is .707.

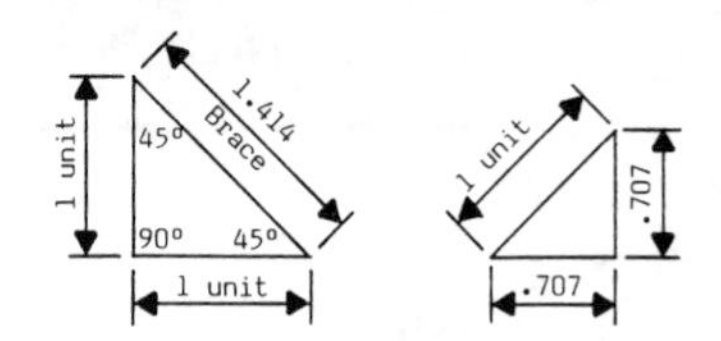

RULE: **To find the length of the Side multiply the sine of a 45° angle times the length of the Brace.**

FORMULA: **S = .707B**

Example
Problem 1. Find the length of the side on the right.

Step 1. Use the formula --------------- S = .707B

Step 2. Substitute the known values into the formula -------------- S = .707 X 3

Step 3. Calculate the formula (do the operations indicated) --------- S =

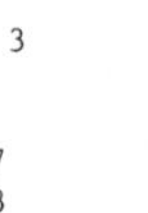

.707
X 3
2.121'

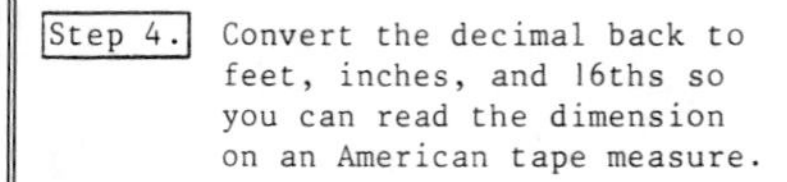

Step 4. Convert the decimal back to feet, inches, and 16ths so you can read the dimension on an American tape measure.

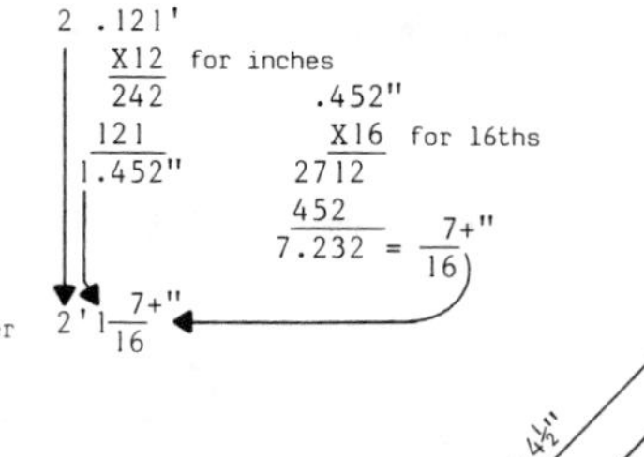

2 .121'
X12 for inches
242
121
1.452"

.452"
X16 for 16ths
2712
452
7.232 = 7+"/16

Answer 2' 1 7+"/16

Example
Problem 2. Find the length of the side on the right.

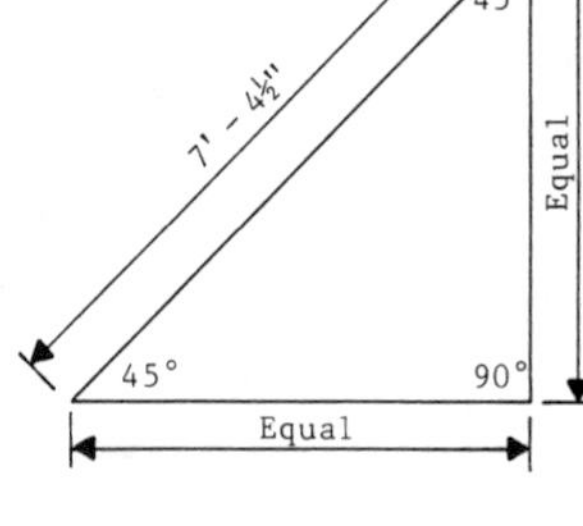

Step 1. Use the formula -------------- S = .707B

Step 2. Substitute the known values into the formula ------------- S = .707 X 7.375'

[Change 7'4½" to decimals of a foot]

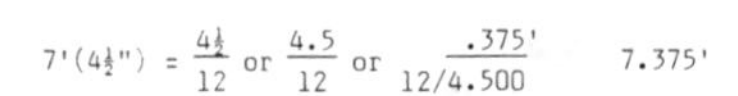

7'(4½") = 4½/12 or 4.5/12 or .375'/(12/4.500) 7.375'

Step 3. Calculate the formula (do the operations indicated) --------- S =

```
   7.375'
X   .707
   51625
 516250
5.214125'
```

Step 4. Convert 5.214' so you can read the dimension on an American tape measure or scale to 1/16 ± of an inch accuracy.

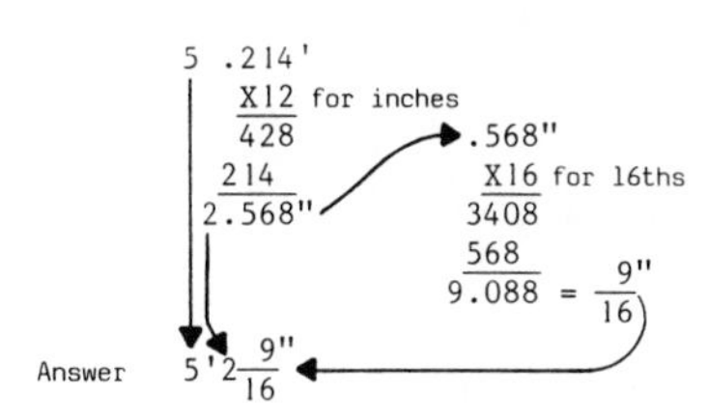

Another Way to Find the Side When the Brace (hypotenuse) is Known

When the diameter of a circle or shaft is known, the side of the square encompassed within it can be found by multiplying the diameter times .707.

Example
Problem 3. Find the length of the side of the square encompassed within the shaft on the right.

? Side

2 3/4"
Diameter

Step 1. Use the formula --------------- S = .707B

Step 2. Substitute the known values into the formula -------------- S = .707 X 2.75"

Step 3. Calculate the formula (do the operations indicated) --------- S =

```
   2.75
X  .707
   1925
  19250
1.94425"
```

Step 4. Convert the decimal 1.944" so you can read the dimension on an American tape measure or scale to 1/32 ± of an inch accuracy.

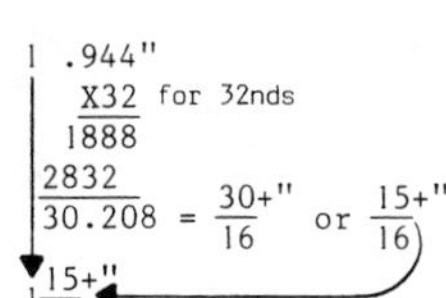

PRACTICAL APPLICATION OF HOW TO FIND THE SIDE WHEN THE BRACE IS KNOWN

In most cases, actual brace dimensions will not be given in decimals. They will be actually measured out (in feet, inches, & 16ths) or be in feet, inches, and 16ths to start with. Therefore, you should be aware of how to change fractions into decimals.

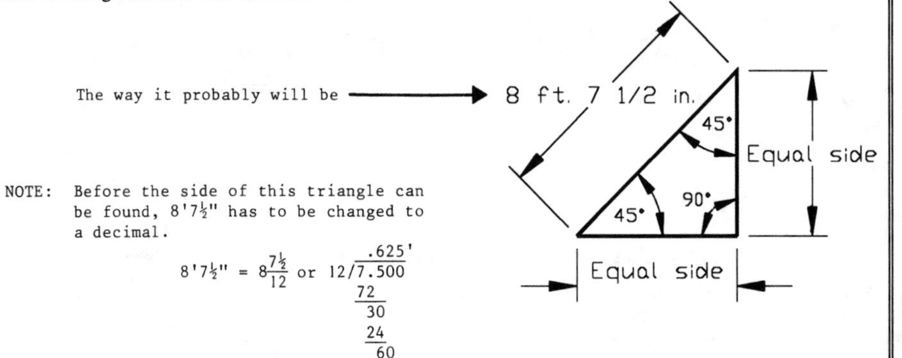

NOTE: Before the side of this triangle can be found, 8'7½" has to be changed to a decimal.

$$8'7\tfrac{1}{2}" = 8\frac{7\frac{1}{2}}{12} \text{ or } 12\overline{)7.500} = .625'$$

```
     .625'
12/7.500
   72
    30
    24
     60
     60
```

Step 1. Change the dimension to a decimal as shown above.

Step 2. Multiply the brace in decimals (8.625') times .707

```
   8.625'
  X .707
   60375
 603750
6.097875'
```

Step 3. Convert the decimal back to feet, inches, and 16ths so you can read the dimension on an American tape measure.

```
6 .098'
   X12
   196
   98
  1.176"
```

$6'1\frac{3}{16}"$ equals the length of each side

Task No. ▸	R2A

Task Number R1 demonstrates how to find the brace when the side of an isosceles right triangle is known. This module demonstrates how to find the side or leg of an isosceles right triangle. In other words, if you know the length of the brace you can find the length of the side. Use this module in conjunction with Task Number R1 because the same examples are used to illustrate better how the rule works.

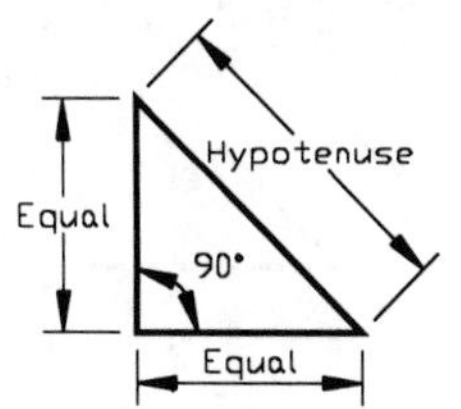

Isosceles Right Triangle

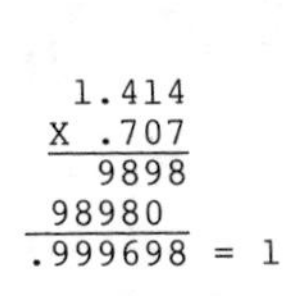

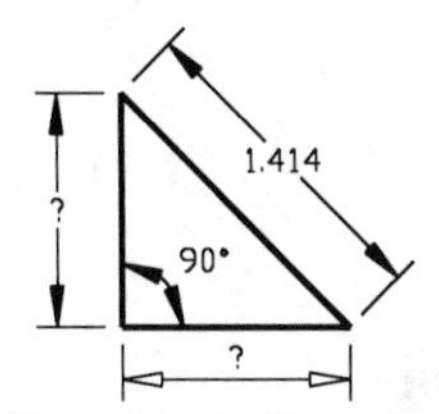

Find the Side when the Hypotenuse is Known

RULE: **To find the length of a Side (leg) of an isosceles right triangle multiply the hypotenuse (Brace) times .707 (.707 is the sine of a 45° angle).**

FORMULA: **S = B.707 (Side = Base X .707)**

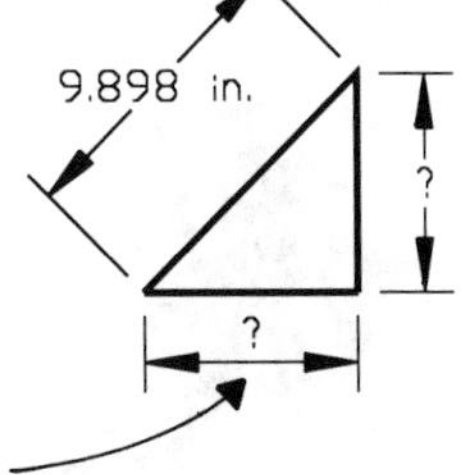

Example
Problem 1. Find the length of side on the right.

Step 1. Use the formula ---------------- S = B.707

Step 2. Substitute the known values into the formula ---------------- S = 9.898" X .707

Step 3. Calculate the formula (do the operations indicated) ----------- S = 6.997886 or 7"

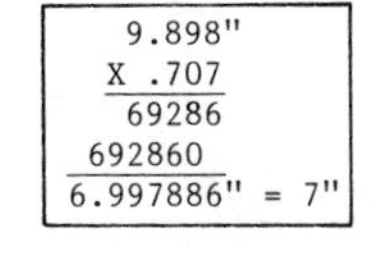

Example
Problem 2. Find the length of the side on the right.

Step 1. Use the formula ---------------- S = B.707

Step 2. Substitute the known values into the formula ---------------- S = 7.07' X .707

Step 3. Calculate the formula (do the operations indicated) ----------- S = 4.99849' = 5'

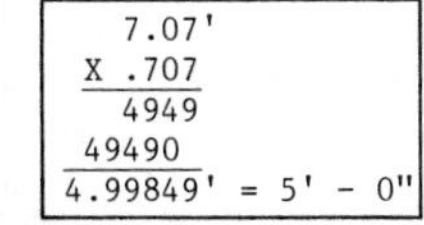

Example
Problem 3. Find the length of the side on the right.

Step 1. Use the formula ---------------- S = B.707

Step 2. Substitute the known values into the formula ---------------- S = 10.4862' X .707

Step 3. Calculate the formula (do the operations indicated) ---------- S = 7.4137434'

Step 4. Convert the decimal so you can read the dimension on an American tape measure.

```
7 .4137'
    X12
   8274
  4137
 4.9644" = 5"
7' - 5"
```

```
   10.4862'
   X .707
   734034
  7340340
 7.4137434' = 7' - 5"
```

Example
Problem 4. Find the length of the side inscribed within the circle shown below.

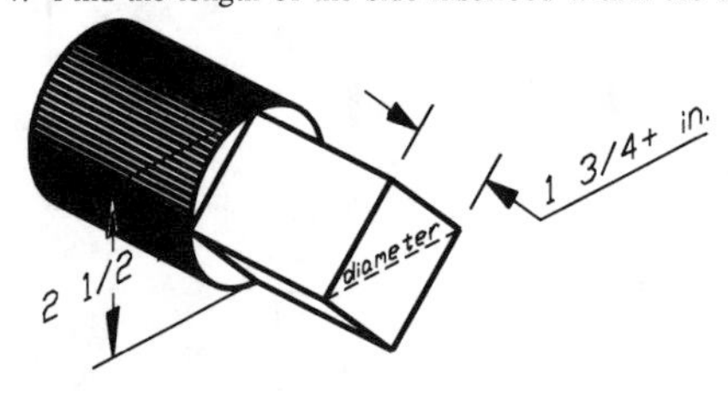

Step 1. Use the formula -------------- S = B.707

Step 2. Substitute the known values into the formula -------------- S = 2.5 X .707

[Change $2\frac{1}{2}$" to a decimal: $2\frac{1}{2}$" = 2.5"]

Step 3. Calculate the formula (do the operations indicated) --------- S =

```
   .707
  X 2.5
   3535
  1414
 1.7675"
```

Step 4. Convert the decimal back to feet, inches, and 16ths, so you can read the dimension on an American tape measure.

```
1 .7675"
    X16   for 16ths
  46050
  7675
 12.2800 = 12+"/16 = 3+"/4
```

$1\frac{3+}{4}$" Length of the side.

Example
Problem 5. Find the length of the side of the square inscribed within the circle shown below.

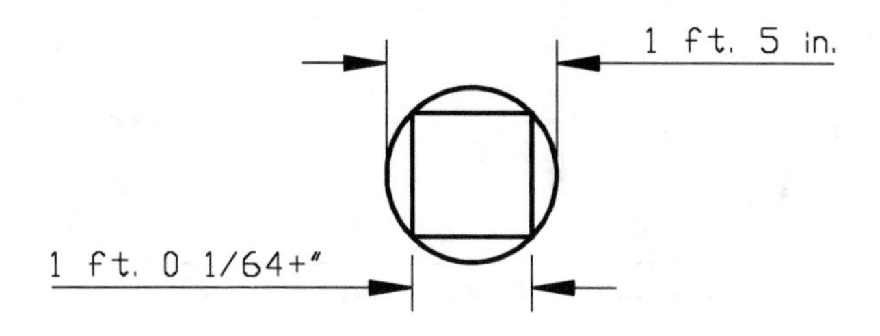

Step 1. Use the formula ------------------- S = B.707

Step 2. Substitute the known values into the formula ------------------- S = 1.416 X .707

[1'5" = $1\frac{5'}{12}$ or (12 ÷ 5 = .416') 1.416']

Step 3. Calculate the formula (do the operations indicated) -------------- S =

```
   1.416'
 X  .707
    9912
   99120
1.001112'
```

Step 4. Convert the decimal back to feet, inches, and 16ths (64ths in this case) so you can read the dimension on an American (English) tape measure.

```
1 .001112'
       X 12  for inches
       2224
      1112
 0.013344  inches
       X 64
      53376
     80064
    .854016

1'0 1/64-"
```

[Or change to all inches and calculate]

```
1'5" = 17"      .707
               X 17
               4949
               707
             12.019"
```

Note: The answer is a little closer this way. To get the anwer closer using feet, multiply by .4166 or .41666.

APPENDIX

English Linear Measure (Length)

1 foot (ft.) = 12 inches (in.)

1 yard (yd.) = 3 feet (ft.)

1 rod (rd.) = 16 ½ ft. or 5 ½ yd.

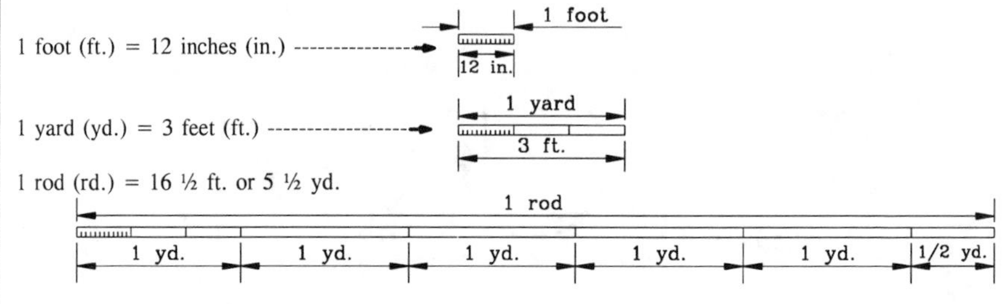

1 mile = 320 rd. (320 X 16.5 ft.)
1 mile = 1760 yd. (320 X 5½ yd.)
1 mile = 5280 ft. (1760 X 3 ft.)

English Surface measure (Area)

1 sq. ft. = 144 sq. in.

1 sq. yd. = 9 sq. ft.

1 sq. rd. = 30¼ sq. yd. (5.5 X 5.5)
1 sq. rd. = 272.25 sq. ft. (16.5 X 16.5)

1 acre = 160 sq. rd.
1 acre = 4840 sq. yd.
1 square mile = 640 square acres
1 section = 640 acres
1 quarter section = 160 acres
1 forty = 40 acres
1 section = 16 forties
1 acre = 43,560 sq. ft.
(A square acre is 208.71' X 208.71')
1 township = 36 sections

Surveyor's Linear Measure:

1 mile = 80 chains
1 chain = 66 ft. (5280 ÷ 80)
1 chain = 100 links
1 link = 7.92 in. (.66 X 12)

Surveyor's Square Measure:

1 sq. chain = 16 sq. rd.
10 sq. chains = 1 acre
1 sq. mile = 640 acres

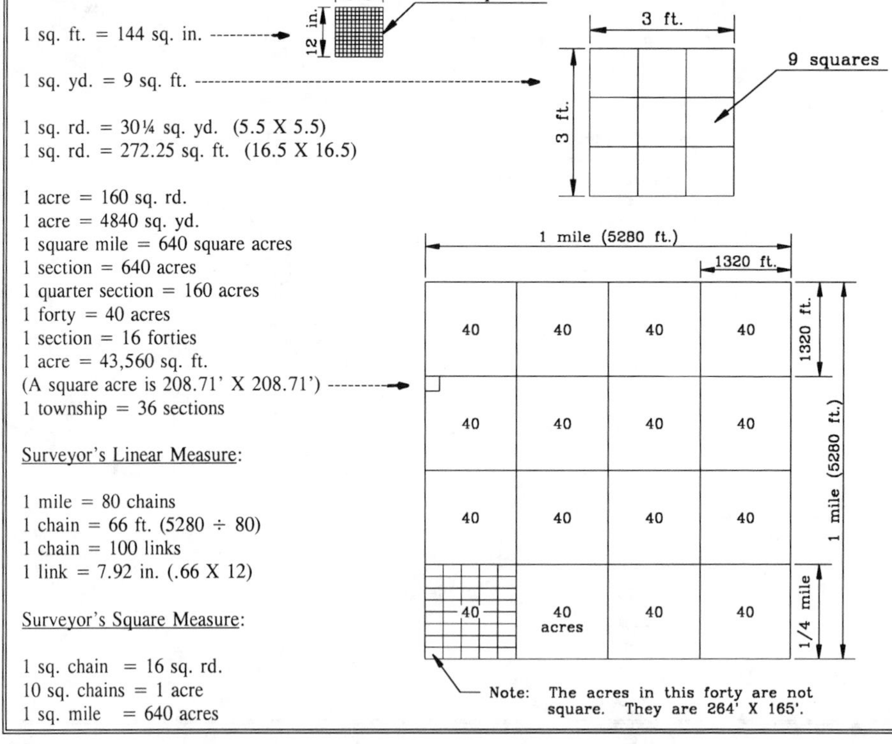

English Cubic Measure (Volume)

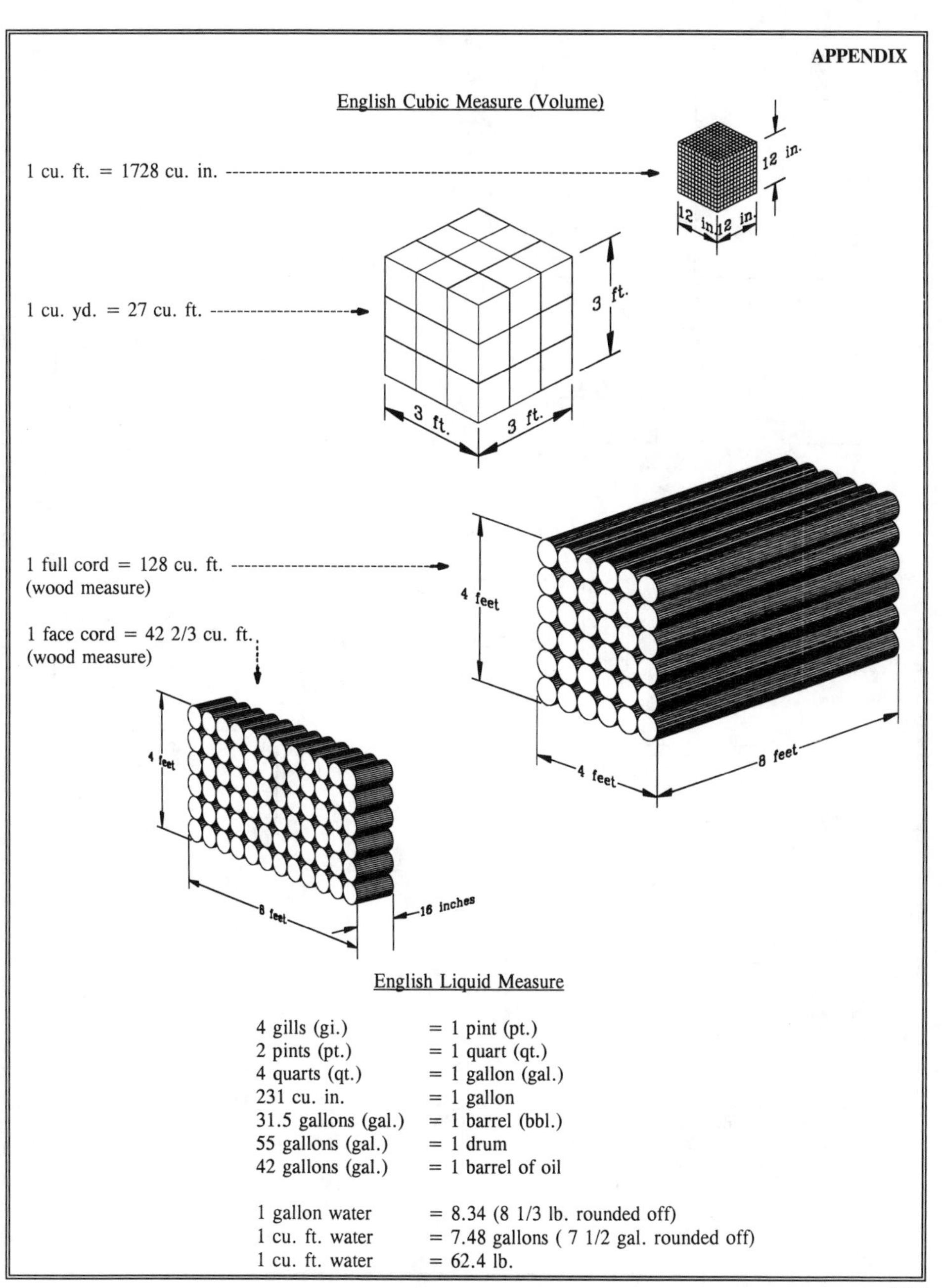

English Liquid Measure

4 gills (gi.)	= 1 pint (pt.)
2 pints (pt.)	= 1 quart (qt.)
4 quarts (qt.)	= 1 gallon (gal.)
231 cu. in.	= 1 gallon
31.5 gallons (gal.)	= 1 barrel (bbl.)
55 gallons (gal.)	= 1 drum
42 gallons (gal.)	= 1 barrel of oil
1 gallon water	= 8.34 (8 1/3 lb. rounded off)
1 cu. ft. water	= 7.48 gallons (7 1/2 gal. rounded off)
1 cu. ft. water	= 62.4 lb.

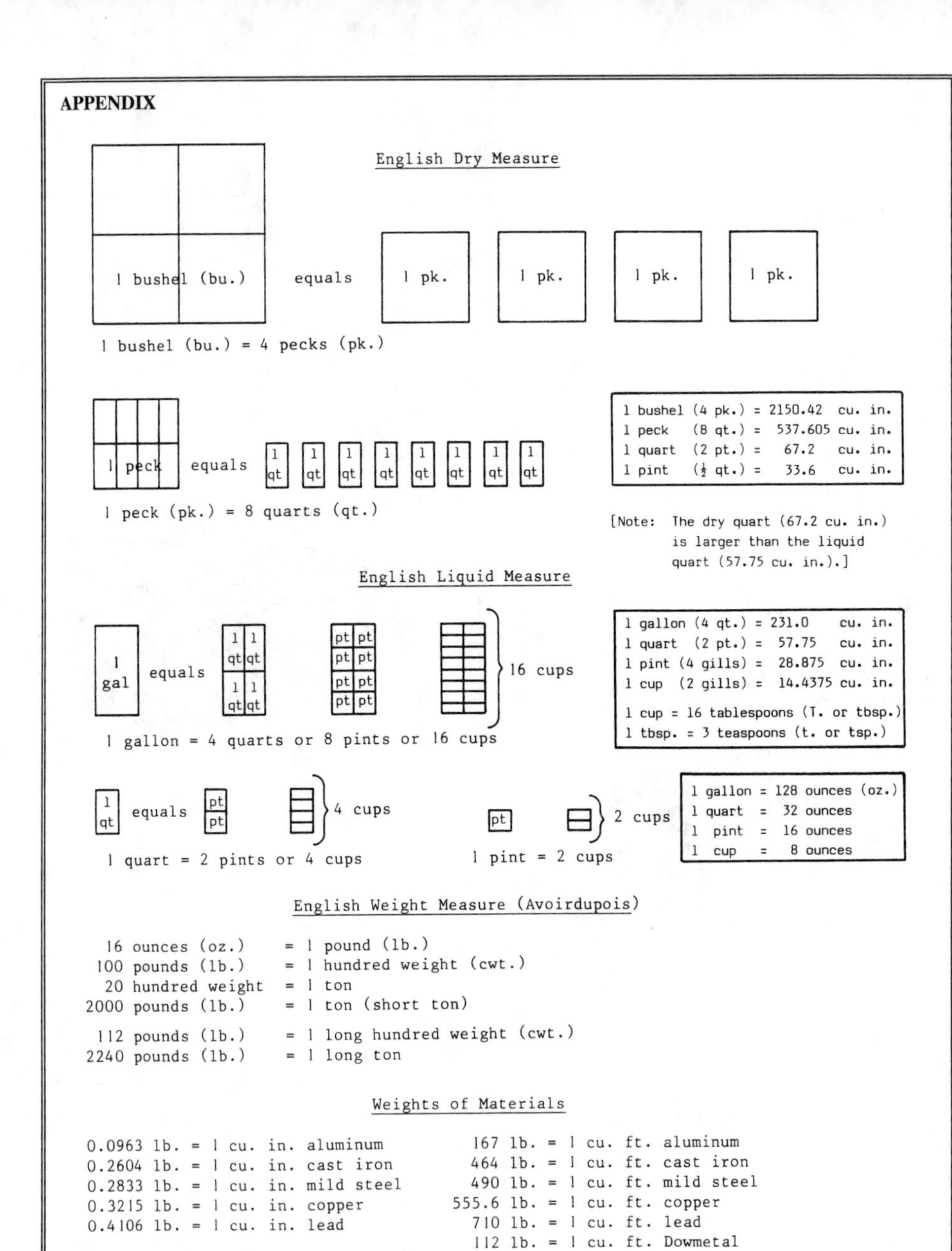

APPENDIX

English Dry Measure

1 bushel (bu.) equals 1 pk. 1 pk. 1 pk. 1 pk.

1 bushel (bu.) = 4 pecks (pk.)

1 peck equals 1 qt 1 qt 1 qt 1 qt 1 qt 1 qt 1 qt 1 qt

1 peck (pk.) = 8 quarts (qt.)

1 bushel	(4 pk.) =	2150.42	cu. in.
1 peck	(8 qt.) =	537.605	cu. in.
1 quart	(2 pt.) =	67.2	cu. in.
1 pint	($\frac{1}{2}$ qt.) =	33.6	cu. in.

[Note: The dry quart (67.2 cu. in.) is larger than the liquid quart (57.75 cu. in.).]

English Liquid Measure

1 gal equals 1 qt 1 qt 1 qt 1 qt; pt pt pt pt pt pt pt pt; 16 cups

1 gallon = 4 quarts or 8 pints or 16 cups

1 gallon	(4 qt.) =	231.0	cu. in.
1 quart	(2 pt.) =	57.75	cu. in.
1 pint	(4 gills) =	28.875	cu. in.
1 cup	(2 gills) =	14.4375	cu. in.

1 cup = 16 tablespoons (T. or tbsp.)
1 tbsp. = 3 teaspoons (t. or tsp.)

1 qt equals pt pt; 4 cups

1 quart = 2 pints or 4 cups

pt; 2 cups

1 pint = 2 cups

1 gallon	=	128 ounces (oz.)
1 quart	=	32 ounces
1 pint	=	16 ounces
1 cup	=	8 ounces

English Weight Measure (Avoirdupois)

16 ounces (oz.)	=	1 pound (lb.)
100 pounds (lb.)	=	1 hundred weight (cwt.)
20 hundred weight	=	1 ton
2000 pounds (lb.)	=	1 ton (short ton)
112 pounds (lb.)	=	1 long hundred weight (cwt.)
2240 pounds (lb.)	=	1 long ton

Weights of Materials

0.0963 lb. = 1 cu. in. aluminum
0.2604 lb. = 1 cu. in. cast iron
0.2833 lb. = 1 cu. in. mild steel
0.3215 lb. = 1 cu. in. copper
0.4106 lb. = 1 cu. in. lead

167 lb. = 1 cu. ft. aluminum
464 lb. = 1 cu. ft. cast iron
490 lb. = 1 cu. ft. mild steel
555.6 lb. = 1 cu. ft. copper
710 lb. = 1 cu. ft. lead
112 lb. = 1 cu. ft. Dowmetal

Time Measure

60 seconds (sec.)	= 1 minute (min.)	
60 minutes (min.)	= 1 hour (hr.)	
24 hours (hr.)	= 1 day (da.)	
7 days	= 1 week (wk.)	
4 weeks (wk.)	= 1 month (mo.)	(4.3 weeks is more accurate)
12 months (mo.)	= 1 year (yr.)	
52 weeks (wk.)	= 1 year	
365 days	= 1 year	
366 days	= 1 leap year	
10 years	= 1 decade	
100 years	= 1 century	

Circle and Angle Measure

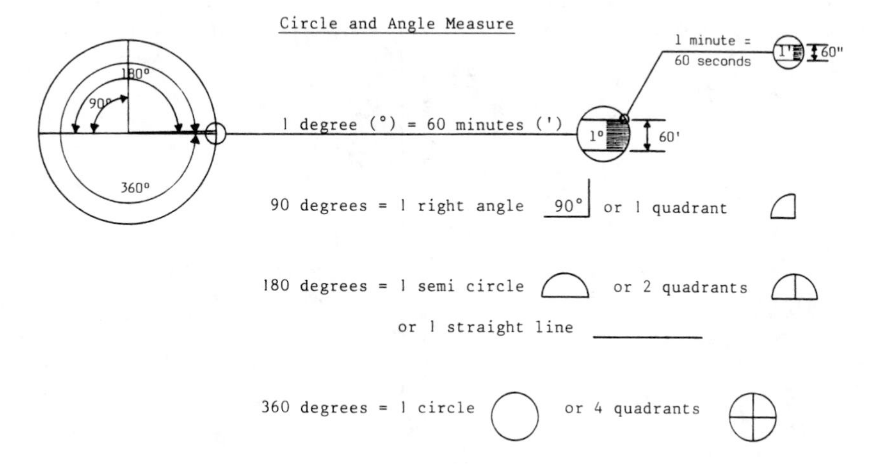

Miscellaneous Measure

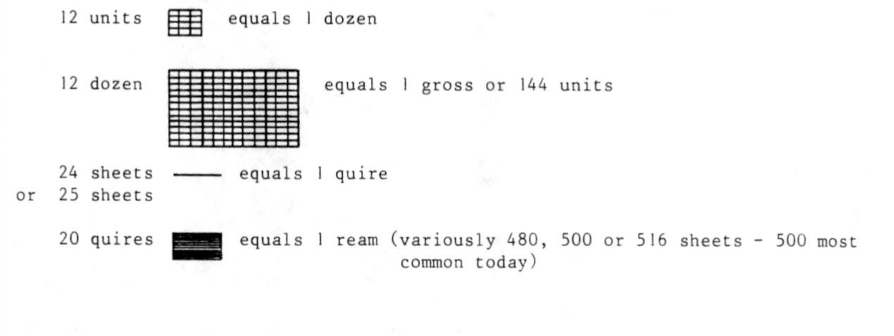

6 feet (ft.) = 1 fathom
6076.1 ft. = 1 nautical mile

APPENDIX

Fahrenheit Temperature	Celsius Temperature
[English (Customary) Measure]	[Metric Measure]

The Fahrenheit scale has 180° (212° - 32°) from the freezing point to the boiling point, whereas the Celsius scale has 100° from the freezing point to the boiling point.

Thus: $1°F = \frac{100°}{180°}C$ or $\frac{5}{9}$ of 1°C

To change Fahrenheit to Celsius, multiply $\frac{5}{9}$ by the Fahrenheit degrees minus 32.

Formula: $C = \frac{5}{9}(°F - 32)$ for greatest accuracy

Or: $C = .555\ (°F - 32)$

Example: Change 72°F to Celsius degrees.

Formula: $C = .555\ (°F - 32)$

$C = .555\ (72° - 32)$

$C = .555 \times (40)$

$C = 22.2°$

Another Example: Change 98.6°F to Celsius degrees.

Formula: $C = \frac{5}{9}(°F - 32)$

$C = \frac{5}{9} \times (98.6 - 32)$

$C = \frac{5}{9} \times (\frac{66.6}{1})$

$C = \frac{333}{9}$

$C = 37°$

[98.6°F or 37°C is normal body temperature.]

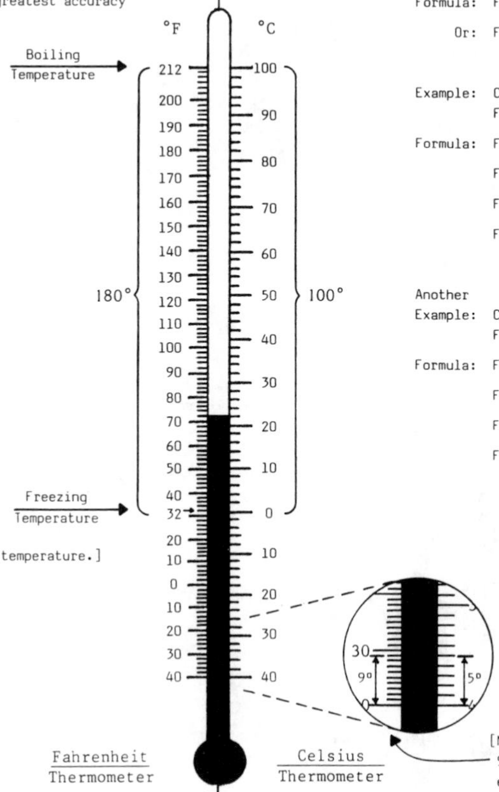

The Celsius scale has 100° from the freezing point to the boiling point, whereas the Fahrenheit scale has 180° (212° - 32°) from the freezing point to the boiling point.

Thus: $1°C = \frac{180°}{100°}F$ or $\frac{9}{5}$ of 1°F

To change Celsius to Fahrenheit, multiply $\frac{9}{5}$ by the Celsius degrees and add 32.

Formula: $F = (\frac{9}{5} \times °C) + 32$

Or: $F = (1.8 \times °C) + 32$

Example: Change 22.2°C to Fahrenheit degrees.

Formula: $F = (1.8 \times °C) + 32$

$F = (1.8 \times 22.2) + 32$

$F = (39.96) + 32$

$F = 71.96°$ or $72°$

Another Example: Change 37°C to Fahrenheit degrees.

Formula: $F = (1.8 \times °C) + 32$

$F = (1.8 \times 37°) + 32$

$F = (66.6) + 32$

$F = 98.6°$

[Note that changing from Celsius degrees to Fahrenheit degrees is the reciprocal (opposite) operation of changing from Fahrenheit degrees to Celsius degrees.]

METRIC MEASURE AND CONVERSION TABLES

Metric Linear Measure
[Length]

Metric Linear Measure
Prefixes, Symbols, and English Linear Equivalents

Multiplication Factor	Prefix	Unit	Symbol	Inches	Feet
1 000 (thousands) = 10^3	kilo	kilometer	km	39370.07874"	3280.839895'
100 (hundreds) = 10^2	hecto	hectometer	hm	3937.007874"	328.0839895'
10 (tens) = 10^1	deka	dekameter	dam	393.7007874"	32.80839895'
1 [Base unit] = 10^0		meter	m	39.37007874"	3.280839895'
.1 (tenths) = 10^{-1}	deci	decimeter	dm	3.937007874"	.3280839895'
.01 (hundredths) = 10^{-2}	centi	centimeter	cm	.3937007874"	.03280839895'
.001 (thousandths) = 10^{-3}	milli	millimeter	mm	.03937007874"	.003280839895'

Summarized Metric Linear Measure Sequence

10 mm = 1 cm
10 cm = 1 dm or 100 mm
10 dm = 1 m or 100 cm or 1000 mm
10 m = 1 dam or 100 dm or 1000 cm or 10000 mm
10 dam = 1 hm or 100 m or 1000 dm or 10000 cm
10 hm = 1 km or 100 dam or 1000 m or 10000 dm

Note: The most common metric linear measure units of measure used are the millimeter, centimeter, meter, and kilometer.]

Metric and English Tape Measure

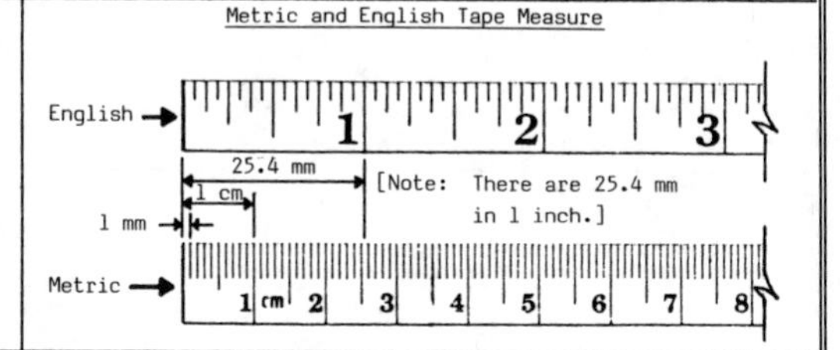

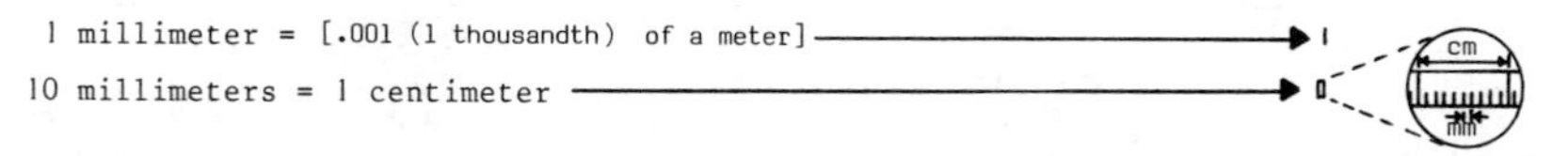

inches/millimeters

1 inch = 25.4 millimeters

1 millimeter = $\frac{1}{25.4}$ or .03937 inch

To change millimeters to inches, multiply millimeters by .03937.
To change inches to millimeters, multiply inches by 25.4.

$$\frac{\text{millimeters} \times .03937}{= \text{inches}} \qquad \frac{\text{inches} \times 25.4}{= \text{millimeters}}$$

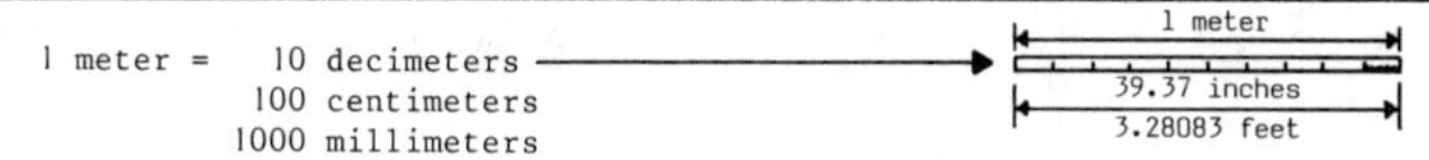

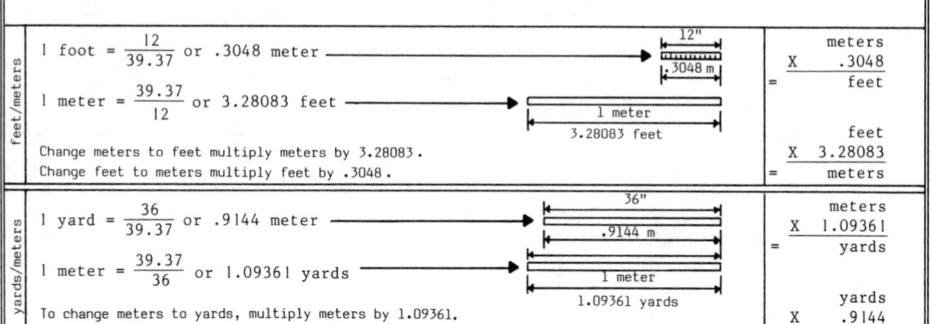

feet/meters

1 foot = $\frac{12}{39.37}$ or .3048 meter

1 meter = $\frac{39.37}{12}$ or 3.28083 feet

Change meters to feet multiply meters by 3.28083.
Change feet to meters multiply feet by .3048.

$$\frac{\text{meters} \times .3048}{= \text{feet}} \qquad \frac{\text{feet} \times 3.28083}{= \text{meters}}$$

yards/meters

1 yard = $\frac{36}{39.37}$ or .9144 meter

1 meter = $\frac{39.37}{36}$ or 1.09361 yards

To change meters to yards, multiply meters by 1.09361.
To change yards to meters, multiply yards by .9144.

$$\frac{\text{meters} \times 1.09361}{= \text{yards}} \qquad \frac{\text{yards} \times .9144}{= \text{meters}}$$

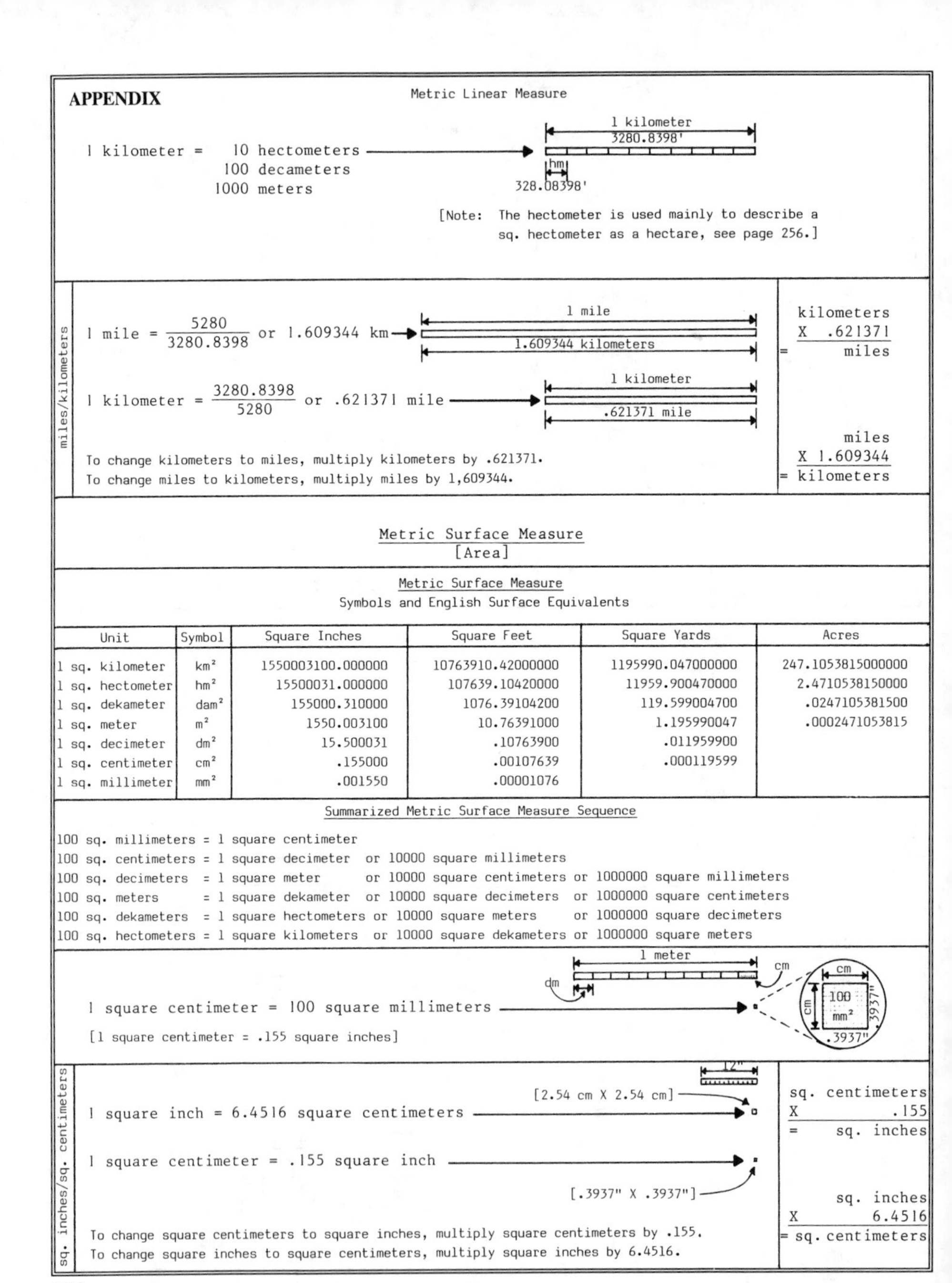

Metric Linear Measure

1 kilometer = 10 hectometers
100 decameters
1000 meters

[Note: The hectometer is used mainly to describe a sq. hectometer as a hectare, see page 256.]

miles/kilometers

$$1 \text{ mile} = \frac{5280}{3280.8398} \text{ or } 1.609344 \text{ km}$$

$$1 \text{ kilometer} = \frac{3280.8398}{5280} \text{ or } .621371 \text{ mile}$$

To change kilometers to miles, multiply kilometers by .621371.
To change miles to kilometers, multiply miles by 1,609344.

kilometers
X .621371
= miles

miles
X 1.609344
= kilometers

Metric Surface Measure
[Area]

Metric Surface Measure
Symbols and English Surface Equivalents

Unit	Symbol	Square Inches	Square Feet	Square Yards	Acres
1 sq. kilometer	km^2	1550003100.000000	10763910.42000000	1195990.047000000	247.1053815000000
1 sq. hectometer	hm^2	15500031.000000	107639.10420000	11959.900470000	2.4710538150000
1 sq. dekameter	dam^2	155000.310000	1076.39104200	119.599004700	.0247105381500
1 sq. meter	m^2	1550.003100	10.76391000	1.195990047	.0002471053815
1 sq. decimeter	dm^2	15.500031	.10763900	.011959900	
1 sq. centimeter	cm^2	.155000	.00107639	.000119599	
1 sq. millimeter	mm^2	.001550	.00001076		

Summarized Metric Surface Measure Sequence

100 sq. millimeters = 1 square centimeter
100 sq. centimeters = 1 square decimeter or 10000 square millimeters
100 sq. decimeters = 1 square meter or 10000 square centimeters or 1000000 square millimeters
100 sq. meters = 1 square dekameter or 10000 square decimeters or 1000000 square centimeters
100 sq. dekameters = 1 square hectometers or 10000 square meters or 1000000 square decimeters
100 sq. hectometers = 1 square kilometers or 10000 square dekameters or 1000000 square meters

1 square centimeter = 100 square millimeters
[1 square centimeter = .155 square inches]

sq. inches/sq. centimeters

1 square inch = 6.4516 square centimeters

1 square centimeter = .155 square inch

To change square centimeters to square inches, multiply square centimeters by .155.
To change square inches to square centimeters, multiply square inches by 6.4516.

sq. centimeters
X .155
= sq. inches

sq. inches
X 6.4516
= sq. centimeters

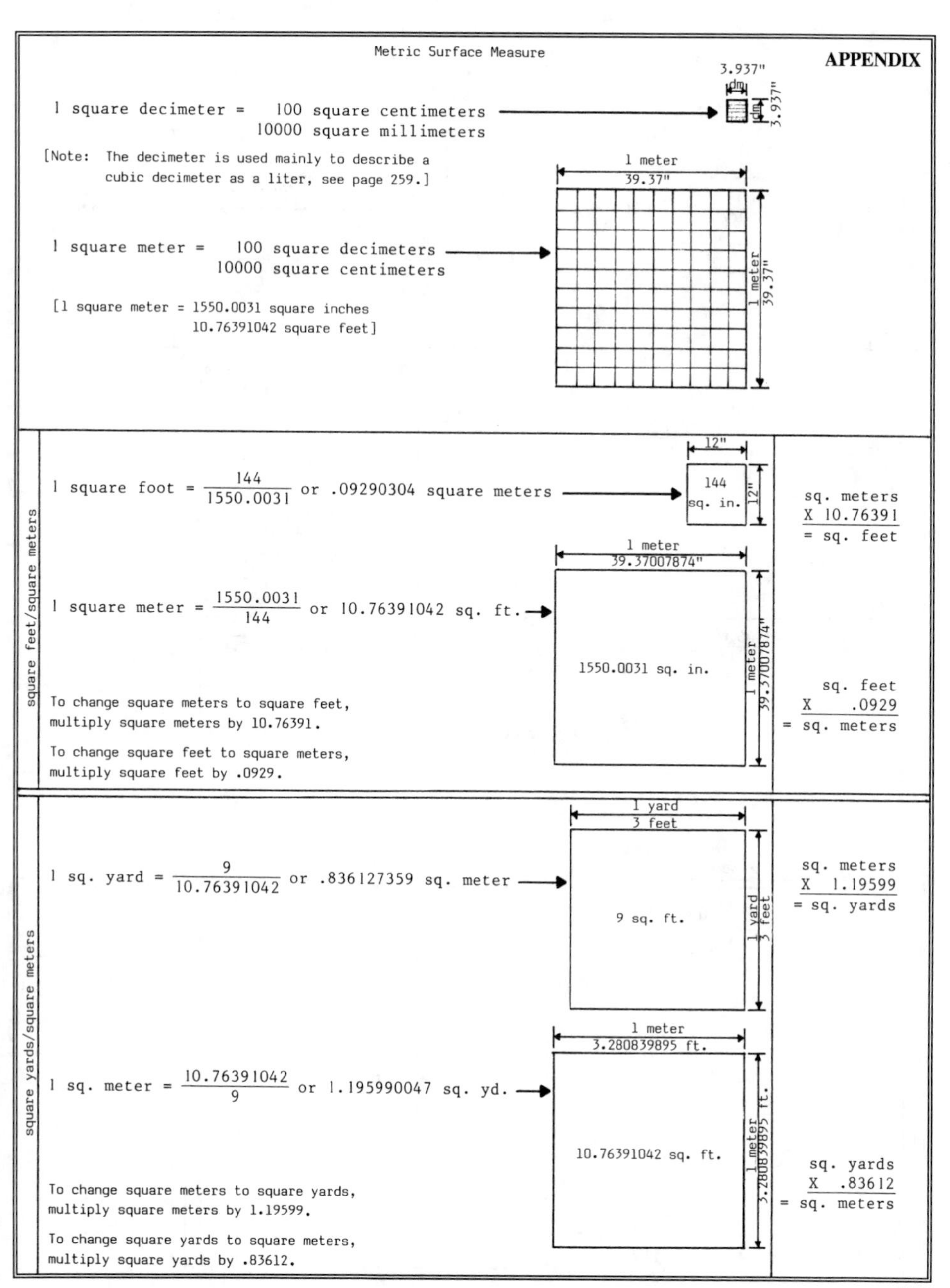
Metric Surface Measure
APPENDIX
1 square decimeter = 100 square centimeters
10000 square millimeters
3.937"
dm
3.937"
[Note: The decimeter is used mainly to describe a cubic decimeter as a liter, see page 259.]
1 meter
39.37"
1 square meter = 100 square decimeters
10000 square centimeters
[1 square meter = 1550.0031 square inches
10.76391042 square feet]
1 meter
39.37"
square feet/square meters
1 square foot = 144/1550.0031 or .09290304 square meters
12"
144 sq. in.
12"
sq. meters
X 10.76391
= sq. feet
1 meter
39.37007874"
1 square meter = 1550.0031/144 or 10.76391042 sq. ft.
1550.0031 sq. in.
1 meter
39.37007874"
To change square meters to square feet, multiply square meters by 10.76391.
To change square feet to square meters, multiply square feet by .0929.
sq. feet
X .0929
= sq. meters
square yards/square meters
1 yard
3 feet
1 sq. yard = 9/10.76391042 or .836127359 sq. meter
9 sq. ft.
1 yard
3 feet
sq. meters
X 1.19599
= sq. yards
1 meter
3.280839895 ft.
1 sq. meter = 10.76391042/9 or 1.195990047 sq. yd.
10.76391042 sq. ft.
1 meter
3.280839895 ft.
To change square meters to square yards, multiply square meters by 1.19599.
To change square yards to square meters, multiply square yards by .83612.
sq. yards
X .83612
= sq. meters

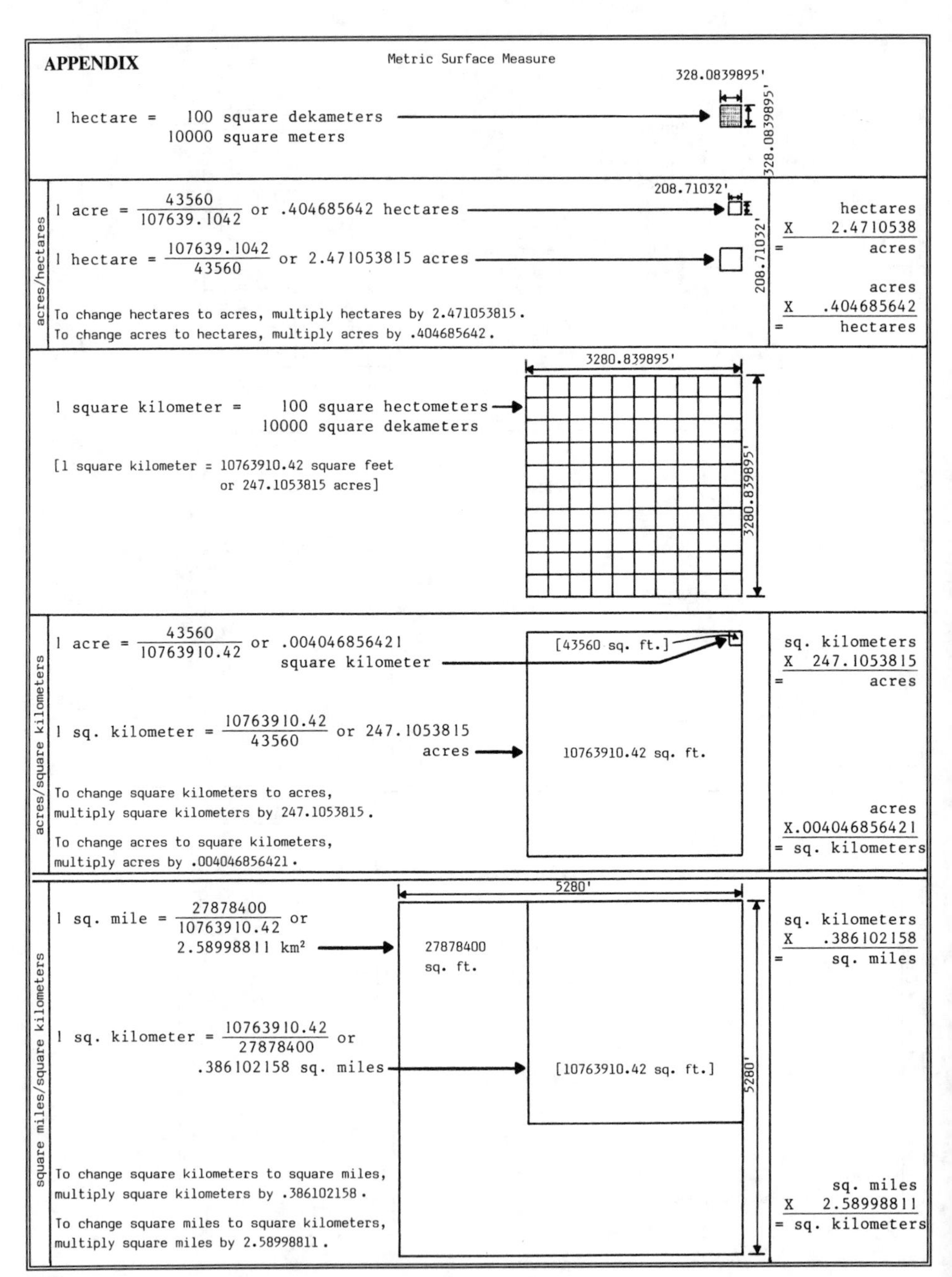
APPENDIX
Metric Surface Measure
328.0839895'
1 hectare = 100 square dekameters
10000 square meters
328.0839895'
acres/hectares
208.71032'
1 acre = 43560 / 107639.1042 or .404685642 hectares
1 hectare = 107639.1042 / 43560 or 2.471053815 acres
208.71032'
To change hectares to acres, multiply hectares by 2.471053815.
To change acres to hectares, multiply acres by .404685642.
hectares
X 2.4710538
= acres
acres
X .404685642
= hectares
3280.839895'
1 square kilometer = 100 square hectometers
10000 square dekameters
[1 square kilometer = 10763910.42 square feet
or 247.1053815 acres]
3280.839895'
acres/square kilometers
1 acre = 43560 / 10763910.42 or .004046856421 square kilometer
[43560 sq. ft.]
1 sq. kilometer = 10763910.42 / 43560 or 247.1053815 acres
10763910.42 sq. ft.
To change square kilometers to acres, multiply square kilometers by 247.1053815.
To change acres to square kilometers, multiply acres by .004046856421.
sq. kilometers
X 247.1053815
= acres
acres
X.004046856421
= sq. kilometers
square miles/square kilometers
5280'
1 sq. mile = 27878400 / 10763910.42 or 2.58998811 km²
27878400 sq. ft.
1 sq. kilometer = 10763910.42 / 27878400 or .386102158 sq. miles
[10763910.42 sq. ft.]
5280'
To change square kilometers to square miles, multiply square kilometers by .386102158.
To change square miles to square kilometers, multiply square miles by 2.58998811.
sq. kilometers
X .386102158
= sq. miles
sq. miles
X 2.58998811
= sq. kilometers

Metric Cubic Measure
[Volume]

Metric Cubic Measure
Symbols and English Volume Equivalents

Unit	Symbol	Cubic Inches	Cubic Feet	Cubic Yards
1 cu. meter	m^3	61023.74409	35.31466672	1.307950619
1 cu. decimeter	dm^3	61.02374409	.03531466672	.001307950619
1 cu. centimeter	cm^3	.06102374409	.00003331466672	
1 cu. millimeter	mm^3	.00006102374409		

Summarized Metric Cubic Measure Sequence

1000 cubic millimeters = 1 cubic centimeter
1000 cubic centimeters = 1 cubic decimeter (liter) or 1000000 cubic millimeters
1000 cubic decimeters = 1 cubic meter or 1000000 cubic centimeters

1 cubic centimeter = 1000 cubic millimeters →

[1 cubic centimeter = .06102374409 cubic inches]

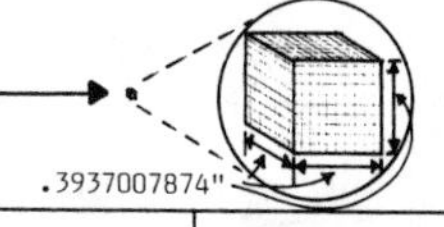

cu. in./cu. centimeters

1 cubic inch = $\frac{1}{.06102374409}$ or 16.387064 cubic centimeters →

1 cubic centimeter = .06102374409 cubic inch →

To change cubic centimeters to cubic inches, multiply cubic centimeters by .06102374409.
To change cubic inches to cubic centimeters, multiply cubic inches by 16.387064.

cu. centimeters
X .06102374409
= cu. inches

cu. inches
X 16.387064
= cu. centimeters

1 cubic decimeter = [1 liter] 1000 cubic centimeters 1000000 cubic millimeters →

3.937007874"

[1 cubic decimeter = 61.02374409 cu. inches]

cubic inches/cubic decimeters

1 cubic inch = $\frac{1}{61.02374409}$ or .016387064 cubic decimeters →

1 cubic decimeter = 61.02374409 cubic inches →

To change cubic decimeters to cubic inches, multiply cubic decimeters by 61.02374409.
To change cubic inches to cubic decimeters, multiply cubic inches by .016387064.

cu. decimeters
X 61.02374409
= cu. inches

cu. inches
X .016387064
= cu. decimeters

cubic feet/cubic decimeters

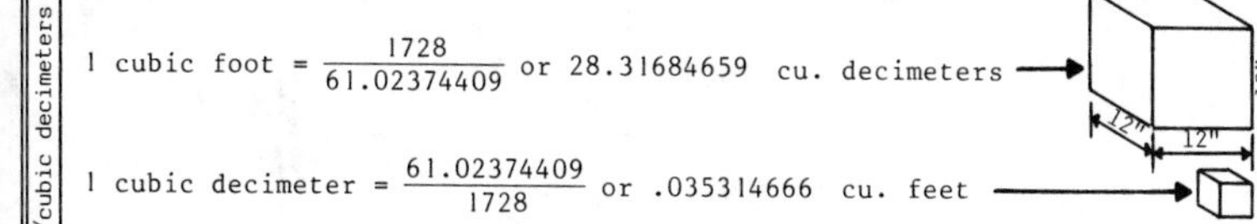

1 cubic foot = $\frac{1728}{61.02374409}$ or 28.31684659 cu. decimeters →

1 cubic decimeter = $\frac{61.02374409}{1728}$ or .035314666 cu. feet →

To change cubic decimeters to cubic feet, multiply cubic decimeters by .035314666.
To change cubic feet to cubic decimeters, multiply cubic feet by 28.31684659.

cu. decimeters
X .035314666
= cu. feet

cu. feet
X 28.31684659
= cu. decimeters

Metric Cubic Measure

1 cubic meter = [1 kiloliter]
1000 cu. decimeters
1000000 cu. centimeters

[1 cubic meter = 61023.7449 cubic inches,
35.31466672 cubic feet,
1.307950619 cubic yards]

39.37007874" × 39.37007874" × 39.37007874"

cubic feet/cubic meters

1 cu. ft. = $\frac{1728}{61023.74409}$ or .028316846 cu. meters

1 cu. meter = $\frac{61023.74409}{1728}$ or 35.31466672 cu. ft.

12" × 12" × 12"

cubic meters
X 35.31466672
= cubic feet

cubic feet
X .028316846
= cubic meters

To change cubic meters to cubic feet, multiply cubic meters by 35.31466672.
To change cubic feet to cubic meters, multiply cubic feet by .028316846.

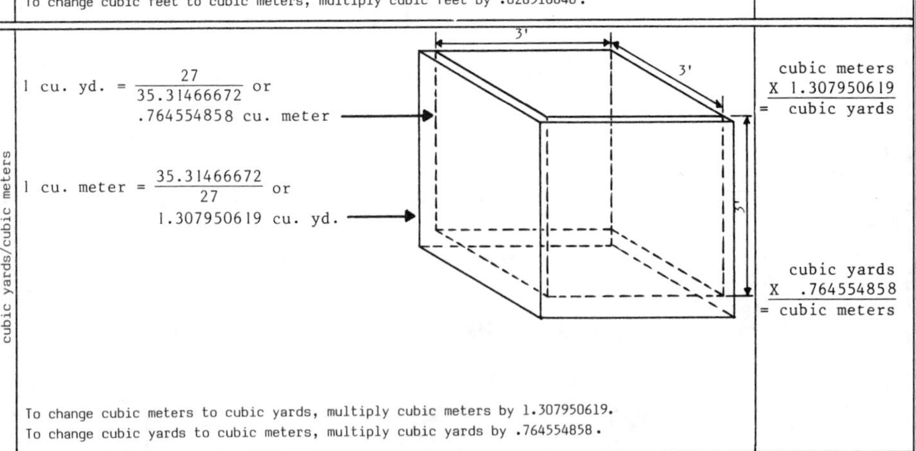

cubic yards/cubic meters

1 cu. yd. = $\frac{27}{35.31466672}$ or .764554858 cu. meter

1 cu. meter = $\frac{35.31466672}{27}$ or 1.307950619 cu. yd.

cubic meters
X 1.307950619
= cubic yards

cubic yards
X .764554858
= cubic meters

To change cubic meters to cubic yards, multiply cubic meters by 1.307950619.
To change cubic yards to cubic meters, multiply cubic yards by .764554858.

Metric Liquid Measure

Metric Liquid Measure
Symbols and English Liquid Equivalents

Unit	Symbol	Ounces	Pints	Quarts	Gallons	cu. in./cu. ft.
1 kiloliter	kl	33814.0227	2113.376419	1056.688209	264.172052	61023.74/35.31466
1 hectoliter	hl	3381.40227	211.3376419	105.6688209	26.4172052	6102.374/3.5314666
1 dekaliter	dal	338.140227	21.13376419	10.56688209	2.64172052	610.2374/.35314666
1 liter [Base]	l	33.8140227	2.113376419	1.056688209	.264172052	61.02374/.03531466
1 deciliter	dl	3.38140227	.2113376419	.1056688209	.0264172052	6.102374/.00353146
1 centiliter	cl	.338140227	.02113376419	.01056688209	.00264172052	.6102374/.00035314
1 milliliter	ml	.0338140227	.002113376419	.001056688209	.000264172052	.0610237/.00003531

Summarized Metric Liquid Measure Sequence

10 milliliters = 1 centiliter
10 centiliters = 1 deciliter or 100 milliliters
10 deciliters = 1 liter or 100 centiliters or 1000 milliliters = [dm^3 or 1000 cm^3]
10 liters = 1 dekaliter or 100 deciliters or 1000 centiliters or 10000 milliliters
10 dekaliters = 1 hectoliters or 100 liters or 1000 deciliters or 10000 centiliters or 100000 milliliters
10 hectoliters = 1 kiloliter or 100 dekaliters or 1000 liters or 10000 deciliters or 100000 centiliters = [$1 m^3$]

1 liter = 1000 milliliters → 33.8140227 ounces
[1 liter = 61.023744 cubic inches]

pints/liters

1 pint = $\frac{16}{33.8140227}$ or .473176473 liter → 16 ounces

1 liter = $\frac{33.8140227}{16}$ or 2.113376419 pints → 33.8140227 ounces

To change liters to pints, multiply liters by 2.113376419.
To change pints to liters, multiply pints by .473176473.

liters × 2.113376419 = pints

pints × .473176473 = liters

quarts/liters

1 quart = $\frac{32}{33.8140227}$ or .946352946 liter → 32 ounces

1 liter = $\frac{33.8140227}{32}$ or 1.056688209 quarts → 33.8140227 ounces

To change liters to quarts, multiply liters by 1.056688209.
To change quarts to liters, multiply quarts by .946352946.

liters × 1.056688209 = quarts

quarts × .946352946 = liters

gallons/liters

1 gallon = $\frac{128}{33.8140227}$ or 3.785411784 liters → 128 ounces

1 liter = $\frac{33.8140227}{128}$ or .264172052 gallon → 33.8140227 ounces

To change liters to gallons, multiply liters by .264172052.
To change gallons to liters, multiply gallons by 3.785411784.

liters × .264172052 = gallons

gallons × 3.785411784 = liters

Metric Mass and Weight Measure

Metric Mass and Weight Measure
Symbols and English Weight Equivalents

Unit	Symbol	Grains	Ounces	Pounds	Tons (short)
1 metric ton	MT		35273.96196	2204.622624	1.102311312
1 quintal	q		3527.396196	220.4622624	.1102311312
1 myriagram	mym		352.7396196	22.04622624	.01102311312
1 kilogram	kg	15432.0	35.27396196	2.204622624	.001102311312
1 hectogram	hg	1543.2	3.527396196	.2204622624	
1 dekagram	dag	154.32	.3527396196	.02204622624	
1 gram	g	15.432	.03527396196	.002204622624	
1 decigram	dg	1.5432	.003527396196		
1 centigram	cg	.15432			
1 milligram	mg	.015432			

Summarized Metric Mass and Weight Sequence

10 milligrams	= 1 centigram				
10 centigrams	= 1 decigram	or 100 milligrams			
10 decigrams	= 1 gram	or 100 centigrams	or 1000 milligrams		
10 grams	= 1 dekagram	or 100 decigrams	or 1000 centigrams	or 10000 milligrams	
10 dekagrams	= 1 hectograms	or 100 grams	or 1000 decigrams	or 10000 centigrams	or 100000 milligrams
10 hectograms	= 1 kilograms	or 100 dekagrams	or 1000 grams	or 10000 decigrams	or 100000 centigrams
10 kilograms	= 1 myriagrams	or 100 hectograms	or 1000 dekagrams	or 10000 grams	or 100000 decigrams
10 myriagrams	= 1 quintal	or 100 kilograms	or 1000 hectograms	or 10000 dekagrams	or 100000 grams
10 quintal	= 1 metric ton	or 100 myriagrams	or 1000 kilograms	or 10000 hectograms	or 100000 dekagrams

1 gram = 10 decigrams → 15.432 grains
100 centigrams
1000 milligrams

[1 gram = .03527396196 ounce or 15.432 grains]

ounces/grams

1 ounce = 28.349523125 grams →

1 gram = $\frac{1}{28.349523125}$ or .03527396196 ounce →

To change grams to ounces, multiply grams by .03527396196.
To change ounces to grams, multiply ounces by 28.349523125.

grams
X .035273961
= ounces

ounces
X 28.3495231
= grams

1 kilogram = 10 hectograms →
100 dekagrams
1000 grams

[1 kilogram = 35.273961196 ounces
or 2.204622624 pounds]

pounds/kilograms

1 pound = $\frac{16}{35.27396196}$ or .453592369 kilograms →

1 kilogram = $\frac{35.27396196}{16}$ or 2.204622622 pounds →

To change kilograms to pounds, multiply kilograms by 2.204622622.
To change pounds to kilograms, multiply pounds by .453592369 .

kilograms
X 2.204622622
= pounds

pounds
X .453592369
kilograms

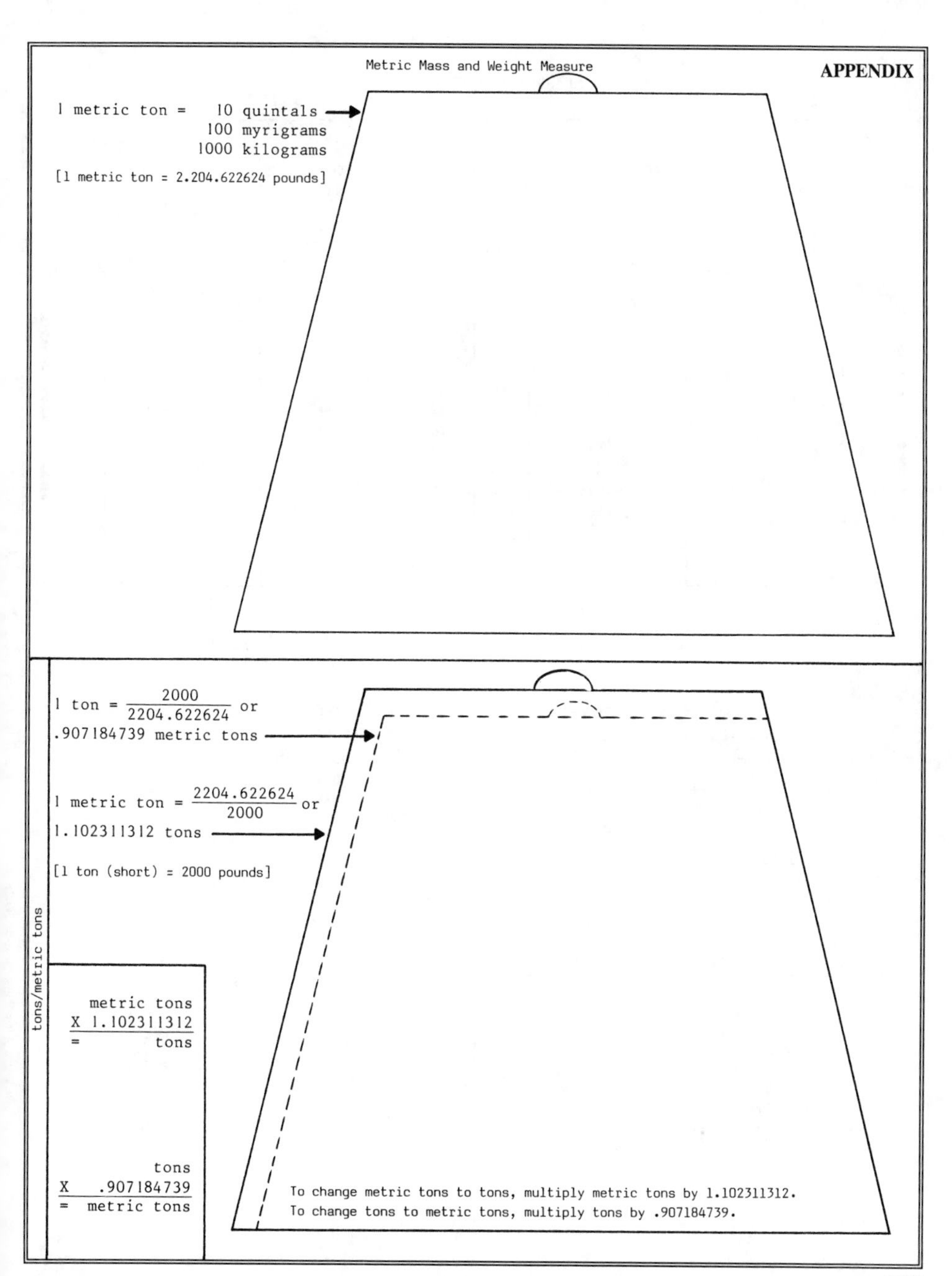

Metric Mass and Weight Measure
1 metric ton = 10 quintals
100 myrigrams
1000 kilograms
[1 metric ton = 2.204.622624 pounds]
tons/metric tons
1 ton = 2000 / 2204.622624 or
.907184739 metric tons
1 metric ton = 2204.622624 / 2000 or
1.102311312 tons
[1 ton (short) = 2000 pounds]
metric tons
X 1.102311312
= tons
tons
X .907184739
= metric tons
To change metric tons to tons, multiply metric tons by 1.102311312.
To change tons to metric tons, multiply tons by .907184739.

METRIC MEASURE AND THE CUSTOMARY CUP

CUP AND FLUID OUNCES TO MILLILITERS

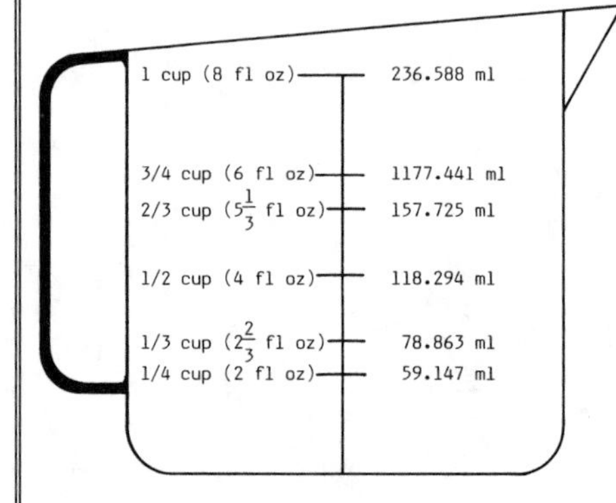

ENGLISH CUP
[CUSTOMARY]

1 cup = 236.588 milliliters (capacity/volume)

1 cup = 16 tablespoons (tbsp.)
1 tbsp. = 3 teaspoons (tsp.)

1 tbsp. = .5 oz (14.786 ml)
1 tsp. = .166 oz (4.928 ml)

How the above calculations were determined.

1 liter	= 1000 milliliters (ml)
1 liter	= 33.8140227 fluid ounces
1 fluid ounce	= 1000 ÷ 33.8140227 or 29.573529 ml
8 fluid ounces	= 29.573529 X 8 or 236.588 ml

1 gallon = 3.785411 liters
1 quart = 946.3527 ml
1 pint = 473.1764 ml
1 cup = 236.5882 ml
1 fl oz = 29.5735 ml
1 tbsp. = 14.7868 ml
1 tsp. = 4.9289 ml

QUARTER LITER TO FLUID OUNCES

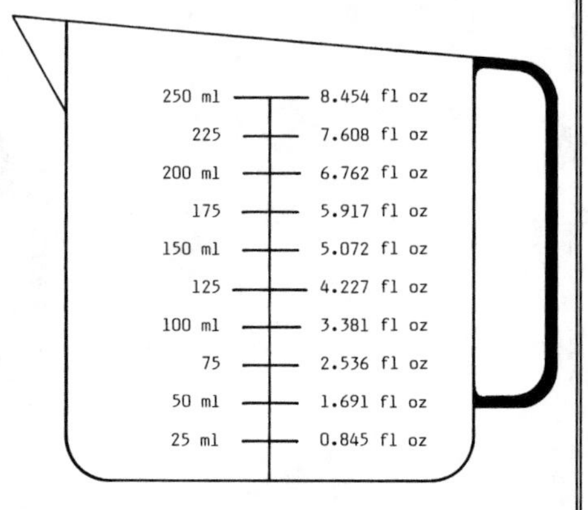

QUARTER LITER CUP
[METRIC]

250 milliliters is approximately 1 tablespoon larger than the English cup.

1 tbsp. = 15 ml
1 tsp. = 5 ml
½ tsp. = 2.5 ml
¼ tsp. = 1.25 ml
} approximate

"This page intentionally blank."

GLOSSARY OF TERMS

ABSOLUTE VALUE: The arithmetic number of the signed number, whether positive or negative.
ABSTRACT NUMBER: A number not applied to a particular object or quantity, e.g. 7, 12, 76.
ACUTE ANGLE: Any angle less than 90°.
ACUTE TRIANGLE: A triangle with all angles less than 90°.
ADDEND: A number that is added to another.
ADDITION: The process of combining numbers to obtain their total.
ADJ: The abbreviation of adjacent used to identify the side adjacent.
ADJACENT ANGLE: Two angles that have the same vertex and a common side.
ADJACENT SIDE: The side adjacent to the hypotenuse in a right triangle (see page 179).
ALGEBRA: A part of mathematics that uses letters to represent numbers in the forms of equations and formulas calculated according to the rules of arithmetic.
ALTITUDE: The perpendicular distance in a geometric figure.
AMOUNT: The sum of the base and the percentage, e.g. 12% of 100 = 12, 100 + 12, or 112 is the amount. In an interest problem, the sum of the principal and interest.
ANGLE: The opening between two meeting lines or any opening between lines that cross each other.
ANGULAR MEASURE: Degrees, minutes, and seconds.
ANSWER: The solution to a problem.
ANTECEDENT: The first term of a ratio.
ARABIC NUMERAL: The most commonly used system of numbering of which ten digits are used, 0 through 9.
ARC: A continuous portion of a curved line, as of a circle.
AREA: The amount of surface a geometric figure covers in the same plane.
ARITHMETIC: The part of mathematics that deals with computing numbers.
AVOIRDUPOIS WEIGHT: Common English weights that use the pound (16 ounces) as the basic unit of measure.
BASE: The whole number on which a percent is reckoned, e.g. 12% of 100 = 12, 100 is the base.
BASE NUMBER: The original number that the rate or percent is applied to.
BASE OF A GEOMETRIC FIGURE: The side on which it seems to stand.
BENCH MARK: A point used as a reference for measurement.
BINARY SYSTEM: A system in which the value of each place is twice that of the place to its right.
BISECT: Divide into two equal parts.
BOARD FEET: The method in which the volume of lumber is measured. A board foot is 12" x 12" x 1".
BRACE: The hypotenuse of an isosceles right triangle. Two of its sides are equal, two angles are 45°, and one angle is 90°.
BUILDERS' LEVEL: A leveling instrument that is mounted on a tripod used to sight horizontal level lines. It can also be used to lay out (measure) horizontal angles.
CALCULATE: Determine by mathematical process.
CALIPER: An instrument used to measure the thickness of an object or distances between surfaces.
CELSIUS: Thermometer scale which has 100° from the freezing point to the boiling point.
CENTI-: Means one hundredth of whichever unit you're measuring using the metric system.
CENTIGRAM: Equals .01 gram in the metric system written as *cg*.
CENTILITER: Equals .01 liter in the metric system written as *cl*.
CENTIMETER: Equals .01 meter in the metric system written as *cm*.
CENTRAL ANGLE: An angle whose vertex is at the center of a circle.
CHORD: A line segment joining two points on a circle.
CIRCLE: A closed plane curve, all points of which are the same distance from a point inside the curve. That point is called its center. There are 360° around a circle.

CIRCLE SECTOR: A geometric figure that includes two equal radii and one part of the circumference, the arc, of a circle.
CIRCUMFERENCE: The distance around a circle.
CIRCUMSCRIBE: To draw a circle around a geometric figure, so as to touch as many points as possible.
COMMON DENOMINATOR: Fractions that have the same denominator.
COMMON FRACTION: Any number of equal parts of a whole.
COMPLEMENT OF AN ANGLE: The angle, which, when added to another angle, equals 90°.
COMPLEX FRACTION: A fraction in which a mixed number or a fraction appears in the numerator or the denominator or both.
COMPOSITE NUMBER: A number that can be divided exactly by other numbers as well as by itself and 1, e.g. 8 can be divided exactly by 2 or 4, and also by 8 and 1.
CONCENTRIC CIRCLE: Circles in the same plane that have a different radii but the same center.
CONE: A geometric figure (solid) whose base is a circle and whose curved surface tapers uniformly to a common point.
CONGRUENT TRIANGLE: Triangles which have exactly the same shape and size.
CONSEQUENT: The second term of a ratio.
CORD: The cubic unit of measure used to measure wood (see full cord or face cord in the appendix, page 367).
COSECANT OF AN ANGLE: The reciprocal (inverted fraction) function of the sine of an angle (see page 171 and 177).
COSINE OF AN ANGLE: A term used to describe one of the six common fraction functions of a triangle in relation to a circle (see page 171 and 177).
COTANGENT OF AN ANGLE: The reciprocal (inverted fraction) function of the tangent of an angle (see page 171 and 177).
CONVERSION: The mathematical process of changing to an equivalent.
CUBE: A geometric figure that is one unit in length, one unit in width, and one unit in height. All six sides are square and equal. Each is parallel to its opposite side and perpendicular to the other four sides.
CUBE ROOT: Whatever number taken as a factor three times, yields the original number, e.g. the cube root of 8 is 2 because 2 x 2 x 2 = 8.
CUBIC UNIT: Used in measuring volume.
CURVE: A set of many points that do not form a straight line.
CYLINDER: A solid figure formed by turning a rectangle about one side as an axis.
DECI-: Means one tenth of whichever unit you're measuring using the Metric System.
DECIGRAM: Equals .1 grams in the metric system, written as *dg*.
DECILITER: Equals .1 liter in the metric system written as *dl.*
DECIMAL: Based on the number 10.
DECIMAL FRACTION: A fraction in which the denominator is a power of 10, e.g. .25 = 25/100.
DECIMAL POINT: The dot to the left of a decimal fraction, or to the right of a whole number.
DECIMETER: Equals .1 meter in the metric system written as *dm.*
DEGREE: The unit of measure used to measure angles.
DEKA-: Means ten times whichever unit you are measuring using the Metric System.
DEKAGRAM: Equals 10 grams in the metric system, written as *dag*.
DEKALITER: Equals 10 liters in the metric system written as *dal.*
DEKAMETER: Equals 10 meters in the metric system written as *dam.*
DENOMINATE NUMBER: Any number that applies to a specific unit of measure, e.g. 5 ft. 2 in., 8 lb. 5 oz., 2 hr. 5 min.
DENOMINATOR: The bottom number of a fraction and is always the divisor.

DIAGONAL: A line that joins two non-consecutive vertices, or a line that extends from one corner to the corner in a four sided figure.

DIAMETER: A line drawn through the center of a circle, going completely across the circle. The diameter of a circle is twice the radius.

DIFFERENCE: The remainder when the percentage is subtracted from the base, e.g. 12% of 100 = 12, 100 - 12 = 88, 88 is the difference.

DIGIT: A symbol used for a whole number from 0 to 9.

DIMENSION: Measurement as in height, length, breadth.

DIRECT PROPORTION: One in which the order of the ratios is the same, e.g. small is to large as small is to large.

DIVIDEND: A number to be divided.

DIVISION: The process of repeated subtraction of one number from another.

DIVISOR: A number by which the dividend is divided.

ELLIPSE: A closed plane surface with two focuses and shaped so that the sum of the distances from any point on the curve to the two focuses is always equal.

EQUAL: Identical in value.

EQUATION: A formal statement of equivalence between mathematical expressions symbolized by the = sign.

EQUIANGULAR: A triangle with all sides equal, or 60°.

EQUILATERAL TRIANGLE: All sides equal.

EQUIVALENT FRACTION: Fractions that are equal, e.g. 3/4 = 6/8.

EXPONENT: The symbol written raised, and to the right of a number indicating its power.

EXPRESSION: A collection of terms combined by either addition or subtraction or both.

EXTERIOR ANGLE: The angle formed by a side of the triangle and the adjacent side extended.

EXTREMES: The outside terms of a proportion.

FACTORS: The numbers that are multiplied to form the product.

FAHRENHEIT: Thermometer scale which has 180° (212° - 32°) from the boiling point to the freezing point.

FORMULA: A shortened version of a mathematical rule.

FRACTION: A number that is more than zero, but less than 1, or if negative, less than 0 but more than negative 1 (-1).

FRUSTUM OF A CONE: The part of a cone left after a uniform top has been cut off.

FRUSTUM OF A PYRAMID: The part of the pyramid left after a uniform top has been cut off.

GEOMETRY: The segment of mathematics involving measurements and relationships of points, lines, solids, and planes.

GRAM: A unit of mass and weight in the metric system written as *g*.

HECTARE: A unit of measure used to measure metric "acres" (or - land areas in the metric system). A hectare is 2.471053815 acres or one hundredth of a square kilometer.

HECTO-: Means one hundred times whichever unit you're measuring in the Metric System

HECTOGRAM: Equals 100 grams in the metric system, written as *hg*.

HECTOLITER: Equals 100 liters in the metric system written as *hl*.

HECTOMETER: Equals 100 meters in the metric system written as *hm*.

HEIGHT OF A POLYGON: The length of the line drawn from the highest point to the base, forming a right angle with the base.

HEPTAGON: A seven sided figure.

HEXAGON: A six sided figure.

HYP: The abbreviation of hypotenuse.

HYPOTENUSE: The side opposite the right angle in a right triangle.
IMPROPER FRACTION: A fraction in which the numerator is larger than the denominator.
INDEX OF THE ROOT: The raised number placed in the radical sign.
INDIRECT PROPORTION: One in which the order of ratio is inverse, e.g. large is to small as small is to large.

INSCRIBE: To draw inside a figure so as to touch in as many places as possible.
INTEGER: A whole number, e.g. 7 and 68.
INTERPOLATION: The process of calculating angle function values that do not come out even to a whole degree(s) plus minute(s). Calculations are taken from the trig tables.
ISOSCELES TRIANGLE: A three sided figure of which two sides are equal.
KILO-: Means one thousand of whichever unit you're measuring in the metric system.
KILOGRAM: Used to measure weight in the metric system written as *kg*.
KILOLITER: Equals 1000 liters in the metric system, written as *kl*.
KILOMETER: Equals 1000 meters in the metric system written as *km*.
LATERAL: Sideways.
LEVEL, BUILDERS': A leveling instrument that is mounted on a tripod and used to sight horizontal level lines. It can also be used to lay out (measure) horizontal angles.
LEVEL, LINE: A leveling instrument that attaches to a taut line and that is used for horizontal leveling (see page 344).
LEVEL, SPIRIT: Hand held leveling instrument used to level angles, level horizontally, and level vertically (see page 344).
LEVEL, TRANSIT: A leveling instrument that is mounted on a tripod and used to sight level lines both horizontally and vertically. It can also be used to measure both horizontal and vertical angles.
LEVEL, WATER: A leveling instrument used to level horizontally by using a clear hose and colored water (optional). It is based on the principal that water seeks its own level.
LINEAR: Relating to a line.
LINEAR MEASURE: The measurement of distance.
LIQUID OR DRY MEASURE: The measurement of capacity.
LITER: The unit of volume in the metric system written as *l*.
LITERAL NUMBER: Letters that represent numbers in a formula.
LOWEST TERM: The lowest possible number that can be divided evenly.
MATHEMATICS: The study of quantities and relationships using numbers and symbols.
MEANS: The inside terms of a proportion.
MEDIAN: The line segment connecting any vertex of a triangle to the midpoint of the opposite side.
METER: The unit of length in the metric system, written as *m*.
METRIC SYSTEM: The decimal system of weights and measures based on the meter, kilogram, and liter.
MICROMETER: An instrument used for measuring minute distances.
MILLI-: Means one thousandth of whichever unit you're measuring in the metric system.
MILLIGRAM: Equals .001 gram in the metric system, written as *mg*.
MILLILITER: Equals .001 liter in the metric system, written as *ml*.
MILLIMETER: Equals .001 meter in the metric system, written as *mm*.
MINUEND: The number which you subtract the subtrahend from. In the equation $50 - 16 = 34$, the minuend is 50.
MIXED NUMBER: A whole number and a fraction written together.
MULTIPLICATION: The process of repeated addition of identical numbers.
MYRIAGRAM: Equals 10,000 grams in the metric system, written as *mym*.
NEGATIVE: Any number less than zero and written with a negative (-) sign.

NUMBER: A word form to reference to one or more that one.
NUMERATOR: The top number of a fraction.
OBLIQUE TRIANGLE: Has no angle that is 90°.
OBTUSE ANGLE: Any angle more than 90°.
OBTUSE TRIANGLE: A triangle with one angle larger than 90°.
OCTAGON: An eight sided plane figure.
OPP: The abbreviation used to identify the side opposite of a right triangle.
PARALLEL: Lines the same distance away from each other at all points, and will never meet.

PARALLELOGRAMS: Quadrilaterals that have opposite sides that are parallel.
PENTAGON: A five sided figure.
PERCENT: Any part of 100 parts.
PERCENTAGE: A part of a whole expressed in hundredths.
PERIMETER: The distance around a figure. The sum of all sides.
PERPENDICULAR: A line that is 90° from another line.
PI: (π) The ratio of the circumference of a circle to its diameter, 3.1416. The circumference of the circle is 3.1416 units longer than the diameter.
PLANE GEOMETRY: The part of mathematics that deals with flat figures and lines. Plane geometry measurements are known as surface measure.
PLUMB BOB: A machined weight used to level vertically.
PLUS: Add.
POLYGONS: Geometric figures that are made of straight lines with closed sides that lie in a plane (flat surface).
POLYHEDRON: A closed geometric figure consisting off four or more polygons and their interiors, all in different planes.
POSITIVE NUMBER: Any number greater than zero and written with a positive (+) sign.
POWER OF A NUMBER: Indicates the number of times that number is a factor when multiplying by itself.
PRIME NUMBER: A number that can be divided exactly, only by itself and the number 1, e.g., 1,2,3,5,7.
PRISM: A solid geometric figure that has equal and parallel bases at each end.
PRODUCT: The number resulting from the multiplication of two or more numbers.
PROPER FRACTION: A fraction in which the numerator is smaller than the denominator.
PROPORTION: An equality of two ratios.
PROTRACTOR: A device used to measure angles in degrees.
PYRAMID: A geometric figure (solid) whose base is a polygon and whose sides are equal triangles that meet in a common point.
QUADRANGLE: The four angles inside a quadrilateral that add up to 360°.
QUADRANT: One quarter of a circle.
QUADRILATERALS: A plane figure with four straight sides.
QUARTER: A term used to signify one quarter (1/4) of a whole unit. To divide a unit into 4 equal parts.
QUINTAL: Equals 1 metric ton in the Metric System written as *q*.
QUOTIENT: The number resulting from dividing one number by another.
RADIAN: Angle measurement in relation to a sector of a circle. The sector is made up of two radii and an arc which are all equal in length. There are 6.283 radians encompassed within a circle.
RADICAL SIGN: $\sqrt{\ }$ Indicates square root.
RADII: Plural of Radius.
RADIUS: A line drawn from the center of a circle to the curved line.

RATE: Number of parts or percent taken in the formula, P = BR (Percentage equals Base times the Rate). 12% of 100 = 12, 12 is the Rate.
RATIO: The quotient of two numbers of the same kind.
RAY: A line starting at one point and continuing forever in the same direction.
RECIPROCAL: One of a pair of numbers (as 3/5, 5/3) whose product is one.
RECTANGLE: Opposite sides are parallel. Adjacent sides are not equal. All angles are 90°.
RECTANGULAR SOLID: A geometric figure that has six rectangular faces that meet to form right angles.
REMAINDER: The final undivided part after division that is less than the divisor.
RHOMBOID: Opposite sides are parallel. Adjacent sides are not equal. No right angles.
RHOMBUS: Opposite sides are parallel. All sides are equal. No right angles.

RIGHT ANGLE: 90° Angle.
RIGHT PRISM: A solid in which the bases or ends are perpendicular to the sides. All the sides are parallelograms.
RIGHT TRIANGLE: Has one angle that is 90°, and two acute angles (less than 90°).
ROD, LEVELING: A measuring stick used to measure when sighting with a builders' level or a level-transit.
ROMAN NUMERALS: A system of numbering in which Roman letters are used as numerals, e.g. I = 1, V = 5, X = 10, etc.
ROOT OF A NUMBER: Whatever number, taken as a factor a given number of times, yields the original number.
RULE OF PYTHAGORAS: In any right triangle, the square of the hypotenuse is equal to the sum of the squares of the other two sides.
SCALENE TRIANGLE: No two sides are equal.
SECANT OF AN ANGLE: The reciprocal (inverted fraction) function of the cosine of an angle (see page 171 and 177).
SECTOR: The part of a circle that lies between two radii.
SEMI-CIRCLE: One-half of a circle. There are 180° around a semi-circle.
SIDE OPPOSITE: The side opposite the hypotenuse in a right triangle (see page 179).
SIDES OF AN ANGLE: The lines that form the angle.
SIGN OF DEGREE: ° (e.g., 1° = 1 degree)
SIGN OF PERCENT: %
SINE OF AN ANGLE: A term used to describe one of the six functions of a triangle in relation to a circle (see page 171 and 177).
SOLID GEOMETRY: The part of mathematics which deals with three-dimensional figures.
SOHCAHTOA: The acronym used to identify a tricks of the trade formula for solving trigonometry problems.
SPHERE: A geometric figure (solid) whose surface at every point is equally distant from the center of the sphere.
SQUARE: Four sided figure of which opposite sides are parallel, all sides are equal, all angles are 90°.
SQUARE MEASURE: A system used in measuring area.
SQUARE ROOT: The number which taken as a factor twice, that is, multiplied by itself, yields the original number, (e.g., the square root of 9 is 3, because 3 multiplied by itself is 9.)
STRAIGHT ANGLE: Straight line.
STRAIGHT LINE: The shortest distance between two points.
SUBSTITUTION: The changing of letters or symbols to numbers in a formula or equation.
SUBTRACTION: The process of deducting one number from another.

SUBTRAHEND: The number that is subtracted from the minuend.
SURFACE: The outer or the topmost boundary of an object.
SUM: The result of adding numbers.
SUPPLEMENT OF AN ANGLE: The angle, which, when added to another angle, equals 180°.
TANGENT: The point at which a line meets an arc of a circle and will not cut it if extended.
TANGENT OF AN ANGLE: A term used to describe one of the six common functions of a triangle in relation to a circle (see page 171 and 177).
TAPE MEASURE: A tape used for measuring inches, feet, yards, etc.
TERMS OF A FRACTION: The numerator and the denominator.
TERMS OF A RATIO: The two numbers being compared to find a ratio.
TETRAGON: Four sided figure.
TIME MEASURE: Seconds, minutes, hours.
TRANSIT (ALSO CALLED A LEVEL-TRANSIT): See level-transit.
TRAPEZIUM: Quadrilateral (figure with four sides that has no two sides parallel.)
TRAPEZOIDS: Quadrilaterals (figures with four sides that have only two sides that are parallel.)

TRIANGLES: Polygons that have three sides. All angles of a triangle add up to 180°.
TRIG: The abbreviation of trigonometry.
TRIGONOMETRY: Triangular measurement. It is a branch of mathematics which studies the relationships between sides and angles of a triangle.
TRIPOD: A mechanical device that is made up of three legs hinged to a base. The base can be used to mount either a builders' level or a level-transit.
VERNIER: A small auxiliary scale made to slide along the main, fixed scale of an instrument to enable smaller intervals of the main scale to be measured.
VERNIER CALIPER: A measuring device consisting of a main scale with a fixed jaw and a sliding jaw with an attached Vernier.
VERNIER MICROMETER: A small auxiliary device on some micrometers.
VERNIER PROTRACTOR: Used to measure angles in degrees and minutes.
VERTEX OF AN ANGLE: The point where the sides meet.
VERTEX OF A POLYGON: A point where two lines meet.
VERTICES: The plural of vertex.
VIAL: The glass or plastic tube part of a spirit level that is filled with alcohol or ether to prevent breakage due to freezing. An air bubble inside correlates with the lines drawn on the vial that represents level when exactly between the lines.
VOLUME: The space taken up by a geometric figure that has three dimensions.
WATER LEVEL: A leveling instrument used to level horizontally by using a clear hose and colored water (optional). It is based on the principal that water seeks its own level.
WIDTH: The term used to signify the breadth or measurement of an object taken a right angles to the length.
WHOLE NUMBER: A whole number or integer is the entire number. It is not a fraction or does not include a fraction of a number.
YARD: A yard is 3 feet or 36 inches.
ZERO: The number that has no value when used alone.

INDEX

A

INDEX

C

INDEX

INDEX

INDEX

Task No. | | Page

Q

R

S

TASK NO. PAGE

INDEX

U

V

INDEX

“This page intentionally blank.”

ENGLISH (CUSTOMARY) MEASURE AND METRIC MEASURE EQUIVALENTS

Linear Measure	Millimeters	Centimeters	Decimeters	Meters	Dekameters	Hectometers	Kilometers
Inch	25.4	2.54	.254	.0254			
Feet	304.8	30.48	3.048	.3048	.03048	.003048	
Yards	914.4	91.44	9.144	.9144	.09144	.009144	
Rods	5029.215	502.9215	50.29215	5.029215	.5029215	.05029215	.005029215
Miles				.00160934	.01609344	.1609344	1.609344

Square Measure	mm^2	cm^2	dm^2	m^2	dam^2	Hectare	km^2
Sq. Inches	645.16	6.4516	.064516	.0064516			
Sq. Feet		929.0304	9.290304	.09290304			
Sq. Yards		8361.27359	83.6127359	.83612735			
Sq. Rods				25.29285	.2529285		
Acres				4046.856	40.46856	.404685	.00404685
Sq. Miles						258.99881	2.5899881

Cubic Measure	mm^3	cm^3	dm^3	m^3	dam^3	hm^3	km^3
Cu. Inches	16387.064	16.387064	.0163870				
Cu. Feet		28316.846	28.316846	.028316846			
Cu. Yards			764.554856	.764554856			

Liquid Measure	Milliliters	Centiliters	Deciliters	Liters	Dekaliters	Hectoliters	Kiloliters
Ounces	29.57352956	2.957352956	.29573529	.029573529			
Pints	473.176473	47.3176473	4.73176473	.473176473			
Quarts	946.352946	94.6352946	9.46352946	.946352946			
Gallons	3785.41178	378.541178	37.8541178	3.78541178	.378541178	.0378541178	.00378541178
Cu. Inches	16.38706	1.638706	.1638706	.01638706			
Cu. Feet				28.3168	2.83168	.283168	.0283168

Mass/Weight Measure	Milligrams	Centigrams	Decigrams	Grams	Dekagrams	Hectograms	Kilograms
Grains	64.4800414	6.4800414	.64800414	.064800414			
Ounces				28.349523125			
Pounds				453.59236	45.359236	4.5359236	.45359236
Tons (short)							907.184739
Tons (long)							1016.046908

APPROXIMATE 16TH-INCH EQUIVALENTS OF 1 TO 26 MILLIMETERS

mm	Dec. In.	16ths	Equivalents
1	0.03937	0.62992	1/16-"
2	0.07874	1.25984	1/16+"
3	0.11811	1.88976	1/8-"
4	0.15748	2.51968	3/16-"
5	0.19685	3.14960	3/16+"
6	0.23622	3.77952	1/4-"
7	0.27559	4.40944	1/4+"
8	0.31496	5.03936	5/16"
9	0.35433	5.66928	3/8-"
10	0.39370	6.29920	3/8+"
11	0.43307	6.92912	7/16"
12	0.47244	7.55904	1/2-"
13	0.51181	8.18896	1/2+"
14	0.55118	8.81888	9/16-"
15	0.59055	9.44880	9/16+"
16	0.62992	10.07870	5/8"
17	0.66929	10.70860	11/16-"
18	0.70866	11.33860	11/16+"
19	0.74803	11.96850	3/4-"
20	0.78740	12.59840	13/16-"
21	0.82677	13.22830	13/16+"
22	0.86614	13.85820	7/8-"
23	0.90551	14.48820	7/8+"
24	0.94488	15.11810	15/16+"
25	0.98425	15.74800	1-"
26	1.02362	16.37790	1+"